T0142113

THE GENUS *YERSINIA*

ADVANCES IN EXPERIMENTAL MEDICINE AND BIOLOGY

Recent Volumes in this Series

Volume 594
MOLECULAR ASPECTS OF THE STRESS RESPONSE: CHAPERONES, MEMBRANES AND NETWORKS
Edited by Peter Csermely and Lászió Vígh

Volume 595
THE MOLECULAR TARGETS AND THERAPEUTIC USES OF CURCUMIN IN HEALTH AND DISEASE
Edited by Bharat B. Aggarwal, Young-Joon Surh, and Shishir Shishodia

Volume 596
MECHANISMS OF LYMPHOCYTE ACTIVATION AND IMMUNE REGULATION XI
Edited by S. Gupta, F.W. Alt, M.D. Cooper, F. Melchers, and K. Rajewsky

Volume 597
TNF RECEPTOR ASSOCIATED FACTORS
Edited by Hao Wu

Volume 598
CURRENT TOPICS IN INNATE IMMUNITY
Edited by J.D. Lambris

Volume 599
OXYGEN TRANSPORT TO TISSUE XXVIII
Edited by David Maguire, Duane F. Bruley, and David K. Harrison

Volume 600
SEMAPHORINS
Edited by Jeroen R. Pasterkamp

Volume 601
IMMUNE MEDIATED DISEASES: FROM THEORY TO THERAPY
Edited by Michael R. Shurin and Yuri S. Smolkin

Volume 602
OSTEOIMMUNOLOGY
Edited by Yongwon Choi

Volume 603
THE GENUS YERSINIA
Edited by Robert D. Perry and Jacqueline D. Fetherston

A Continuation Order Plan is available for this series. A continuation order will bring delivery of each new volume immediately upon publication. Volumes are billed only upon actual shipment. For further information please contact the publisher.

THE GENUS *YERSINIA*:
From Genomics to Function

Edited by

Robert D. Perry and Jacqueline D. Fetherston

Department of Microbiology, Immunology, and Molecular Genetics University of Kentucky Lexington, KY, USA

Robert D. Perry
Department of Microbiology,
Immunology, and Molecular Genetics
University of Kentucky
800 Rose Street
Lexington, KY 40536
USA
rperry@email.uky.edu

Jacqueline D. Fetherston
Department of Microbiology,
Immunology, and Molecular Genetics
University of Kentucky
800 Rose Street
Lexington, KY 40536
USA

ISBN 978-1-4939-3891-9 ISBN 978-0-387-72124-8 (eBook)
DOI 10.1007/978-0-387-72124-8

Printed on acid-free paper.

9 8 7 6 5 4 3 2 1

springer.com

Contributors

Frederico Abath
Departamento de Microbiologia
Centro de Pesquisas Aggeu Magalhães
Cidade Universitaria
Recife, Brazil

Arwa Abu Khweek
Department of Microbiology, Immunology and Molecular Genetics
University of Kentucky
Lexington, KY, USA

Moshe Aftalion
Department of Biochemistry and Molecular Genetics
Israel Institute for Biological Research
Ness-Ziona, Israel

Alzira Almeida
Departamento de Microbiologia
Centro de Pesquisas Aggeu Magalhães
Cidade Universitaria
Recife, Brazil

Brad Anderson
Laboratory of Genetics
University of Wisconsin-Madison
Madison, WI, USA

Deborah Anderson
Department of Microbiology
University of Chicago
Chicago, IL, USA

Andrey P. Anisimov
Laboratory for Plague Microbiology, Department of Infectious Diseases
State Research Center for Applied Microbiology and Biotechnology
Obolensk, Moscow Region, Russia

Sonia Arafah
Faculté de Médecine Henri Warembourg
Institut Pasteur de Lille
Université Lille II
Lille, France

Guangchun Bai
Wadsworth Center
New York State Department of Health
Albany, NY, USA

Irina V. Bakhteeva
State Research Center for Applied Microbiology and Biotechnology
Obolensk, Moscow Region, Russia

Sergey V. Balakhonov
Antiplague Research Institute of Siberia and Far East
Irkutsk, Russia

Scott Bearden
Bacterial Zoonoses Branch
Centers for Disease Control and Prevention
Fort Collins, CO, USA

Raphael Ber
Department of Biochemistry and Molecular Genetics
Israel Institute for Biological Research
Ness-Ziona, Israel

Frederick R. Blattner
Laboratory of Genetics
University of Wisconsin-Madison
Madison, WI, USA

Alexander G. Bobrov
Department of Microbiology, Immunology and Molecular Genetics
University of Kentucky
Lexington, KY, USA

Katja Böhme
Institut für Mikrobiologie
Technische Universität Braunschweig
Braunschweig, Germany

Christine G. Branger
Center for Infectious Diseases
and Vaccinology
The Biodesign Institute
Arizona State University
Tempe, AZ, USA

Jeanette E Bröms
Department of Medical Countermeasures
Division of NBC Defence
Swedish Defence Research Agency
Umeå, Sweden

Robert R. Brubaker
Department of Microbiology
and Molecular Genetics
Michigan State University
East Lansing, MI, USA

Valerie Burland
Laboratory of Genetics
University of Wisconsin-Madison
Madison, WI, USA

Olga V. Bystrova
N.D. Zelinsky Institute of Organic
Chemistry
Russian Academy of Sciences
Moscow, Russia

Eric Cabot
Laboratory of Genetics
University of Wisconsin-Madison
Madison, WI, USA

Elisabeth Carniel
Yersinia Research Unit
National Reference Laboratory and WHO
Collaborative Center for *Yersinia*
Institut Pasteur
Paris, France

Patrick Chain
Biosciences and Biotechnology Division
Lawrence Livermore National Laboratory
Livermore, CA, USA

Viviane Chenal-Francisque
Yersinia Research Unit
National Reference Laboratory and WHO
Collaborative Center for *Yersinia*
Institut Pasteur
Paris, France

Sara Cohen
Department of Biochemistry and Molecular
Genetics
Israel Institute for Biological Research
Ness-Ziona, Israel

Claire Cornelius
Department of Microbiology
University of Chicago
Chicago, IL, USA

Yujun Cui
Institute of Microbiology
and Epidemiology
Academy of Military Medical Sciences
Beijing, China

Roy Curtiss III
Center for Infectious Diseases
and Vaccinology
The Biodesign Institute
Arizona State University
Tempe, AZ, USA

Catherine Daniel
Laboratoire des Bactéries Lactiques
et Immunité des Muqueuses
Institut Pasteur de Lille
Lille, France

Andrew J. Darwin
Department of Microbiology
New York University School of Medicine
New York, NY, USA

Svetlana V. Dentovskaya
Laboratory for Plague Microbiology,
Department of Infectious Diseases
State Research Center for Applied
Microbiology and Biotechnology
Obolensk, Moscow Region, Russia

Petra Dersch
Institut für Mikrobiologie
Technische Universität Braunschweig
Braunschweig, Germany

Rodrigue Dessein
Faculté de Médecine
Institut Pasteur de Lille
Université de Lille
Lille, France

Joëlle Dewulf
Laboratoire des Bactéries Lactiques
et Immunité des Muqueuses
Institut Pasteur de Lille
Lille, France

Solveig K. Ericsson
Department of Medical Countermeasures
Division of NBC Defence
Swedish Defence Research Agency
Umeå, Sweden

Alexandra Farias
Departamento de Microbiologia
Centro de Pesquisas Aggeu Magalhães
Cidade Universitaria
Recife, Brazil

Suleyman Felek
Department of Biologic and Materials
Science
University of Michigan School of Dentistry
Ann Arbor, MI, USA

Jacqueline D. Fetherston
Department of Microbiology, Immunology
and Molecular Genetics
University of Kentucky
Lexington, KY, USA

Richard ffrench-Constant
School of Biological Sciences
University of Exeter
Falmouth, United Kingdom

Claire Flamez
Faculté de Médecine Henri Warembourg
Institut Pasteur de Lille
Université Lille II
Lille, France

Yehuda Flashner
Department of Biochemistry and Molecular
Genetics
Israel Institute for Biological Research
Ness-Ziona, Israel

Benoit Foligné
Faculté de Médecine
Institut Pasteur de Lille
Université de Lille
Lille, France

Åke Forsberg
Department of Medical Countermeasures
Division of NBC Defence
Swedish Defence Research Agency
Umeå, Sweden

Emilio Garcia
Biosciences and Biotechnology Division
Lawrence Livermore National Laboratory
Livermore, CA, USA

Beth A. Garvy
Department of Microbiology, Immunology
and Molecular Genetics
Department of Internal Medicine
University of Kentucky
Lexington, KY, USA

Jeremy Glasner
Laboratory of Genetics
University of Wisconsin-Madison
Madison, WI, USA

Jon Goguen
Department of Molecular Genetics
and Microbiology
University of Massachusetts Medical
School
Worcester, MA, USA

Andrey Golubov
Wadsworth Center
New York State Department of Health
Albany, NY, USA

Olivier Gorgé
Division of Analytical Microbiology
Centre d'Etudes du Bouchet
Vert le Petit, France

John M. Greene
Health Research Systems
SRA International, Inc.
Rockville, MD, USA

Ibtissem Grissa
Institut de Génétique et Microbiologie
Université Paris-Sud
Orsay, France

David Gur
Department of Biochemistry and Molecular Genetics
Israel Institute for Biological Research
Ness-Ziona, Israel

Johanna Haiko
Department of Biological
and Environmental Sciences
University of Helsinki
Helsinki, Finland

Mohamad A. Hamad
Department of Microbiology
University of Colorado Health Science Center
Denver, CO, USA

Thomas Hampton
Health Research Systems
SRA International, Inc.
Rockville, MD, USA

Michelle Hares
School of Biological Sciences
University of Exeter
Falmouth, United Kingdom

Jürgen Heesemann
Max von Pettenkofer Institute for Hygiene and Medical Microbiology
Ludwig Maximillians University
Munich, Germany

Ann Kathrin Heroven
Institut für Mikrobiologie
Technische Universität Braunschweig
Braunschweig, Germany

Stewart Hinchliffe
Department of Infectious and Tropical Diseases
London School of Hygiene and Tropical Medicine
London, United Kingdom

B. Joseph Hinnebusch
Laboratory of Zoonotic Pathogens
Rocky Mountain Laboratories
Hamilton, MT, USA

Sarah Howard
Department of Infectious and Tropical Diseases
London School of Hygiene and Tropical Medicine
London, United Kingdom

Clayton Jarrett
Laboratory of Zoonotic Pathogens
Rocky Mountain Laboratories
Hamilton, MT, USA

Sabrina S. Joseph
Department of Microbiology
and Immunology
University of Miami Miller School
of Medicine
Miami, FL, USA

Olga Kirillina
Department of Microbiology, Immunology and Molecular Genetics
University of Kentucky
Lexington, KY, USA

Stefan D. Knight
Department of Molecular Biology
Swedish University of Agricultural Science
Uppsala, Sweden

Yuriy A. Knirel
N.D. Zelinsky Institute of Organic Chemistry
Russian Academy of Sciences
Moscow, Russia

Nina A. Kocharova
N.D. Zelinsky Institute of Organic Chemistry
Russian Academy of Sciences
Moscow, Russia

Timo K. Korhonen
General Microbiology
Faculty of Biosciences
University of Helsinki
Helsinki, Finland

Eric S. Krukonis
Department of Microbiology and Immunology
University of Michigan School of Medicine
Ann Arbor, MI, USA

Maini Kukkonen
General Microbiology
Faculty of Biosciences
University of Helsinki
Helsinki, Finland

Kaarina Lähteenmäki
General Microbiology
Faculty of Biosciences
University of Helsinki
Helsinki, Finland

Dorothy Lang
Biosciences and Biotechnology Division
Lawrence Livermore National Laboratory
Livermore, CA, USA

Frank Larimer
Life Sciences Division
Oak Ridge National Laboratory
Oak Ridge, TN, USA

Moa Lavander
Department of Medical Countermeasures
Division of NBC Defence
Swedish Defence Research Agency
Umeå, Sweden

Nilma Leal
Departamento de Microbiologia
Centro de Pesquisas Aggeu Magalhães
Cidade Universitaria
Recife, Brazil

Alexandre J.L. Leclercq
Yersinia Research Unit
National Reference Laboratory and WHO Collaborative Center for *Yersinia*
Institut Pasteur
Paris, France

Chrono Lee
Department of Molecular Genetics and Microbiology
University of Massachusetts Medical School
Worcester, MA, USA

Yanjun Li
Institute of Microbiology and Epidemiology
Academy of Military Medical Sciences
Beijing, China

Luther Lindler
Department of Bacterial Diseases
Walter Reed Army Institute of Research
Silver Spring, MD, USA

Buko Lindner
Research Center Borstel
Center for Medicine and Biosciences
Borstel, Germany

Paul Liss
Laboratory of Genetics
University of Wisconsin-Madison
Madison, WI, USA

Leandro Lobo
General Microbiology
Faculty of Biosciences
University of Helsinki
Helsinki, Finland

Stephanie Malfatti
Biosciences and Biotechnology Division
Lawrence Livermore National Laboratory
Livermore, CA, USA

Emanuelle Mamroud
Department of Biochemistry and Molecular Genetics
Israel Institute for Biological Research
Ness-Ziona, Israel

Michaël Marceau
Faculté de Médecine Henri Warembourg
Institut Pasteur de Lille
Université Lille II
Lille, France

Bob Mau
Laboratory of Genetics
University of Wisconsin-Madison
Madison, WI, USA

Kathleen A. McDonough
Wadsworth Center
New York State Department of Health
Albany, NY, USA

Shirly Mildiner-Earley
Molecular Microbiology and Pathogenesis
Washington University
St. Louis, MO, USA

Virginia L. Miller
Molecular Microbiology and Pediatrics
Washington University
St. Louis, MO, USA

Scott A. Minnich
Department of Microbiology, Molecular Biology, and Biochemistry
University of Idaho
Moscow, ID, USA

Brian S. Murphy
Department of Internal Medicine
University of Kentucky
Lexington, KY, USA

Eric Neeno-Eckwall
Laboratory of Genetics
University of Wisconsin-Madison
Madison, WI, USA

Heinrich Neubauer
Friedrich-Loeffler Institute
German Reference Center for Human and Animal Brucellosis
Jena, Germany

Matthew L. Nilles
Department of Microbiology and Immunology
School of Medicine and Health Sciences
University of North Dakota
Grand Forks, ND, USA

Mark F. Oellerich
Max von Pettenkofer Institute for Hygiene and Medical Microbiology
Ludwig Maximillians University
Munich, Germany

Marcos A. Oliveira
Department of Pharmaceutical Sciences
Feik School of Pharmacy
University of the Incarnate Word
San Antonio, TX, USA

Ning Pan
Department of Molecular Genetics and Microbiology
University of Massachusetts Medical School
Worcester, MA, USA

Evgeniy A. Panfertsev
Laboratory for Plague Microbiology, Department of Infectious Diseases
State Research Center for Applied Microbiology and Biotechnology
Obolensk, Russia

Kenneth D. Parrish
Advanced Implant and Perio Center
Louisville, KY, USA

Janice Pata
Wadsworth Center
New York State Department of Health
Albany, NY, USA

Chandra N. Patel
Hematology/Cardiology
Bayer HealthCare LLC
Berkeley, CA, USA

Nicole T. Perna
Laboratory of Genetics
University of Wisconsin-Madison
Madison, WI, USA

Robert D. Perry
Department of Microbiology, Immunology
and Molecular Genetics
University of Kentucky
Lexington, KY, USA

Gerald B. Pier
Channing Laboratory
Brigham and Women's Hospital
Harvard Medical School
Boston, MA, USA

Gregory V. Plano
Department of Microbiology
and Immunology
University of Miami Miller School
of Medicine
Miami, FL, USA

Mikhail E. Platonov
State Research Center for Applied
Microbiology and Biotechnology
Obolensk, Moscow Region, Russia

Guy Plunkett III
Laboratory of Genetics
University of Wisconsin-Madison
Madison, WI, USA

Sabine Poiret
Laboratoire des Bactéries Lactiques
et Immunité des Muqueuses
Institut Pasteur de Lille
Lille, France

Bruno Pot
Laboratoire des Bactéries Lactiques
et Immunité des Muqueuses
Institut Pasteur de Lille
Lille, France

David Pot
Health Research Systems
SRA International, Inc.
Rockville, MD, USA

Christine Pourcel
Institut de Génétique et Microbiologie
Université Paris-Sud
Orsay, France

Michael B. Prentice
Department of Microbiology, Department
of Pathology
University College Cork
Cork, Ireland

Yu Qiu
Laboratory of Genetics
University of Wisconsin-Madison
Madison, WI, USA

Lauriane Quenee
Department of Microbiology
University of Chicago
Chicago, IL, USA

Alexander Rakin
Max von Pettenkofer Institute for Hygiene
and Medical Microbiology
Munich, Germany

Päivi Ramu
General Microbiology
Faculty of Biosciences
University of Helsinki
Helsinki, Finland

Isabelle Ricard
Faculté de Médecine Henri Warembourg
Institut Pasteur de Lille
Université Lille II

Lille, France Harold N. Rohde
Department of Microbiology, Molecular Biology, and Biochemistry
University of Idaho
Moscow, ID, USA

Jason A. Rosenzweig
Division of Math Science and Technology
Nova Southeastern University
Ft. Lauderdale, FL, USA

Lisa M. Runco
Department of Molecular Genetics and Microbiology
State University of New York Stony Brook
Stony Brook, NY, USA

Michael Rusch
Laboratory of Genetics
University of Wisconsin-Madison
Madison, WI, USA

Kurt Schesser
Department of Microbiology and Immunology
Miller School of Medicine
University of Miami
Miami, FL, USA

Olaf Schneewind
Department of Microbiology
University of Chicago
Chicago, IL, USA

Florent Sebbane
Faculté de Médicine Henri Warembourg
Institut National de la Santé et de la Recherche Médicale Unité 801
Université de Lille II
Lille, France

Sof'ya N. Senchenkova
N.D. Zelinsky Institute of Organic Chemistry
Russian Academy of Sciences
Moscow, Russia

Avigdor Shafferman
Department of Biochemistry and Molecular Genetics
Israel Institute for Biological Research
Ness-Ziona, Israel

Rima Z. Shaikhutdinova
State Research Center for Applied Microbiology and Biotechnology
Obolensk, Moscow Region, Russia

Matthew Shaker
Health Research Systems
SRA International, Inc.
Rockville, MD, USA

Lorie Shaull
Health Research Systems
SRA International, Inc.
Rockville, MD, USA

Panna Shetty
Health Research Systems
SRA International, Inc.
Rockville, MD, USA

Chuan Shi
Health Research Systems
SRA International, Inc.
Rockville, MD, USA

Michel Simonet
Faculté de Médecine Henri Warembourg
Institut Pasteur de Lille
Université Lille II
Lille, France

Mikael Skurnik
Department of Bacteriology and Immunology
Haartman Institute
University of Helsinki
Helsinki, Finland

Stephen T. Smiley
Trudeau Institute
Saranac Lake, NY, USA

Eric Smith
Wadsworth Center
New York State Department of Health
Albany, NY, USA

Yajun Song
Institute of Microbiology
and Epidemiology
Academy of Military Medical Sciences
Beijing, China

Gerlane Souza
Departamento de Microbiologia
Centro de Pesquisas Aggeu Magalhães
Cidade Universitaria
Recife, Brazil

Ida Steinberger-Levy
Department of Biochemistry and Molecular Genetics
Israel Institute for Biological Research
Ness-Ziona, Israel

Susan C. Straley
Department of Microbiology, Immunology
and Molecular Genetics
University of Kentucky
Lexington, KY, USA

Daniel Sturdevant
Research Technology Section
Rocky Mountain Laboratories
Hamilton, MT, USA

Marjo Suomalainen
General Microbiology
Faculty of Biosciences
University of Helsinki
Helsinki, Finland

Tat'yana E. Svetoch
Laboratory for Plague Microbiology,
Department of Infectious Diseases
State Research Center for Applied
Microbiology and Biotechnology
Obolensk, Russia

David G. Thanassi
Department of Molecular Genetics
and Microbiology
State University of New York Stony Brook
Stony Brook, NY, USA

Nicholas R. Thomson
The Pathogen Sequencing Unit
Wellcome Trust Genome Campus
The Wellcome Trust Sanger Institute
Cambridge, United Kingdom

Galina M. Titareva
State Research Center for Applied
Microbiology and Biotechnology
Obolensk, Moscow Region, Russia

Gabriela Torrea
Yersinia Research Unit
National Reference Laboratory and WHO
Collaborative Center for *Yersinia*
Institut Pasteur
Paris, France

Hien Tran-Winkler
Institut für Mikrobiologie
Technische Universität Braunschweig
Braunschweig, Germany

Konrad Trülzsch
Max von Pettenkofer Institute for Hygiene
and Medical Microbiology
Ludwig Maximillians University
Munich, Germany

Viveka Vadyvaloo
Laboratory of Zoonotic Pathogens
Rocky Mountain Laboratories
Hamilton, MT, USA

Baruch Velan
Department of Biochemistry and Molecular Genetics
Israel Institute for Biological Research
Ness-Ziona, Israel

Gilles Vergnaud
Institut de Génétique et Microbiologie
Université Paris-Sud
Orsay, France

Ritva Virkola
General Microbiology
Faculty of Biosciences
University of Helsinki
Helsinki, Finland

Kimberly A. Walker
Molecular Microbiology and Pathogenesis
Washington University
St. Louis, MO, USA

Nick Waterfield
Department of Biology and Biochemistry
University of Bath
Bath, United Kingdom

Benita Westerlund-Wikström
General Microbiology
Faculty of Biosciences
University of Helsinki
Helsinki, Finland

Jon Whitmore
Health Research Systems
SRA International, Inc.
Rockville, MD, USA

Mary Wong
Health Research Systems
SRA International, Inc.
Rockville, MD, USA

Patricia Worsham
Bacteriology Division
United States Army Medical Research
Institute of Infectious Diseases
Fort Detrick, MD, USA

Brian W. Wortham
Department of Pharmaceutical Sciences
University of Kentucky
Lexington, KY, USA

Brendan W. Wren
Department of Infectious and Tropical
Diseases
London School of Hygiene and Tropical
Medicine
London, United Kingdom

Christine R. Wulff
Department of Microbiology, Immunology
and Molecular Genetics
University of Kentucky
Lexington, KY, USA

Ruifu Yang
Institute of Microbiology
and Epidemiology
Academy of Military Medical Sciences
Beijing, China

Glenn M. Young
Department of Food Science
and Technology
University of California, Davis
Davis, CA, USA

Eran Zahavy
Department of Infectious Diseases
Israel Institute for Biological Research
Ness-Ziona, Israel

Sam Zaremba
Health Research Systems
SRA International, Inc.
Rockville, MD, USA

Ayelet Zauberman
Department of Biochemistry and Molecular
Genetics
Israel Institute for Biological Research
Ness-Ziona, Israel

Dongsheng Zhou
Institute of Microbiology
and Epidemiology
Academy of Military Medical Sciences
Beijing, China

Preface

Picture 1. Logo for the 9th International Symposium on *Yersinia*. Logo design by Cesar Ibanez, Web Designer, American Society for Microbiology.

The 9th International Symposium on *Yersinia* was held in Lexington, Kentucky, USA on October 10-14, 2006. Over 250 *Yersinia* researchers from 18 countries gathered to present and discuss their research. In addition to 37 oral presentations, there were 150 poster presentations. This Symposium volume is based on selected presentations from the meeting and contains both reviews and research articles. It is divided into six topic areas: 1) genomics; 2) structure and metabolism; 3) regulatory mechanisms; 4) pathogenesis and host interactions; 5) molecular epidemiology and detection; and 6) vaccine and antimicrobial therapy development. Consequently, this volume covers a wide range of current research areas in the *Yersinia* field.

Robert D. Perry

Jacqueline D. Fetherston

Department of Microbiology, Immunology,
and Molecular Genetics
University of Kentucky
Lexington, KY
USA

Acknowledgments

We thank the members of the Scientific Advisory Board: Andrey Anisimov (Obelensk), José Bengoechea (Palma de Mallorca), James Bliska (Stony Brook), Elisabeth Carniel (Paris), Virginia Miller (St. Louis), Heinrich Neubauer (Jena), Mikael Skurnik (Helsinki), George Smirnov (Moscow), Hans Wolf-Watz (Umeå), and Brendan Wren (London), for the speaker suggestions, for organizational suggestions, and for reviewing submitted abstracts.

A number of companies and institutions generously supported the symposium: Battelle, Fisher Scientific, Pfizer, the Great Lakes Research Center for Excellence in Biodefense (GLRCEB), the Southeastern Research Center for Excellence in Biodefense (SERCEB), and the University of Kentucky. The Symposium could not have been held without their support.

Practical organization of the meeting was professionally performed by the American Society for Microbiology, Department of Meetings and Industry Relations – a special thanks go to Traci Williams and Evangelos Koutalas for their efforts and organizational skills.

Andrey Anisimov kindly provided many of the photographs from the meeting that are in this volume.

Contents

Part I - Genomics

Picture 2. Luther Lindler introduces the Genomics Session. Photo by A. Anisimov.

Picture 3. Andrey Anisimov, Yuriy Knirel, Valentine Fedorova, Rima Shaikhutdinova, and Svetlana, Dentovskaya at the posters. Photo from A. Anisimov.

1
Comparative Genome Analyses of the Pathogenic Yersiniae Based on the Genome Sequence of *Yersinia enterocolitica* Strain 8081

Nicholas R. Thomson[1], Sarah Howard[2], Brendan W. Wren[2], and Michael B. Prentice[3,4]

[1] The Pathogen Sequencing Unit, Wellcome Trust Genome Campus, The Wellcome Trust Sanger Institute, nrt@sanger.ac.uk
[2] Department of Infectious and Tropical Diseases, London School of Hygiene and Tropical Medicine, London
[3] Department of Microbiology, University College Cork
[4] Department of Pathology, University College Cork

Abstract. This chapter represents a summary of the findings from the *Yersinia enterocolitica* strain 8081 whole genome sequence and the associated microarray analysis. Section 1 & 2 provide an introduction to the species and an overview of the general features of the genome. Section 3 identifies important regions within the genome which highlight important differences in gene function that separate the three pathogenic *Yersinia*s. Section 4 describes genomic loci conferring important, species-specific, metabolic and virulence traits. Section 5 details extensive microarray data to provide an overview of species-specific core *Y. enterocolitica* gene functions and important insights into the intra-species differences between the high, low and non-pathogenic *Y. enterocolitica* biotypes.

1.1 Introduction

Yersinia enterocolitica represents a key link in our understanding of how the pathogenic members of the *Yersinia* genus have evolved to produce diverse clinical manifestations. The disease potential of the human pathogenic *Yersinia* ranges from gastroenteritis for *Y. enterocolitica* and *Yersinia pseudotuberculosis*, which are primarily enteropathogens, to bubonic plague caused by *Yersinia pestis* (Perry and Fetherston 1997). It is estimated that *Y. enterocolitica* and *Y. pseudotuberculosis* diverged within the last 200 million years and that *Y. pestis* is a clone of *Y. pseudotuberculosis* that has emerged within the last 1,500–20,000 years (Achtman et al. 2004; Achtman et al. 1999; Wren 2003).

Since splitting from *Y. pseudotuberculosis, Y. enterocolitica* has evolved into a genetically and biochemically heterogeneous collection of organisms that has been divided into six biotypes differentiated by biochemical tests (1A, 1B, 2, 3, 4 and 5) (Wauters et al. 1987). These *in vitro* biotypes can be placed into three distinct lineages based on pathogenic potential: a mostly non-pathogenic group (biogroup 1A); a weakly pathogenic group that is unable to kill mice (biogroups 2 to 5); and a highly pathogenic, mouse-lethal group (biogroup 1B) (Mcnally et al. 2004; Prentice et al.

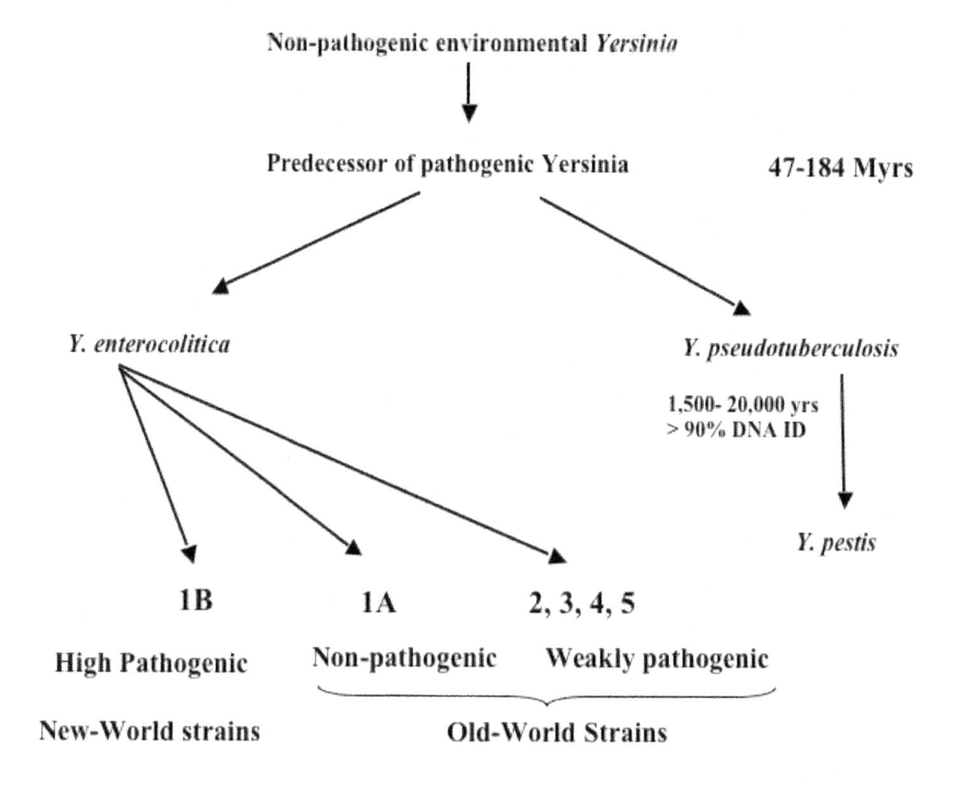

Fig. 1. A basic model describing the evolution of the pathogenic *Yersinia* (adapted from (Wren 2003)).

1991; Van Noyen et al. 1981; Wauters et al. 1987). These biogroups also form geographically distinct groups with biotype1B being most frequently isolated in North America (termed the 'New-World' strains), whereas biogroups 2-5 predominate in Europe and Japan (termed the 'Old-World' strains) (Schubert et al. 2004) (Fig. 1).

Representatives of the two other human pathogenic *Yersinia* species, *Y. pseudotuberculosis* strain IP32953 (referred to as *Y. pseudotuberculosis*), and *Y. pestis* (strains CO92 [biovar Orientalis], KIM10+ [biovar Mediaevalis], and 91001 [biovar Microtis]), have been sequenced (Chain et al. 2004; Deng et al. 2002; Parkhill et al. 2001; Song et al. 2004). Consequently, the three pathogenic *Yersinia* represent an ideal genus to study bacterial pathogenesis and the evolution of virulence (Wren 2003).

In this chapter we have condensed the whole genome sequence analysis of *Y. enterocolitica* strain 8081 biotype 1B (serotype 0:8). We provide examples of ancestral gene functions which appear to have been lost following the divergence of the pathogenic *Yersinia* from their last common ancestor. In addition, we have highlighted regions that appear to have been acquired by *Y. enterocolitica* strain 8081 and define it at the species through to the strain level. For a more complete analysis of the 8081 genome sequence and its comparative analysis to other *Y. entrocolitica* strains refer to (Thomson et al. 2006) and (Howard et al. 2006).

1.2 *Y. enterocolitica* 8081 Chromosome

The characteristics of the *Y. enterocolitica* chromosome are very similar to those of *Y. pestis* and *Y. pseudotuberculosis* (Table 1). The most notable differences lie in the numbers of insertion-sequence (IS) elements. Although *Y. enterocolitica* possesses fewer in total than the other yersiniae, their diversity is greater with 15 IS families in *Y. enterocolitica* compared to 4 and 5 in *Y. pseudotuberculosis* and *Y. pestis* (CO92), respectively.

Y. enterocolitica also possesses far fewer pseudogenes than *Y. pestis,* which is thought to have >140 (Parkhill et al. 2001). The recent expansion of a few types of IS element in *Y. pestis* and the accumulation of so many pseudogenes is thought to reflect a marked change in lifestyle (associated with specific plasmid-acquisition events) (Chain et al. 2004; Parkhill et al. 2001). Conversely, this also implies that *Y. enterocolitica* and *Y. pseudotuberculosis* have been stably maintained in a consistent niche.

Although general characteristics of the *Y. enterocolitica* genome are similar to those of *Y. pseudotuberculosis* and *Y. pestis*, these figures disguise considerable

Table 1. Properties of all the published *Yersinia* genomes

Property	Y. entero-colitica 8081	Y. pestis CO92[a]	Y. pestis KIM10+[b]	Y. pestis 91001[c]	Y. pseudo-tuberculosis IP32953[d]
Size	4,615,899	4,653,726	4,600,755	4,595,065	4,744,671
G+C content	47.27%	47.64%	47.64%	47.65%	47.61%
Number of CDSs	4,037	4,012	4,198	4037	3,974
Coding density	83.8%	83.8%	86%	81.6%	82.5%
Ave. gene size	968 bp	998 bp	940 bp	966 bp	998 bp
rRNA operons	7	6	7	7	7
tRNA	81	70	73	72	85
Pseudogenes[e]	67	149	54	141	62
IS elements	60	139	122	109	20
Prophage regions	4	4	3	ND	5

[a](Parkhill et al. 2001), [b](Deng et al. 2002), [c](Song et al. 2004), [d](Chain et al. 2004), [e]Figures taken from original publication. ND - not determined. Taken from (Thomson et al. 2006).

variation in gene repertoire. A comparison of orthologous gene sets shared between *Y. enterocolitica, Y. pestis* (strain CO92), and *Y. pseudotuberculosis* (Fig. 2) showed a core set of 2,747 CDSs shared by all, as well as a significant number of CDSs being unique to *Y. enterocolitica* (~29%), *Y. pseudotuberculosis* (~9%), or *Y. pestis* (~11%).

Perhaps the biggest surprise from the genome was the number of CDS's shared exclusively between *Y. enterocolitica* and either *Y. pseudotuberculosis* or *Y. pestis.*

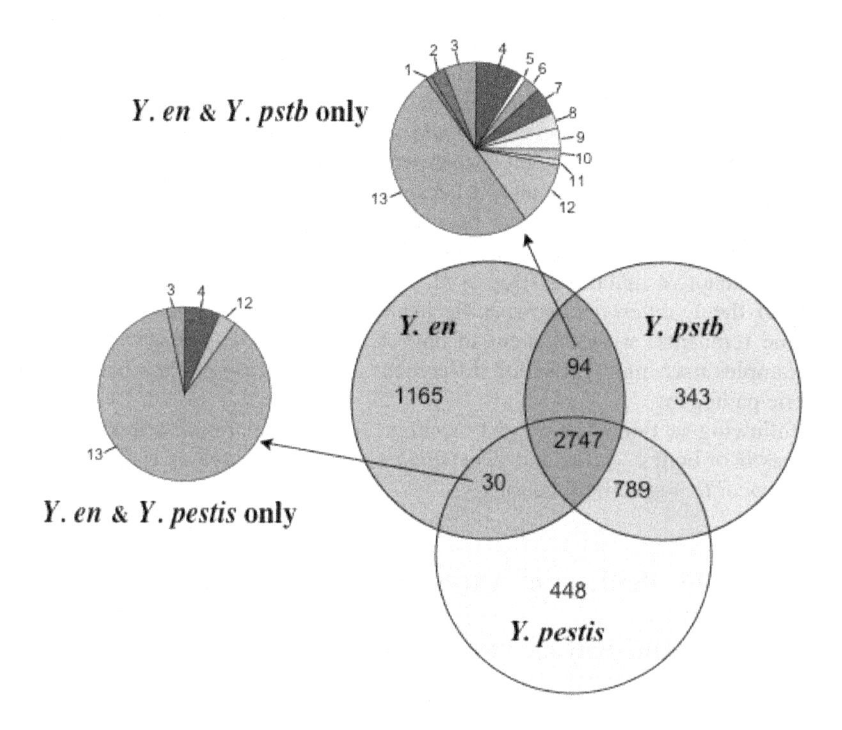

Fig. 2. Distribution of orthologous CDSs in *Y. enterocolitica* 8081, *Y. pestis* CO92, and *Y. pseudotuberculosis* IP32953. The Venn diagram shows the number of genes unique or shared between two other *Yersinia* species. The associated pie charts show the breakdown of the functional groups assigned for CDSs in relevant sections of the Venn diagram. Number code for the pie charts is as follows: pathogenicity and virulence [1]; general regulation [2]; and miscellaneous function [3]; conserved hypothetical proteins [4]; chemotaxis and motility [5]; protective responses [6]; transport and binding proteins [7]; adaptations to atypical conditions [8]; synthesis and modification of macromolecules [9]; central intermediary metabolism [10]; energy metabolism [11]; periplasmic/exported/lipoproteins [12]; laterally acquired (including prophage CDSs) [13]. *Y. en, Y. enterocolitica* strain 8081; *Y. pstb, Y. pseudotuberculosis* strain IP32953; *Y. pestis, Y. pestis* strain CO92.

Close inspection of these CDS's showed that most were prophage-related (Fig. 2). Moreover all those shared exclusively between *Y. enterocolitica* and *Y. pestis,* which were not of phage origin were *in silico* artifacts, accounted for by differences in annotation. However, this was not the explanation for those shared between *Y. pseudotuberculosis* and *Y. enterocolitica,* which fell into a range of functional categories. These CDSs were interesting because it is highly unlikely that both *Y. pseudotuberculosis* and *Y. enterocolitica* independently acquired these functions since the divergence of *Y. pseudotuberculosis* and *Y. pestis*. Therefore the most parsimonious explanation for this was that these functions have probably been lost by *Y. pestis* since diverging from *Y. pseudotuberculosis*.

An investigation of the genomic context of CDSs shared exclusively between *Y. pseudotuberculosis* and *Y. enterocolitica* and the corresponding regions in the other *Yersinia* revealed that in several instances there were remnants of these regions in *Y. pestis*. These CDSs are thought to represent ancestral functions important for an enteric lifestyle, but which subsequently became redundant for *Y. pestis*. On the other hand, the loss of function may have been selectively advantageous, given the increased virulence potential of *Y. pestis,* that if true would make these further examples of pathoadaptive mutations (Day et al. 2001). Similar observations were made for some of the *Y. enterocolitica*-specific loci except that in this instance deletion scars (gene remnants) were apparent in both *Y. pestis* and *Y. pseudotuberculosis*. These examples may highlight subtle differences in the disease process between the two enteric pathogens.

The following sections will present examples of ancestral *Yersinia* functions lost from *Y. pestis* or both *Y. pestis* and *Y. pseudotuberculosis,* as well as examples from the *Y. enterocolitica*-specific functions.

1.3 Evidence for the Loss of Ancestral *Yersinia* Gene Functions

1.3.1 The Methionine-Salvage Pathway

One example of an entire metabolic pathway retained by *Y. enterocolitica* and *Y. pseudotuberculosis,* but apparently lost by *Y. pestis*, is the methionine-salvage pathway. The methionine-salvage pathway recycles the sulphur-containing compound, methylthioadenosine (MTA), formed during spermidine and spermine synthesis, and as a byproduct of *N*-acylhomoserine lactone production. MTA is recycled back to methionine, which can be further metabolised to produce S-adenosylmethionine, an essential reactant in several methylation reactions (Sekowska et al. 2004).

The methionine-salvage pathways appear to be intact and are encoded by seven CDSs in one locus (*mntA-E, mtnK* and *mtnU*) and a single CDS at an unlinked site (*mtnN*) in both *Y. enterocolitica* and *Y. pseudotuberculosis*. In *Y. pestis,* all of the CDSs encoded in the *mtnK–mtnU* locus are missing (presumably deleted). However, the *mtnN* gene has been retained in *Y. pestis* and remains intact and in the same genetic context as the *Y. enterocolitica mtnN* gene. It is known that in nutrient-rich environments and in the presence of low concentrations of dioxygen, facultatively anaerobic bacteria, such as *Escherichia coli,* simply convert MTA into methylthioribose, using MtnN, and excrete it from the cell. This is likely to be the case for *Y. pestis,* too, since growth outside of the nutrient-rich environment of the host is unnecessary for its current lifestyle.

1.3.2 The Cellulose Operon

Evidence for more extensive gene loss is illustrated by the cellulose *(cel)* biosynthetic operon. The genes encoding this pathway have been apparently lost from both *Y. pseudotuberculosis* and *Y. pestis,* leaving deletion scars. The *Y. enterocolitica cel* operon is highly similar in gene content and sequence to that carried by most *Salmonella*. The only remaining *cel* CDS in *Y. pseudotuberculosis* and *Y. pestis* is *bcsZ,* encoding endo-1,4-β-glucanase. Although *bcsZ* appears intact in *Y. pseudotuberculosis* (YPTB3837), the *Y. pestis bcsZ* orthologue carries a frameshift mutation (Fig. 3.). An identical mutation is present in the *bcsZ* genes in all of the sequenced *Y. pestis* isolates.

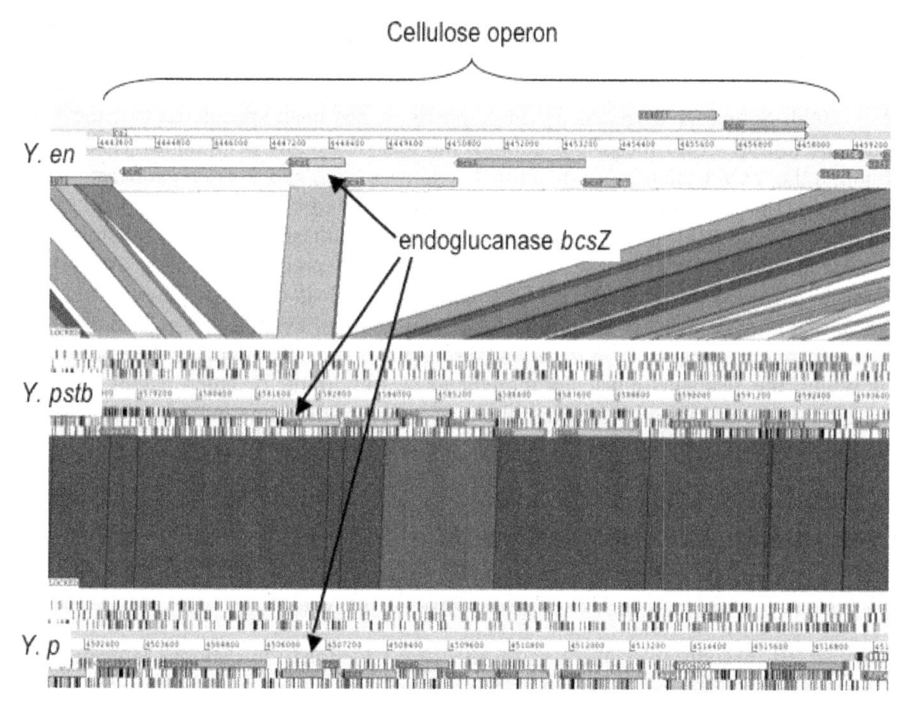

Fig. 3. Comparison of the *cel* locus of *Y. enterocolitica* strain 8081 (*Y. en*), Y. pseudotuberculosis strain IP32953 (*Y. pstb*), Y. pestis strain CO92 (*Y. p*). ACT comparison of amino-acid matches between the complete six-frame translations (computed using TBLASTX) of representatives of the three sequenced *Yersinia* species (http://www.sanger.ac.uk/Software/ACT). The grey shaded bars spanning between the genomes represent individual TBLASTX matches. CDS are marked as shaded boxes positioned on the grey DNA lines: The scale is marked in bps.

Salmonella produce cellulose in concert with thin aggregative fimbriae to form an inert and highly hydrophobic extracellular matrix. It has been suggested that the protection afforded by this matrix increases retention time of the bacterium in the gut (Zogaj et al. 2001). Cellulose production is presumably redundant for *Y. pestis* in its new lifestyle. However, why this operon should have been lost by *Y. pseudotuberculosis* is unclear and may reflect niche differences within the enteric environments of the two enteropathogenic *Yersinia* species.

1.3.3 Loss of Function in the *Yersinia* Core Functions

In addition to the loss of complete biochemical pathways, the *Y. enterocolitica* sequence revealed more subtle examples of loss of function in *Y. pestis*. All the pathogenic yersiniae possess a cluster of 13 CDSs on a genomic island displaying a lower G + C content denoted *Yersinia* Genomic Island 1 (YGI-1). YGI-1 is related to *tad* loci (tight adherence), present in diverse bacterial and archaeal species, including *Actinobacillus actinomycetemcomitans* (Schreiner et al. 2003).

The *tad* locus of *A. actinomycetemcomitans,* a human pathogen causing endocarditis and periodontitis, has been shown to be important for virulence by encoding the biosynthesis and transport of pili involved in tight, nonspecific adherence (Kachlany et al. 2000; Schreiner et al. 2003). In *Y. pestis,* it has been speculated that the *tad* genes are important for the colonisation of the flea (Kachlany et al. 2000). However, although the YGI-1 islands are intact in *Y. enterocolitica* and *Y. pseudotuberculosis,* in *Y. pestis* the essential pilin gene, *flp,* has been deleted by the insertion of IS*1541* elements and *rcpA* (pilin secretion protein) carries a frameshift mutation (identical in all sequenced *Y. pestis* strains). Moreover, since it is predicted that the Tad pilus would be exposed on the surface of the cell, like the loss of YadA (Rosqvist et al. 1988), this may be another example of a key mutational event that was selected for by the change in lifestyle of *Y. pestis*. Consequently, far from being an adaptation to life within the flea, this cluster is thought to be important for enteropathogenicity. The global loss of the pilin gene and the possession of the same mutation in *rcpA* in *Y. pestis* suggests that this event occurred soon after speciation.

1.4 *Y. enterocolitica* Unique Functions

1.4.1 Hydrogenases and Other Metabolic Functions

As expected in addition to the loss of ancestral functions the genome sequence of *Y. enterocolitica* also revealed that there had been significant gene accretion since *Y. enterocolitica* and *Y. pseudotuberculosis* diverged. Some of these species-specific loci significantly broaden the metabolic capability of this bacterium and were shown by microarray analysis to be characteristic of the species. These include two [NiFe]-containing hydrogenase complexes Hyd-2 and Hyd-4.

The ability to exploit locally generated hydrogen as a source of energy is known to be essential for colonisation of the gut, and for the virulence of enteric bacteria such as *Salmonella* and *Helicobacter* (Maier 2005; Maier et al. 2004; Olson and

Maier 2002). The two *Y. enterocolitica* hydrogenase gene clusters *hyf* and *hyb* are extremely compact, encoding all of the CDSs essential for Hyd-4 and Hyd-2 functioning and maturation. In other enteric bacteria these gene functions tend to be distributed over several different loci dispersed throughout their genome. There is no evidence of the loci in the *Y. pestis* and *Y. pseudotuberculosis* genomes. Coupled with the loci's genetic compactness, this may suggest that they have been acquired by *Y. enterocolitica,* despite the absence of obvious mobility genes in these clusters.

Other metabolic capabilities potentially relevant to enteric survival and apparently acquired by a single horizontal transfer event include the capacity for cobalamin synthesis and propanediol utilisation conferred by the divergent *cob-pdu* operon. This closely resembles the *cob-pdu* operon horizontally acquired by *Salmonella* and lost from *E. coli* (Lawrence and Roth 1996).

1.4.2 The Plasticity Zone

The PZ is the largest region of species-specific genomic variation found within the *Y. enterocolitica* 8081 genome and accounts for ~16% of the *Y. enterocolitica* unique CDSs (an ~199 kb locus extending from 3,761,922–3,960,673 bps and encoding 186 CDSs [Fig. 4]). The PZ is unlikely to have been acquired during a single event and is more likely to have arisen through a series of independent insertions at this site. Several discrete functional units are identifiable within this region, some of which are known to be mobile or sporadically distributed in other bacteria, and some of which are flanked by repeat sequences (Fig. 4).

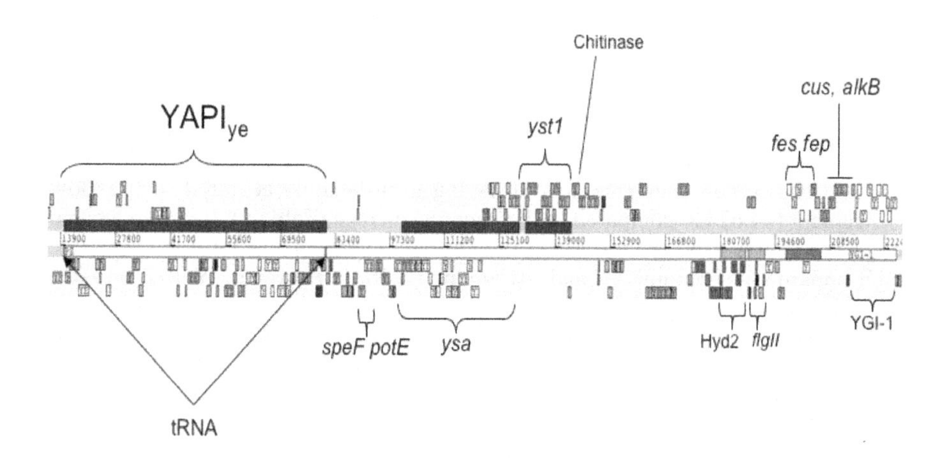

Fig. 4. The *Y. enterocolitica* strain 8081 PZ and YGI-1 genomic loci. An Artemis (http://www.sanger.ac.uk/Software/Artemis) view showing the genetic composition of the PZ locus. CDS are marked on their translation line as shaded boxes above and below the forward and reverse strands of DNA lines (grey): The scale is marked in bps.

These functional units within the PZ include a region highly similar to the *Y. pseudotuberculosis* adhesion pathogenicity island (Collyn et al. 2004) denoted $YAPI_{ye}$. The *Yersinia* YAPI islands are members of a broader family of genomic islands found in a diverse set of bacteria, and include the *Salmonella typhi* pathogenicity island, SPI-7 (Pickard et al. 2003; Collyn et al. 2006; Mohd-Zain et al. 2004). $YAPI_{ye}$ carries a type IV pilus operon, characteristic of this family of islands and shown to be important for virulence in *Y. pseudotuberculosis* (Collyn et al. 2004). In addition, $YAPI_{ye}$ also encodes a possible hemolysin (YE3454), a toxin/antitoxin system (YE3480 and YE3481), and an arsenic-resistance operon (YE3472–YE3475).

Other functions encoded on the PZ include the second *Y. enterocolitica* type three secretion system (T3SS) and general secretion pathway (GSP)–like system, Ysa and Yst, respectively (Foultier et al. 2002; Haller et al. 2000; Iwobi et al. 2003). Both of these secretion systems are known to be important for pathogenicity. The *Y. enterocolitica* PZ also carries several other gene clusters capable of conferring survival benefits in the gut or wider environment. These include the Hyd-2 biosynthetic operon (discussed above) and genes encoding products with possible roles in osmoprotection and osmoregulation, metal uptake and resistance, including the betaine/proline transporter, ProP and the ferric enterochelin operon *fepBDGC fes* and *fepA* (Schubert et al. 1999).

1.4.3 Additional Genomic Islands

In addition to the PZ, *in silico* analysis and microarray experiments showed that there were other potential genomic islands (GI). These included an island, denoted YGI-2, which encodes 13 CDS highly conserved as a unit in a wide range of the *Enterobacteriaceae* including the phytopathogen *Erwinia carotovora* subsp. *atroseptica* the pathogenic *E. coli:* enteroheamorrhagic *E. coli* 0157:H7 and uropathogenic *E. coli* CFT073 as well as the probiotic *E. coli* strain Nissle. Although, notably this island is missing from *E. coli* K12.

YGI-2 has certain characteristics of having been laterally acquired such as low G+C content (44.62 %) compared to the genome average (Table 1). It is also located alongside a tRNA gene (tRNA Asp), a common integration site for both lysogenic phage and mobile genomic islands (Campbell 2003). However, YGI-2 and the analogous loci in the other Enterobacteriaceae lack obvious mobility functions.

YGI-2 appears to encode biosynthesis, modification and export of an outer membrane anchored glyco-lipo-protein. Functional analysis of the CDS in this cluster suggest that the lipid moiety has a complex structure with one or more double bonds and hydroxyl groups and may in fact be O-methylated by the product of YE0894. The presence of an acyl transferase (YE0897) and a bifunctional glycosyl transferase/acyl transferase (YE0903) suggests that a sugar group is also acylated and attached like the lipid moiety. It is possible that YE0905 is responsible for the transport of the lipoprotein and likewise YE906 and YE0900 are involved in the export of the glycolipid. It is not possible to predict the true function of such a product, but there are very weak similarities to surfactins encoded by other bacteria.

Other interesting GI's also include two proposed integrated plasmid elements that, like YGI-2, have an atypical G + C content. The first, YGI-3, is inserted alongside the stable RNA *ssrA* gene (located at 1097155–1116114 bps) and flanked by 14 bp direct repeats and the second element, YGI-4 (located at 1308551–1323148 bps) is inserted into a gene, leaving an intact copy on one side and a partially duplicated copy on the other side of the element. Although YGI-3 and YGI-4 carry a small number of genes associated with conjugation and transfer, their true mode of transfer and whether they confer any possible selective benefits to the host is unclear.

1.5 Microarray Analysis of *Y. enterocolitica*

1.5.1 Phylogenomic Analysis of *Y. enterocolitica*

Using a 8081 DNA microarray containing all chromosomal 4037 CDSs we applied comparative phylogenomics (whole genome comparisons of microbes using DNA microarrays combined with Bayesian phylogenies) to investigate a diverse collection of *Y. enterocolitica* strains representing non pathogenic (37 biotype 1A isolates), low pathogenicity (49 isolates; biotypes 2-4) and highly pathogenic (8 biotype1B isolates) strains. The analysis confirmed three distinct statistically supported clusters comprising of a non-pathogenic clade, a low pathogenic clade and a highly pathogenic clade. Surprisingly, a larger than expected number of genetic differences (125 CDSs), were found in all highly pathogenic biotype 1B strains but absent in the other clades (Howard et al. 2006). These included several previously uncharacterised CDSs that may encode novel virulence determinants including a hemolysin, metalloprotease and type III secretion effector protein. Additionally 27 CDSs were identified which were present in all low pathogenicity strains but absent from the non-pathogenic 1A isolates (Howard et al. 2006).

Recently, several effectors proteins secreted by the chromosomally encoded Ysa T3SS were defined in a strain derived from the sequenced genome *Y. enterocolitica* 8081 1B serotype O:8 (Matsumoto and Young 2006). To investigate whether genes encoding the effectors identified in 8081 were unique to this strain or were also present in other biotype 1B isolates, we searched DNA microarray data obtained from eight US *Y. enterocolitica* biotype 1B strains comprising a range of different serotypes (Table 2) (Howard et al. 2006). With the exception of *yspY* and *yspP* the remaining suite of Ysa effectors were present in all 1B isolates, indicating that although these genes are scattered throughout the chromosome of 8081 it is unlikely that each of these strains has acquired these genes independently.

1.5.2 Variation in Prophage and GIs

Consistent with observations made from many other bacterial genomes the single largest source of large scale genetic variation between *Y. enterocolitica* isolates were prophage, none of which were conserved in the non- (biotype 1A) or mildly pathogenic (biotypes 2, 3, and 4) *Y. enterocolitica*. Although, perhaps more surprising was

Table 2. Distribution of chromosomally encoded effectors secreted by the Ysa T3SS within eight US *Y. enterocolitica* biotype 1B strains determined by comparative DNA microarray analysis

Designated name	Gene[a]	Reference	Strains containing Ysa effector protein(s) determind by DNA microarray analysis
YspA	YE3534	Foultier *et al*. (2002)	8/8 strains tested
YspB	YE3537	Foultier *et al*. (2002)	8/8 strains tested
YspL	YE0113	Matsumoto and Young (2006)	8/8 strains tested
YspN	YE3543	Matsumoto and Young (2006)	8/8 strains tested
YspC	YE3536	Foultier *et al*. (2002)	8/8 strains tested
YspY	YE1357	Matsumoto and Young (2006)	5/8 strains tested
YspF	YE1986	Matsumoto and Young (2006)	8/8 strains tested
YspP	YE4194	Matsumoto and Young (2006)	4/8 strains tested
YspE	YE0115	Matsumoto and Young (2006)	ND
YspD	YE3535	Foultier *et al*. (2002)	8/8 strains tested
YspI	YE2444	Matsumoto and Young (2006)	8/8 strains tested
YspK	YE2447	Matsumoto and Young (2006)	8/8 strains tested

[a] *Y. enterocolitica* 8081 systematic gene name. ND no data.

that one of the *Y. enterocolitica* 8081 prophage, ϕYE200, was fully or partially represented in all other biotype 1B isolates. This phage is notable since it appears to be a degenerate P2-family prophage which carries the restriction/modification enzyme *Yen*I within a low G+C pocket of its sequence.

In general there was also considerable variation in the complement and structure of genomic islands detected in *Y. enterocolitica* strain 8081 compared to the other *Y. enterocolitica* isolates analysed. The exception being some of the important species-specific metabolic loci identified in the *in silico* analysis. These data confirmed that *cel*, *hyb*, *hyf* and *cob-pdu* (Thomson et al. 2006) were detected in all *Y. enterocolitica* biotypes tested and so are likely to represent key factors for niche adaptation by this enteropathogen.

The most extreme variation within a single locus was seen in the PZ, sub-regions of which showed significant heterogeneity within biotypes 1B, 1A, 2,3 and 4 with notable sub-regions of the PZ showing clear biotype-specific delineations, consistent with it being a region of hypervariability (a summary of these differences is shown in Table 3 and described below). Two regions within the PZ were common to all strains; the hydrogenase 2 cluster (*hyb*) and a locus predicted to encode SpeF and PotE. As previously reported, the chromosomally encoded T3SS, Ysa, and Yst1, were only detected in *Y. enterocolitica* biotype 1B strains.

Similarly YAPI$_{ye}$ was only detected in *Y. enterocolitica* 8081. However, regions common to all members of this family of GIs (encoding a type IV pilus system) were detected in one other biotype 1B isolate indicating a distinct but related island is present in this strain (Thomson et al. 2006). Furthermore, the microarray analysis confirmed that the siderophore uptake and utilization system YE3615-YE3622, was present in biotype 1A, 1B, 2 and 3 isolates but absent from biotype 4 strains (Schubert et al. 1999). This pattern of variability is consistent with the notion that the PZ is composed of multiple independently acquired units and offers potential for a PCR-based screen.

The other GI's mentioned above also showed marked variation: YGI-2 was found to be restricted to biotypes 1A and 1B only. YGI-4 was restricted to only a small number of other *Y. enterocolitica* 1B strains (3 out of 8 1B isolates) and YGI-3 was

Table 3. Summary of biotype specific regions within the PZ determined using comparative DNA microarray analysis of 94 strains of *Y. enterocolitica*

Region within PZ	Locus/Gene name(s)	Biotype 1B	Biotype 1A	Biotype 2	Biotype 3	Biotype 4
YE3450-YE3515	$YAPI_{ye}$	Present[$]	Absent	Absent	Absent	Absent
YE3526-YE3532	speF, potE	Present	Present	Present	Present	Present
YE3533-YE3580	ysa, yst1	Present	Absent	Absent	Absent	Absent
YE3611-YE3614	flgII	Present	Absent	Absent	Absent	Absent
YE3615-YE3622	fep, fes	Present	Present	Present	Present	Absent

[$] Entire $YAPI_{ye}$ locus only present in biotype 1B strain 8081.

unique to the sequenced *Y. enterocolitica* 1B strain 8081. Conversely the YGI-1, predicted to encode the *tad* genes, was shown to be present in all of the pathogenic *Y. enterocolitica* biotypes tested (1B, 2-4) but absent from the non-pathogenic 1A biotypes tested (37 strains). This reinforces the view that this locus is important for enteropathogenicity (Howard et al. 2006; Thomson et al. 2006).

1.6 Summary

The genome of *Y. enterocolitica* revealed important insights into gene loss and acquisition that have occurred since these yersiniae diverged. Essentially these could be defined as: loci that were *Y. enterocolitica*–specific, but which showed evidence of having been present and subsequently deleted from both *Y. pestis* and *Y. pseudotuberculosis*; those that were present in *Y. enterocolitica* and *Y. pseudotuberculosis* but which had been lost from *Y. pestis*; and the functions appearing to be truly species-specific, presumably acquired since divergence. Therefore as well as defining functions that are likely to be important for enteropathogenicity (being no longer important to *Y. pestis*) the loss of ancestral functions also highlight regions which, whilst still important to *Y. enterocolitica,* are clearly no longer essential for the enteric lifestyle of *Y. pseudotuberculosis.* Therefore this predicts that the niche and growth dynamics within the host are significantly different between *Y. enterocolitica* and *Y. pseudotuberculosis*.

Microarray analysis confirmed that flexible loci such as the PZ are important sites of variation encompassing regions known to be important for virulence or niche adaptation and showing marked difference in their species, biotype and strain distribution.

Interestingly, patterns of gene loss and acquisition seen when comparing the *Yersinia* are replicated in other bacterial pathogens such as *S. typhi* and *Salmonella paratyphi* A which, like *Y. pestis*, have become acute human pathogens whilst their relatives have remained essentially as enteropathogens. This implies that members of the yersiniae and salmonellae have found common solutions to niche adaptation by gene acquisition and loss, perhaps even occupying similar metabolic niches.

1.7 References

Achtman, M., Morelli, G., Zhu, P., Wirth, T., Diehl, I., Kusecek, B., Vogler, A.J., Wagner, D.M., Allender, C.J., Easterday, W.R., Chenal-Francisque, V., Worsham, P., Thomson, N.R., Parkhill, J., Lindler, L.E., Carniel, E. and Keim, P. (2004) Microevolution and history of the plague bacillus, *Yersinia pestis*. Proc. Natl. Acad. Sci. USA 101, 17837-17842.

Achtman, M., Zurth, K., Morelli, G., Torrea, G., Guiyoule, A. and Carniel, E. (1999) *Yersinia pestis,* the cause of plague, is a recently emerged clone of *Yersinia pseudotuberculosis*. Proc. Natl. Acad. Sci. USA 96, 14043-14048.

Campbell, A. (2003) Prophage insertion sites. Res Microbiol. 154, 277-282.

Chain, P.S., Carniel, E., Larimer, F.W., Lamerdin, J., Stoutland, P.O., Regala, W.M., Georgescu, A.M., Vergez, L.M., Land, M.L., Motin, V.L., Brubaker, R.R., Fowler, J., Hinnebusch, J., Marceau, M., Medigue, C., Simonet, M., Chenal-Francisque, V., Souza, B., Dacheux, D., Elliott, J.M., Derbise, A., Hauser, L.J. and Garcia, E. (2004) Insights into the evolution of *Yersinia pestis* through whole-genome comparison with *Yersinia pseudotuberculosis*. Proc. Natl. Acad. Sci. USA 101, 13826-13831.

Collyn, F., Billault, A., Mullet, C., Simonet, M. and Marceau, M. (2004) YAPI, a new *Yersinia pseudotuberculosis* pathogenicity island. Infect. Immun. 72, 4784-4790.

Collyn, F., Guy, L., Marceau, M., Simonet, M. and Roten, C.A. (2006) Describing ancient horizontal gene transfers at the nucleotide and gene levels by comparative pathogenicity island genometrics. Bioinformatics 22, 1072-1079.

Day, W.A., Jr., Fernandez, R.E. and Maurelli, A.T. (2001) Pathoadaptive mutations that enhance virulence: genetic organization of the *cadA* regions of *Shigella* spp. Infect. Immun. 69, 7471-7480.

Deng, W., Burland, V., Plunkett, G., 3rd, Boutin, A., Mayhew, G.F., Liss, P., Perna, N.T., Rose, D.J., Mau, B., Zhou, S., Schwartz, D.C., Fetherston, J.D., Lindler, L.E., Brubaker, R.R., Plano, G.V., Straley, S.C., McDonough, K.A., Nilles, M.L., Matson, J.S., Blattner, F.R. and Perry, R.D. (2002) Genome sequence of *Yersinia pestis* KIM. J. Bacteriol. 184, 4601-4611.

Foultier, B., Troisfontaines, P., Muller, S., Opperdoes, F.R. and Cornelis, G.R. (2002) Characterization of the *ysa* pathogenicity locus in the chromosome of *Yersinia enterocolitica* and phylogeny analysis of type III secretion systems. J. Mol. Evol. 55, 37-51.

Haller, J.C., Carlson, S., Pederson, K.J. and Pierson, D.E. (2000) A chromosomally encoded type III secretion pathway in *Yersinia enterocolitica* is important in virulence. Mol. Microbiol. 36, 1436-1446.

Howard, S.L., Gaunt, M.W., Hinds, J., Witney, A.A., Stabler, R. and Wren, B.W. (2006) Application of comparative phylogenomics to study the evolution of *Yersinia enterocolitica* and to identify genetic differences relating to pathogenicity. J. Bacteriol. 188, 3645-3653.

Iwobi, A., Heesemann, J., Garcia, E., Igwe, E., Noelting, C. and Rakin, A. (2003) Novel virulence-associated type II secretion system unique to high-pathogenicity *Yersinia enterocolitica*. Infect. Immun. 71, 1872-1879.

Kachlany, S.C., Planet, P.J., Bhattacharjee, M.K., Kollia, E., DeSalle, R., Fine, D.H. and Figurski, D.H. (2000) Nonspecific adherence by *Actinobacillus actinomycetemcomitans* requires genes widespread in bacteria and archaea. J. Bacteriol. 182, 6169-6176.
Lawrence, J.G. and Roth, J.R. (1996) Evolution of coenzyme B12 synthesis among enteric bacteria: evidence for loss and reacquisition of a multigene complex. Genetics 142, 11-24.
Maier, R.J. (2005) Use of molecular hydrogen as an energy substrate by human pathogenic bacteria. Biochem. Soc. Trans. 33, 83-85.
Maier, R.J., Olczak, A., Maier, S., Soni, S. and Gunn, J. (2004) Respiratory hydrogen use by *Salmonella enterica* serovar Typhimurium is essential for virulence. Infect. Immun. 72, 6294-6299.
Matsumoto, H. and Young, G.M. (2006) Proteomic and functional analysis of the suite of Ysp proteins exported by the Ysa type III secretion system of *Yersinia enterocolitica* Biovar 1B. Mol. Microbiol. 59, 689-706.
McNally, A., Cheasty, T., Fearnley, C., Dalziel, R.W., Paiba, G.A., Manning, G. and Newell, D.G. (2004) Comparison of the biotypes of *Yersinia enterocolitica* isolated from pigs, cattle and sheep at slaughter and from humans with yersiniosis in Great Britain during 1999-2000. Lett. Appl. Microbiol. 39, 103-108.
Mohd-Zain, Z., Turner, S.L., Cerdeno-Tarraga, A.M., Lilley, A.K., Inzana, T.J., Duncan, A.J., Harding, R.M., Hood, D.W., Peto, T.E. and Crook, D.W. (2004) Transferable antibiotic resistance elements in *Haemophilus influenzae* share a common evolutionary origin with a diverse family of syntenic genomic islands. J. Bacteriol. 186, 8114-8122.
Olson, J.W. and Maier, R.J. (2002) Molecular hydrogen as an energy source for *Helicobacter pylori*. Science 298, 1788-1790.
Parkhill, J., Wren, B.W., Thomson, N.R., Titball, R.W., Holden, M.T., Prentice, M.B., Sebaihia, M., James, K.D., Churcher, C., Mungall, K.L., Baker, S., Basham, D., Bentley, S.D., Brooks, K., Cerdeno-Tarraga, A.M., Chillingworth, T., Cronin, A., Davies, R.M., Davis, P., Dougan, G., Feltwell, T., Hamlin, N., Holroyd, S., Jagels, K., Karlyshev, A.V., Leather, S., Moule, S., Oyston, P.C., Quail, M., Rutherford, K., Simmonds, M., Skelton, J., Stevens, K., Whitehead, S. and Barrell, B.G. (2001) Genome sequence of *Yersinia pestis*, the causative agent of plague. Nature. 413, 523-527.
Perry, R.D. and Fetherston, J.D. (1997) *Yersinia pestis*--etiologic agent of plague. Clin. Microbiol. Rev. 10, 35-66.
Pickard, D., Wain, J., Baker, S., Line, A., Chohan, S., Fookes, M., Barron, A., Gaora, P.O., Chabalgoity, J.A., Thanky, N., Scholes, C., Thomson, N., Quail, M., Parkhill, J. and Dougan, G. (2003) Composition, acquisition, and distribution of the Vi exopolysaccharide-encoding *Salmonella enterica* pathogenicity island SPI-7. J. Bacteriol. 185, 5055-5065
Prentice, M.B., Cope, D. and Swann, R.A. (1991) The epidemiology of *Yersinia enterocolitica* infection in the British Isles 1983-1988. Contrib. Microbiol. Immunol. 12, 17-25.
Rosqvist, R., Skurnik, M. and Wolf-Watz, H. (1988) Increased virulence of *Yersinia pseudotuberculosis* by two independent mutations. Nature 334, 522-524.
Schreiner, H.C., Sinatra, K., Kaplan, J.B., Furgang, D., Kachlany, S.C., Planet, P.J., Perez, B.A., Figurski, D.H. and Fine, D.H. (2003) Tight-adherence genes of *Actinobacillus actinomycetemcomitans* are required for virulence in a rat model. Proc. Natl. Acad. Sci. USA 100, 7295-7300.
Schubert, S., Fischer, D. and Heesemann, J. (1999) Ferric enterochelin transport in *Yersinia enterocolitica*: molecular and evolutionary aspects. J. Bacteriol. 181, 6387-6395.
Schubert, S., Rakin, A. and Heesemann, J. (2004) The Yersinia high-pathogenicity island (HPI): evolutionary and functional aspects. Int. J. Med. Microbiol. 294, 83-94.
Sekowska, A., Denervaud, V., Ashida, H., Michoud, K., Haas, D., Yokota, A. and Danchin, A. (2004) Bacterial variations on the methionine salvage pathway. BMC Microbiol. 4, 9.

Song, Y., Tong, Z., Wang, J., Wang, L., Guo, Z., Han, Y., Zhang, J., Pei, D., Zhou, D., Qin, H., Pang, X., Zhai, J., Li, M., Cui, B., Qi, Z., Jin, L., Dai, R., Chen, F., Li, S., Ye, C., Du, Z., Lin, W., Yu, J., Yang, H., Huang, P. and Yang, R. (2004) Complete genome sequence of *Yersinia pestis* strain 91001, an isolate avirulent to humans. DNA Res. 11, 179-197.

Thomson, N.R., Howard, S., Wren, B.W., Holden, M.T., Crossman, L., Challis, G.L., Churcher, C., Mungall, K., Brooks, K., Chillingworth, T., Feltwell, T., Abdellah, Z., Hauser, H., Jagels, K., Maddison, M., Moule, S., Sanders, M., Whitehead, S., Quail, M.A., Dougan, G., Parkhill, J. and Prentice, M.B. (2006) The complete genome sequence and comparative genome analysis of the high pathogenicity *Yersinia enterocolitica* strain 8081. PLoS Genet 2, e206.

Van Noyen, R., Vandepitte, J., Wauters, G. and Selderslaghs, R. (1981) *Yersinia enterocoli tica*: its isolation by cold enrichment from patients and healthy subjects. J. Clin. Pathol. 34, 1052-1056.

Wauters, G., Kandolo, K. and Janssens, K. (1987) Revised biogrouping scheme of *Yersinia enterocolitica*. Contrib. Microbiol. Immunol. 9, 14-21.

Wren, B.W. (2003) The yersiniae–a model genus to study the rapid evolution of bacteria pathogens. Nat. Rev. Microbiol. 1, 55-64.

Zogaj, X., Nimtz, M., Rohde, M., Bokranz, W. and Romling, U. (2001) The multi-cellular morphotypes of *Salmonella typhimurium* and *Escherichia coli* produce cellulose as the second component of the extracellular matrix. Mol. Microbiol. 39, 1452-1463.

Picture 4. Nicholas Thomson presents the whole genome analysis of *Y. enterocolitica* 8081. Photograph by A. Animisov.

2

Pestoides F, an Atypical *Yersinia pestis* Strain from the Former Soviet Union

Emilio Garcia[1], Patricia Worsham[2], Scott Bearden[3], Stephanie Malfatti[1], Dorothy Lang[1], Frank Larimer[4], Luther Lindler[5], and Patrick Chain[1]

[1] Biosciences and Biotechnology Division, Chemistry Materials and Life Sciences Directorate, Lawrence Livermore National Laboratory, garcia12@llnl.gov
[2] United States Army Medical Research Institute of Infectious Diseases
[3] Bacterial Zoonoses Branch, Centers for Disease Control and Prevention
[4] Oak Ridge National Laboratory
[5] Department of Bacterial Diseases, Walter Reed Army Institute of Research

Abstract. Unlike the classical *Yersinia pestis* strains, members of an atypical group of *Y. pestis* from Central Asia, denominated *Y. pestis* subspecies *caucasica* (also known as one of several pestoides types), are distinguished by a number of characteristics including their ability to ferment rhamnose and melibiose, their lack of the small plasmid encoding the plasminogen activator (*pla*) and pesticin, and their exceptionally large variants of the virulence plasmid pMT (encoding murine toxin and capsular antigen). We have obtained the entire genome sequence of *Y. pestis* Pestoides F, an isolate from the former Soviet Union that has enabled us to carryout a comprehensive genome-wide comparison of this organism's genomic content against the six published sequences of *Y. pestis* and their *Y. pseudotuberculosis* ancestor. Based on classical glycerol fermentation (+ve) and nitrate reduction (+ve) *Y. pestis* Pestoides F is an isolate that belongs to the biovar antiqua. This strain is unusual in other characteristics such as the fact that it carries a non-consensus V antigen (*lcrV*) sequence, and that unlike other Pla$^-$ strains, Pestoides F retains virulence by the parenteral and aerosol routes. The chromosome of Pestoides F is 4,517,345 bp in size comprising some 3,936 predicted coding sequences, while its pCD and pMT plasmids are 71,507 bp and 137,010 bp in size respectively. Comparison of chromosome-associated genes in Pestoides F with those in the other sequenced *Y. pestis* strains reveals differences ranging from strain-specific rearrangements, insertions, deletions, single nucleotide polymorphisms, and a unique distribution of insertion sequences. There is a single ~7 kb unique region in the chromosome not found in any of the completed *Y. pestis* strains sequenced to date, but which is present in the *Y. pseudotuberculosis* ancestor. Taken together, these findings are consistent with Pestoides F being derived from the most ancient lineage of *Y. pestis* yet sequenced.

2.1 Introduction

Most of our knowledge of *Yersinia pestis* at the genomic level has been obtained from studies on strains derived from "classical" isolates of the Americas, Africa and some from Asia that share some broad phenotypic and virulence properties. However, atypical strains from the territories encompassed by the former Soviet Union harbor a number of clearly distinct *Y. pestis* strains described extensively in the recent review by Anisimov et al. (2004). Standard biochemical characteristic, their specific animal host range, virulence properties as well as numerical taxonomy (Martinevskii 1969) have helped classify these widely diverse strains of *Y. pestis*. The term "pestoides", in particular, was first coined by Martinevskii to describe *Y. pestis* isolates from natural foci located in the Transcaucasian highland, in the Mountain Altai and Transbikalian regions. The term was later adopted by scientists in the US to designate strains derived from the former Soviet Union (FSU) (Worsham and Roy 2003). Pestoides F, the strain whose genomic characteristics are being presented in this work, is one of a number of "pestoides" strains (known as subspecies *caucasica* in the FSU) originally described by Worsham and colleagues that is characterized by its atypical biochemical characteristics, its lack of pPCP (one of the unique virulence plasmids of the *Y. pestis* group), the presence of an enlarged pMT plasmid, possession of unusual animal host specificity and full virulence by parenteral and aerosol route in the mouse model (Worsham and Hunter 1998; Worsham and Roy 2003).

In recent studies by Achtman et al. (2005), the pestoides isolates have been placed as early offshoots in phylogenetic trees, together with another recently characterized avirulent Y. pestis strain, 91001, which has been given the designation of subspecies Microtus. These studies however, were based on single nucleotide polymorphisms found among the strains available at the time. The result is that isolates closely related to those whose genomes have been sequenced can be firmly placed on the phylogenetic tree, yet distinct isolates that belong to independent non-sequenced lineages can be only loosely placed. We have determined the complete sequence of *Y. pestis* Pestoides F, an isolate we have determined belongs to the FSU *caucasica* subspecies, and have performed preliminary analyses that firmly place this group as the oldest lineage sequenced to date.

2.2 Materials and Methods

Completed whole genome sequence was derived from a standard genome shotgun approach, deep (~10-fold) sequencing of 3 differently sized libraries (average sizes of 3 kb, 7 kb and 40 kb). Genome closure was performed by a combination of PCR-sequencing and directed primer walking off of clones. Genome assembly was verified with the properly assembled clone end-sequences as well as by PCR where physical gaps remained. All repeat sequences surpassing the length of sequencing reads (~600 bp) were separately assembled to assure correct sequence and assembly.

Whole genome sequence alignments were performed using MUMmer (Kurtz et al. 2004) and/or BLAST (Altschul et al. 1997) and visualized using ACT. Separate gene alignments were conducted using BLAST and/or CLUSTALW (Thompson et al. 1994).

2.3 Results

The *Y. pestis* Pestoides F genome has been sequenced to completion and is avail able for download from our Lawrence Livermore Laboratory web site (http://www.llnl.gov/bio/groups/genomics_virulence/Yersinia/YersiniaGenomics.html) as well as from our sister site at the Oak Ridge National Laboratory Computational Biology site, as part of the Department of Energy's Joint Genome Institute (https://maple.lsd.ornl.gov/microbial/ypes_1570/). The genome of Pestoides F consists of a single chromosome of 4,517,345 bp and only two of the three virulence plasmids typical of *Y. pestis* strains, pCD and pMT at 71,507 bp and 137,010 bp in size respectively.

The pCD1 plasmid is essentially identical to that of CO92 but contains a non-consensus *lcrV* (Fig. 1), a gene encoding the V antigen, an important factor involved in immunosuppresion of the host during *Yersinia* infection. In contrast, the pMT plasmid of Pestoides F is substantially larger (137,010 bp in size) than those of the classical *Yersinia* (at roughly 96-100 kb) and is quite similar to that of the pFra plasmid reported by Golubov et al. (2004), although it contains a number of single nucleotide polymorphisms and rearrangements.

As in the other *Y. pestis* and *Y. pseudotuberculosis* genomes (Parkhill et al. 2001; Deng et al. 2002; Song et al. 2004; Chain et al. 2004) the chromosome of Pestoides F displays a large number of rearrangements compared to all other *Y. pestis* chromosomes. There are a total of 106 IS elements (of the four main types: IS*100*, IS*1541*, IS*285*, IS*1661*) in the Pestoides F chromosome, which are associated with most of the rearrangement events observed between *Y. pestis* (as well as *Y. pseudotuberculosis*) genomes. In addition to the repeated IS elements, there are a few other observed instances of repeated sequences, most prominently seven copies of the rRNA operon (3 copies of the 16S-tRNA-tRNA-23S-5S, and 4 copies of the 16S-tRNA-23S-5S), as is observed in most *Y. pestis* strains (only 6 are present in the KIM sequence, due to a presumed deletion between IS*100* elements).

The Pestoides F strain has been characterized biochemically as an isolate of the Antiqua biovar, despite its classification as a pestoides isolate (in the *caucasica* subspecies). We examined the genes responsible for this classification, as identified by Motin et al. (2001) and Achtman et al. (2005), and have found wild type versions of *glpD*, which in Orientalis strains harbors a deletion responsible for the glycerol fermentation negative phenotype, as well as wild type *napA*, which in Medievalis strains harbors one of two mutations responsible for the nitrate reduction negative phenotype. Of special note is a ~7 kb unique region not found in any other *Y. pestis* strain but shared with the enteropathogen ancestor, *Y. pseudotuberculosis* and several of the non-pathogenic Yersiniae (*Y. intermedia*, *Y. mollaretii* and *Y. frederiksenii*). This region has been found to encode a number of genes involved in various enzymatic activities (apolipoprotein acyl transferase, aminotransferase, aldolase, methylthioribose kinase).

```
Y. pseudo   LSVRGRGSLN IPTMGRDQEN QR*CEGKLFN MIRAYEQNPQ HFIEDLEKVR  50
Pestoides   .......... .......... ..*....... .......... ..........
Microtus    .......... .......... ..*....... .......... ..........
CO92,Kim    .......... .......... ..*....... .......... ..........
Antiqua     .......... .......... ..*....... .......... ..........

Y. pseudo   VEQLTGHGSS VLEELVQLVK DKNIDISIKY DPRKDSEVFA NRVITDDIEL  100
Pestoides   .......... .......... .......... .......... ..........
Microtus    .......... .......... .......... .......... ..........
CO92,KIM    .......... .......... .......... .......... ..........
Antiqua     .......... .......... .......... .......... ..........

Y. pseudo   LKKILAYFLP EDAILKGGHY DNQLQNGIKR VKEFLESSPN TQWELRAFMA  150
Pestoides   .......... .......... .......... .......... ..........
Microtus    .......... .......... .......... .......... ..........
CO92, KIM   .......... .......... .......... .......... ..........
Antiqua     .......... .......... .......... .......... ..........

Y. pseudo   VIHFSLTADR IDDDILKVIV DSMNHHGDAR SKLREELAEL TAELKIYSVI  200
Pestoides   .M........ .......... .......... .......... ..........
Microtus    .M........ .......... .......... .......... ..........
CO92, KIM   .M........ .......... .......... .......... ..........
Antiqua     .M........ .......... .......... .......... ..........

Y. pseudo   QAEINKHLSS GGTINIHDKS INLMDKNLYG YTDEEIFKAS AEYKILEKMP  250
Pestoides   .......... S......... .......... .......... ..........
Microtus    .......... S......... .......... .......... ..........
CO92, KIM   .......... S......... .......... .......... ..........
Antiqua     .......... S......... .......... .......... ..........

Y. pseudo   QTTIQEGETE KKIVSIKNFL ESEKKRTGAL GNLKDSYSYN KDNNELSHFA  300
Pestoides   .....VDGS. .......D.. G..N...... ....N..... ..........
Microtus    .....VDGS. .......D.. G..N...... ....N..... ..........
CO92, KIM   .....VDGS. .......D.. G..N...... ....N..... ..........
Antiqua     .....VDGS. .......D.. G..N...... ....N..... ..........

Y. pseudo   TTCSDKSRPL NDLVSQKTTQ LSDITSRFNS AIEALNRFIQ KYDSVMQRLL  350
Pestoides   .......... .......... .......... .......... ..........
Microtus    .......... .......... .......... .......... ..........
CO92, KIM   .......... .......... .......... .......... ..........
Antiqua     .......... .......... .......... .......... ..........

Y. pseudo   DDTSGK*HEV IMQQETTDTQ EYQLAMESFL KGGGTIA 387
Pestoides   ...R*      x......... .......... .......
Microtus    ...R*      x......... .......... .......
CO92, KIM   ......*... .......... .......... .......
Antiqua     ......*... .......... .......... .......
```

Fig. 1. Alignment of the *lcrV* gene of various *Y. pestis* strains. Dots indicate conserved amino acids; asterisks indicate a stop codon. A 16 nt deletion at the carboxyl end of Pestoides F and Microtus results in a shorter protein and a frameshift.

A recently acquired panel of *Y. pestis* isolates have been obtained from the National Center for Disease Control of the Republic of Georgia and, based on both phenotypic properties and an IS*100*-based genotyping (Motin et al. 2002), have been placed as being closest in properties to Pestoides F. We therefore compared plasmid composition, unique regions and pattern of gene inactivation between Pestoides F and the Georgian isolates. It was determined that in each case, the Pestoides F characteristics closely resemble those of the Georgian strains. This is in agreement with Pestoides F belonging to the *Y. pestis* subspecies *caucasica* in the classification used for strains from the former Soviet Union (Anisimov et al. 2004), and together with the above results, suggests that this group may belong to one of the oldest lineages of *Y. pestis*.

2.4 Conclusion

Detailed genome comparisons among closely related bacterial strains offer an incomparable opportunity to ascertain important evolutionary relationships that are more difficult, or impossible, to obtain by other means. The genome sequence of Pestoides F has been determined and found to be similar in content to other *Y. pestis* strains with a number of important exceptions. The comparison between this newly-sequenced atypical *Y. pestis* strain and those from previously sequenced members of this group has led us to conclude that Pestoides F likely belongs to the oldest linage of *Y. pestis* thus far sequenced. Facts that support this conclusion include the lack of pPCP plasmid that encodes the plasminogen activator and pesticin (presumably not acquired by this isolate) and the "unique" region of the genome of Pestoides F compared with all other completed *Y. pestis* that is shared with *Y. pseudotuberculosis* (the *Y. pestis* progenitor). Together with the observed genome rearrangement and IS element abundance/patterns these findings strongly suggest that Pestoides F derives from the most ancient lineage of *Y. pestis* yet studied.

2.5 Acknowledgments

This work was performed under the auspices of the U.S. Department of Energy by the University of California, Lawrence Livermore National Laboratory under contract No. W-7405-Eng-48. This work was financially supported by the Department of Homeland Security, Office of Science, Chemical and Biological Division.

2.6 References

Achtman, M., Morelli, G., Zhu, P., Wirth, T., Diehl, I., Kusecek, B., Vogler, A.J., Wagner, D.M., Allender, C.J., Easterday, W.R., Chenal-Francisque, V., Worsham, P., Thomson, N.R., Parkhill, J., Lindler, L.E., Carniel, E. and Keim, P. (2004) Microevolution and history of the plague bacillus, *Yersinia pestis*. PNAS 101, 17837-17842.

Altschul, S.F., Madden, T.L., Schaffer, A.A., Zhang, J., Zhang, Z., Miller, W. and Lipman, D.J. (1997) Gapped BLAST and PSI-BLAST: a new generation of protein database search programs. Nucleic Acids Res. 25, 3389-3402.

Anisimov, A.P., Lindler, L.E. and Pier, G.B. (2004) Intraspecific diversity of *Yersinia pestis*. Clin. Microbiol. Rev. 17, 434-464.

Chain, P.S.G., Carniel, E., Larimer, F.W., Lamerdin, J., Stoutland, P.O., Regala, W.M., Georgescu, A.M., Vergez, L.M., Land, M.L., Motin, V.L., Brubaker, R.R., Fowler, J., Hinnebusch, J., Marceau, M., Medigue, C., Simonet, M., Chenal-Francisque, V., Souza, B., Dacheux, D., Elliott, J.M., Derbise, A., Hauser, L.J. and Garcia, E. (2004) Insights into the evolution of *Yersinia pestis* through whole-genome comparison with *Yersinia pseudotuberculosis*. PNAS 101, 13826-13831.

Deng, W., Burland, V., Plunkett, G., III, Boutin, A., Mayhew, G.F., Liss, P., Perna, N.T., Rose, D.J., Mau, B., Zhou, S., Schwartz, D.C., Fetherston, J.D., Lindler, L.E., Brubaker, R.R., Plano, G.V., Straley, S.C., McDonough, K.A., Nilles, M.L., Matson, J.S., Blattner, F.R. and Perry, R.D. (2002) Genome Sequence of *Yersinia pestis* KIM. J. Bacteriol. 184, 4601-4611.

Golubov, A., Neubauer, H., Nolting, C., Heesemann, J. and Rakin, A. (2004) Structural Organization of the pFra Virulence-Associated Plasmid of Rhamnose-Positive *Yersinia pestis*. Infect. Immun. 72, 5613-5621.

Kurtz, S., Phillippy, A., Delcher, A. L., Smoot, M. Shumway, M., Antonescu, C. and Salzberg, S. L. (2004) Versatile and open software for comparing large genomes. Genome Biol. 5, R12.

Martinevsky, I.L. (1969) *Biology and genetic features of plague and plague-related microbes [in Russian]*. Moscow: Meditsina Press, Moscow, USSR.

Song, Y., Tong, Z., Wang, J., Wang, L., Guo, Z., Han, Y., Zhang, J., Pei, D., Zhou, D., Qin, H., Pang, X., Han, Y., Zhai, J., Li, M., Cui, B., Qi, Z., Jin, L., Dai, R., Chen, F., Li, S., Ye, C., Du, Z., Lin, W., Wang, J., Yu, J., Yang, H., Wang, J., Huang, P. and Yang, R. (2004) Complete genome sequence of *Yersinia pestis* strain 91001, an isolate avirulent to humans. DNA Res. 11, 179-197.

Thompson, J.D., Higgins, D.G. and Gibson, T.J. (1994) CLUSTAL W: improving the sensitivity of progressive multiple sequence alignment through sequence weighting, position-specific gap penalties and weight matrix choice. Nucleic Acids Res. 22, 4673-4680.

Worsham, P.L. and Hunter, M. (1998) Characterization of Pestoides F, an atypical strain of *Yersinia pestis*. Med Microbiol. 6(Suppl.II), 34-35.

Worsham, P.L. and Roy, C. (2003) Pestoides F, a *Yersinia pestis* strain lacking plasminogen activator, is virulent by the aerosol route. Adv. Exp. Med. Biol. 529, 129-131.

Picture 5. Emilio Garcia presents the Pestoides F genome analysis. Photograph by A. Anisimov.

3
Variability of the Protein Sequences of LcrV Between Epidemic and Atypical Rhamnose-Positive Strains of *Yersinia pestis*

Andrey P. Anisimov, Evgeniy A. Panfertsev, Tat'yana E. Svetoch, and Svetlana V. Dentovskaya

State Research Center for Applied Microbiology and Biotechnology, Obolensk, anisimov@obolensk.org

Abstract. Sequencing of *lcrV* genes and comparison of the deduced amino acid sequences from ten *Y. pestis* strains belonging mostly to the group of atypical rhamnose-positive isolates (non-*pestis* subspecies or pestoides group) showed that the LcrV proteins analyzed could be classified into five sequence types. This classification was based on major amino acid polymorphisms among LcrV proteins in the four "hot points" of the protein sequences. Some additional minor polymorphisms were found throughout these sequence types. The "hot points" corresponded to amino acids 18 (Lys → Asn), 72 (Lys → Arg), 273 (Cys → Ser), and 324-326 (Ser-Gly-Lys → Arg) in the LcrV sequence of the reference *Y. pestis* strain CO92. One possible explanation for polymorphism in amino acid sequences of LcrV among different strains is that strain-specific variation resulted from adaptation of the plague pathogen to different rodent and lagomorph hosts.

3.1 Introduction

LcrV (V antigen) of the medically significant yersiniae is a multifunctional protein involved in modification of innate immune response to these pathogens as well as in regulation and translocation of Yop effectors of the type III secretion system (Brubaker 2003). Moreover, LcrV is a major protective antigen and a principal component of the modern anti-plague subunit vaccines which are currently under development (Titball and Williamson 2004).

Recent studies also revealed some heterogeneity in the sequences of the *lcrV* genes of different origin (Adair et al. 2000; Anisimov, Lindler, and Pier 2004; Hakansson et al. 1993; Motin et al. 1992; Price et al. 1989; Roggenkamp et al. 1997; Sing et al. 2005; Snellings, Popek, and Lindler 2001; Song et al. 2004) especially in *Y. enterocolitica* (Hakansson et al. 1993; Sing et al. 2005; Snellings et al. 2001). Although major types of LcrV antigens of *Y. pestis* and *Y. pseudotuberculosis* are cross-protective (Motin et al. 1994), certain variations in the sequence of LcrV resulted in the reduction of the cross-protectivity against some yersiniae strains (Anisimov et al. 2004; Roggenkamp et al. 1997; Une and Brubaker 1984).

In this study we determined sequences of LcrV of ten strains of *Y. pestis* and compared them with those from the sequenced *lcrV* genes of this pathogen. Most of the strains used for our sequencing experiments belonged to atypical rhamnose-positive isolates of *Y. pestis* avirulent to guinea pigs and humans (Anisimov et al. 2004; Song et al. 2004). They are generally known as pestoides group, which according to the classification adopted in Former Soviet Union countries are referred as non-main subspecies of *Y. pestis* (Anisimov et al. 2004). In another classification, this category of strains is referred as isolates of the microtus biovar (Song et al. 2004). Accordingly, the isolates of the typical biovars such as antiqua, medievalis and orientalis we designated here as epidemic strains to emphasize their historic role in plague pandemics.

3.2 Materials and Methods

The strains of *Y. pestis* analyzed in this study belonged to five "subspecies" circulating in the Eurasian natural plague foci and differed in their epidemiological significance (Table 1; Anisimov et al. 2004). All *lcrV* genes were sequenced by using primers LcrVF (5'-CAGCCTCAACATCCCTACGA-3'), LcrVFI (5'-GCAAAAT GGCATCAAGCGAG-3'), and LcrVR (5'-TGTCTGTCGTCTCTTGTTGC-3') and compared with the data available at GenBank/EMBL/DDBJ (accession numbers, M26405, AF167309, AF167310.1, AE017043.1, CP000311).

3.3 Results and Discussion

3.3.1 Comparative Analysis of the V Antigen Sequence Heterogeneity

To address the genetic variation among LcrV antigens of *Y. pestis* strains of different origin and epidemiological significance, we determined the complete sequence of *lcrV* genes for ten strains. We found that only three sequences were identical to the predominant *lcrV* sequence-type initially reported by Price et al. (1989). Two of the isolates which showed this type of LcrV (I-1996 and I-2638, see Table 1) belonged to the epidemic type of *Y. pestis* strains, biovar antiqua, and one isolate (I-2836, Table 1) was atypical. Furthermore, nucleotide sequence analysis of the amplified *lcrV* alleles indicated that the sizes of the *lcrV* genes ranged from 975 to 981 nucleotides in length (encoding 324 to 326 amino acids). The LcrV proteins analyzed can be classified into sequence types A - biovar microtus strain 91001; subspecies *caucasica* strains: Pestoides F, 1146, C-585 (324 amino acids), B - subspecies *pestis* biovar antiqua strains: Antiqua, I-1996, I-2638, biovar medievalis strain KIM, biovar orientalis strain CO92; subspecies *ulegeica* strain I-2836 (326 amino acids), B/C - subspecies *ulegeica* strain I-2422 (326 amino acids), C - subspecies *hissarica* strain A-1728; subspecies *caucasica* strain C-582 (324 amino acids), and D - biovar antiqua strain Angola (326 amino acids); subspecies *altaica* strains: I-3455 (326 amino acids), I-2359 (325 amino acids). This classification was based on appearance of major amino acid polymorphisms among LcrV antigens in the four "hot points" of the protein sequences. Some additional minor polymorphisms were found through out these sequence types. The "hot points" correspond to amino acids 18 (Lys →

Table 1. *Y. pestis* strains used in these studies

Strain	Geographical origin[a]	Biovar/ subspecies[b]	Main host	*lcrV* accession number	Reference
C-585	Transcaucasian highland (foci #4-6)	antiqua / *caucasica*	*M. arvalis*		This study
C-582	Transcaucasian highland (foci #4-6)	antiqua / *caucasica*	*M. arvalis*	DQ489557.1	This study
1146	Transcaucasian highland (foci #4-6)	antiqua / *caucasica*	*M. arvalis*		This study
I-1996	Trans-Baikal focus #38	antiqua / *pestis*	*Citellus dauricus*		This study
I-2638	Mongun-Taigin focus #37	antiqua / *pestis*	*Citellus undulatus*		This study
A-1728	Gissar focus #34, Tadji-kistan, Uzbekistan	medievalis / *hissarica*	*Microtus carruthersi*	DQ489552.1	This study
I-2422	Northeast Mongolia, Gobi Desert	medievalis / *ulegeica*	*O. pricei*	DQ489554.1	This study
I-2836	Northeast Mongolia, Gobi Desert	medievalis / *ulegeica*	*Ochotona pricei*	DQ489553.1	This study
I-2359	Mountain-Altai focus #36	medievalis / *altaica*	*O. pricei*	DQ489556.1	This study
I-3455	Mountain-Altai focus #36	medievalis / *altaica*	*O. pricei*	DQ489555.1	This study

For information on [a]geographical location of plague natural foci and [b]biovar-subspecies interrelations see ref. (Anisimov et al. 2004).

Asn), 72 (Lys → Arg), 273 (Cys → Ser), and 324-326 (Ser-Gly-Lys → Arg) in the LcrV sequence of the reference epidemic strain CO92:

type A	Lys_{18},	Lys_{72},	Cys_{273},	Arg_{324};
type B	Lys_{18},	Lys_{72},	Cys_{273},	$Ser\text{-}Gly\text{-}Lys_{324\text{-}326}$;
type B/C	Asn_{18},	Arg_{72},	Cys_{273},	$Ser\text{-}Gly\text{-}Lys_{324\text{-}326}$;
type C	Asn_{18},	Arg_{72},	Ser_{273},	Arg_{324};
type D	Asn_{18},	Arg_{72},	Ser_{273},	$Ser\text{-}Gly\text{-}Lys_{324\text{-}326}$.

The LcrV of the type B group is a predominant variant among *Y. pestis* strains sequenced so far. It has been found in 23 *Y. pestis* strains belonging to all three biovars of the epidemic *Y. pestis* isolates originated from Asia, Africa and Americas as well as in one representative of *ulegeica* subspecies from Mongolia (Adair et al. 2000; Motin et al. 1992; Price et al.; this study). This type of LcrV is the most homologous to LcrV from *Y. pseudotuberculosis* strains of 1b and 3 serovars available in the GenBank/EMBL/DDBJ (Bergman et al. 1991; Motin et al. 1992). The strains possessing LcrV type B may have "universal virulence" (epidemic strains are pathogenic to many mammals) or be pathogenic only for a few rodent and lagomorph species including their natural host (atypical, subspecies *ulegeica*) (Anisimov et al. 2004). It is unlikely that the "selective virulence" of the atypical rhamnose-positive

Y. pestis strains is a result of variations in the LcrV amino-acid sequence, although other LcrV types were found exclusively in the atypical strains circulating in the populations of *Microtus* species (types A and C) or *O. pricei* (types B/C and D). Data about the natural host of the strain Angola, type D are not available. The strains carrying the same LcrV types are circulating in geographically distant natural plague foci: type A – foci #4-6 (Transcaucasian highland, Armenia, Azerbaijan, Georgia) and focus L (Inner Mongolia, China); type C – foci #4-6 (Transcaucasian highland, Armenia, Azerbaijan, Georgia) and focus #34 (Hissarian Ridge, Tadjikistan, Uzbekistan); type D – focus #36 (Mountain Altai, Russia) and Angola (Africa).

The deletion that was caused by two direct repeats (ATGACACG) at the 3' terminus of *lcrV* gene (Song et al. 2004) is characteristic of the *Y. pestis* types A and C (this study) as well as to the majority of LcrV variants from *Y. enterocolitica* (GenBank/EMBL/DDBJ accession numbers, X96796.1, X96797.1, X96798.1, X96799.1, X96800.1, X96801.1 (Roggenkamp et al. 1997), AF102990.1 (Hakansson et al. 1993), AF336309.1 (Snellings et al. 2001), and AY150843.2). Moreover, other amino-acid replacements such as amino acids 18 (Lys → Asn), 72 (Lys → Arg), 273 (Cys → Ser), are specific only for *Y. pestis*.

Taken together, these observations demonstrate that LcrV can display polymorphism in size and amino acid sequence among atypical rhamnose-positive *Y. pestis* strains. The presence of a modified LcrV protein apparently have not altered lethality of these strains for mice and their natural hosts, since atypical rhamnose-positive *Y. pestis* strains were reported to be highly virulent for these animal species (Anisimov et al. 2004; Song et al. 2004). However, the influence of LcrV sequence polymorphism on the "selective virulence" (host range) of these strains needs to be further investigated.

3.3.2 Conclusion

This study demonstrated that the LcrV antigens analyzed could be classified into five sequence types, according to the appearance of major amino acid polymorphisms among LcrV proteins in the four "hot points" of the proteins with some minor polymorphisms found throughout these sequence types. One possible explanation for variation in amino acid sequences among different strains is that strain-specific variation might result from adaptation for continued transmission of the plague pathogen within different rodent and lagomorph species. Further explorations into functional activity of different *Y. pestis* LcrV sequence types will reveal the biological significance of this phenomenon.

3.4 Acknowledgments

This work was performed within the framework of the International Science and Technology Center (ISTC) Project #2426. The authors are grateful to Dr. Vladimir L. Motin for helpful discussions and comments.

3.5 References

Adair, D.M., Worsham, P.L., Hill, K.K., Klevytska, A.M., Jackson, P.J., Friedlander, A.M., and Keim, P. (2000) Diversity in a variable-number tandem repeat from *Yersinia pestis*. J. Clin. Microbiol. 38, 1516-1519.

Anisimov, A.P., Lindler, L.E. and Pier, G.B. (2004) Intraspecific diversity of *Yersinia pestis*. Clin. Microbiol. Rev. 17, 434-464.

Bergman, T., Hakansson, S., Forsberg, A., Norlander, L., Macellaro, A., Backman, A., Bolin, I., and Wolf-Watz, H. (1991) Analysis of the V antigen lcrGVH-yopBD operon of Yersinia pseudotuberculosis: evidence for a regulatory role of LcrH and LcrV. J. Bacteriol. 173 (5), 1607-1616.

Brubaker, R.R. (2003) Interleukin-10 and inhibition of innate immunity to yersiniae: roles of Yops and LcrV (V antigen). Infect. Immun. 71, 3673-3681.

Hakansson, S., Bergman, T., Vanooteghem, J.C., Cornelis, G., and Wolf-Watz, H. (1993) YopB and YopD constitute a novel class of *Yersinia* Yop proteins. Infect. Immun. 61, 71-80.

Motin, V.L., Nakajima, R., Smirnov, G.B., and Brubaker, R.R. (1994) Passive immunity to yersiniae mediated by anti-recombinant V antigen and protein A-V antigen fusion peptide. Infect. Immun. 62, 4192-4201.

Motin, V.L., Pokrovskaya, M.S., Telepnev, M.V., Kutyrev, V.V., Vidyaeva, N.A., Filippov, A.A., and Smirnov, G.B. (1992) The difference in the *lcrV* sequences between *Y. pestis* and *Y. pseudotuberculosis* and its application for characterization of *Y. pseudotuberculosis* strains. Microb. Pathog. 12, 165-175.

Price, S.B., Leung, K.Y., Barveand, S.S., and Straley, S.C. (1989) Molecular analysis of *lcrGVH*, the V antigen operon of *Yersinia pestis*. J. Bacteriol. 171, 5646-5653.

Roggenkamp, A., Geiger, A.M., Leitritz, L., Kessler, A., and Heesemann, J. (1997) Passive immunity to infection with *Yersinia* spp. mediated by anti-recombinant V antigen is dependent on polymorphism of V antigen. Infect. Immun. 65, 446-451.

Sing, A., Reithmeier-Rost, D., Granfors, K., Hill, J., Roggenkamp, A., and Heesemann, J. (2005) A hypervariable N-terminal region of *Yersinia* LcrV determines Toll-like receptor 2-mediated IL-10 induction and mouse virulence. Proc. Natl. Acad. Sci. USA. 102, 16049-16054.

Snellings, N.J., Popek, M., and Lindler, L.E. (2001) Complete DNA sequence of *Yersinia enterocolitica* serotype O:8 low-calcium-response plasmid reveals a new virulence plasmid-associated replicon. Infect. Immun. 69, 4627-4638.

Song, Y., Tong, Z., Wang, J., Wang, L., Guo, Z., Han, Y., Zhang, J., Pei, D., Zhou, D., Qin, H., Pang, X., Han, Y., Zhai, J., Li, M., Cui, B., Qi, Z., Jin, L., Dai, R., Chen, F., Li, S., Ye, C., Du, Z., Lin, W., Wang, J., Yu, J., Yang, H., Wang, J., Huang, P., and Yang, R. (2004) Complete genome sequence of *Yersinia pestis* strain 91001, an isolate avirulent to humans. DNA Res. 11, 179-197.

Titball, R.W. and Williamson, E.D. (2004) *Yersinia pestis* (plague) vaccines. Expert. Opin. Biol. Ther. 4, 965-973.

Une, T. and Brubaker, R.R. (1984) Roles of V antigen in promoting virulence and immunity in yersiniae. J. Immunol. 133, 2226-2230.

4

A New Asset for Pathogen Informatics – the Enteropathogen Resource Integration Center (ERIC), an NIAID Bioinformatics Resource Center for Biodefense and Emerging/Re-emerging Infectious Disease

John M. Greene[1], Guy Plunkett III[2], Valerie Burland[2], Jeremy Glasner[2], Eric Cabot[2], Brad Anderson[2], Eric Neeno-Eckwall[2], Yu Qiu[2], Bob Mau[2], Michael Rusch[2], Paul Liss[2], Thomas Hampton[1], David Pot[1], Matthew Shaker[1], Lorie Shaull[1], Panna Shetty[1], Chuan Shi[1], Jon Whitmore[1], Mary Wong[1], Sam Zaremba[1], Frederick R. Blattner[2], and Nicole T. Perna[2]

[1] Health Research Systems, SRA International, Inc., john_greene@sra.com
[2] Genome Center of Wisconsin, University of Wisconsin-Madison, ntperna@wisc.edu

Abstract. ERIC (Enteropathogen Resource Information Center) is one of the National Institute of Allergy and Infectious Diseases (NIAID) Bioinformatics Resource Centers for Biodefense and Emerging/Re-emerging Infectious Disease. ERIC serves as a comprehensive information resource for five related pathogens: *Yersinia enterocolitica*, *Yersinia pestis*, diarrheagenic *E. coli*, *Shigella* spp., and *Salmonella* spp. ERIC integrates genomics, proteomics, biochemical and microbiological information to facilitate the interpretation and understanding of ERIC pathogens and select related non-pathogens for the advancement of diagnostics, therapeutics, and vaccines. ERIC (www.ericbrc.org) is evolving to provide state-of-the-art analysis tools and data types, such as genome sequencing, comparative genomics, genome polymorphisms, gene expression, proteomics, and pathways as well as expertly curated community genome annotation. Genome sequence and genome annotation data and a variety of analysis and tools for eight strains of *Yersinia enterocolitica* and *Yersinia pestis* pathogens (*Yersinia pestis* biovars Mediaevalis KIM, Mediaevalis 91001, Orientalis CO92, Orientalis IP275, Antiqua Angola, Antiqua Antiqua, Antiqua Nepal516, and *Yersinia enterocolitica* 8081) and two strains of *Yersinia pseudotuberculosis* (*Yersinia pseudotuberculosis* IP32953 and IP31758) are currently available through the ERIC portal. ERIC seeks to maintain a strong collaboration with the scientific community so that we can continue to identify and incorporate the latest research data, tools, and training to best meet the current and future needs of the enteropathogen research community. All tools and data developed under this NIAID contract will be freely available. Please contact info@ericbrc.org for more information.

4.1 Introduction

The Enteropathogen Resource Integration Center (ERIC) was formed over two years ago and is available online at www.ericbrc.org. ERIC is a partnership between personnel at the Genome Center of the University of Wisconsin, Madison, and at SRA International in an effort to combine top-notch science with professional, disciplined software and systems engineering. ERIC focuses on the integration of data from five closely related pathogens: *Yersinia pestis*, *Yersinia enterocolitica*, *Shigella* spp., *Salmonella* spp., and diarrheagenic *E. coli*.

4.1.1 The BRC Program

In July 2004, NIAID funded a new bioinformatics effort intended to integrate the vast amount of genomic and other biological data already available, as well as being produced by the ramp-up in biodefense research. Eight Bioinformatics Resource Centers for Biodefense and Emerging/Re-Emerging Infectious Disease were funded to provide the research community working on a selected group of pathogens access to integrated genomics data to facilitate the discovery and development of novel therapeutics, vaccines, and diagnostics for these pathogenic organisms. The initial term of this program is for five years, and altogether this is likely the largest bioinformatics effort to date dedicated to human pathogens. Pathogenic species of biodefense and special public health interest have been classified by NIAID into three high-priority categories (Category A, B, and C; www3.niaid.nih.gov/Biodefense/bandc_priority.htm), based on their relative potentials for causing morbidity or mortality from disease in case of bio-warfare.

Among the primary goals of the BRCs is to provide users with easy Web access (and other types of user interfaces such as APIs or web services) to genomic and related data on these pathogens. Such data includes the genome sequences of multiple strains of these organisms and related plasmids, protein sequences, annotations, microarray data, epitopes, SNPs, proteomic data, and epidemiological data—as much data as possible to allow global and comparative analyses. Eventually, the BRCs will evolve to contain data relevant to host-pathogen interactions. A full list of the BRCs is available on a central portal designed to link the BRCs, www.brc-central.org, and at www.niaid.nih.gov/dmid/genomes/brc/default.htm, the NIAID program site.

All data and software produced under the contracts are freely available to the research community. The BRC program has a policy to publicly release any type of new data placed in the BRCs within 6-12 months from data deposition and upon publication. All data contributed to the BRCs is to be attributed as to source, and each BRC is required to have Data Transfer and Data Access Agreements.

A major emphasis of the BRC program is outreach to the research community to determine and serve their needs, and to make them aware of these new Centers' substantial resources. In addition to direct interactions with researchers, each BRC has a Scientific Working Group (SWG) of approximately ten experts on their assigned pathogens' biology, evolutionary genomics, biodefense, and bioinformatics who help provide vision for the BRCs, expand community ties, and represent the needs of the scientific community working on these pathogens to the BRC staffs.

4.1.2 ERIC – Enteropathogen Resource Integration Center

The Enteropathogen Resource Integration Center (ERIC) focuses on enteropathogens, including *Yersinia enterocolitica*, and the closely related *Yersinia pestis* (see Table 1) with a meticulous, disciplined focus on the annotation and curation of genes and gene families, which will be discussed in detail in Section 4.4 below.

Another focal point for ERIC comes from the realization that such a bioinformatics resource center will add the most value by integrating the disparate data tightly, such that a researcher can find out everything the resource has on a gene (for an example of a biological entity), than by simply providing a collection of disconnected tools. To accomplish this, the user interfaces for these tools and databases must be both scientist-friendly and work well together. Interacting closely with the enterobacterial research community has and will be critical to provide a truly useful resource to the community.

Finally, we also recognized that we can perform some strategic bioinformatics analyses both as starting points for further analysis and as examples of how to use ERIC. The first of these will be identification of lists of genes common to these enterobacteria, and of those genes unique to each species. We have also initiated collaborations with NIAID's Immune Epitope Database and Proteomics Resource Centers programs to identify targeted subsets of genes that are relevantly expressed and immunologically activating to identify candidates for vaccines and diagnostics.

To help improve usability, we chose to use a true Web Portal framework to allow more flexibility, configurability, and customization. ERIC also makes extensive use of Open Source software, which allows the engineering team more flexibility to modify existing tools for better integration. ERIC is in continuous and accelerating development, but we still retain the flexibility to work with users to change ERIC's priorities on short notice to better serve researchers.

4.2 Organisms in ERIC

4.2.1 Yersiniae

At present, ERIC contains seven isolates of *Yersinia pestis*, and one isolate of *Yersinia enterocolitica*. Table 1 displays the isolates/strains and sequencing status.

4.2.2 Reference Genomes for Comparative Genomics

In addition, ERIC has information on related genomes that were not explicitly covered in the contract, but that are essential for full coverage for comparative genomics. These genomes benefit from our annotation strategy emphasizing orthology (cf. Section 4.4.1), using the community specialist knowledge to add annotations to their genes of interest. In addition to *E.coli* K-12, of special note are *Yersinia pseudotuberculosis* IP32953 and IP31758 in this reference set.

4.3 ERIC Portal

The ERIC web site is a true web portal system, and is evolving to allow customization by the end user of which portlets should appear on his home page. ERIC will make use of a "data warehouse" approach to store both contributed pathogen data (annotations, sequence, microarray, proteomics, etc.) and data from external sources, enabling better integration of inputs from multiple sources.

Table 1. *Yersiniae* in ERIC and Genomic Sequence Status

Organism	Isolate/Strain	Genomic Sequence Status
Yersinia enterocolitica	8081	Sequencing complete
Yersinia pestis biovar Mediaevalis	KIM	Sequencing complete
Yersinia pestis biovar Mediaevalis	91001	Sequencing complete
Yersinia pestis biovar Orientalis	CO92	Sequencing complete
Yersinia pestis biovar Orientalis	IP275	Draft assembly
Yersinia pestis biovar Antiqua	Angola	Draft assembly
Yersinia pestis biovar Antiqua	Antiqua	Sequencing complete

4.3.1 ASAP

ERIC's Web portal currently centers on its pathogen annotation system, ASAP (A Systematic Annotation Package for community analysis of genomes), originally developed in the laboratory of co-investigator Nicole Perna at the University of Wisconsin – Madison to acquire, store, update, and distribute genome sequence data and functional characterization (Glasner et al. 2006). ASAP is a relational database and web interface designed to facilitate ongoing community annotation of genomes and tracking of information as genome projects move from preliminary data collection through post-sequencing functional analysis. This system's functions allow versioning of genomes and careful annotation of genes and other genome features (e.g. insertion elements or pseudogenes) with evidence codes. Annotation is performed by six on-staff ERIC annotators/curators, and by direct annotation by members of the scientific community, a critical part of our ongoing efforts.

4.3.2 GBrowse

GBrowse is used in ERIC for genome viewing, with direct links to ERIC's gene annotations. GBrowse is Open Source software under development as part of the Generic Model Organism System Database Project (GMOD) (Stein et al. 2002). For the end user, features of the browser include the ability to scroll and zoom through arbitrary regions of a genome, to enter a region of the genome by searching for a landmark or performing a full text search of all features, and the ability to enable and disable tracks and change their relative order and appearance (color, shape, and size). The user can upload private annotations to view them in the context of the public ones, and publish those annotations to the community. In addition, URLs can be attached to features, and customized tracks can be added to the display for new data types. GBrowse has been adapted by a number of the BRCs due to its flexible configuration and easy integration, and therefore will promote interoperability. The software, its documentation, and support are available at www.gmod.org.

4.3.3 Mauve

For comparative genomics analysis, the ERIC system makes integrated use of Mauve, developed at the University of Wisconsin (Darling et al. 2004). Mauve represents the first alignment system that integrates analysis of large-scale evolutionary events with traditional multiple sequence alignment, to discover common elements in subsets of the aligned sequences. Unlike other such programs, Mauve can handle more than two genomes at a time, and is more capable of handling genomic rearrangements, deletions, and insertions.

Aligning eight enterobacterial genomes demonstrates that several chromosomal rearrangements can be seen among the organisms, including numerous inversions. Regions of sequence unique to subsets of the organisms are easily identified. Mauve features an interactive display allowing functions such as zoom (allowing Mauve to also be of value in SNP location and prediction), re-centering on a shared Locally Collinear Block (LCB, a homologous region of sequence shared by two or more of the genomes under study which does not contain any rearrangements of that homologous sequence), and retrieving LCB coordinates and alignments. ERIC also provides access to some pre-aligned genome comparisons (for example, comparison of four *Yersiniae*).

4.3.4 mAdb Microarray Database

For microarray analysis, we are leveraging seven years of development of the mAdb (for microArray database) system – a scalable, modular, enterprise-level system for both storage of microarray data and analysis over the Web which was developed by the National Cancer Institute (NCI) in collaboration with NIH's Center for Information Technology and SRA International. It is currently supporting nearly 1,500 registered users at NIH and their collaborators worldwide. NCI's mAdb now contains over 63,000 microarray experiments, making it one of the largest microarray databases in existence.

The mAdb system requires only a web browser for the end user. Composite images and quantitated data files are uploaded to the system over the web. The composite image file is used to display individual array spot images upon request, and the quantitated data file is parsed into Oracle 10g tables. Investigators are given the authority to determine access privileges to their projects, allowing data privacy while still enabling data sharing with collaborators. The ERIC mAdb system can currently accept two-color array data quantitated with several common array scanners, as well as Affymetrix data; other scanner/array outputs can be easily added. Users can filter the spots based on numerous spot quality parameters to create reusable datasets, from which subsets can be created by the application of additional filters and/or analysis tools. A history of data subset filtering is maintained to track what has been done.

The array data display page is highly configurable, and each feature on the array links to the relevant details of that gene in our system. mAdb is MIAME compliant, with MAGE-ML import/export under development. Among the large number of features and analysis tools currently included in the system are agglomerative hierarchical, K-means, and self-organizing map (SOM) gene and/or array clustering; Principal Components Analysis (PCA) and Multidimensional Scaling (MDS); the PAM (Prediction Analysis of Microarrays) classifier from the Tibshirani group at Stanford (Tibshirani et al. 2002; to our knowledge, this is the first web-based version of their command line tool); array group assignment and group statistics (mean, median, and standard deviation) for each group; Boolean comparison of datasets; T-test, Wilcoxon Rank Sum, ANOVA, and Kruskal-Wallis statistical analyses; and the SAM (Significance Analysis of Microarrays) method (Tucker et al. 2001). There is also a Pathways/Ontology Summary report (currently summarizing BioCarta, KEGG, and GO). To support additional external data analysis needs, data can be exported to Excel, tab-delimited files, or other analysis tools.

4.4 *Yersinia* Genome Annotations and Curation

ERIC provides continuously updated versions of all complete and many draft *Yersinia* genomes, plus user interfaces to query, download and contribute to the ongoing annotation projects. This section provides an overview of the process and insights into how these annotations differ from what a user may download from other sources. Three key principles are useful to keep in mind:

1) ERIC annotations are continuously changing, with input from both dedicated personnel and the greater community instantaneously available to all users.
2) All annotations are accompanied by supporting evidence and reviewed over time by expert curators to allow users to assess the logic behind each label.
3) Annotated features from one genome are connected to related features from other genomes to leverage information across enterobacteria.

4.4.1 Overall Strategy

Most genomes enter the system starting from the sequence and annotation deposited in GenBank by the primary sequencing center. Pairs of coordinates delimiting the boundaries of annotated features are enumerated, along with a set of feature qualifiers that contain the core annotation information, like gene names and functions. These features and annotations are ingested into the ASAP database (Glasner et al. 2003; Glasner et al. 2006) and are immediately available through the ERIC system. Initially, the annotations are presented exactly as they appear in the GenBank record and are tagged in ASAP as supported by the evidence "Published Annotation: GenBank Accession Number XXXXXX". They are also marked as "Approved" to confirm that they accurately reflect the information present in the published GenBank annotation. It is important to note that approval indicates only accuracy with reference to the evidence cited. An important goal for ERIC is to steadily replace these annotations based on weak evidence with annotations linked to the best direct experimental evidence available in the literature.

The features annotated in GenBank are also the starting point for comparative analyses across genomes. Reciprocal BLASTP searches and comparison-specific filtering are typically used to generate an initial set of predicted orthologs between each new genome and every other genome in the ERIC system. This strategy is known to be error prone for members of multigene families where close paralogs in either genome can lead to both false positives and false negatives. To combat these errors, automated processing is limited to cases where there is a single best match (lowest E-value) and no alternative matches of comparable significance. These automated predictions can be reviewed and rejected or supplemented by manual curation at any point. Extensive conservation within groups, for example, among the *Yersinia*, makes it possible to achieve higher sensitivity and specificity in ortholog prediction through genome-scale alignment. ERIC's curators are currently working to use the integrated Mauve alignment system (described above in Section 4.3.3) to confirm and augment the BLASTP-based ortholog predictions. Additional searches are conducted against the complete GenPept and InterPro databases. All search results are stored and presented for users to view and use in annotation updates.

Once a genome annotation project has been established in the ERIC system, dedicated curators and community annotators can correct and augment annotations and add supporting evidence based on published literature, additional bioinformatics analyses, or by inheriting annotations from related features in another ERIC genome. Each new annotation is accompanied by both the annotator contact information and the supporting evidence cited. These annotations are immediately available for all users, including other annotators, and for all public genomes, the entire research community. The annotations will be tagged as "Uncurated" to indicate that no designated expert curator has reviewed the annotation for accuracy relative to the evidence referenced. Even records contributed by ERIC-designated annotators are reviewed by a second curator. Approved records are tagged conspicuously for users while rejected records are hidden from default views but remain accessible for tracking the history of each annotated feature. Curators also add and delete features, or adjust annotated coordinates of features. These changes are also immediately accessible to users, and amended feature lists and boundaries are used in subsequent updates of BLAST and InterPro searches (Fig. 1).

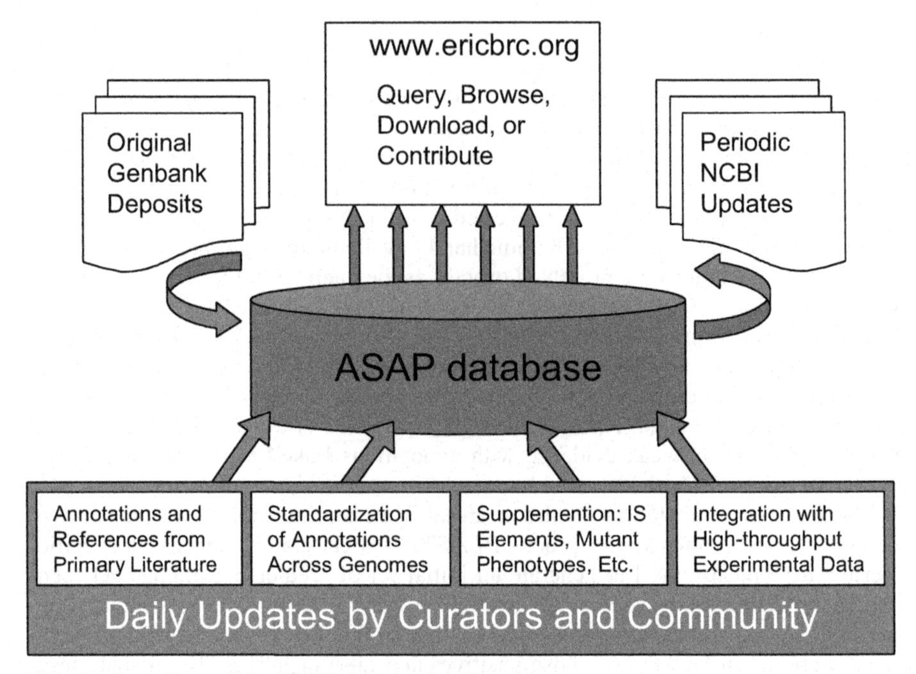

Fig. 1. Overview of the ERIC Annotation Process.

In prokaryotes a "pseudogene" is a gene that is disrupted in the particular strain or isolate whose genome was sequenced. They are recognized as such by comparison to a related organism where the wild-type or "ancestral" state is seen. Pseudogenes are distinguished from missense mutations, where the gene is still intact but may have altered functionality; disruption can be due to in-frame stop codons, frameshifts, the insertion of IS elements, prophages, or islands, deletions, and more complex rearrangements. These features are of particular interest, having been implicated in the evolution of pathogens and their adaptation to new niches. In the context of evolution and comparative genomics, is a given gene present or absent in a particular genome, and is it intact or disrupted? In a given pathogen, are any partial genes actually expressed? Is the consequence of a pseudogene something other than the straightforward loss of function? Pseudogenes are inconsistently annotated in current genomes, making these issues harder to address. In current GenBank entries, a pseudogene may be annotated as a CDS (with a /pseudo qualifier), a gene (with no underlying CDS), or even misc_feature. As part of our internal standardization of annotations, the feature type pseudogene has been introduced in ERIC, at the same hierarchical level as CDS, rRNA, tRNA, misc_RNA, etc. In addition to standardizing existing pseudogene annotations, we are examining groups of related genomes to delineate previously unrecognized pseudogenes.

The following sections summarize the ERIC curators' updates to the *Yersinia pestis* genomes published to date.

4.4.2 *Yersinia pestis* Annotation Summary

ERIC's curators are actively updating the annotations of all published *Yersinia pestis* genomes. The first *Y. pestis* genomes (strains CO92 and KIM) were published and deposited in GenBank in 2002 and the most recent *Y. pestis* genomes have emerged this year (Parkhill et al. 2001; Deng et al. 2002; Song et al. 2004; Chain et al. 2006). Curators have reviewed PubMed records to augment annotations based on literature published on *Y. pestis* since 2002. Curators also inspected and resolved many inconsistencies across genomes. Efforts have centered on strain CO92, with the intention of propagating these updates across all orthologous genes in all *Yersinia* genomes in 2007. Thus, here we focus on a more detailed description of the current state of the CO92 annotation.

A total of 41,000 annotations were imported from the original GenBank deposit. Curator review has led to rejection of 5,479 of these records. An additional 22,111 lines of (approved or uncurated) annotation have been added during the ERIC project. Table 2 shows the distribution of annotation types included in this update, though exact numbers are continuously in flux. The annotations have been subdivided into four categories to facilitate discussion. Those in the "Source" category are related to the ERIC approach to integrating all replicons as a single genome. In the case of *Y. pestis* CO92, this includes three plasmids (pCD1, pMT1, and pPCP1) in addition to the chromosome. When users query the ERIC annotations for CO92, they have the option to restrict searches to a single replicon, but the default is to search across all four. The new "Source" annotations correspond to unified cross-references to the organism and details about the replicon identities.

The "Identifiers" category includes a variety of annotation types that link the CO92 entry in ERIC to other resources. For example, over 3,800 "db_xref" annotations are the corresponding identifiers for each protein in the UniProt/SwissProt database (Wu et al. 2006). The "EnteroFam" annotations link these CO92 genes to sets conserved across a broader collection of enterobacteria that will allow automated propagation of annotation information across these highly conserved genes to minimize the effort required to maintain and standardize the annotations of multiple related genomes. The "locus tag" annotations are a key unique identifier that has changed usage at NCBI since deposition of the original entry. This revision brings the CO92 entry into compliance with current standards. The "name" and "synonym" annotations are among the most important updates, because these alternate forms of gene names are a common way for users to query the data. The use of a single primary "name" while allowing infinite "synonym" annotations allows us to provide alternate routes for finding a given gene. Given the extensive experimental research on enterobacteria, it is perhaps not surprising to find that some genes have multiple (up to eight added to a single CO92 gene) synonyms in the literature.

All annotations gathered in the "Role" category are related to the physiological role, cellular structure, or function of the gene product. The "product" category reflects the primary descriptor, with updates derived from a variety of sources including primary literature and related gene products from other ERIC genomes or reference organisms, such as the highly curated *Escherichia coli* K-12 genomes (Riley, Abe et al. 2006). Many of the "EC number" entries showing the official Enzyme Commission designation for enzymatic activity were obtained by mapping the corresponding entries from the UniProt database. Particularly notable are the three types of GO, or Gene Ontology, annotations. GO annotations use a controlled vocabulary to describe the biological role, molecular function, and location of gene products and are increasingly important for systematic analyses of high-throughput experimental data (Gene Ontology Consortium 2006). Many of these CO92 annotations were collected by reviewing the InterPro search results for matches to the TIGRFams (Haft et al. 2003) and mapping associated GO annotations, but others were added by curators using other evidence. In each case, users see the evidence types and links to the specific evidence references alongside the annotations. For example, a GO annotation that cites an Evidence Type like "Inferred by Sequence Similarity" and cites "TIGRFams" as a Reference will also include a link to the TIGRFam entry in the CMR database that serves as the primary TIGRFam repository. Although many of the annotation types added in ERIC can be included in updated deposits to NCBI, the integration of supporting evidence is only found via ERIC.

Table 2. Annotations Added to *Y. pestis* CO92 by ERIC curators

Annotation Category	Annotation Type	Count of New ERIC Annotations
Source	genome context	4
	molecule type	4
	nucleic acid type	4
	organism	3
	plasmid	3
	strain	4
Identifiers	db xref	8,423
	EnteroFam	1,615
	locus tag	4,203
	name	669
	protein id	197
	synonym	986
Role	alternate product name	75
	EC number	718
	function	56
	GO biological process	1,256
	GO cellular component	551
	GO molecular function	1,190
	product	659
	pseudo	108
Additional Information	comment	235
	genetic interaction	12
	insertion sequence	144
	molecular interaction	27
	mutant phenotype	148
	note	674
	overexpression phenotype	3
	physical properties	12
	regulation	104
	structure	13
	subfeatures	10
	transl except	1
All Annotations	Total	22,111

The remaining annotation types, grouped in the category "Additional Information" include a variety of structured and free text records. The "note" field corresponds to the standard "/note" field attached to many GenBank annotations, distinct from the "comment" annotation which has no counterpart in a GenBank entry. This distinction allows annotators to target certain general descriptive information for inclusion in updates to NCBI genome deposits. Other annotation types in this category were designed for ERIC to house specialized information. These include "mutant phenotype", "overexpression phenotype", "genetic interaction", and "molecular interaction", each of which is populated with structured information manually curated from primary literature. The "regulation" annotations describe both genetic and external (acetate, pH, etc.) factors that control expression of the annotated gene product. Additional "physical property" and "structure" annotations describe biochemical characterizations of (usually) protein products.

In total, 988 annotations from the CO92 genome cite PubMed ID's as a reference. As a general rule, this is the optimal reference because it links the annotation directly to peer-reviewed primary literature. While some of these annotations are derived from direct experimentation with *Y. pestis* strain CO92 (137), others reflect experiments conducted with different *Yersinia* (431) or more distantly related enterobacteria (420). An optional field provided on the annotation input form allows ERIC curators to record the organism that is the primary focus of the reference. This option is particularly useful when the evidence cited is sequence similarity, but it also allows propagation of experimental evidence with the cautionary note that the organism in which the experiment was conducted may differ from the organism in which the annotation is being examined.

Collectively, these updates to CO92 in ERIC provide a new standard for *Y. pestis* genome annotation. Wherever appropriate, these updates will be leveraged across all existing and upcoming *Y. pestis* genomes in the ERIC system through the use of orthologs. These substantial updates will be gathered together to create new genome entries for deposition in NCBI databases and at BRC Central, where all eight BRC's maintain comparable records for their target organisms. It is important to note again that the annotations available from the main ERIC portal are continuously updated and include supporting evidence. This means that regardless of frequency of deposition elsewhere, ERIC's records reflect the most up-to-date and complete annotations and curation available.

4.4.3 *Y. pestis* Comparisons: Orthologs, Alignment, Pseudogenes, and Insertion Sequences

Propagation of annotation updates to the CO92 genome to other *Y. pestis* genomes is expected to have substantial returns because of the high degree of similarity among these genomes (Parkhill et al. 2001; Deng et al. 2002; Chain et al. 2004; Song et al. 2004; Chain et al. 2006). Differences between genomes are also potentially useful markers for sub-typing *Y. pestis* strains. ERIC allows examination of such variations in several ways including curated orthologs lists, genome alignments, and standardized mobile element predictions.

Users can actively query predicted orthologs in the ERIC database, by selecting a reference genome and one or more genomes for comparison. Results are displayed showing all genes from the reference genome and any predicted orthologs in the comparison genomes. For example, a query to retrieve orthologs of the newly reannotated CO92 protein coding genes from *Y. pestis* strains KIM and 91001 plus *Y. pseudotuberculosis* IP32953 returns a list of the 4,193 genes in CO92 and corre-

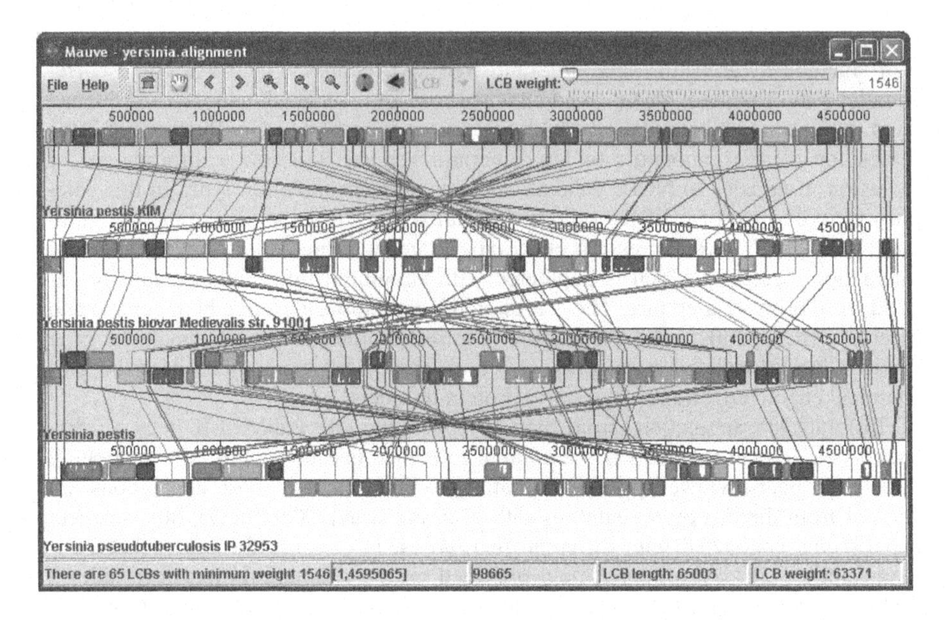

Fig. 2. A Mauve alignment of four *Yersinia* genomes.

sponding entries from the other genomes. The results can be browsed interactively or downloaded in tabular form for further analysis. In this comparison, simple sorting reveals that there are 3,588 predicted orthologs in *Y. pestis* KIM, 3,676 in *Y. pestis* 91001 and 3,455 in *Y. pseudotuberculosis* IP32953. By copying the list of FeatureIDs for *Y. pestis* CO92 genes not found in any of the other genomes and pasting it back into the ERIC annotation query form, users can obtain a complete set of annotations for strain-specific genes. Future development of the ERIC system will expand integrated user interfaces for conducting these types of tasks.

While the ERIC predicted ortholog sets are a good starting point for comparisons, they are likely to be conservative predictions and underestimate overall genome similarity. The Mauve alignment tool provides another means of comparison (Fig. 2). A DNA sequence alignment conducted with optimized Mauve parameters is linked on every detailed annotation page. This link launches the interactive Mauve visualization tool, highlighting the annotated feature of interest in the genome from which the query was initiated. Aligned regions of other *Yersinia* genomes are displayed as additional panels. Users can zoom in to the nucleotide alignment level or zoom out for an overview of the similarity across genomes and the boundaries of regions that have undergone rearrangements characteristic of *Y. pestis* genomes.

Many of the rearrangements are bounded by repetitive elements, specifically Insertion Sequences (IS), which are relatively abundant in *Yersinia* genomes compared to many other enterobacteria. Diversity in structure and sequence of IS element families and continuing discovery of new elements make accurate prediction and enumeration an ongoing challenge, as does the presence of partial elements. Annotation style and consistency of predictions in the original GenBank deposits is quite variable. ERIC curators are actively updating IS predictions for all *Yersinia* genomes. To date, predictions have been augmented and corrected for all chromosomes and the plasmids are in progress. The analysis includes sequence similarity searches against the complete IS finder database (Siguier et al. 2006) and filtering that retains all intact or partial (>10%) elements. Manual review and resolution of differences between new and existing predictions produced a standardized set summarized in Table 3. Updated coordinates for these features and annotations of the IS family names are can be searched by choosing "repeat_region" as the feature type and "insertion sequence" as the annotation type on query forms.

4.4.4 Community Participation in *Yersinia* Annotation

As noted in several sections above, the ASAP annotation system component of ERIC was created both to allow and encourage community annotation—based on the principle that researchers working on a given organism will know it better than any dedicated annotator could. It is somewhat paradoxical that most of the research community agrees that annotation is very important, but that so few researchers actually participate. The ERIC team will be happy to provide training on the use of the ASAP system, either over the Web, or if there is a cluster of interested researchers, in person. Even if you are expert in only one gene or system of the *Yersiniae*, your contribution is vitally important. At the least, we invite you to look at the annotations relevant to your research in ERIC, and if you have something to add or correct, send an e-mail to us via the provided link to the ERIC annotation team.

4.5 Upcoming Features of ERIC

4.5.1 Text Mining

SRA is an industry leader in natural language processing (NLP)-based text mining, which can find relationships in unstructured text, and is staffed by a dedicated group of linguists and software engineers. SRA's powerful NetOwl® suite of products has routinely placed very highly in Government text mining evaluations such as the Automated Content Extractions (ACE), run by NIST, where NetOwl® has been the top scorer in entity and relationship extraction for the past three years.

ERIC intends to use these tools to extract semantic relationships from the literature for the BRC Program. These programs can perform multilingual information extraction, text clustering, and text summarization – this is not just keyword searching. They have been extensively used by numerous commercial and government clients in areas as diverse as financial services and intelligence applications, and it is likely that they will perform as well in analyzing the biological literature. Discovering heretofore unrecognized relationships in the biomedical literature may be key to

Table 3. Insertion Sequence Annotations

Genome	Existing GenBank Annotations	Chromosomal			
		Total IS	Complete IS	Coordinate Updates	Partial IS
Y. pestis 91001	Yes, updated	127	101	80	26
Y. pestis KIM	Yes, updated	139	109	3	30
Y. pestis CO92	Yes, reformatted	144	130	3	14
Y. pestis Antiqua	No	189	152	n/a	37
Y. pestis Nepal 516	No	144	115	n/a	29
Y. pseudotuberculosis IP32953	No	32	20	n/a	12
Totals for *Yersinia*		775	627	86	148

designing experiments which will lead to the identification of targets for vaccines, therapeutics, and diagnostics.

Initial research and development of the NetOwl® software directed to enteropathogen based literature has focused on the extraction of gene, gene product, and operon functions. The system has been trained to extract gene to role relationships (e.g. "These results suggest that virK function is an essential virulence determinant.") and gene mutation to phenotype relationships (e.g. "The katF mutant has significantly reduced virulence in mice."), information that is highly relevant in the functional annotation genome annotation. Simple yet effective presentation of extracted biological relationships to our research community will be critical to the even more rapid and effective capture of known facts in a central and usable forum. Development of these interfaces is underway through close interaction with future users of the system. All indications are that automatic extraction of information from literature will be of use to the *Yersinia* (and broader enteropathogen) community.

4.5.2 Other Tools and Databases

Also under development for incorporation into the ERIC portal is pathway software capable of being overlaid with 'omics' data, such as gene expression, proteomics, metabolomics, etc. ERIC will also be capable of handling the analysis of genome polymorphism and SNP data as it becomes available, and will add tools to better visualize comparative genome hybridization (CGH) data. Proteomics data, both in terms of mass spectrometric-derived data and protein-protein interactions, are being and will be respectively stored in ERIC. In addition, ERIC will also integrate phenotypic data, including details on pathogenicity and virulence.

As noted above, in all cases, the emphasis will be on thorough integration of this data, such that wherever a user enters the portal, he or she will be able to "drill down" easily to all of the relevant information on that biological entity and to be able to query across data types.

4.6 Conclusions

ERIC has been and is being developed to serve diverse sets of researchers, including pathogen experts in *Yersinia* as well as other enteropathogens, researchers in comparative genomics and microbial evolution, biodefense researchers, and developers of therapeutics, vaccines, and diagnostics. We believe the resource center will become the pre-eminent source of information on the genomics of the *Yersiniae*. However, to achieve this goal will require the participation of the research community, by actively using this resource, by contributing community annotation, and by providing suggestions for its improvement and feedback on its features. Such input is always welcome at info@ericbrc.org.

4.7 Acknowledgements

We wish to thank our Scientific Working Group (Drs. Robert Perry, Emilio Garcia, Fiona Brinkman, Tom Cebula, Stan Maloy, Stephen Calderwood, Shelley Payne, Will Gilbert, Julian Davies, Mary Lipton, Valerie de Crecy-Lagard, and Jim Kaper) for their excellent counsel and ideas. We also thank our program officer, Valentina Di Francesco, for her contributions both to the BRC program and to ERIC. J.M.G wishes to thank Dr. Susan Castillo, Michael Fultz, Dr. Timothy Cooke, and Gio Patterson for their support and counsel. This project has been funded in whole or in part with Federal funds from the National Institute of Allergy and Infectious Diseases, National Institutes of Health, Department of Health and Human Services, under Contract No. HHSN266200400040C.

4.8 References

Chain, P. S., Carniel, E., Larimer, F. W., Lamerdin, J., Stoutland, P. O., Regala, W. M., Georgescu, A. M., Vergez, L. M., Land, M. L., Motin, V. L., Brubaker, R. R., Fowler, J., Hinnebusch, J.. Marceau, M., Medigue, C., Simonet, M., Chenal-Francisque, V., Souza, B., Dacheux, D., Elliott, J. M., Derbise, A., Hauser, L. J. and Garcia, E. (2004). Insights into the evolution of *Yersinia pestis* through whole-genome comparison with *Yersinia pseudotuberculosis*. PNAS 101, 13826-13831.

Chain, P. S., Hu, P., Malfatti, S. A., Radnedge, L., Larimer, F., Vergez, L. M., Worsham, P., Chu, M. C. and Andersen, G. L. (2006). Complete genome sequence of *Yersinia pestis* strains Antiqua and Nepal516: evidence of gene reduction in an emerging pathogen. J. Bacteriol. 188, 4453-4463.

Darling, A. C., Mau, Blattner, F. R. and Perna, N. T. (2004). Mauve: multiple alignment of conserved genomic sequence with rearrangements. Genome Res 14, 1394-1403.

Deng, W., Burland, V., Plunkett, 3rd, Boutin, A., Mayhew, G. F., Liss, P., Perna, N. T., Rose, D. J., Mau, B., Zhou, S., Schwartz, D. C., Fetherston, J. D., Lindler, L. E., Brubaker, R. R., Plano, G. V., Straley, S. C., McDonough, K. A., Nilles, M. L., Matson, J. S., Blattner, F. R. and Perry, R. D. (2002). Genome sequence of *Yersinia pestis* KIM. J. Bacteriol. 184, 4601-4611.

Gene Ontology Consortium (2006). The Gene Ontology (GO) project in 2006. Nucleic Acids Res. 34(Database issue), D322-326.

Glasner, J. D., Liss, P., Plunkett, 3rd, G., Darling, A., Prasad, T., Rusch, M., Byrnes, A., Gilson, M., Biehl, B., Blattner, F. R. and Perna , N. T. (2003). ASAP, a systematic annotation package for community analysis of genomes. Nucleic Acids Res. 31, 147-151.

Glasner, J. D., Rusch, M., Liss, P., Plunkett, 3rd, G., Cabot, E. L., Darling, A., Anderson, B. D., Infield-Harm, P., Gilson, M. C. and Perna, N. T. (2006). ASAP: a resource for annotating, curating, comparing, and disseminating genomic data. Nucleic Acids Res. 34(Database issue), D41-D45.

Greene J.M., Asaki, E., Bian, X., Bock, C., Castillo, S.O., Chandramouli, G., Martell, R., Meyer, K., Ruppert, T., Sundaram, S., Tomlin, J., Yang, L. and Powell, J. I. (2003). The NCI/CIT microArray database (mAdb) system - bioinformatics for the management and analysis of Affymetrix and spotted gene expression microarrays. AMIA Annu. Symp. Proc. 2003, 1066.

Haft, D. H., Selengut, J. D. and White, O. (2003). The TIGRFAMs database of protein families. Nucleic Acids Res. 31, 371-373.

Parkhill, J., B. Wren, W., Thomson, N. R., Titball, R. W., Holden, M. T., Prentice, M. B., Sebaihia, M., James, K. D., Churcher, C., Mungall, K. L., Baker, S., Basham, D., Bentley, S. D., Brooks, K. , Cerdeno-Tarraga, A. M., Chillingworth, T., Cronin, A., Davies, R. M., Davis, P., Dougan, G., Feltwell, T., Hamlin, N., Holroyd, S., Jagels, K., Karlyshev, A. V., Leather, S., Moule, S., Oyston, P. C., Quail, M., Rutherford, K., Simmonds, M., Skelton, J., Stevens, K., Whitehead, S. and Barrell, B. G. (2001). Genome sequence of *Yersinia pestis*, the causative agent of plague. Nature 413, 523-527.

Riley, M., Abe, T., Arnaud, M. B., Berlyn, M. K., Blattner, F. R., Chaudhuri, R. R., Glasner, J. D., Horiuchi, T., Keseler, I. M., Kosuge, T., Mori, H., Perna, N. T., Plunkett, 3rd, G., Rudd, K. E., Serres, M. H., Thomas, G. H., Thomson, N. R., Wishart, D. and Wanner, B. L. (2006). *Escherichia coli* K-12: a cooperatively developed annotation snapshot–2005. Nucleic Acids Res. 34, 1-9.

Siguier, P., Perochon, J., Lestrade, L., Mahillon, J. and Chandler, M. (2006). ISfinder: the reference centre for bacterial insertion sequences. Nucleic Acids Res. 34(Database issue), D32-D36.

Song, Y., Tong, Z., Wang, J., Wang, L., Guo, Z., Han, Y., Zhang, J., Pei, D., Zhou, D., Qin, H., Pang, X., Zhai, J., Li, M., Cui, B., Qi, Z., Jin, L., Dai, R., Chen, F., Li, S., Ye, C., Du, Z., Lin, W., Yu, J., Yang, H., Huang, P. and Yang, R. (2004). Complete genome sequence of *Yersinia pestis* strain 91001, an isolate avirulent to humans. DNA Res. 11, 179-197.

Stein, L. D., Mungall, C., Shu, S., Caudy, M., Mangone, M., Day, A., Nickerson, E., Stajich, J.E., Harris, T.W., Arva, A. and Lewis, S. (2002) The generic genome browser: a building block for a model organism system database. Genome Res. 12, 1599-1610.

Tibshirani R., Hastie, T., Narasimhan, B. and Chu, G. (2002) Diagnosis of multiple cancer types by shrunken centroids of gene expression. PNAS 99, 6567-6666672.

Tusher, V. G., Tibshirani, R. and Chu, G. (2001). Significance analysis of microarrays applied to the ionizing radiation response. PNAS. 98, 5116-5121.

Wu, C. H., Apweiler, R., Bairoch, A., Natale, D. A., Barker, W. C., Boeckmann, B., Ferro, S., Gasteiger, E., Huang, H., Lopez, R., Magrane, M., Martin, M. J., Mazumder, R., O'Donovan, C., Redaschi, N. and Suzek, B. (2006). The Universal Protein Resource (UniProt): an expanding universe of protein information. Nucleic Acids Res. 34(Database issue), D187-D-191.

Picture 6. Attendees at an oral presentation. Photograph by A. Anisimov.

Part II – Structure and Metabolism

Picture 7. Robert Brubaker during his intermediary metabolism and LCR talk. Photograph by A. Anisimov.

Picture 8. Attendees at an oral presentation. Photograph by R. Perry.

5
My Life with *Yersinia*

Mikael Skurnik

Department of Bacteriology and Immunology, Haartman Institute, University of Helsinki, mikael.skurnik@helsinki.fi

Abstract. This review is based on the opening lecture I was honored to give during the 9th International Symposium on Yersinia in Lexington, Kentucky in October 2006. I present some topics that have been close to my interest during the past 25 years with some historical anecdotes. For example, how detection of intervening sequences in *Yersinia enterocolitica* rDNA genes resulted in development of microbial diagnostic applications. How the adhesin YadA was detected and named and what do we know of its function now? What was the first pseudogene sequenced in *Yersinia pestis*? I will also discuss *Yersinia* lipopolysaccharide, bacteriophages and serum resistance mechanisms which we have worked on lately.

5.1 My Entry to the World of *Yersinia*

I received my MSc degree in 1977 majoring in biochemistry but it took still three years before I started working with *Yersinia*. In 1976 I had started working on cancer immunology with my father, a general practitioner who in his spare time studied cancer immunology and this happened at the Department of Medical Microbiology, University of Oulu, Finland. As it happened, he succumbed to pancreatic carcinoma in December 1976 but I stayed in the department and continued the work practically on my own. In 1980 I finally realized that with my resources I had no chances to develop an immunological cure for cancer. This approach has not been successful even in big laboratories.

At this time, the microbiologist Tapio Nurmi had joined the department and was responsible for the *Yersinia* diagnostics in the clinical routine laboratory. In May 1980 we met in the corridor and he showed the paper of Don Zink (Zink et al. 1980) on the plasmid-mediated tissue invasiveness in *Yersinia enterocolitica* and said that it could be worthwhile to have a look if the Finnish *Y. enterocolitica* isolates also carry the plasmid. As I had some practical experience on plasmid isolation and my cancer immunology research was not leading anywhere I took over the task.

We had a nice collection of Finnish strains available and I started the plasmid isolations, which was not so straightforward due to the big size of the virulence plasmid, pYV. At the same time Timo Vesikari's group in Tampere, Finland, had also started studies on *Y. enterocolitica* virulence based on their previous experience on invasion experiments and tissue cultures. We decided to join forces and this resulted in publication of our paper in 1981 (Vesikari et al. 1981). The net result was that, yes, Finnish *Y. enterocolitica* isolates do carry pYV and it is associated with adherence to and toxicity for Hep-2 cells, calcium dependency and autoagglutination. Strains that had lost pYV adhered less but became more invasive to Hep-2 cells.

From there on, I continued my work on *Y. enterocolitica* with the idea that this could be the subject of my PhD thesis. At that time, and still in Finland, a PhD thesis was based on a minimum of 4-5 articles published in peer-reviewed international journals where the candidate should be the first author in most and the single author in some papers. So this gave me a goal and framework for the coming years.

The readers should also appreciate that at that time in Finland people got married and had children at a younger age than nowadays. When I started my life with *Yersinia* I already had 2 children, my wife was studying and in 1981 I also started medical studies with the illusion of becoming more competitive in the application of researcher positions. Until then I only had had short-term contracts. During the medical studies I earned our living by working on an hourly basis in the routine clinical laboratory doing *Yersinia* and *Chlamydia* diagnostics and used electron microscopy (EM) for viral diagnostics of stools.

5.2 PhD Thesis and Dissertation

Since I already had a degree in biochemistry I had, in the medical studies, some loose time which I used to work on my first first-author articles on *Yersinia* (Skurnik 1984; Skurnik et al. 1983). The scientific approach in these papers was not very sophisticated and I was lucky to get them published, however, the process was very educational. In the fimbria work (Skurnik 1984), I already had the idea that pYV might have something to do with the bacterial surface which is the interface between the bacterium and the host. Without the special EM staining techniques to see close to the bacterial surface I missed both the YadA-fibrils and the type III secretion system injectisomes.

5.2.1 Autoagglutination

I had in parallel started to study the autoagglutination (AA) phenomenon (Laird and Cavanaugh 1980). In spring 1981 I found that AA bacteria express a prominent high molecular mass protein which we now know as YadA. The same protein was also identified by Ingrid Bölin and Hans Wolf-Watz from *Yersinia pseudotuberculosis* who published it in 1982 (Bölin et al. 1982); it was not noticed by Dan Portnoy and Stanley Falkow (Portnoy et al. 1981). Dan Portnoy later told me that he missed YadA since he threw the YadA-band away with the dirty-looking top of the SDS-polyacrylamide gel that he cleaned for the photograph. Comparison of the laboratory note books revealed that Ingrid had seen the YadA-band a week earlier than I.

Anyway, I submitted my findings on "Characterization of autoagglutination phenomenon of *Y. enterocolitica* O:3" to the Journal of Bacteriology in March 1983 and got it returned with the advice that just seeing a band in SDS-PAGE in pYV-positive autoagglutinated bacteria does not prove that the protein is responsible for the phenomenon, I should provide, for example, genetic evidence. This was a good point and showed how unequipped I was at that time for bacterial genetics. In my basic

Table 1. Selected list of participants in the EMBO course 1983; *Yersinia* people in bold

Organizers	Staffan Normark, Bernt-Eric Uhlin, Sven Bergström, Hans Wolf-Watz, Ingrid Bölin, Roland Rosqvist, Åke Forsberg
Teachers	**Daniel Portnoy**, Werner Göbel, Magdalene So, Michael Koomey, Catharine Svanborg-Edén, Alex Gabain, Joel Belasco, Wim Gaastra, John Mekalanos, **Stanley Falkow**, Richard Goldstein, Thomas Grundström, Jörg Hacker
Students	Alan Barbour, Bianca Colonna, Mauro Nicoletti, Gunna Christiansen, Torkel Wadström, Mo-Quen Klinkert, Arlette Michaud, Shulamit Michaeli, John Swanson, Airi Palva, Rino Rappuoli, **Mikael Skurnik**

biochemistry education, bacterial genetics was poorly represented in the curriculum thus leaving me to learn the background and possibilities of bacterial genetics throughout my career. As a first step, I was very lucky to get accepted to an EMBO practical course on "Molecular Genetics of Bacterial Pathogenesis" organized in Umeå, Sweden (August 6-20, 1983).

5.2.2 The EMBO Course

Surprisingly many of the present day top names in bacterial pathogenesis participated in the EMBO course either as students or teachers. I have collected some of the names in Table 1.

During the course I learned and tried to absorb the basics of bacterial genetics and the strategies that could be used to study bacterial pathogenesis, that were delivered to us by the leading gurus in the field. During the course I also performed a couple of experiments, outside the course program, using *yadA::*Tn*5* mutants isolated by Ingrid Bölin and could demonstrate that YadA was responsible for AA. The manuscript revised with this data and co-authored by Hans Wolf-Watz (Hasse) and Ingrid Bölin was then published in the Journal of Bacteriology (Skurnik et al. 1984).

I returned back to Oulu from the EMBO course with a bunch of bacterial strains from Hasse's laboratory and full of ideas. I also took a break in the medical studies (see above) to concentrate full time on the PhD thesis. In the end I never returned to the medical studies. I started my first real experiments in bacterial genetics by trying to generate a transposon insertion library into *Y. enterocolitica* O:3 strain using the bacteriophage P1clm,clr100:Tn*5* delivery system that had been elegantly used for *Y. pseudotuberculosis* and *Yersinia pestis*, however, without success. Later I learned that *Y. enterocolitica* O:3 does not have the P1 receptor, the terminal glucose of the LPS core present in *Escherichia coli, Y. pestis* and *Y. pseudotuberculosis.* Those experiments, however, planted the seed for searching for transducing yersiniophages that I discuss later.

One of the projects that I started for the PhD thesis was based on trying to identify the pYV-encoded proteins of *Y. enterocolitica* O:3. For that reason I immunized several rabbits with formalin-killed autoagglutinated bacteria and also with a long immunization protocol using live bacteria starting with pYV-negative followed by

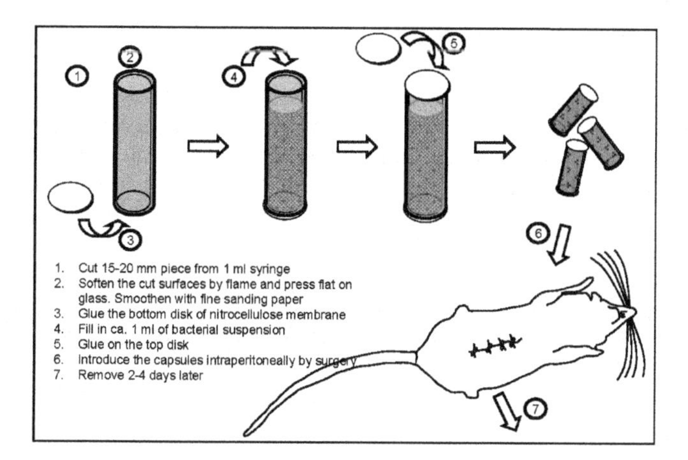

Fig. 1. Use of intraperitoneal capsules to study *in vivo* expression of pYV-encoded proteins.

increasing doses of pYV-positive bacteria. The results of these experiments "Expression of antigens encoded by the virulence plasmid of *Yersinia enterocolitica* under different growth conditions" were then published in Infection and Immunity (Skurnik 1985a). In this last paper I made use of intraperitoneal capsules (Fig. 1). In the capsules I could introduce ca. 10^8 bacteria into *in vivo* conditions and recover the bacteria for biochemical studies.

Although I have never counted myself as a plague scientist, I have made comparisons between *Y. pestis* and *Y. pseudotuberculosis/Y. enterocolitica* whenever possible. Thus I also added *Y. pestis* into the capsules. Using Western blotting, where I designed rolling bottles to incubate membranes, and anti-*Y. enterocolitica* or anti-*Y. pestis* antisera I could detect the expression of pYV-encoded proteins, now known as Yops, by both *Y. enterocolitica* and *Y. pestis* in the intraperitoneal capsules (Skurnik 1985a). For the latter this was first evidence for the *in vivo* expression of the Yops.

5.2.3 Dissertation

After completing the above-mentioned 5 articles (Skurnik 1984; Skurnik 1985a; Skurnik et al. 1984; Skurnik et al. 1983; Vesikari et al. 1981) I thought that it would be enough for a PhD thesis, however, finally one of the papers (Skurnik et al. 1983) was replaced by another one (Skurnik and Poikonen 1986a) where I used immunohistochemistry to follow wild type *Y. enterocolitica* O:3 bacteria in orally infected rats during the first days of infection. As primary antibodies I had a monoclonal antibody (D66) against YadA (Skurnik and Poikonen 1986b) and the polyclonal rabbit antisera mentioned above against formalin-killed and live O:3 bacteria. Finally, based on these five papers I wrote the PhD thesis and defended it publicly on June 25, 1985 (Skurnik 1985b).

5.3 Postdoc in Umeå, Sweden

After obtaining my PhD degree I had an opportunity to work for almost 2 years as a post-doc with Hasse in Umeå. Somehow Hasse was able to scrape together a little money and I got also some funding so that I could move with my family to Sweden.

There we decided that my project would be to clone the autoagglutination protein gene, i.e. the *yadA* gene. In Western blotting using the Mab D66 I had seen that the YadA-band size varies between different *Y. enterocolitica* and *Y. pseudotuberculosis* strains (Skurnik et al. 1986b) and that *Y. pestis* does not express it at all. Ingrid Bölin had expressed YadA from a cosmid clone (Bölin and Wolf-Watz 1984) and crudely mapped its 5'-end to BamHI fragment 10 of pIB1 (the virulence plasmid of *Y. pseudotuberculosis* strain YPIII/pIB1). She had also isolated a *yadA::*Tn*5* mutant which was as virulent in mice as the wild type strain. This was a reason for Hasse and Ingrid not to continue with this protein and so Ingrid and others in the lab took up other pYV-proteins which we now know as Yops (Bölin et al. 1985). Thus when I arrived in Umeå the YadA field was open for me. Since I had seen YadA in almost all the virulent strains that I had studied I thought it must have a function since nature does not keep a gene for nothing.

5.3.1 The *yadA* Gene

We decided that I would clone the *yadA* gene from both *Y. pseudotuberculosis* and *Y. enterocolitica* serotype O:8 and O:3 strains. Additionally, since the BamHI map of the *Y. pestis* virulence plasmid and that of pIB1 in the fragment 10 region were identical we also included *Y. pestis* into the cloning project. I cloned ClaI-fragment libraries into pBR322 from the four virulence plasmids, isolated the *yadA*-carrying clones and expressed the YadA-protein in *E. coli.* The cloning and expression results were published with a notation that "the DNA region of plasmid pYV019 (*Y. pestis* plasmid) that corresponded to the structural gene of YadA was also conserved, even though YadA was not expressed from this plasmid. The inability to express YadA may be due to minor changes in the structural gene, to defects in its regulation, or both" (Bölin et al. 1988).

The next major task was to subclone the genes into M13mp18/19 vectors for sequencing which I did manually setting up sequencing reactions using reverse transcriptase as a DNA polymerase (Skurnik and Wolf-Watz 1989). Altogether, the detailed characterization of the *yadA* genes; cloning, sequencing, Northern blottings, promoter mappings by primer extension and first trials to understand the protein using computer-based prediction programs (using the early versions of the GCG program suite (Devereux et al. 1984) or the in-house developed Geneus program package (Harr et al. 1985)), took most of my post-doctoral period and continued after my return to Finland in June 1987.

Based on my post-doc period we published 3 papers (Bölin et al. 1988; Rosqvist et al. 1988; Skurnik et al. 1989). In addition to learning how to do science we had time for other activities in Umeå. One was lunch-time jogging 2-3 times a week with a group of microbiologists which Hasse and I crowned with our first marathon in Stockholm in 1987. The other was the birth of my daughter in December 1986.

5.3.2 I Still Don't Work with *Y. pestis*; First Sequence of Its Pseudogene

The *yadA* gene of *Y. pestis* turned out to be silenced by a single nucleotide deletion; instead of 9 adenosines in *Y. pseudotuberculosis* there were only 8 in the *Y. pestis* gene leading to a frame shift mutation and premature stop codon. This apparently also affected the half-life of the mRNA since in Northern blotting I could not detect any *yadA* signal from *Y. pestis* (Skurnik et al. 1989). These findings and plasmid constructs were useful when combined with Roland Rosqvist's work on invasin leading to the publication of the infamous Nature article on "Increased virulence of *Yersinia pseudotuberculosis* by two independent mutations" (Rosqvist et al. 1988). I will return to the structure and function studies on YadA later in this review.

5.4 Move to Turku, Starting Own Group

In Umeå I got an offer to join the reactive arthritis group of Paavo and Auli Toivanen at the Department of Medical Microbiology, University of Turku, Finland. My task was to bring bacterial genetics expertise to the group that consisted mostly of immunologists and clinicians. In Turku I set up the DNA laboratory and got involved in different projects, some on topics directly or indirectly related to reactive or rheumatoid arthritis, others on genetics and mechanisms of antibiotic resistance and on the use of PCR in microbial diagnostics.

In *Yersinia* research, I continued with the YadA-project (see below, section 8), and started to supervise the PhD thesis work of Anna-Mari Viitanen, MD, who tried to find a relationship between HLA-B27 and *Y. enterocolitica* at the DNA level based on the molecular mimicry hypothesis. Some of the Yops and also YadA were suspected to play a role in molecular mimicry and therefore she ended up cloning and characterizing the *lcrE/yopN* operon from both *Y. enterocolitica* O:3 and *Y. pseudotuberculosis* YPIII/pIB1 (Forsberg et al. 1991; Viitanen et al. 1990). Based on the *lcrE* sequence Anna-Mari set up a PCR-method for the detection of *Y. enterocolitica* DNA from synovial samples of reactive arthritis patients (Viitanen et al. 1991) and defended her PhD thesis in 1990 (Viitanen 1990).

I started new projects with *Y. enterocolitica* lipopolysaccharide (see section 6) and with a potentially arthritogenic molecule which turned out to be the β–subunit of urease (Gripenberg-Lerche et al. 2000; Mertz et al. 1994; Skurnik et al. 1993). Finally, I isolated bacteriophages from the Turku City sewage to obtain transducing phages between *E. coli* and *Y. enterocolitica*. In 1988 Reija Venho was employed by my laboratory as a technician and she was instrumental in all the projects until 2003 when I moved to Helsinki. One of the first projects where she contributed was the IVS-project.

5.5 IVS, the Intervening Sequences and Microbial Diagnostics

During Northern blotting, total bacterial RNA is separated by denaturing agarose gel electrophoresis. When analyzing RNA isolated from the three *Yersinia* species

(Skurnik et al. 1989) I noticed that the 23S rRNA band was absent in *Y. enterocolitica* O:3 or partially absent in *Y. enterocolitica* O:8 while it was intact in *Y. pseudotuberculosis* and *Y. pestis* (Fig. 2).

We started to elucidate this phenomenon in the genus *Yersinia,* located the cleavage site by hybridizations and RNA-sequencing and sequenced over the cleavage site in the 23S rDNA genes. We found out that in certain *Y. enterocolitica* and *Yersinia bercovieri* strains a few or all of the seven 23S rDNA genes contained an IVS located between nucleotides 1171 and 1172 (the *E. coli* 23S rRNA numbering, Fig. 2). As we were almost ready to submit our findings on this phenomenon an extensive report on intervening sequences (IVS) in *Salmonella* was published in Cell (Burgin et al. 1990). We were fortunate to publish our finding in Molecular Microbiology (Skurnik and Toivanen 1991). Since then IVSs have been identified in many other bacterial taxa and they map to certain hotspots in the 23S rDNA. The function of IVSs, if any, has not been solved.

The work with the IVSs raised my interest to read the rRNA literature. At that time Carl Woese had just published his review on bacterial evolution (Woese 1987) largely based on the 16S rRNA and rDNA sequence comparisons. As I was working in close association with a clinical microbiology laboratory we became interested in the possible use of rDNA sequences in bacterial diagnostics. To that end I designed, based on the available 23S rRNA sequences, a primer pair (MS-37&MS-38) that we tested against all available strains in our own and all the neighboring culture collections. The primer pair was fantastic, it worked with all strains representing different bacterial phyla, and most importantly, with all clinically important bacteria including spirochetes and *Chlamydiae* (Kotilainen et al. 1998). In addition, the PCR was sensitive.

Therefore our routine laboratory started to offer universal bacterial PCR to the clinicians to detect bacterial DNA from patients with difficult symptoms and later on we also started using rDNA sequencing in the identification of bacteria (Jalava et al. 1995; Jalava et al. 1996; Jalava et al. 2001; Rantakokko-Jalava et al. 2000).

5.6 Lipopolysaccharide Research

I started to study *Yersinia* LPS, its genetics and role in virulence after my move to Turku. Similar to many turns in my life, I did not plan it, the project just dropped into my lap as I happened to be in the right place. The reactive arthritis immunologists had generated monoclonal antibodies A6 and 2B5 specific for the LPS O-antigen (O-ag) and core of *Y. enterocolitica* O:3 (Pekkola-Heino et al. 1987). In a Monday evening progress meeting the question, what could they be used for, was asked. Without knowing much of LPS and even less of its genetics I proposed that we should use it to clone and express the LPS gene in *E. coli.* A literature search in 1988 revealed that such had been done in Peter Reeves's laboratory for *Vibrio cholerae* and *Salmonella typhimurium* (Brahmbhatt et al. 1986; Manning et al. 1986).

Thus when Ayman Al-Hendy, an eager Egyptian MD, came to Turku to work for his PhD, I gave him the LPS cloning project. He made a genomic library of *Y. enterocolitica* O:3 in pBR322 and managed to get four Mab A6 positive colonies on

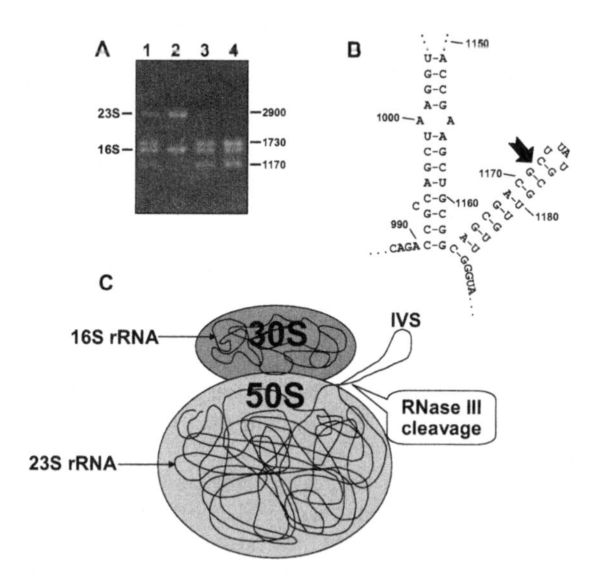

Fig. 2. IVS in 23S rDNA gene cause cleavage of the 23s rRNA. Panel A. Ethidium bromide stained denaturing agarose gel electrophoresis showing total RNA from *Y. bercovieri* (lane 1), *Shigella flexneri* (lane 2), *Y. enterocolitica* O:8 (lane 3) and *Y. enterocolitica* O:3 (lane 4). Panel B. Location of the cleavage site in the hairpin loop 1164-1185 of *E. coli* 23S rRNA sequence. Panel C. Possible mechanism of cleavage. The 30S and 50S subunits including their 16S and 23S rRNAs are drawn schematically. The IVS sequence forms a hairpin loop outside the 50S ribosomal body and is accessible to RNase III cleavage.

colony immunoblotting. Two of the clones were characterized and carried about 11 and 12 kb inserts (Al-Hendy et al. 1991b). Later he also screened the library using Mab 2B5 and isolated a core-expressing clone (Al-Hendy et al. 1991a). The latter clone has not been studied further but it is very likely carrying the outer core (OC) gene cluster that we identified later.

5.6.1. *Yersinia* LPS Plays a Role in Virulence

There were reports associating the O-ags of *Salmonella, E. coli, Shigella* and *Pseudomonas aeruginosa* with virulence (Binns et al. 1985; Engels et al. 1985; Mäkelä et al. 1988; Okamura et al. 1983; Porat et al. 1987) and the O-ag of *Y. enterocolitica* with resistance to the alternative complement pathway (Wachter and Brade 1989). Therefore we wanted to examine whether the O-ag of *Y. enterocolitica* O:3 played any role in virulence. To this end we isolated a rough mutant, YeO3-R2, using an

Table 2. STM-screens pick LPS-mutants in *Yersinia*

Bacterial species	Infection route	*N* mutants			References
		LPS mutants	Total attenuated	Total screened	
Y. ent O:8	Intraperitoneal	16	55	2016	1
Y. pstb O:3	Orogastric	3	13	960	2
Y. pstb O:3	Intravenous	10	31	603	3

1. (Darwin and Miller 1999)
2. (Mecsas et al. 2001)
3. (Karlyshev et al. 2001)

O-ag specific bacteriophage ϕYeO3-12 (Pajunen et al. 2000). By oral infection in mice, YeO3-R2 was about 50-fold less virulent than the wild type parental strain (Al-Hendy et al. 1992).

Later on we expanded our LPS studies to serotype O:8 O-ag and identified the OC gene cluster in serotype O:3 (Skurnik et al. 1995; Zhang and Skurnik 1994; Zhang et al. 1996). We demonstrated that these also play a role in virulence (Skurnik et al. 1999; Zhang et al. 1997). In recent signature-tagged-mutagenesis (STM) screens (Table 2) with *Y. enterocolitica* and *Y. pseudotuberculosis*, a large number of hits accumulated in LPS-genes further strengthening the importance of LPS structure to virulence.

5.6.2. LPS Structure Variation

Structurally LPS is divided into three components: lipid A, core and O-polysaccharide (OPS or O-ag). The endotoxic activity is dependent on the lipid A part while the core and the OPS are believed to play a role in outer membrane (OM) integrity and providing correct surface topology for the function of 'true' virulence factors. Recent studies have revealed that the LPS is a dynamic structure that responds to environmental signals. The effect of temperature on LPS structure has been observed in all pathogenic *Yersinia,* and the first noticed effect was down-regulation of OPS expression at 37°C. Since all known virulence factors are in one way or another regulated by temperature, it is of interest to study temperature-regulation.

The number and structure of the lipid A fatty acids and the substitutions of the 1- and 4'-positions in the glucosamine disaccharide can vary. Variation can also occur in the inner and outer core regions and in the OPS. In *Yersiniae* these structures appear to be under tight genetic control which modulates the OM properties and functions which are critical to pathogenesis.

The endotoxic activity of LPS depends on the lipid A acylation pattern such that hexa-acylated lipid A is the most endotoxic and tetra-acylated almost non-toxic. Structural studies have shown variation in the *Y. pestis* LPS structure. In *Y. pestis* grown at 37°C, lipid A is tetra-acylated and penta- and hexa-acylated at 25°C

(Bengoechea et al. 2003; Bengoechea et al. 1998; Hitchen et al. 2002; Kawahara et al. 2002; Knirel et al. 2006; Knirel et al. 2005a; Knirel et al. 2005b; Rebeil et al., 2004). This has raised speculations that, *in vivo*, *Y. pestis* would have tetra-acylated LPS which would allow it to grow in the blood to very high numbers (>10^8/ml) without killing the host by endotoxic shock. This would secure enough bacteria in the flea blood meal for efficient transmission to a new host. On the other hand, a *Y. pestis* strain with constitutively hexa-acylated lipid A appears to be severely attenuated (Montminy et al. 2006) thus extending the importance of LPS regulation to the early phases of infection where hexa-acylated LPS functions as a triggering signal for innate immunity.

5.6.3 Y. enterocolitica O:3 LPS

Our first papers on O:3 LPS created a fruitful collaboration with Joanna Radziejewska-Lebrecht (University of Silesia, Poland) and Otto Holst (Research Center Borstel, Germany) who are experts in carbohydrate structures. Central to elucidating the functions of the LPS gene products is to determine the structures of the final biosynthetic products expressed by wild type and mutant strains.

The LPS of *Y. enterocolitica* serotype O:3 shows some structural peculiarities which make it an interesting study object. Its homopolymeric OPS is composed of β1,2-linked 6-deoxy-L-altrose residues (Hoffman et al. 1980) and is the serotype-specific O-ag. A single OPS is ca. 60-70 monomers long and it is linked to the inner core of LPS. A hexasaccharide is also linked to the inner core and gives the LPS a branched structure (Radziejewska-Lebrecht et al. 1998; Skurnik et al. 1999; Skurnik et al. 1995). We call the hexasaccharide the OC although strictly speaking it is not similar to the OC region present in *E. coli* and *Salmonella.* Our recent studies have revealed that, under some circumstances, the enterobacterial common antigen (ECA) may also be linked to the inner core (Radziejewska-Lebrecht et al. 2003). Down-regulation of the OPS expression in *Y. enterocolitica* O:3 at 37°C has been long known (Al-Hendy et al. 1991a; Kawaoka et al. 1983a; Kawaoka et al. 1983b; Ogasawara et al. 1985; Wartenberg et al. 1983) but the effect of temperature on the other LPS components has been little studied.

5.6.3.1 The LPS Gene Clusters of *Y. enterocolitica* O:3

LPS is a very complex polysaccharide structure with a wide variety of sugar specificities and glycosidic linkages. In bacteria, the genes involved in LPS biosynthesis form gene clusters. The LPS genes encode NDP-sugar biosynthetic enzymes, glycosyltransferases forming the glycosidic bonds between the sugar residues, the O-unit polymerases and the machinery required to translocate the structure to the OM.

The O-ag and the OC gene clusters (Fig. 3) have independent locations in the *Y. enterocolitica* O:3 genome. They consist of 8 and 9 genes, respectively. The functional categories for the gene products are indicated based on similarity predictions; we have solid experimental evidence for the functions of WbcP and Gne, for the

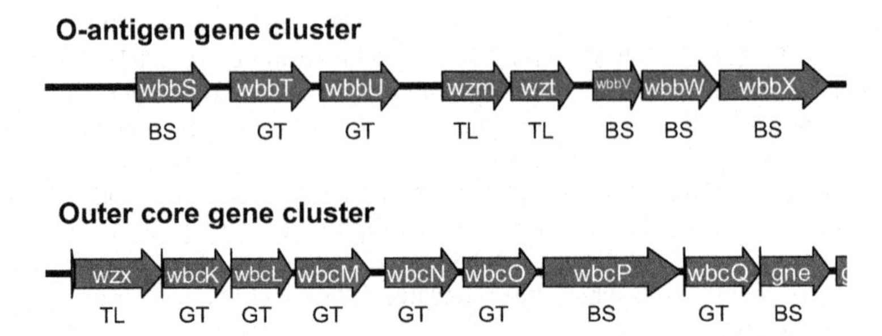

Fig. 3. The LPS gene clusters of *Y. enterocolitica* O:3. The functions of the gene products are indicated: BS, NDP-sugar biosynthesis; GT, glycosyltransferase; TL, translocation.

remaining we have either indirect evidence to support the function or we are working to generate the data.

5.6.3.2 *Y. enterocolitica* O:3 OC-Specific Bacteriophage, Enterocoliticin and Mab 2B5

Surface structures of bacteria are targets for bacteriophages. We have isolated several bacteriophages from the Turku City sewage with different specificities for *Y. enterocolitica* O:3. One of the phages is ϕR1-37, specific for the OC (Kiljunen et al. 2005). After isolation of a new phage it is usual to study its host range. ϕR1-37 is able to infect a wide set of *Y. enterocolitica* serotypes including O:3, O:5,27, O:6,31, O:9, O:25,26,44, O:41,43 and O:50, also one *Yersinia intermedia* serotype O:52,54 strain and surprisingly *Y. pseudotuberculosis* serotype O:9 strains (Kiljunen et al. 2005; Skurnik et al. 1995). This indicates that the phage receptor is present in these other strains and for *Y. enterocolitica* we know that this is the OC structure and that the OC gene cluster is present in the strains. However, in *Y. pseudotuberculosis* O:9 structure of the receptor seems to consist, at least in part, of its O-ag since a phage-resistant mutant has lost it (Kiljunen et al. 2005). Furthermore, OC gene cluster specific PCRs gave negative results with *Y. pseudotuberculosis* O:9 DNA.

Enterocoliticin is a phage tail-like bacteriocin that Eckhard Strauch characterized (Strauch et al. 2001) and we later showed that it also uses the OC structure as a receptor (Strauch et al. 2003). Mab 2B5 mentioned above (Pekkola-Heino et al. 1987) is also specific for the OC (Biedzka-Sarek et al. 2005). Thus we have serendipitously obtained three different biotools specific for the OC. Our preliminary data from truncated OC constructs indicate that Mab 2B5 and enterocoliticin do not tolerate any OC truncation while ϕR1-37 accepts a tri- or tetrasaccharide OC stump. We are currently elucidating the structural details of the receptors.

Bacteriophage ϕR1-37 offered us another surprise; it is one of the largest phages known and in its DNA deoxyuridine (dU) has completely replaced thymidine (Kiljunen et al. 2005). It will be interesting to find out how it accomplishes this.

5.6.4 Y. enterocolitica O:8 LPS

Shortly after Ayman Al-Hendy, Lijuan Zhang joined my group originally recruited to set up PCR-methods to study the viral etiology of rheumatoid arthritis (Zhang et al. 1993). Having learned to master the PCR-technique she was eager to learn molecular biology methods and as a training project she took up sequencing of the *Y. enterocolitica* O:3 O-ag gene cluster (Zhang et al. 1993) and then turned to serotype O:8 O-ag gene cluster which she characterized and showed that a rough O:8 strain is attenuated (Zhang et al. 1997; Zhang et al. 1994; Zhang et al. 1996). She also laid the foundation for regulation studies that José Antonio Bengoechea (Josean) took up when he started in 1998 as a post-doc. In a very nice series of papers Josean exposed the complex regulatory networks involved in the control of the O-ag expression in *Y. enterocolitica* O:8 (Bengoechea et al. 2004; Bengoechea and Skurnik 2000; Bengoechea et al. 2002).

A remarkable finding was that for pathogenicity it is not enough to have the O-ag expressed on the surface, it is also critical how it is expressed. O:8 bacteria use temperature to control the amount of the O-ag but they also need to control its chain length to retain full virulence (Bengoechea et al. 2004; Najdenski et al. 2006; Najdenski et al. 2003). In preliminary mouse experiments with a set of luciferase reporter strains we noticed that bacteria shut off O-ag expression when they reside in the spleen or liver while in the Peyer's patches the expression is on. We don't know what the *in vivo* signals are that regulate expression.

5.7 LPS Genes in *Yersinia* are Between the *hemH* and *gsk* Genes

The O-ag gene cluster of O:8 and the OC gene cluster of O:3 both were flanked with the *hemH* and *gsk* genes. In 1996, partial sequences of the O-ag gene clusters of *Y. pseudotuberculosis* serotype O:2a and O:5a were available and in these strains the cluster was located downstream of the *hemH* gene. I designed a degenerate pair of primers based on aligned *hemH* and *gsk* sequences and set up long range PCR to find out whether this was a rule in *Yersinia* including *Y. pestis.* On March 30, 1996 I ran the long range PCR where I observed that *Y. pestis* DNA gave a ca. 20 Kb product indicating that also it carries an O-ag gene cluster, although it must be non-functional as *Y. pestis* does not express the O-ag (Fig. 4A). This made me curious as it also had evolutionary implications towards the origin of *Y. pestis;* at that time we only speculated that it had evolved from *Y. pseudotuberculosis* but there was no hard evidence for that. In order to get funding to look at the question, I turned to Finnish Scientific Advisory Board for Defense and proposed that we would develop a *Y. pestis* – specific PCR based on the differences within the O-ag gene clusters between *Y. pestis* and *Y. pseudotuberculosis.* After waiting for another year the funding was granted and we started in 1998. I received the *Y. pseudotuberculosis* reference strains of Hiroshi Fukushima from Elisabeth Carniel and soon found out that the serotype O:1b O-ag gene cluster was most similar to that of *Y. pestis.* Sequence analysis revealed that five of the 17 genes in the O-ag gene cluster of *Y. pestis* are pseudogenes explaining the lack of O-ag expression (Skurnik et al. 2000).

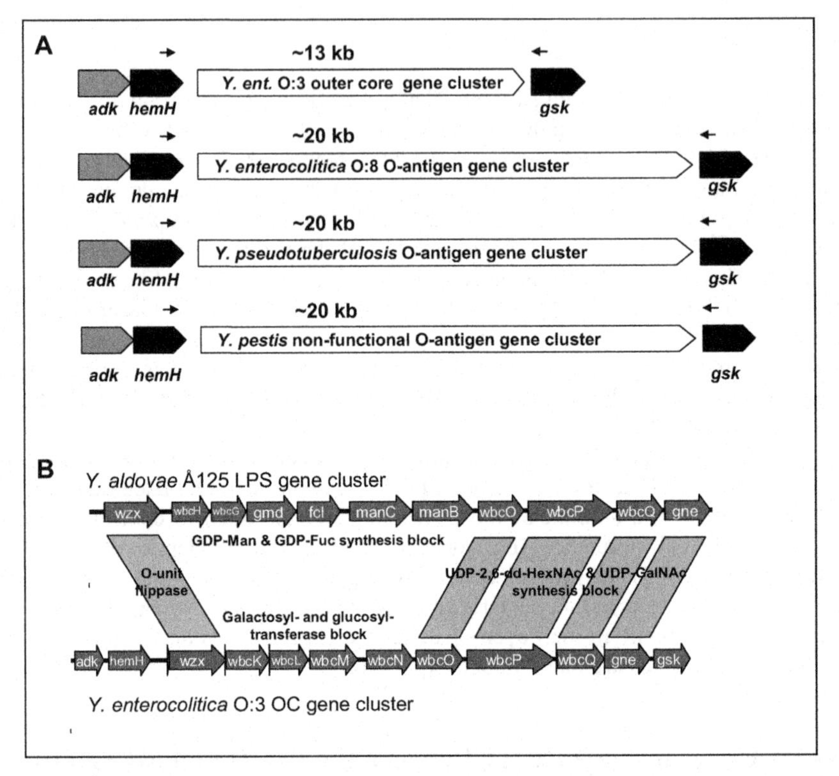

Fig. 4. LPS gene clusters in *Yersinia* are located between the *hemH* and the *gsk* genes. Panel A. Schematic representation of four LPS-gene clusters. The small arrows indicate the primers used in long range PCR. Panel B. Comparison of the gene clusters of *Yersinia aldovae* strain Å125 (GenBank accession no. AJ871364) and *Y. enterocolitica* O:3 outer core (accession no. Z47767). The gene cluster graphics were generated with the ggnVIEW sequence file viewer (http://colibase.bham.ac.uk/cgi-bin/fileprepare.cgi).

As most of the sequencing was done by primer walking and we had promised to develop a diagnostic PCR for *Y. pestis* we selected gene-specific primer pairs from them and analyzed all *Y. pseudotuberculosis* reference strains for the presence of the *Y. pestis* –specific genes.

The results revealed that in addition to O:1b, serotypes O:11 and O:14 shared 15 of the 17 genes and that the gene repertoire sharing dropped blockwise paralleling the sugar composition of the O-units known for some of the serotypes. In general, in the O-ag clusters a gene block dedicated to the biosynthesis of GDP-mannose and to specific mannosyltransferases is physically tightly organized to ensure that during horizontal gene transfer the whole block is transferred as one. Such a case is illustrated in Fig. 4B. Sequence alignments of the *Y. pseudotuberculosis* O-ag cluster genes, especially *wzx* and *wzz*, revealed scars of ancient recombination events indicating that these genes have been recombination hotspots during the evolution of new O-ag specificities.

In any case, the high similarity between the O-ag gene clusters of *Y. pestis* and some of the *Y. pseudotuberculosis* serotypes made the development of the *Y. pestis* – specific PCR problematic. We managed to do it although it was not fully satisfactory. However, in parallel we designed a multiplex PCR for easy O-ag-genotyping of *Y. pseudotuberculosis* strains (Bogdanovich et al. 2003).

5.8 Function and Structure Studies on YadA

I will now return back to YadA for a while. After accomplishing the nucleotide sequence of the *Y. pseudotuberculosis yadA* gene I started to think of the structure-function relationships of YadA. I had the amino acid sequence at hand, but what to do with it.

In November 1986 I wrote to Tom Blundell (Birkbeck College, London) about YadA (at that time known as Yop1):

"This protein ... may play a role during the infection caused by some *Yersinia* strains...The role of Yop1 could be attachment of bacteria on the gut epithelium, and thus help the bacteria in colonization of the intestine. Yop1 was originally associated with ... ... autoagglutination. Yop1 is located in the OM of the bacteria where it forms very thin hair-like structures, fimbrillae, with a diameter of 2 nm and length of 100 nm (this refers to Kapperud paper (Kapperud, Namork and Skarpeid 1985)). In SDS-PAGE Yop1 is seen in two bands, >200 kD and subunit form 45-50 kD. If the sample is cooked extensively, almost all >200 kD form turns into the subunit form.

I made some calculations based on the assumption that a protein of 50 kD would take a volume of ca. 100 cubic nm. The volume of the above mentioned fimbrill would be ca. 400 cubic nm. This implicates that one fimbrill is composed of four subunits, and that in fact the >200 kD band seen in SDS-PAGE represents intact fimbrillae. The forces keeping the subunits together seem to be quite strong, as they resist the denaturing effect of the SDS-PAGE sample buffer and cooking for 5 min...." (Then I continue telling that I have the deduced amino acid sequence of YadA available and that) "I'd like to know how the four subunits bind together to form the fimbrillae, ... I imagine that in your laboratory you have advanced computer programs for this kind of problems ..."

Tom replied that "ultimately a high resolution analysis will be required. Can you produce enough to crystallise? ... The prediction of the structure and the nature of the subunit interactions is a difficult task." Indeed it has taken a long time to finally get to the YadA structure but even with the structure, we have surprisingly little details of the functional mechanisms.

Table 3. Characteristics and phenotypes associated with YadA

Characteristic or phenotype	References
Expression regulated by temperature and VirF, not by [Ca^{2+}]	1
Fibrils, tags or lollipops seen in EM	2
YadA forms a trimer & trimeric autotransporter	3
Binds to extracellular matrix (ECM) proteins: collagen, immobilized fibronectin and laminin; binds to intestinal submucosa and mucin	4
Mediates adhesion to tissue culture cells	5
Functions as an invasin	6
Mediates complement resistance	7
Associated with reactive arthritis in a rat model	8

1. (Lambert de Rouvroit et al. 1992; Skurnik and Toivanen 1992)
2. (Hoiczyk et al. 2000; Kapperud et al. 1985; Zaleska et al. 1985)
3. (Gripenberg-Lerche et al. 1995; Roggenkamp et al. 2003)
4. (Emödy et al. 1989; Paerregaard et al. 1991a; Paerregaard et al. 1991b; Schulze-Koops et al. 1993; Schulze-Koops et al. 1992; Schulze-Koops et al. 1995; Skurnik et al. 1994; Tamm et al. 1993; Tertti et al. 1992)
5. (Bukholm et al. 1990; Heesemann and Grüter 1987)
6. (Bliska et al. 1993; Heise and Dersch 2006; Yang and Isberg 1993)
7. (Balligand et al. 1985; China et al. 1993; Pilz et al. 1992)
8. (Gripenberg-Lerche et al. 1995; Gripenberg-Lerche et al. 1994).

5.8.1 YadA is a Multifunctional Protein

In 1990 during a Keystone meeting in Frisco, Colorado, to differentiate it from true released Yops the protein which had been called P1, autoagglutination protein, Yop1 and YopA, was named as YadA, the *Yersinia* adhesin A, to accommodate its stickiness to everything. After its identification different YadA-associated phenomena had started to accumulate in the literature. As these phenomena have been reviewed recently (El Tahir and Skurnik 2001) I will just list a few in Table 3.

5.8.2 YadA has a C-terminal Membrane Anchor

As mentioned above, purification of YadA for X-ray studies was the advice I got from Tom Blundell. I made several extraction and ammonium sulphate precipitation experiments and obtained small amounts of the protein. But when I read from the paper of Thomas Michiels (Michiels et al. 1990) that a segment typical of membrane-associated structures was predicted to span amino acids 83-104 of YadA I thought that this part was the membrane anchor. I designed, with Anu Tamm, a PCR-based protocol to delete exactly those 22 amino acids (aa) from the structure. To our disappointment the 22-aa deleted version of YadA was not secreted into the culture supernatant; the protein was still membrane-associated and monoclonal anti-YadA antibodies could detect the protein on the bacterial surface (Tamm et al. 1993).

The other possible membrane anchor predicted from the amino acid sequence was the hydrophobic C-terminus and indeed YadA constructs with C-terminal deletions did not reach the bacterial surface. Since we now know that YadA is a trimeric autotransporter (Linke et al. 2006; Roggenkamp et al. 2003) the explanation is that these deletions destroyed the translocator β-barrel. None of these constructs gave any easy solution for the purification of YadA.

5.8.3 The 22-aa Deletion Mutant is Avirulent

We noticed that the 22-aa deletion abolished YadA-mediated autoagglutination and this prompted us to look for other YadA-associated phenomena such as binding to ECM proteins (Table 3). To that end we sent the constructs to Timo Korhonen in Helsinki. The 22-aa deletion mutant had lost collagen-binding ability but retained laminin- and fibronectin binding. The deletion did not affect serum resistance. Since serotype O:3 is not lethal to mice we constructed the 22-aa YadA deletion in a serotype O:8 strain which turned out to be completely avirulent in orally infected mice (Tamm et al. 1993), and was non-arthritogenic in a rat model of reactive arthiritis (Gripenberg-Lerche et al. 1995). Soon Andreas Roggenkamp from Jürgen Heesemann's laboratory reported that histidines 156 and 159 of $YadA_{YeO:8}$ were involved in collagen-binding and that their collagen-binding mutant was also avirulent (Roggenkamp et al. 1995). This mutant, however, had not lost the autoagglutination property in contrast to the 22-aa deletion mutant.

5.8.4 Epitope Mapping of YadA

After Anu Tamm had finished the YadA-project it was taken over by Yasmin El Tahir, a Sudanese veterinarian who found herself working with me because she followed her to-become husband Ahmed to Turku and Finland. She used a lot of time trying different methods to purify YadA or C-terminally truncated constructs for crystallography but without much success. Instead her most important contribution was the epitope-mapping of YadA. The rationale behind the mapping experiment was that we had three different rabbit antisera against *Y. enterocolitica* O:3 or gel-extracted YadA. All antisera reacted strongly with YadA in immunoblotting but only the one against gel-extracted YadA inhibited the YadA-mediated collagen-binding (Emödy et al. 1989; Schulze-Koops et al. 1992). Identification of the YadA epitopes specific for the latter antiserum would help us map the collagen-binding sites of YadA. For epitope mapping we initially used an overlapping set of 16-mer peptides biotinylated at one end to allow their mobilization onto streptavidin-coated plates, and later 4-8-mer peptides for fine-mapping. We identified altogether 12 antigenic epitopes of which 8 were specific for the collagen-binding inhibitory antiserum (El Tahir et al. 2000). To provide evidence that these epitopes would be relevant to collagen-binding we could either try to inhibit collagen-binding by synthetic individual peptides or by site-directed mutagenesis of the individual epitopes. The former approach was tried but was completely unsuccessful, the latter was more informative. By inspecting the sequences of the epitopes we noticed that certain of them shared an NSVAIG--S motif. That motif with little variation was present actually in several repeats and now we know that these motifs can be aligned as shown in Fig. 5.

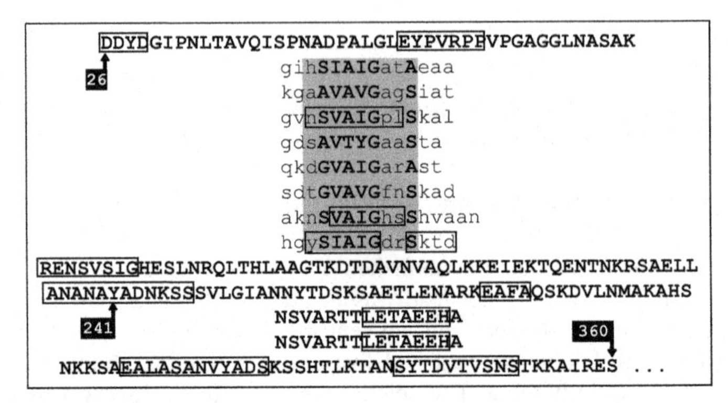

Fig. 5. Structurally important sequences of $YadA_{YeO:3}$. Shown is the sequence from aa 26 to 360. The identified antigenic epitopes are boxed. The NSVAIG--S motifs of YadA are aligned and indicated by grey background. Also marked is Y241, the last residue in the crystallized head domain (Nummelin et al. 2002).

Since collagen is a regular repeating structure we anticipated that also the binding motifs on YadA could be repeating and that the NSVAIG--S motif could form part of it. To test this hypothesis we mutagenized several of the motifs by changing the hydrophobic VAI aa-triplets to hydrophilic amino acids of similar size. To our satisfaction most of our motif-mutants lost collagen-binding either totally or partially. Thus we predicted that the NSVAIG--S motif would be responsible for collagen-binding (El Tahir et al. 2000). This turned out to be a wrong prediction.

5.8.5 Crystal Structure of the Head Domain of YadA

Adrian Goldman was recruited to Turku Centre for Biotechnology in 1992 to set up an X-ray crystallography laboratory and I immediately introduced him to YadA. My group moved to the Centre for Biotechnology in 1993 so we were in close contact but it still took 3-4 years before anything really started to happen. Heli Nummelin was a chemist that joined Adrian's group in 1997 and soon took up the YadA challenge. Based on the work that Yasmin had done and the structural domain predictions published in the excellent paper from Jürgen Heesemann's group (Hoiczyk et al. 2000) Pauli Ollikka, then a post-doc in my laboratory, made a few C-terminally truncated constructs where most of the predicted stalk was removed. We were safe to do this as the collagen-binding domain mapped to the N-terminal head-domain of YadA (Westerlund-Wikström et al. 1997). One of the truncated constructs (expressing aa 26-241, Fig. 5) produced a good yield of collagen binding product that could be crystallized (Nummelin et al. 2002) and the structure was solved at 1.55 Å resolution (Nummelin et al. 2004). Without going into any details about the beautiful structure which can be seen in the publication, a few points deserve discussion.

The NSVAIG--S motifs identified by Yasmin turned out to be located in the hydrophobic core of the head domain and therefore inaccessible to collagen-binding.

```
       -- --  -   +           - +        +   + -  +   +     -++ +  -               + +-       -
          -- -                -     -                                   +            +  +      -
            --    +           - - +                                         -        +  +      -
Ypstb  EEPE=GN====R=S=======V==K==VGLYPAKPILRQENPKLPPRGPQGPEKKRARLAEAIQ=Q=L=======R==DPY========== 107
YeO:3  ____DDYDGIPNLTAVQISPNADPALGLEY______________________________[PVRPPVPGAG]GLNASAKGIHSIAIGATAEAA 82
YeO:8  ____NNDEV__HF==========DSHVVI________________________________FQ=AAEAL==T==L==S=====V==S==== 79
```

Fig. 6. Alignment of the N-terminal aa sequences of YadA of *Y. pseudotuberculosis* O:3 (top, aa 26-117), *Y. enterocolitica* O:3 (middle, aa 26-82) and *Y. enterocolitica* O:8 (bottom, aa 26-79). The charged aa are indicated on top. The other sequences are compared to the middle sequence, identical aa are indicated by =, gaps in alignment by _, and different aa by letters. The YadA$_{YeO:3}$ aa 52-61 loop is boxed.

Instead they are crucial to the head-domain structure. Introduction of charged aa into the core most likely distorts the compact structure by bringing charged residues to face each other. This results in repulsion between the monomers and swelling of the head cylinder that destroys the collagen-binding surface of YadA. Site-directed mutagenesis of surface-located aa showed that most likely a large surface of YadA participates in collagen binding (Nummelin et al. 2004). We still are missing details of the binding mechanisms as there have been problems in obtaining co-crystals with synthetic collagen-like peptides or native collagen fragments.

On the top of the head domain cylinder a loop formed by aa 52-61 was not stable and its structure could not be solved. Very interestingly, alignment of the YadA aa sequences of *Y. enterocolitica* O:3, O:8 and *Y. pseudotuberculosis* O:3 reveals that in this part of YadA the sequences differ largely (Fig. 6).

Amino acids 29-81 of YadA$_{YeO:8}$ are responsible for neutrophil binding, while the extra loop in YadA$_{Ypstb}$ is responsible for invasiveness (Heise et al. 2006). Thus bacteria seem to decorate YadA with functionally different surface domains. Deletion of the extra loop in YadA$_{Ypstb}$ changed the binding properties of YadA to ECM components (Heise et al. 2006).

5.9 Serum Resistance of *Y. enterocolitica* O:3

5.9.1 The Complement System

Serum resistance is a property that most invasive animal pathogens possess to survive in the host. The serum and tissue fluids contain defense mechanisms of innate immunity of which the complement system is the most powerful. The complement system is a set of ca. 30 plasma proteins which, upon recognition of a foreign structure, initiate a multistep enzymatic amplification cascade. This leads to the formation of membrane-attack-complexes (MACs) that kill target cells by making holes in the membranes.

Three different complement activation pathways are known: (i) classical pathway (CP) activated by the binding of C1 to an antigen-antibody complex, (ii) lectin pathway (LP) activated by the binding of mannan binding lectin to a microbial surface polysaccharide, and (iii) alternative pathway (AP) activated when a circulating C3b binds directly to the microbial surface. All these pathways lead to the formation of C3-convertase; CP and LP to C4b2a, composed of C4b and C2a and AP to C3bBb, composed of C3b and Bb. C3-convertases, by cleaving C3, generate huge amounts of active C3b molecules on site that are able to covalently attach to target surfaces nearby to generate more C3-convertases, or when binding to a C3-convertase change it to C5-convertase. At this level the three pathways converge; upon formation of C5b by C5-convertase, the cascade leads to MAC.

Complement activation is tightly controlled to prevent self-destruction. Regulation takes place at three critical levels:

1. C1-inhibitor keeps C1 in an inactive form in solution
2. C3-convertase activity is regulated by decay accelerating factors or by specific proteins like C4 binding protein (C4Bp) and factor H which directly inhibit the convertase or render it accessible to factor I that inactivates C3b by cleavage that produces iC3b
3. C9 polymerization that is inhibited e.g. by host cell surface proteins CD59 or a homologous restriction factor.

To survive the complement attack pathogenic bacteria have developed several different strategies. These include thick protective capsules or long OPSs that prevent insertion of MAC into the bacterial membrane. Specific serum resistance factors can either inhibit complement activation by interfering in a critical activation step or by diverting the complement attack away from the bacterium. In addition, some bacteria camouflage themselves under host-specific structures such as sialic acid.

5.9.2 Serum Resistance of *Y. enterocolitica*

In 1982 Chik Pai and in 1983 Rafael Martinez, Bob Perry and Jürgen Heesemann first associated serum resistance of *Y. enterocolitica* to the presence of pYV and growth at 37°C (Heesemann et al. 1983; Martinez 1983; Pai and DeStephano 1982; Perry and Brubaker 1983). I also confirmed these findings (Skurnik et al. 1984). Chiesa, however, found that some pYV-negative derivatives were serum-resistant in their hands (Chiesa and Bottone 1983). In 1985 Balligand from Guy Cornelis's laboratory was first to associate YadA with serum resistance (Balligand et al. 1985). Wachter and Brade were the first to study complement activation in more detail by analyzing C3 and C9 consumption from serum and C3 deposition onto bacterial surfaces (Wachter et al. 1989)

I started to dissect the serum resistance of *Y. enterocolitica* O:3 using strains that expressed YadA and LPS in different combinations and in 1991 submitted a full manuscript to Infection and Immunity on our findings. I wrote that "these combinations established separate roles for YadA and O-ag, and revealed a third, chromosomally coded serum resistance factor, Trf, which was only expressed at 37°C. Yops did not play a role in serum resistance." We also concluded that serum resistance in *Y. enterocolitica* is a multifactorial property; YadA binds a soluble regulatory factor from serum, most likely factor H, and in CP activation, O-ag would be the target for natural antibodies. The manuscript was rejected based on many reasons, one reviewer writing that "the manuscript itself is so poorly written that I find it offensive of the Author to submit a manuscript in such a state" (that was rather depressing), another was that there were no statistics on any of the data. One shortcoming also was that the experiments were done with a single individual's serum. Anyway, later that same year I resubmitted the manuscript but even then I got it back rejected. After that the serum resistance project hibernated in my laboratory for almost ten years.

Meanwhile, the field had made some progress. In 1992 Pilz et al. studied the role of YadA in the AP resistance of *Y. enterocolitica* O:3. They were able to show that YadA-expressing bacteria bound less MAC than YadA-negative bacteria and that the effect was vastly enhanced if the bacteria were opsonized before the serum incubation (Pilz et al. 1992). The difference was apparent only after about a 60 min incubation in Mg-EGTA treated serum. They also showed that YadA-expressing bacteria convert C3b to iC3b more rapidly that YadA-negative bacteria and offered factor H binding by YadA as a possible mechanism. In 1993 Bernard China from the laboratory of Guy Cornelis showed that YadA-expressing bacteria fix less C3b on their surface than YadA-negative bacteria and was the first to demonstrate, using goat anti-factor H antibodies, that factor H from serum binds to the YadA-expressing bacteria and, in affinity blotting, to the YadA-band (China et al. 1993).

The role of Ail in serum resistance was first reported by Jim Bliska in 1992, followed by Dorothy Pierson (Bliska and Falkow 1992; Pierson and Falkow 1993). Later on, using site-directed mutagenesis, Virginia Miller reported that certain membrane loop residues of Ail are involved in serum resistance (Miller et al. 2001). However, the role of Ail in complement resistance has not been elucidated.

5.9.3 Resuscitation of the Serum Resistance Project

During the hibernation we identified the OC gene cluster of *Y. enterocolitica* O:3. OC plays a role in virulence and in resistance to cationic antimicrobial peptides but had no direct role in complement resistance (Skurnik et al. 1999). Meanwhile we constructed an *ail* mutant in *Y. enterocolitica* O:3 as I suspected that the above mentioned third serum resistance factor Trf would be identical to Ail. Our goal was to generate a full set of strains expressing all possible combinations of YadA, Ail, O-ag and OC. Most of these strains were constructed in the fall of 2001 when Marta Biedzka-Sarek started her MSc thesis work in my laboratory as a Sokrates/Erasmus exchange student and she soon completed construction of the missing ones. In total we studied 24 isogenic strains.

The first task was to perform the serum killing assays in a statistically acceptable manner, *i.e.,* from triplicate individual cultures in two parallel killing assays. We measured the CP and AP mediated killing in 66.7% serum at 30 and 120 min time points. Based on the results (Biedzka-Sarek et al. 2005) we could see several serum resistance phenotypes which we arbitrarily divided into five classes (Table 4). This division was based on killing kinetics, *i.e.*, survivors at 30 min (rapid <30%, delayed

Table 4. Serum resistance phenotypes of *Y. enterocolitica* O:3 grown at 37°C

Resistance phenotype (*n* strains)	YadA-, Ail-, O-ag-, and OC-expression phenotype (*n* strains)
Rapid sensitive (8)	----, Oag, OC, AilOag(2), OagOC, AilOagOC(2)
Rapid intermediate (6)	Ail, YadA, AilOag, YadAOag, YadAOC, YadAOagOC
Delayed sensitive (1)	AilOC
Delayed intermediate (5)	Ail (2), AilOC, YadAAil, YadAAilOagOC
Delayed resistant (4)	YadAAil, YadAAilOag (2), YadAAilOC

>30%) and the final survivors at 120 min (sensitive <1%, intermediate 1-20% and resistant >20%).

Even though the cutoffs are artificial (demonstrated by the fact that different mutants with the same phenotype, such as Ail, AilOC, YadAAil or AilOag, are classified in two resistance phenotypes) it helps in handling and interpreting the results (Biedzka-Sarek et al. 2005).

A surprising result is that the wild type strain (YadAAilOagOC) is not the most resistant; the most resistant strain is the one missing the O-ag, followed by strains missing OC or both O-ag and OC. This most likely reflects the fact that we are observing killing *in vitro;* in real life the wild type strain has the freedom to regulate its factors during the infection to its best advantage.

Another conclusion worth mentioning is that a strain must express both YadA and Ail to reach the highest resistance, losing one or the other drops immediately the resistance level. This illustrates the multifactorial nature of the resistance in *Y. enterocolitica* O:3. As a rule, expression of Ail delays killing, however, in combination with O-ag, the Ail-function is masked. The LPS components alone without YadA or Ail are not able to confer any resistance to complement. We also noticed that logarithmic phase bacteria are more resistant than stationary phase bacteria.

5.9.4 Dissection of the Complement Resistance Mechanisms

The complement system is a complex dynamic system and therefore difficult to dissect. Traditionally people have used purified complement components and studied their interactions with the bacteria or with isolated resistance factors. Alternatively, specific antibodies for individual complement components have been used to monitor, for example, deposition of C1, C3 or C5 onto bacteria.

Our first target was C3 (Biedzka-Sarek et al. 2005). Upon activation of complement and also when complement activation is inhibited, different forms of C3 (such as C3b, iC3b and covalently bound C3b) can be detected in the reaction (Vogel et al. 1997). We studied the deposition of C3 forms on the bacteria at 3, 15 and 30 min time points (Biedzka-Sarek et al. 2005). In general, C3 deposition could not be correlated with a serum resistance phenotype. On the contrary, the presence of O-ag prevented early C3b and covalently bound C3b deposition. On the other hand, prominent accumulation of iC3b took place in most resistant strains. In conclusion, the C3 deposition on *Y. enterocolitica* O:3 appears to be more complex than for example to

Neisseria meningitidis (Vogel et al. 1997). With this approach we can only observe a snapshot of the net result of C3b deposition, however, we have difficulties differentiating between the covalent binding of C3b to different bacterial surface components, its inactivation to iC3b, or its function as part of C3 and C5 convertases.

The other components we have looked at thus far are factor H and MAC. While setting up reliable methods to monitor these components we have noticed that the complement activation reaction is extremely rapid and that in some cases all the relevant activity has taken place within a couple of minutes. We have confirmed that factor H binds to YadA-expressing *Y. enterocolitica* O:3 but also to some other variants of the 24 isogenic strains we have tested.

At the same time we have also started to dissect the molecular basis of YadA-mediated serum resistance. Based on the report, again from Jürgen Heesemann's laboratory, where they identified the stalk of YadA as the serum resistance determinant (Roggenkamp et al. 2003), Marta Biedzka-Sarek has now constructed 30 short internal deletions in the stalk and analyzed their effect on factor H binding and on serum resistance. Several deletions in different parts of the stalk affect factor H binding or serum resistance or both and these do not always correlate, indicating that the interaction is complex. The deletion analysis of the stalk is tricky due to its trimeric coiled-coil structure; a deletion in one part of the stalk may cause structural constraints to a distant part and thereby cause a phenotype.

In addition, participation of factors such as other complement components, serum proteins and bacterial structures must be elucidated quantitatively and qualitatively to get a comprehensive understanding of the interactions taking place between the bacteria and the serum components. This highly dynamic situation involving several actors on both sides will still need more of our attention before we know what is really going on.

5.10 Acknowledgements

I consider myself a fortunate person. I have had a possibility to do research and I was somehow selected to work with *Yersinia.* This has given me an opportunity to interact with many *Yersinia* people in many different roles: as collaborator, coauthor, meeting organizer, meeting participant, supervisor, anonymous reviewer or anonymously reviewed, student and most importantly as a friend. I thank you all.

My work reported in this chapter has been supported by several funding agencies over the years including the European Union, the Academy of Finland, the Finnish Funding Agency for Technology and Innovation, and the Finnish Scientific Advisory Board for Defense. Visit also the homepages http://www.hi.helsinki.fi/yersinia/.

5.11 References

Al-Hendy, A., Toivanen, P. and Skurnik, M. (1991a). The effect of growth temperature on the biosynthesis of *Yersinia enterocolitica* O:3 lipopolysaccharide: temperature regulates the transcription of the *rfb* but not of the *rfa* region. Microb. Pathog. 10, 81-86.

Al-Hendy, A., Toivanen, P. and Skurnik, M. (1991b). Expression cloning of *Yersinia enterocolitica* O:3 *rfb* gene cluster in *Escherichia coli* K12. Microb. Pathog. 10, 47-59.

Al-Hendy, A., Toivanen, P. and Skurnik, M. (1992). Lipopolysaccharide O side chain of *Yersinia enterocolitica* O:3 is an essential virulence factor in an orally infected murine model. Infect. Immun. 60, 870-875.

Balligand, G., Laroche, Y. and Cornelis, G. (1985). Genetic analysis of a virulence plasmid from a serogroup 9 *Yersinia enterocolitica* strain: role of outer membrane protein P1 in resistance to human serum and autoagglutination. Infect. Immun. 48, 782-786.

Bengoechea, J.A., Brandenburg, K., Arraiza, M.D., Seydel, U., Skurnik, M. and Moriyon, I. (2003). Pathogenic *Yersinia enterocolitica* strains increase the outer membrane permeability in response to environmental stimuli by modulating lipopolysaccharide fluidity and lipid A structure. Infect. Immun. 71, 2014-2021.

Bengoechea, J.A., Brandenburg, K., Seydel, U., Díaz, R. and Moriyón, I. (1998). *Yersinia pseudotuberculosis* and *Yersinia pestis* show increased outer membrane permeability to hydrophobic agents which correlates with lipopolysaccharide acyl-chain fluidity. Microbiology. 144, 1517-1526.

Bengoechea, J.A., Najdenski, H. and Skurnik, M. (2004). Lipopolysaccharide O antigen status of *Yersinia enterocolitica* O:8 is essential for virulence and absence of O antigen affects the expression of other *Yersinia* virulence factors. Mol. Microbiol. 52, 451-469.

Bengoechea, J.A. and Skurnik, M. (2000). Temperature-regulated efflux pump / potassium antiporter system mediates resistance to cationic antimicrobial peptides in *Yersinia*. Mol. Microbiol. 37, 67-80.

Bengoechea, J.A., Zhang, L., Toivanen, P. and Skurnik, M. (2002). Regulatory network of lipopolysaccharide O-antigen biosynthesis in *Yersinia enterocolitica* includes cell envelope-dependent signals. Mol. Microbiol. 44, 1045-1062.

Biedzka-Sarek, M., Venho, R. and Skurnik, M. (2005). Role of YadA, Ail, and lipopolysaccharide in serum resistance of *Yersinia enterocolitica* serotype O:3. Infect. Immun. 73, 2232-2244.

Binns, M.M., Vaughan, S. and Timmis, K.N. (1985). 'O'-antigens are essential virulence factors of *Shigella sonnei* and *Shigella dysenteriae* 1. Zentralbl. Bakteriol. Mikrobiol. Hyg. [B]. 181, 197-205.

Bliska, J.B., Copass, M.C. and Falkow, S. (1993). The *Yersinia pseudotuberculosis* adhesin YadA mediates intimate bacterial attachment to and entry into HEp-2 cells. Infect. Immun. 61, 3914-3921.

Bliska, J.B. and Falkow, S. (1992). Bacterial resistance to complement killing mediated by the Ail protein of *Yersinia enterocolitica*. Proc. Natl. Acad. Sci. USA. 89, 3561-3565.

Bogdanovich, T., Carniel, E., Fukushima, H. and Skurnik, M. (2003). Use of O-antigen gene cluster-specific PCRs for the identification and O-genotyping of *Yersinia pseudotuberculosis* and *Yersinia pestis*. J. Clin. Microbiol. 41, 5103-5112.

Brahmbhatt, H.N., Quigley, N.B. and Reeves, P.R. (1986). Cloning part of the region encoding biosynthetic enzymes for surface antigen (O-antigen) of *Salmonella typtimurium*. Mol. Gen. Genet. 203, 172-176.

Bukholm, G., Kapperud, G. and Skurnik, M. (1990). Genetic evidence that the *yopA* gene-encoded *Yersinia* outer membrane protein Yop1 mediates inhibition of the anti-invasive effect of interferon. Infect. Immun. 58, 2245-2251.

Burgin, A.B., Parodos, K., Lane, D.J. and Pace, N.R. (1990). The excision of intervening sequences from Salmonella 23S ribosomal RNA. Cell. 60, 405-414.

Bölin, I., Forsberg, Å., Norlander, L., Skurnik, M. and Wolf-Watz, H. (1988). Identification and mapping of the temperature-inducible plasmid-encoded proteins of *Yersinia* spp. Infect. Immun. 56, 343-348.

Bölin, I., Norlander, L. and Wolf-Watz, H. (1982). Temperature-inducible outer membrane protein of *Yersinia pseudotuberculosis* and *Yersinia enterocolitica* is associated with the virulence plasmid. Infect. Immun. 37, 506-512.

Bölin, I., Portnoy, D. and Wolf-Watz, H. (1985). Expression of the temperature-inducible outer membrane proteins of *Yersiniae*. Infect. Immun. 48, 234-240.

Bölin, I. and Wolf-Watz, H. (1984). Molecular cloning of the temperature-inducible outer membrane protein 1 of *Yersinia pseudotuberculosis*. Infect. Immun. 43, 72-78.

Chiesa, C. and Bottone, E.J. (1983). Serum resistance of Yersinia enterocolitica expressed in absence of other virulence markers. Infect. Immun. 39, 469-472.

China, B., Sory, M.P., Nguyen, B.T., Debruyere, M. and Cornelis, G.R. (1993). Role of the YadA protein in prevention of opsonization of *Yersinia enterocolitica* by C3b molecules. Infect. Immun. 61, 3129-3136.

Darwin, A.J. and Miller, V.L. (1999). Identification of *Yersinia enterocolitica* genes affecting survival in an animal host using signature-tagged transposon mutagenesis. Mol. Microbiol. 32, 51-62.

Devereux, J., Haeberli, P. and Smithies, O. (1984). A comprehensive set of sequence analysis programs for the VAX. Nucleic Acid Res. 12, 387-395.

El Tahir, Y., Kuusela, P. and Skurnik, M. (2000). Functional mapping of the *Yersinia enterocolitica* adhesin YadA. Identification of eight NSVAIG--S motifs on the amino-terminal half of the protein involved in collagen binding. Mol. Microbiol. 37, 192-206.

El Tahir, Y. and Skurnik, M. (2001). YadA, the multifaceted *Yersinia* adhesin. Int. J. Med. Microbiol. 291, 209-218.

Emödy, L., Heesemann, J., Wolf-Watz, H., Skurnik, M., Kapperud, G., O'Toole, P. and Wadström, T. (1989). Binding to collagen by *Yersinia enterocolitica* and *Yersinia pseudotuberculosis*: evidence for *yopA*-mediated and chromosomally encoded mechanisms. J. Bacteriol. 171, 6674-6679.

Engels, W., Endert, J., Kamps, M.A. and van Boven, C.P. (1985). Role of lipopolysaccharide in opsonization and phagocytosis of *Pseudomonas aeruginosa*. Infect. Immun. 49, 182-189.

Forsberg, Å., Viitanen, A.M., Skurnik, M. and Wolf-Watz, H. (1991). The surface-located YopN protein is involved in calcium signal transduction in *Yersinia pseudotuberculosis*. Mol. Microbiol. 5, 977-986.

Gripenberg-Lerche, C., Skurnik, M. and Toivanen, P. (1995). Role of YadA-mediated collagen binding in arthritogenicity of *Yersinia enterocolitica* serotype O:8: Experimental studies with rats. Infect. Immun. 63, 3222-3226.

Gripenberg-Lerche, C., Skurnik, M., Zhang, L.J., Söderström, K.O. and Toivanen, P. (1994). Role of YadA in arthritogenicity of *Yersinia enterocolitica* serotype O:8: Experimental studies with rats. Infect. Immun. 62, 5568-5575.

Gripenberg-Lerche, C., Zhang, L., Ahtonen, P., Toivanen, P. and Skurnik, M. (2000). Construction of urease-negative mutants of *Yersinia enterocolitica* serotypes O:3 and O:8: role of urease in virulence and arthritogenicity. Infect. Immun. 68, 942-947.

Harr, R., Fällman, P., Häggström, M., Wahlström, L. and Gustafsson, P. (1985). GENEUS, a computer system for DNA and protein sequence analysis containing information retrieval system for EMBL data library. Nucleic Acids Res. 14, 273-284.

Heesemann, J. and Grüter, L. (1987). Genetic evidence that the outer membrane protein YOP1 of *Yersinia enterocolitica* mediates adherence and phagocytosis resistance to human epithelial cells. FEMS Microbiol. Lett. 40, 37-41.

Heesemann, J., Keller, C., Morawa, R., Schmidt, N., Siemens, H.J. and Laufs, R. (1983). Plasmids of human strains of *Yersinia enterocolitica*: molecular relatedness and possible importance for pathogenesis. J. Infect. Dis. 147, 107-115.

Heise, T. and Dersch, P. (2006). Identification of a domain in *Yersinia* virulence factor YadA that is crucial for extracellular matrix-specific cell adhesion and uptake. Proc. Natl. Acad. Sci. USA. 103, 3375-3380.

Hitchen, P.G., Prior, J.L., Oyston, P.C., Panico, M., Wren, B.W., Titball, R.W., Morris, H.R. and Dell, A. (2002). Structural characterization of lipo-oligosaccharide (LOS) from *Yersinia pestis*: regulation of LOS structure by the PhoPQ system. Mol. Microbiol. 44, 1637-1650.

Hoffman, J., Lindberg, B. and Brubaker, R.R. (1980). Structural studies of the O-specific side-chains of the lipopolysaccharide from *Yersinia enterocolitica* Ye 128. Carbohydr. Res. 78, 212-214.

Hoiczyk, E., Roggenkamp, A., Reichenbecher, M., Lupas, A. and Heesemann, J. (2000). Structure and sequence analysis of *Yersinia* YadA and *Moraxella* UspAs reveal a novel class of adhesins. EMBO J. 19, 5989-5999.

Jalava, J., Kotilainen, P., Nikkari, S., Skurnik, M., Vänttinen, E., Lehtonen, O.P., Eerola, E. and Toivanen, P. (1995). Use of the polymerase chain reaction and DNA sequencing for detection of *Bartonella quintana* in the aortic valve of a patient with culture-negative infective endocarditis. Clin. Infect. Dis. 21, 891-896.

Jalava, J., Mantymaa, M.L., Ekblad, U., Toivanen, P., Skurnik, M., Lassila, O. and Alanen, A. (1996). Bacterial 16S rDNA polymerase chain reaction in the detection of intra-amniotic infection. British Journal of Obstetrics and Gynaecology. 103, 664-669.

Jalava, J., Skurnik, M., Toivanen, A., Toivanen, P. and Eerola, E. (2001). Bacterial PCR in the diagnosis of joint infection. Ann. Rheum. Dis. 60, 287-289.

Kapperud, G., Namork, E. and Skarpeid, H.-J. (1985). Temperature-inducible surface fibrillae associated with the virulence plasmid of *Yersinia enterocolitica* and *Yersinia pseudotuberculosis*. Infect. Immun. 47, 561-566.

Karlyshev, A.V., Oyston, P.C., Williams, K., Clark, G.C., Titball, R.W., Winzeler, E.A. and Wren, B.W. (2001). Application of high-density array-based signature-tagged mutagenesis to discover novel *Yersinia* virulence-associated genes. Infect. Immun. 69, 7810-7819.

Kawahara, K., Tsukano, H., Watanabe, H., Lindner, B. and Matsuura, M. (2002). Modification of the structure and activity of lipid A in Yersinia pestis lipopolysaccharide by growth temperature. Infect. Immun. 70, 4092-8.

Kawaoka, Y., Otsuki, K. and Tsubokura, M. (1983a). Growth temperature-dependent variation in the bacteriophage-inactivating capacity and antigenicity of *Yersinia enterocolitica* lipopolysaccharide. J. Gen. Microbiol. 129, 2739-2747.

Kawaoka, Y., Otsuki, K. and Tsubokura, M. (1983b). Serological evidence that *Yersinia enterocolitica* lipopolysaccharide produced during growth *in vivo* resembles that produced during growth *in vitro* at 25 °C. J. Gen. Microbiol. 129, 2749-2751.

Kiljunen, S., Hakala, K., Pinta, E., Huttunen, S., Pluta, P., Gador, A., Lönnberg, H. and Skurnik, M. (2005). Yersiniophage φR1-37 is a tailed bacteriophage having a 270 kb DNA genome with thymidine replaced by deoxyuridine. Microbiology. 151, 4093-4102.

Knirel, Y.A., Dentovskaya, S.V., Senchenkova, S.N., Shaikhutdinova, R.Z., Kocharova, N.A. and Anisimov, A.P. (2006). Structural features and structural variability of the lipopolysaccharide of *Yersinia pestis*, the cause of plague. J. Endotoxin Res. 12, 3-9.

Knirel, Y.A., Lindner, B., Vinogradov, E., Shaikhutdinova, R.Z., Senchenkova, S.N., Kocharova, N.A., Holst, O., Pier, G.B. and Anisimov, A.P. (2005a). Cold temperature-induced modifications to the composition and structure of the lipopolysaccharide of *Yersinia pestis*. Carbohydr. Res. 340, 1625-30.

Knirel, Y.A., Lindner, B., Vinogradov, E.V., Kocharova, N.A., Senchenkova, S.N., Shaikhutdinova, R.Z., Dentovskaya, S.V., Fursova, N.K., Bakhteeva, I.V., Titareva, G.M., Balakhonov, S.V., Holst, O., Gremyakova, T.A., Pier, G.B. and Anisimov, A.P. (2005b).

Temperature-dependent variations and intraspecies diversity of the structure of the lipopolysaccharide of *Yersinia pestis*. Biochemistry. 44, 1731-43.

Kotilainen, P., Jalava, J., Meurman, O., Lehtonen, O.-P., Rintala, E., Seppälä, O.-P., Eerola, E. and Nikkari, S. (1998). Diagnosis of meningococcal meningitis by broad-range bacterial PCR with cerebrospinal fluid. J. Clin. Microbiol. 36, 2205-9.

Laird, W.J. and Cavanaugh, D.C. (1980). Correlation of autoagglutination and virulence in *yersiniae*. J. Clin. Microbiol. 11, 430-432.

Lambert de Rouvroit, C., Sluiters, C. and Cornelis, G.R. (1992). Role of the transcriptional activator, VirF, and temperature in the expression of the pYV plasmid genes of *Yersinia enterocolitica*. Mol. Microbiol. 6, 395-409.

Linke, D., Riess, T., Autenrieth, I.B., Lupas, A. and Kempf, V.A. (2006). Trimeric autotransporter adhesins: variable structure, common function. Trends Microbiol. 14, 264-270.

Manning, P.A., Heuzenroeder, M.W., Yeadon, J., Leavesley, D.I., Reeves, P.R. and Rowley, D. (1986). Molecular cloning and expression in *Escherichia coli* K-12 of the O antigens of the Inaba and Ogawa serotypes of the *Vibrio cholerae* O1 lipopolysaccharides and their potential for vaccine development. Infect. Immun. 53, 272-277.

Martinez, R.J. (1983). Plasmid-mediated and temperature-regulated surface properties of *Yersinia enterocolitica*. Infect. Immun. 41, 921-930.

Mecsas, J., Bilis, I. and Falkow, S. (2001). Identification of attenuated *Yersinia pseudotuberculosis* strains and characterization of an orogastric infection in BALB/c mice on day 5 postinfection by signature-tagged mutagenesis. Infect. Immun. 69, 2779-2787.

Mertz, A.K.H., Daser, A., Skurnik, M., Wiesmüller, K.H., Braun, J., Appel, H., Batsford, S., Wu, P.H., Distler, A. and Sieper, J. (1994). The evolutionarily conserved ribosomal protein L23 and the cationic urease β-subunit of *Yersinia enterocolitica* O:3 belong to the immunodominant antigens in *Yersinia*-triggered reactive arthritis: Implications for autoimmunity. Mol. Med. 1, 44-55.

Michiels, T., Wattiau, P., Brasseur, R., Ruysschaert, J.-M. and Cornelis, G. (1990). Secretion of Yop proteins by *Yersiniae*. Infect. Immun. 58, 2840-2849.

Miller, V.L., Beer, K.B., Heusipp, G., Young, B.M. and Wachtel, M.R. (2001). Identification of regions of Ail required for the invasion and serum resistance phenotypes. Mol. Microbiol. 41, 1053-62.

Montminy, S.W., Khan, N., McGrath, S., Walkowicz, M.J., Sharp, F., Conlon, J.E., Fukase, K., Kusumoto, S., Sweet, C., Miyake, K., Akira, S., Cotter, R.J., Goguen, J.D. and Lien, E. (2006). Virulence factors of *Yersinia pestis* are overcome by a strong lipopolysaccharide response. Nature Immunolology. 7, 1066-73.

Mäkelä, P.H., Hovi, M., Saxen, H., Valtonen, M. and Valtonen, V. (1988). *Salmonella*, complement and mouse macrophages. Immunol. Lett. 19, 217-222.

Najdenski, H., Golkocheva, E., Kussovski, V., Ivanova, E., Manov, V., Iliev, M., Vesselinova, A., Bengoechea, J.A. and Skurnik, M. (2006). Experimental pig yersiniosis to assess attenuation of *Yersinia enterocolitica* O:8 mutant strains. FEMS Immunol. Med. Microbiol. 47, 425-435.

Najdenski, H., Golkocheva, E., Vesselinova, A., Bengoechea, J.A. and Skurnik, M. (2003). Proper expression of the O-antigen of lipopolysaccharide is essential for the virulence of *Yersinia enterocolitica* O:8 in experimental oral infection of rabbits. FEMS Immunol. Med. Microbiol. 38, 97-106.

Nummelin, H., El Tahir, Y., Ollikka, P., Skurnik, M. and Goldman, A. (2002). Expression, purification and crystallization of a collagen-binding fragment of *Yersinia* adhesin YadA. Acta Crystallographia Section D: Biological Crystallography. 58, 1042-1044.

Nummelin, H., Merckel, M.C., Leo, J.C., Lankinen, H., Skurnik, M. and Goldman, A. (2004). The *Yersinia* adhesin YadA collagen-binding domain structure is a novel left-handed parallel beta-roll. EMBO J. 23, 701-711.

Ogasawara, M., Granfors, K., Kono, D.H., Hill, J.L. and Yu, D.T.Y. (1985). A *Yersinia enterocolitica* serotype 0:3 lipopolysaccharide-specific monoclonal antibody reacts more strongly with bacteria cultured at room temperature than those cultured at 37°C. J. Immunol. 135, 553-559.

Okamura, N., Nagai, T., Nakaya, R., Kondo, S., Murakami, M. and Hisatsune, K. (1983). Hela cell invasiveness and O-antigen of *Shigella flexneri* as separate and prerequisite attributes of virulence to evoke keratoconjunctivitis in guinea pigs. Infect. Immun. 39, 505-513.

Paerregaard, A., Espersen, F., Jensen, O.M. and Skurnik, M. (1991a). Interactions between *Yersinia enterocolitica* and rabbit ileal mucus: Growth, adhesion, penetration, and subsequent changes in surface hydrophobicity and ability to adhere to ileal brush border membrane vesicles. Infect. Immun. 59, 253-260.

Paerregaard, A., Espersen, F. and Skurnik, M. (1991b). Role of *Yersinia* outer membrane protein YadA in adhesion to rabbit intestinal tissue and rabbit intestinal brush border membrane vesicles. APMIS. 99, 226-232.

Pai, C.H. and DeStephano, L. (1982). Serum resistance associated with virulence in *Yersinia enterocolitica*. Infect. Immun. 35, 605-611.

Pajunen, M., Kiljunen, S. and Skurnik, M. (2000). Bacteriophage ϕYeO3-12, specific for *Yersinia enterocolitica* serotype O:3, is related to coliphages T3 and T7. J. Bacteriol. 182, 5114-5120.

Pekkola-Heino, K., Viljanen, M.K., Ståhlberg, T.H., Granfors, K. and Toivanen, A. (1987). Monoclonal antibodies reacting selectively with core and O-polysaccharide of *Yersinia enterocolitica* O:3 lipopolysaccharide. APMIS. 95, 27-34.

Perry, R.D. and Brubaker, R.R. (1983). Vwa^+ phenotype of *Yersinia enterocolitica*. Infect. Immun. 40, 166-171.

Pierson, D.E. and Falkow, S. (1993). The *ail* gene of *Yersinia enterocolitica* has a role in the ability of the organism to survive serum killing. Infect. Immun. 61, 1846-1852.

Pilz, D., Vocke, T., Heesemann, J. and Brade, V. (1992). Mechanism of YadA-mediated serum resistance of *Yersinia enterocolitica* serotype O3. Infect. Immun. 60, 189-195.

Porat, R., Johns, M.A. and McCabe, W.R. (1987). Selective pressures and lipopolysaccharide subunits as determinants of resistance of clinical isolates of Gram-negative bacilli to human serum. Infect. Immun. 55, 320-328.

Portnoy, D.A., Moseley, S.L. and Falkow, S. (1981). Characterization of plasmids and plasmid-associated determinants of *Yersinia enterocolitica* pathogenesis. Infect. Immun. 31, 775-782.

Radziejewska-Lebrecht, J., Kasperkiewicz, K., Skurnik, M., Brade, L., Steinmetz, I., Swierzko, A.S. and Muszynski, A. (2003). ECA-antibodies in antisera against R mutants of *Yersinia enterocolitica* O:3. Adv. Exp. Med. Biol. 529, 215-218.

Radziejewska-Lebrecht, J., Skurnik, M., Shashkov, A.S., Brade, L., Rozalski, A., Bartodziejska, B. and Mayer, H. (1998). Immunochemical studies on R mutants of *Yersinia enterocolitica* O:3. Acta Biochimica Polonica. 45, 1011-1019.

Rantakokko-Jalava, K., Nikkari, S., Jalava, J., Eerola, E., Skurnik, M., Meurman, O., Ruuskanen, O., Alanen, A., Kotilainen, E., Toivanen, P. and Kotilainen, P. (2000). Direct amplification of rRNA genes in diagnosis of bacterial infections. J. Clin. Microbiol. 38, 32-39.

Rebeil, R., Ernst, R.K., Gowen, B.B., Miller, S.I. and Hinnebusch, B.J. (2004). Variation in lipid A structure in the pathogenic yersiniae. Mol Microbiol. 52, 1363-73.

Roggenkamp, A., Ackermann, N., Jacobi, C.A., Truelzsch, K., Hoffmann, H. and Heesemann, J. (2003). Molecular analysis of transport and oligomerization of the Yersinia enterocolitica adhesin YadA. J Bacteriol. 185, 3735-44.

Roggenkamp, A., Neuberger, H.-R., Flügel, A., Schmoll, T. and Heesemann, J. (1995). Substitution of two histidine residues in YadA protein of *Yersinia enterocolitica* abrogates collagen binding, cell adherence and mouse virulence. Mol. Microbiol. 16, 1207-1219.

Rosqvist, R., Skurnik, M. and Wolf-Watz, H. (1988). Increased virulence of *Yersinia pseudotuberculosis* by two independent mutations. Nature. 334, 522-525.

Schulze-Koops, H., Burkhardt, H., Heesemann, J., Kirsch, T., Swoboda, B., Bull, C., Goodman, S. and Emmrich, F. (1993). Outer membrane protein YadA of enteropathogenic *Yersiniae* mediates specific binding to cellular but not plasma fibronectin. Infect. Immun. 61, 2513-2519.

Schulze-Koops, H., Burkhardt, H., Heesemann, J., von der Mark, K. and Emmrich, F. (1992). Plasmid-encoded outer membrane protein YadA mediates specific binding of enteropathogenic *Yersiniae* to various types of collagen. Infect. Immun. 60, 2153-2159.

Schulze-Koops, H., Burkhardt, H., Heesemann, J., von der Mark, K. and Emmrich, F. (1995). Characterization of the binding region for the *Yersinia enterocolitica* adhesin YadA on types I and II collagen. Arthritis Rheum. 38, 1283-1289.

Skurnik, M. (1984). Lack of correlation between the presence of plasmids and fimbriae in *Yersinia enterocolitica* and *Yersinia pseudotuberculosis*. J. Appl. Bact. 56, 355-363.

Skurnik, M. (1985a). Expression of antigens encoded by the virulence plasmid of *Yersinia enterocolitica* under different growth conditions. Infect. Immun. 47, 183-190.

Skurnik, M., (1985b). Studies on the virulence plasmids of *Yersinia* species. PhD Thesis, University of Oulu, Oulu, 61+39 pp.

Skurnik, M., Batsford, S., Mertz, A., Schiltz, E. and Toivanen, P. (1993). The putative arthritogenic cationic 19-kilodalton antigen of *Yersinia enterocolitica* is a urease β-subunit. Infect. Immun. 61, 2498-2504.

Skurnik, M., Bölin, I., Heikkinen, H., Piha, S. and Wolf-Watz, H. (1984). Virulence plasmid-associated autoagglutination in *Yersinia* spp. J. Bacteriol. 158, 1033-1036.

Skurnik, M., El Tahir, Y., Saarinen, M., Jalkanen, S. and Toivanen, P. (1994). YadA mediates specific binding of enteropathogenic *Yersinia enterocolitica* to human intestinal submucosa. Infect. Immun. 62, 1252-1261.

Skurnik, M., Nurmi, T., Granfors, K., Koskela, M. and Tiilikainen, A.S. (1983). Plasmid associated antibody production against *Yersinia enterocolitica* in man. Scand. J. Inf. Dis. 15, 173-177.

Skurnik, M., Peippo, A. and Ervelä, E. (2000). Characterization of the O-antigen gene clusters of *Yersinia pseudotuberculosis* and the cryptic O-antigen gene cluster of *Yersinia pestis* shows that the plague bacillus is most closely related to and has evolved from *Y. pseudotuberculosis* serotype O:1b. Mol. Microbiol. 37, 316-330.

Skurnik, M. and Poikonen, K. (1986a). Experimental intestinal infection of rats by *Yersinia enterocolitica* O:3. A follow-up study with specific antibodies to the virulence plasmid specified antigens. Scand. J. Inf. Dis. 18, 355-364.

Skurnik, M. and Poikonen, K. (1986b). Monoclonal antibody to the autoagglutination protein P1 of *Yersinia*. In: *Protein-carbohydrate interactions in biological systems*. D.L. Lark (Ed.). Academic Press, New York. p. 355-357.

Skurnik, M. and Toivanen, P. (1991). Intervening sequences (IVSs) in the 23S ribosomal RNA genes of pathogenic *Yersinia enterocolitica* strains. The IVSs in *Y. enterocolitica* and *Salmonella typhimurium* have common origin. Mol. Microbiol. 5, 585-593.

Skurnik, M. and Toivanen, P. (1992). LcrF is the temperature-regulated activator of the *yadA* gene of *Yersinia enterocolitica* and *Yersinia pseudotuberculosis*. J. Bacteriol. 174, 2047-2051.

Skurnik, M., Venho, R., Bengoechea, J.-A. and Moriyón, I. (1999). The lipopolysaccharide outer core of *Yersinia enterocolitica* serotype O:3 is required for virulence and plays a role in outer membrane integrity. Mol. Microbiol. 31, 1443-1462.

Skurnik, M., Venho, R., Toivanen, P. and Al-Hendy, A. (1995). A novel locus of *Yersinia enterocolitica* serotype O:3 involved in lipopolysaccharide outer core biosynthesis. Mol. Microbiol. 17, 575-594.

Skurnik, M. and Wolf-Watz, H. (1989). Analysis of the *yopA* gene encoding the Yop1 virulence determinants of *Yersinia* spp. Mol. Microbiol. 3, 517-529.

Strauch, E., Kaspar, H., Schaudinn, C., Damasko, C., Konietzny, A., Dersch, P., Skurnik, M. and Appel, B. (2003). Analysis of enterocoliticin, a phage tail-like bacteriocin. In: *The Genus Yersinia: entering the functional genomic era.* M. Skurnik, K. Granfors and J.A. Bengoechea (Eds). Kluwer Academic/Plenum Publishers, New York. p. 249-251.

Strauch, E., Kaspar, H., Schaudinn, C., Dersch, P., Madela, K., Gewinner, C., Hertwig, S., Wecke, J. and Appel, B. (2001). Characterization of enterocoliticin, a phage tail-like bacteriocin, and its effect on pathogenic *Yersinia enterocolitica* strains. Appl. Environ. Microb. 67, 5634-5642.

Tamm, A., Tarkkanen, A.M., Korhonen, T.K., Kuusela, P., Toivanen, P. and Skurnik, M. (1993). Hydrophobic domains affect the collagen-binding specificity and surface polymerization as well as the virulence potential of the YadA protein of *Yersinia enterocolitica.* Mol. Microbiol. 10, 995-1011.

Tertti, R., Skurnik, M., Vartio, T. and Kuusela, P. (1992). Adhesion protein YadA of *Yersinia* species mediates binding of bacteria to fibronectin. Infect. Immun. 60, 3021-3024.

Wachter, E. and Brade, V. (1989). Influence of surface modulations by enzymes and monoclonal antibodies on alternative complement pathway activation by *Yersinia enterocolitica.* Infect. Immun. 57, 1984-1989.

Wartenberg, K., Knapp, W., Ahamed, N.M., Widemann, C. and Mayer, H. (1983). Temperature-dependent changes in the sugar and fatty acid composition of lipopolysaccharide from *Yersinia enterocolitica* strains. Int. J. Med. Microb. 253, 523-530.

Vesikari, T., Nurmi, T., Mäki, M., Skurnik, M., Sundqvist, C., Granfors, K. and Grönroos, P. (1981). Plasmids in *Yersinia enterocolitica* serotypes O:3 and O:9: correlation with epithelial cell adherence in vitro. Infect. Immun. 33, 870-876.

Westerlund-Wikström, B., Tanskanen, J., Virkola, R., Hacker, J., Lindberg, M., Skurnik, M. and Korhonen, T.K. (1997). Functional expression of adhesive peptides as fusions to *Escherichia coli* flagellin. Protein Engin. 10, 1319-1326.

Viitanen, A.-M., 1990. Genetic studies on the pathogenetic determinants of *Yersinia*-triggered reactive arthritis. PhD thesis. Annales Universitatis Turkuensis, Ser.D, 64. University of Turku, Turku, 132 pp.

Viitanen, A.-M., Arstila, T.P., Lahesmaa, R., Granfors, K., Skurnik, M. and Toivanen, P. (1991). Application of the polymerase chain reaction and immunofluorescence techniques to the detection of bacteria in *Yersinia*-triggered reactive arthritis. Arthritis Rheum. 34, 89-96.

Viitanen, A.-M., Toivanen, P. and Skurnik, M. (1990). The *lcrE* gene is part of an operon in the lcr region of *Yersinia enterocolitica* O:3. J. Bacteriol. 172, 3152-3162.

Woese, C.R. (1987). Bacterial evolution. Microbiol. Rev. 51, 221-271.

Vogel, U., Weinberger, A., Frank, R., Muller, A., Kohl, J., Atkinson, J.P. and Frosch, M. (1997). Complement factor C3 deposition and serum resistance in isogenic capsule and lipooligosaccharide sialic acid mutants of serogroup B Neisseria meningitidis. Infect. Immun. 65, 4022-4029.

Yang, Y. and Isberg, R.R. (1993). Cellular internalization in the absence of invasin expression is promoted by the *Yersinia pseudotuberculosis yadA* product. Infect. Immun. 61, 3907-3913.

Zaleska, M., Lounatmaa, K., Nurminen, M., Wahlström, E. and Mäkelä, P.H. (1985). A novel virulence-associated cell surface structure composed of 47-kd protein subunits in *Yersinia enterocolitica.* EMBO J. 4, 1013-1018.

Zhang, L., Al-Hendy, A., Toivanen, P. and Skurnik, M. (1993). Genetic organization and sequence of the *rfb* gene cluster of *Yersinia enterocolitica* serotype O:3: Similarities to the dTDP-L-rhamnose biosynthesis pathway of *Salmonella* and to the bacterial polysaccharide transport systems. Mol. Microbiol. 9, 309-321.

Zhang, L., Nikkari, S., Skurnik, M., Ziegler, T., Luukkainen, R., Möttönen, T. and Toivanen, P. (1993). Detection of herpesviruses by polymerase chain reaction in lymphocytes from patients with rheumatoid arthritis. Arthritis Rheum. 36, 1080-1086.

Zhang, L., Radziejewska-Lebrecht, J., Krajewska-Pietrasik, D., Toivanen, P. and Skurnik, M. (1997). Molecular and chemical characterization of the lipopolysaccharide O-antigen and its role in the virulence of *Yersinia enterocolitica* serotype O:8. Mol. Microbiol. 23, 63-76.

Zhang, L. and Skurnik, M. (1994). Isolation of an R^- M^+ mutant of *Yersinia enterocolitica* serotype O:8 and its application in construction of rough mutants utilizing mini-Tn*5* derivatives and lipopolysaccharide-specific phage. J. Bacteriol. 176, 1756-1760.

Zhang, L., Toivanen, P. and Skurnik, M. (1996). The gene cluster directing O-antigen biosynthesis in *Yersinia enterocolitica* serotype O:8: Identification of the genes for mannose and galactose biosynthesis and the gene for the O-antigen polymerase. Microbiology. 142, 277-288.

Zink, D.L., Feeley, J.C., Wells, J.G., Vanderzant, C., Vickery, J.C., Roof, W.D. and O'Donovan, G.A. (1980). Plasmid-mediated tissue invasiveness in *Yersinia enterocolitica*. Nature. 283, 224-6.

Picture 9. Mikael Skurnik gives the opening lecture. Photograph by A. Anisimov.

6
Structure and Assembly of *Yersinia pestis* F1 Antigen

Stefan D. Knight

Department of Molecular Biology, Swedish University of Agricultural Sciences, Stefan. Knight@molbio.slu.se

Abstract. Most Gram negative pathogens express surface located fibrillar organelles that are used for adhesion to host epithelia and/or for protection. The assembly of many such organelles is managed by a highly conserved periplasmic chaperone/usher assembly pathway. During the last few years, considerable progress has been made in understanding how periplasmic chaperones mediate folding, targeting, and assembly of F1 antigen subunits into the F1 capsular antigen. In particular, structures representing snapshots of several of the steps involved in assembly have allowed us to begin to draw a detailed molecular-level picture of F1 assembly specifically, and of chaperone/usher-mediated assembly in general. Here, a brief summary of these new results will be presented.

6.1 Introduction

Gram negative bacteria can grow hair-like structures referred to as pili (from the Latin word for 'hair') or fimbriae (from the Latin word for 'fringe') arranged in a multitude of 'hairstyles' ranging from soft, long, wavy hair, to afro style (Fig. 1). The individual hairs may be rather rigid, cylindrical structures with a central hole (e.g. *Escherichia coli* type-1 and P pili), or thinner, more flexible fibers (e.g. *Yersinia pestis* pH6 antigen and F1 capsular antigen). Bacterial hair is not mainly for looks but rather to provide adhesion of bacteria to target tissues during infection, and/or protection. For example, type-1 and P pili mediate adhesion of uropathogenic *E. coli* (UPEC) to mannose-containing receptors in the bladder epithelium and to galactose-containing receptors in the kidney respectively, whereas the F1 antigen of *Y. pestis* provides protection against phagocytosis.

In spite of the large variation in appearance, a large group of bacterial pili/fimbriae share the same underlying fiber structure, consisting of a string of non-covalently linked immunoglobulin (Ig)-like modules (Knight et al. 2000; Sauer et al. 2000a; Sauer et al. 2000b). This linear fiber structure is assembled from monomeric incomplete Ig modules by a periplasmic chaperone together with an outer membrane usher. The final fiber morphology is determined by secondary interactions between fiber subunits, allowing some to coil into rigid helical structures such as the pilus rods or to remain as relatively flexible extended linear fibers. Fiber assembly chaperone/usher systems are abundant, and often in the order of five to ten such systems can be predicted in complete genomes of Gram negative bacteria. Although the majority of these remain to be studied, the products of several different systems are known and have been characterized. For example, UPEC frequently express Dr adhesins and type-1, P, F1C, and S pili, all of which are known to play important

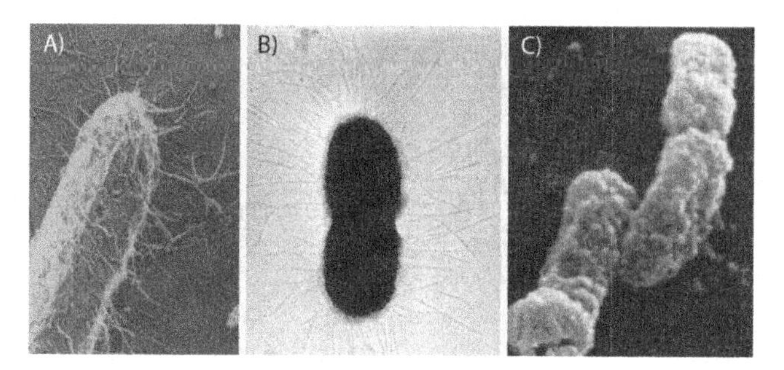

Fig. 1. Examples of bacterial 'hair styles.' A) Type-1 piliated *E. coli*. Type 1 pili are composite structures with a relatively rigid 8 nm wide rod tipped by a thin (2 nm) more flexible tip fibrillum. B) AAF III fimbriated *E. coli* covered with long, flexible, 3-5 nm wide fimbriae. C) *Y. peistis* F1 capsular antigen. No individual fibers are visible. A kindly provided by Dr. John Heuser. B reprinted from (Bernier et al. 2002) with permission from the publisher. C reprinted from (Chen and Elberg 1977) with permission from the publisher.

roles in the pathogenesis of urinary tract infections (Berglund and Knight 2003; Mulvey 2002). *Y. pestis* can express at least two antiphagocytic/adhesive fibrillar structures, the pH6 antigen and the F1 capsular antigen (MacIntyre et al. 2004).

As will be described briefly in the following sections, recent structures representing snapshots of several of the steps involved in assembly of F1 antigen from Caf1 subunits (Fig. 2) have shed new light on the details of chaperone/usher-mediated assembly of F1 antigen in particular and of fibrillar surface structures in general.

6.2 F1 Antigen

F1 antigen, which is expressed exclusively by *Y. pestis*, is produced efficiently at temperatures of 35-37°C to cover the bacterial surface with a massive amount of material forming a gelatinous antiphagocytic capsule (for a recent extensive review of F1 antigen see MacIntyre et al., 2004). At temperatures of 26°C and below, as in e.g. the flea gut, F1 capsule is not expressed. The capsule, which consists of high-molecular weight polymers built from a single 15.6 kDa protein subunit (Caf1) (Galyov et al. 1990), might easily grow to extend to twice the diameter of the bacterial cell. The capsule material is soluble in water, and large amounts can be shed from the bacterial cell surface into the surroundings. Whereas the free Caf1 subunit is only marginally stable (Zavialov et al. 2005), the capsule is extremely stable and remains associated in 0.5% (w/v) SDS at 75°C, and is resistant to proteolysis by trypsin or proteinase K.

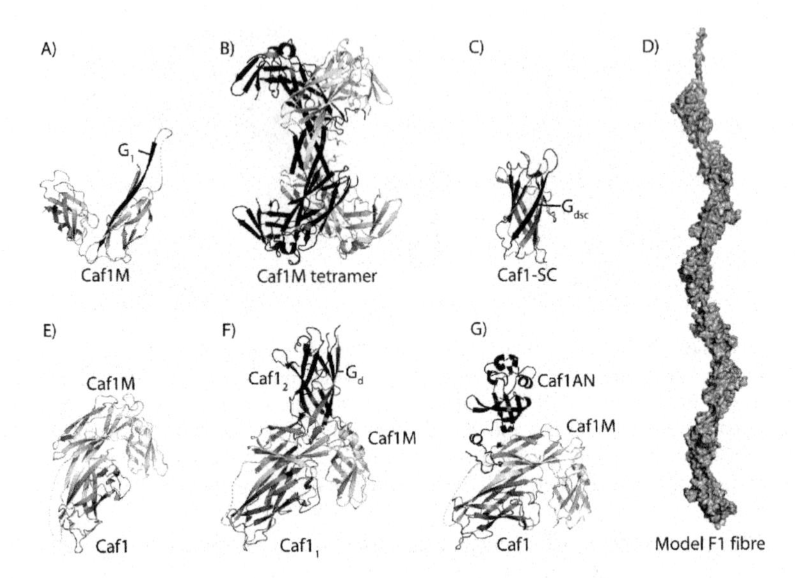

Fig. 2. Structure models representing snapshots of F1 antigen assembly. A) Monomeric Caf1M. B) Caf1M tetramer. C) Caf1-SC. D) F1 fiber. E) Caf1M:Caf1 pre-assembly complex. F) Caf1M:Caf1:Caf1 ternary complex. G) Caf1A_N:Caf1M:Caf1 complex. The fiber model (D) was generated by consecutive extension of the minimal fiber observed in the Caf1M: Caf1:Caf1structure. The monomeric Caf1M structure shown in (A) is the structure of the chaperone as observed in the Caf1M:Caf1:Caf1 complex. Figures were generated using Pymol (DeLano, W.L. The PyMOL Molecular Graphics System (2002), b http://www.pymol.org).

It has been estimated that whilst feeding, a flea carrying *Y. pestis* can inject in the order of 11,000 to 24,000 non-encapsulated *Y. pestis* cells into the host (Perry and Fetherston 1997). Following injection, *Y. pestis* are readily and rapidly phagocytosed. While uptake by neutrophils leads to efficient killing of bacteria, those that are taken up by monocytes or non-activated macrophages can survive and multiply. During this early intracellular stage of infection, a number of 'virulence determinants,' including the antiphagocytic F1 antigen, are expressed. Following release from phagocytic cells, F1 antigen and other antiphagocytic factors dramatically limit phagocytosis (Du et al. 2002). The role of F1 antigen in pathogenesis however remains unclear, since F1 negative strains are as virulent as those expressing F1 in a number of animal models (Davis et al. 1996; Drozdov et al. 1995; Du et al. 1995; Friedlander et al. 1995; Worsham et al. 1995). Nonetheless, owing to its surface location, existence in most *Y. pestis* strains, essentially absolute sequence conservation, and high immunogenicity, much interest has been invested in the F1 subunit, Caf1, as a subunit vaccine, either alone or in combination with V antigen (Titball and Williamson 2001; Titball and Williamson 2004).

6.3 Structure and Assembly of F1 Antigen

6.3.1 Structure of F1 Antigen

Although no defined structure of F1 has been visualized by electron microscopy, recent genetic and structural studies bear out that F1 capsule consists of a tangle of thin (~2 nm) high-molecular weight linear Caf1 fibers (Zavialov et al. 2003; Zavialov et al. 2002). The Caf1 subunit is constructed essentially as an Ig-like β-sandwich, but with a circular permutation that positions the sequence corresponding to the seventh, C-terminal, Ig β-strand (strand G of a canonical Ig domain) at the N-terminus of the polypeptide sequence (Choudhury et al. 1999; Sauer et al. 1999; Sauer et al. 2002; Zavialov et al. 2003) (Fig. 3A). In a typical Ig fold, the 'top' edge of the sandwich, defined by the A and F strands, is capped by the C-terminal G strand, which is hydrogen bonded to the F strand and provides hydrophobic residues to the core of the fold. No structure for isolated Caf1 has been obtained, but structures of Caf1 bound to Caf1M chaperone (Fig. 2E), or incorporated into F1 fibers (Fig. 2F), show that in the absence of a seventh β strand, this edge of the β-sandwich remains uncapped, and a closed hydrophobic core is not formed. Instead, a deep hydrophobic cleft is created on the surface of the molecule. The hydrophobic effect drives the folding of globular proteins by forcing hydrophobic side chains together in a hydrophobic core, shielded from the surrounding water. In Caf1 however, owing to the absence of a seventh, C-terminal, G strand, the polypeptide chain simply cannot fold in such a way as to create a shielded hydrophobic core, explaining the instability of free subunit.

The N-terminal extension of Caf1, which does not take part in the globular fold of Caf1 and which is flexible in solution, carries a β strand motif of alternating hydrophobic and hydrophilic residues. Deletions or mutations in this N-terminal region blocks assembly of Caf1 subunits into F1 fibers (Zavialov et al. 2003; Zavialov et al. 2002). The structure of a Caf1M-Caf1-Caf1 ternary complex (Zavialov et al. 2003; Zavialov et al. 2005) (Fig. 2F), with the minimal *Y. pestis* F1-antigen fiber (Caf1-Caf1) bound to the Caf1M chaperone, revealed that fiber subunits are linked together by insertion of the N-terminal extension of one subunit into the hydrophobic cleft of the second subunit (Fig. 3B). The inserted N-terminal segment adopts a β-strand conformation running parallel to the F strand, hence completing the Ig fold of the subunit. This mode of binding, termed donor strand complementation (DSC), had previously been predicted for type-1 and P pilus fibers (Choudhury et al. 1999; Sauer et al. 1999; Sauer et al. 2002), and is likely to be present in all surface polymers assembled through a chaperone/usher pathway. The resulting linear fiber is composed of globular modules each having an intact Ig topology generated by DSC. In this fiber, each Ig module is made from two polypeptide chains, with the G strand being provided *in trans* (because the N-terminal segment of one subunit is donated to fulfill the role of the Ig-fold G strand in a second subunit, the N-terminal sequence is also referred to as the G_d (d for donor) sequence). A short (4 residues) linker between the Ig modules might allow for considerable flexibility of the fiber, consistent with the appearance of F1 capsular

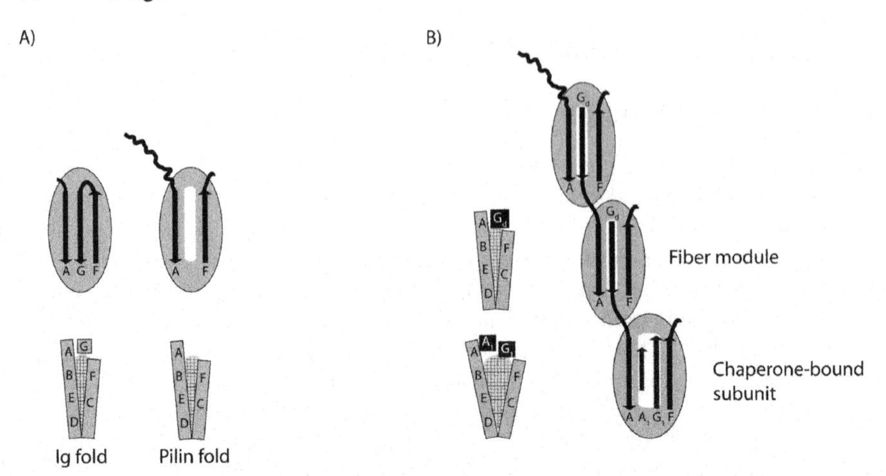

Fig. 3. Schematic illustration of A) the relation between the Ig and the F1 subunit ('pilin') fold, and B) of donor strand complementation before (subunit bound to chaperone; below) and after (in fiber module; above) donor strand exchange. The ellipsoids represent the β-sandwich of the Ig/pilin fold as viewed down at the AF edge of the sandwich; the rectangles represent β sheets and strands (as labeled) viewed edge on.

material not as linear fibers but rather as an amorphous mass in electron micrographs (Fig. 1C).

In spite of their non-covalent nature, and in contrast to free Caf1, or Caf1 bound to Caf1M chaperone, the fiber Ig modules are extremely stable. Whereas Caf1 dissociates from the chaperone: subunit complex and simultaneously melts already at 45°C, the melting temperature for the fiber module is 87.5°C (Zavialov et al. 2005). The enthalpy of melting, determined using differential scanning microcalorimetry, is 840±50 kJ/mol, close to the maximum observed for globular proteins (Makhatadze and Privalov 1995). By a circular permutation in which the donor sequence was fused at the C-terminus rather than at the N terminus as in wt Caf1, a self-complemented monomeric, complete Ig-fold, Caf1 subunit (Caf1-SC) with a similarly high stability as that of the F1 fiber module was constructed (Zavialov et al. 2005). The folding free energy estimated from reversible unfolding of Caf1-SC in guanidinium chloride at 37°C is in the range 70-80 kJ/mol. This should be compared to the typical range of 20-60 kJ/mol maxiumu stability for stable proteins in physiological conditions (Makhatadze and Privalov, 1995). Crystallization and structure determination of Caf1-SC (Fig. 2C) shows that its structure is essentially identical to the structure of the F1 fiber module (Anatoly Dubnovitsky, Anton Zavialov, Stefan Knight, unpublished). The high stability and conserved structure of self-complemented Caf1 makes this construct an interesting candidate for development of an F1 antigen-based monomeric subunit vaccine that would circumvent many of the problems caused by using polymers of variable and unspecified size in vaccine formulations. However, disruption of linear epitopes as well as subtle differences in structure and stability in a self-complemented construct as compared to wt antigen may have deleterious effects on antigen efficacy. An example of this is provided by the recent report that a self-complemented monomeric Caf1 vaccine was significantly less immunogenic and protective in BALB/c mice than recombinant F1 fiber antigen (Chalton et al. 2006).

6.3.2 Caf1M Chaperone Keeps Caf1 Subunits On-Pathway Through Donor Strand Complementation

Biogenesis of stable polymeric surface fibers such as those of the F1 capsular antigen poses many challenges to the Gram negative bacterial cell. It must be able to protect the unstable and highly aggregative fiber subunits from aggregation and proteolytic degradation during their transport from the site of production in the cytoplasm, across the inner membrane and the periplasm, to the site of assembly at the outer membrane of the cell. Having reached the outer membrane, subunit assembly must be controlled to form the desired polymeric structure, which then must be secreted to the cell surface. The Sec secretion system (Stathopoulos et al. 2000) is used for the initial cytoplasmic and inner membrane laps of this journey, while a dedicated periplasmic chaperone/usher pathway that is used for assembly of a wide range of structurally diverse adhesive organelles, takes over following import of subunits into the periplasm (Hung and Hultgren 1998; Sauer et al. 2000b; Soto and Hultgren 1999; Thanassi and Hultgren 2000). Chaperone/usher genes involved in the assembly of F1 antigen are organized in the *caf* gene cluster on the *Y. pestis*-specific pFra plasmid (Karlyshev et al. 1994). In the cytoplasm, Caf1 subunit is expressed as a preprotein with an N-terminal export signal and transported across the plasma membrane using the general secretion (Sec) pathway. In the periplasm, the Caf1M chaperone, together with the outer membrane usher Caf1A, handle the subsequent events that lead to assembly of Caf1 monomers into surface located F1 capsular antigen (Fig. 4). Periplasmic chaperones such as Caf1M are steric chaperones that bind to fiber subunits as they emerge in the periplasm, ensure their correct folding, and deliver the folded subunits to the usher where they are assembled into fibrillar polymeric structures such as pili, fimbriae, or capsules. In the absence of chaperone, subunit folding is slow and leads to a marginally stable and aggregation prone structure, whereas in the presence of chaperone, stable and soluble chaperone: subunit complexes are rapidly formed. In contrast to the many ATP-dependent cellular chaperones, the periplasmic chaperones do not require an input of external energy for subunit release, and organelle assembly is independent of cellular energy.

As all of the periplasmic chaperones (Knight et al. 2000), Caf1M is an L-shaped molecule with two Ig-like domains joined at ~90° angle, with a large cleft between the two domains (Fig. 2A). The F_1 and G_1 β-strands in the 1st, N-terminal, domain are connected by a long and flexible loop that protrudes like a handle from the body of the domain. The beginning of the G_1 β-strand harbors a conserved motif of hydrophobic residues that is critical for subunit binding. FGL chaperones (Hung et al. 1996) such as Caf1M have a relatively long F_1-G_1 loop (hence FGL for FG loop long) with a five-residue hydrophobic subunit binding motif that is used to bind subunits destined for assembly into simple polymeric structures composed of only one or two subunit types, whereas the FGS chaperones used for assembly of composite pilus structures such as type-1 or P pili have a relatively short F_1-G_1 loop

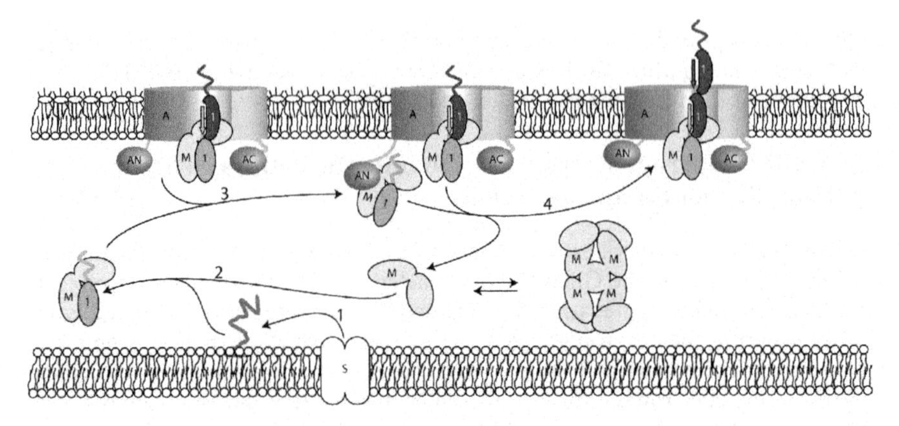

Fig. 4. Schematic illustration of chaperone/usher-assisted assembly of F1 antigen. Cafl (1) enters the periplasm via the Sec system (S) and transiently remains associated with the inner membrane (step 1). In the periplasm, most free Caf1M (M) is in a tetrameric form that protects the chaperone from proteolysis and aggregative off-pathway reactions. Monomeric Caf1M (M) binds newly translocated, partially unfolded Cafl subunits to form soluble and stable chaperone: subunit complexes (step 2). Binary Caf1M:Cafl pre-assembly complexes are targeted to the outer membrane usher Caf1A (A) where they initially interact with the N-terminal periplasmic usher domain (AN) (step 3). At the usher, the incoming Caf1M:Cafl complex attacks the Caf1M:Cafl complex capping the base of the growing fiber (step 4). The usher catalyses DSE in which the capping chaperone at the base is released, the N-terminal G_d donor strand of the attacking subunit is inserted into the polymerization cleft of the subunit at the base to form a new fiber module, and a new chaperone-capped subunit is added at the base. Both the N-terminal (AN) and the C-terminal (AC) periplasmic usher domains are required for DSE to proceed. The folding energy released in the DSE step drives assembly. Concomitantly with assembly, the growing fiber is 'pushed out' and secreted to the bacterial surface. Since the chaperone is too large to pass through the usher pore, capping of the base of the fiber anchors the fiber to the outer membrane.

(FG loop short) with a three-residue motif. FGL chaperones are also distinguished from FGS chaperones by a longer sequence at the N-terminus (N-terminal extension) and a disulphide link bridging the F_1 and G_1 strands.

Cafl binds in the cleft between the two Caf1M domains with its C-terminal carboxyl group anchored by Arg20 and Lys139 at the bottom of the subunit-binding cleft (Fig. 2E). These two residues are strictly conserved in all the periplasmic chaperones and are crucial for chaperone function. As described above, the incomplete fold of Cafl precludes the formation of a closed hydrophobic core, leaving a surface-exposed hydrophobic cleft between the two sheets at the AF edge of the subunit (Fig. 3A). In Caf1M:Cafl complexes (Fig. 2E-G) (Zavialov et al. 2003), the subunit A and F edge strands are hydrogen bonded to the A_1 and G_1 strands of the chaperone (Fig. 3B), which creates a super-barrel of β strands from both the subunit and the chaperone. The chaperone thus caps the hydrophobic polymerization cleft, thereby preventing both premature assembly and unspecific subunit aggregation, and protecting subunits from proteolytic degradation. In the super-barrel, five large hydrophobic residues from the G_1 donor strand of the chaperone are inserted between the two sheets of the subunit β-sandwich and become an integral part of the bound subunit hydrophobic core. It should be noted, that in contrast to the DSC interaction between subunits in the F1 fiber, where the subunit G_d donor strand is inserted antiparallel to the F strand to create a complete canonical Ig module, the chaperone G_1 donor strand is inserted parallel to the F strand.

The hydrophobic subunit binding motifs (A_1 and G_1 strands) of Caf1M are predicted to be solvent exposed in subunit-free chaperone molecules and to contain large unstructured segments (residues 1-12 and 103-132) (Zavialov et al. 2003), and hence the chaperone is expected to show significant unspecific binding. In contrast to ATP-powered chaperones, periplasmic chaperones cannot actively release substrate proteins, which means that failure to recognize a correct partner may lead to long-term blockage of productive pathways, threatening cellular biogenesis. We recently found that subunit-free Caf1M exists predominantly as a tetramer (A Zavialov and SD Knight, manuscript). A 2.9 Å-resolution crystal structure of the Caf1M tetramer (Fig. 2B) reveals that each of the four molecules contributes its subunit binding sequences to form an eight-stranded hetero-sandwich with a well-packed phenylalanine rich hydrophobic core. This packing of extended subunit binding sequences into the hetero-domain protects chaperone molecules against enzymatic proteolysis, and prevents unspecific binding of the chaperone to itself (aggregation) and possibly also to other molecules in the crowded environment of the periplasm.

6.3.3 Assembly of Subunits Proceeds Through Donor Strand Exchange and is Driven by Folding Energy

Caf1M chaperones deliver folded Caf1 subunits to the Caf1A usher. At the usher, chaperone: subunit interactions must be replaced by subunit: subunit interactions. Hence, the A_1 and G_1 strands of the chaperone that cap the subunit at the base of the growing structure, on the periplasmic side of the usher, must be released to allow the N terminal sequence of the next subunit to be inserted into the polymerization cleft (Fig. 4). This exchange process, called donor strand exchange (DSE) (Choudhury et al. 1999; Sauer et al. 1999; Sauer et al. 2002; Zavialov et al. 2003), can occur even in the absence of usher as evidenced by the accumulation of series of low molecular weight Caf1 polymers in the periplasm of Caf1A negative bacteria expressing Caf1M and Caf1, or *in vitro* following incubation of Caf1M:Caf1 complex (Zavialov et al. 2002). However, compared to usher mediated assembly, this process is slow and inefficient, suggesting that DSE assembly is catalyzed by the usher.

Two basic models for DSE have been suggested (Zavialov et al. 2003). In the first model, the chaperone bound to the subunit at the base of a growing fiber is first released, followed by ordering and insertion of the N-terminal G_d donor sequence of an attacking chaperone: subunit complex into the unoccupied polymerization cleft. The second model involves concerted release of the chaperone G_1 donor strand and insertion of the subunit G_d donor strand in a zip-out-zip-in mechanism. The observation of a transient quaternary complex between two chaperone: subunit complexes during assembly of type-1 pili (Vetsch et al. 2006) strongly argues for the second model.

No energy input from external sources is required to convert periplasmic chaperone: subunit pre-assembly complexes into free chaperone and secreted fibers (Jacob-Dubuisson et al. 1994). At the same time, the extensive and well packed interface between chaperone and subunit in the Caf1M:Caf1 pre-assembly complex (~3600 Å^2), compared to the much smaller interface involved in forming the F1 fiber module (~2250 Å^2) (Zavialov et al. 2003), would appear to suggest that the pre-assembly complex is more stable than the fiber module, and hence that assembly cannot occur spontaneously. Structural and energetic analysis of a series of chaperone: subunit complexes provided the answer to this apparent paradox (Zavialov et al. 2003; Zavialov et al. 2005). Comparison of the structure of Caf1 bound to Caf1M, and its structure in the F1 fiber module, revealed a large conformational difference. In the Caf1M:Caf1 super-barrel, the chaperone G_1 donor strand occupies the polymerization cleft, with large hydrophobic residues from the G_1 donor strand inserted between the two sheets of the subunit β-sandwich, preventing them from contacting each other (Fig. 3B). For this reason, the chaperone-bound conformation is referred to as the 'open' or 'expanded' conformation. Molecular dynamics simulations (Zavialov et al. 2005), predict that the open conformation is not stable and would not be maintained in solution. In contrast to the chaperone donor residues, the much smaller donor residues in the subunit N-terminal G_d donor segment do not intercalate between the two sheets of the subunit β-sandwich, allowing close contact between the two sheets in the fiber module (Fig. 3B). Therefore, the fiber conformation is also referred to as the 'closed' or 'condensed' conformation. The observed difference between open and closed conformations, involving a rearrangement and condensation of the subunit hydrophobic core, suggested that periplasmic chaperones trap subunits in a high-energy molten globule-like folding intermediate state, and that folding energy released on final folding of subunit into the fiber module could drive assembly (Zavialov et al. 2003). Using differential scanning microcalorimetry, we found that the Caf1M:Caf1 complex is not as stable as previously thought, in spite of the extensive interface (ΔH at 37°C ~275 kJ/mol; T_m 45°C) (Zavialov et al. 2005). The association constant for the complex has not been measured directly but is expected to be significantly lower than the 2 x 10^8 M^{-1} measured for binding of Caf1M to the base of a fiber (fiber capping) (Zav'yalov et al. 1997). The relatively low stability is explained by the fact that much of the binding energy is needed to stabilize the bound high-energy conformation of Caf1. We also discovered that the F1 fiber module formed following DSE is exceptionally stable (ΔH at 37°C ~395 kJ/mol; T_m 87.5°C). From the thermodynamic data, a lower limit value for the association constant for subunit-subunit binding in the fiber module of 10^{14} M^{-1} was estimated, rating this as one of the tightest known protein-protein and protein-ligand interactions (Brandts and Lin 1990). The large difference in stabilities between Caf1M:Caf1 pre-assembly complex and F1 fiber module creates a free energy potential that drives fiber formation.

Several observations suggest that periplasmic chaperones target and bind subunits in an unfolded or at least partially unfolded state. The high efficiency of chaperone/usher-mediated assembly in vivo (Jacob-Dubuisson et al. 1994) suggests that this process cannot rely on the slow self-folding of subunits (Vetsch et al. 2002). Recently, Vetsch et al. (Vetsch et al. 2004) verified that the FimC chaperone involved in assembly of type 1 pili indeed binds to unfolded pilin subunits. Chaperone binding was shown to increase the rate of folding by a factor of 100.

To achieve binding of unfolded subunits, the periplasmic chaperone has to recognize a common feature of the (partially) unfolded conformations. The extensive interactions between the hydrophobic cores of the N-terminal domain of the chaperone and the subunit observed in the structures of chaperone: subunit complexes suggests that the chaperone might recognize and attract hydrophobic core residues that are exposed in unfolded subunits. The surface exposed hydrophobic patch created by the bulky hydrophobic side chains in the G_1 donor strand of free chaperones (Hung et al. 1996; Soto et al. 1998) might attract (partially) unfolded subunits and provide a template onto which the subunit core can condense, facilitating folding. At the same time however, the subunit is trapped in an open, activated, conformation because of intercalation of the large chaperone donor residues in the subunit hydrophobic core. The resulting meta-stable complex provides a convenient substrate for assembly.

6.3.4 The Usher: A First Glimpse of Usher-Mediated Donor Strand Exchange

For DSE and secretion to proceed, the usher must interact with chaperone-subunit complexes in the periplasm to facilitate dissociation of the chaperone and polymerization of subunits. Growing polymer must then be translocated across the OM to the cell surface. The ushers are large (80-90 kDa) porin-like integral OM proteins (Henderson et al. 2004; Nishiyama et al. 2003; Thanassi et al. 2002). Recent results (Li et al. 2004; So and Thanassi 2006) show that both the PapC (P pilus) and the FimD (type 1 pilus) ushers form 7x10 nm^2 homodimers with a ~2 nm pore in the middle area of each monomer. Such a pore is wide enough to allow translocation of folded structural subunits or their polymers through the OM. The twin-pore structure is similar to the previously reported electron microscopy structures of the Tom40 complex, a mitochondrial outer membrane protein translocase (Ahting et al. 2001). Functional studies of the PapC and FimD ushers have shown that they contain N- and C-terminal periplasmic domains important for initial binding of chaperone-subunit complexes in the periplasm and subsequent pilus assembly (Barnhart et al. 2003; Ng et al. 2004; Nishiyama et al. 2005; Nishiyama et al. 2003; So and Thanassi 2006; Thanassi et al. 2002). Analysis of the full-length ushers and their N- or C-terminal deletion mutants suggested that initial binding of chaperone-subunit complexes occurs to the N-terminal usher domain, while pilus assembly is assisted by the C-terminal domain.

Recently, we solved the crystal structure of the N-terminal periplasmic domain of the Caf1A usher (A_N) bound to the Caf1M:Caf1 pre-assembly complex (A. Dubnovitsky, A. Zavialov, S.D. Knight; unpublished) (Fig. 2G). The structure of this complex, as well as that of a similar complex from the type-1 pilus (Fim) (Nishiyama et al. 2005) system, shows that the ushers have a chaperone binding surface formed by the folded core of the usher domain and by an extended N-terminal 'tail' of the usher. As we predicted already in 1996 (Hung et al. 1996), the usher recognizes a patch of conserved hydrophobic residues on the 'back' of the chaperone N-terminal domain. Whereas in the Fim complex there are also some interactions between the usher N-terminal tail and the chaperone-bound subunit, no such interactions are present in the Caf1 complex. This might reflect the need to distinguish several different subunit types and to assemble these in a particular sequence in the more complex Fim system but not in the Caf1 system.

6.4 Conclusion and Outlook

During the last few years, considerable progress has been made in understanding how periplasmic chaperones guide and control the assembly of intrinsically aggregative protein subunits into ordered fibrillar structures. By providing a folding platform consisting of a pair of template β-strands (A_1 and G_1) and large hydrophobic donor residues, periplasmic chaperones promote subunit folding and partition subunits away from non-productive aggregation pathways. Subunit folding onto this platform results in chaperone donor residues being incorporated into the core of the subunit and formation of a fused super-barrel with the subunit in an open, activated high-energy conformation. Following dissociation of this activated subunit from the chaperone, folding is completed to form a condensed hydrophobic core as observed for the F1 fiber module. By arresting subunit folding and trapping subunits in a molten globule-like high-energy conformation, the chaperones preserve folding energy that can drive assembly even when chaperone: subunit interactions are more extensive than subunit: subunit interactions in the fiber. In contrast to the rather detailed understanding of the periplasmic chaperones, very little is known about the assembly process *per se* and the workings of the outer membrane usher. A first glimpse of how the usher recognizes chaperone: subunit complexes has provided some hints about how specificity and ordered assembly of complex structures is achieved. Having reached the end of the beginning of the chaperone/usher story, in the following years, hopefully new structures of ushers and of complexes between ushers and chaperones and subunits will allow us to begin to write the end of the story.

6.5 Acknowledgments

Work in my laboratory is supported by grants from the Swedish Research Council.

6.6 References

Ahting, U., Thieffry, M., Engelhardt, H., Hegerl, R., Neupert, W. and Nussberger, S. (2001) Tom40, the pore-forming component of the protein-conducting tom channel in the outer membrane of mitochondria. J Cell Biol 153, 1151-1160.

Barnhart, M. M., Sauer, F. G., Pinkner, J. S. and Hultgren, S. J. (2003) Chaperone-subunit-usher interactions required for donor strand exchange during bacterial pilus assembly. J Bacteriol 185, 2723-2730.

Berglund, J. and Knight, S. D. (2003) Structural basis for bacterial adhesion in the urinary tract. Adv Exp Med Biol 535, 33-52.

Bernier, C., Gounon, P. and Le Bouguenec, C. (2002) Identification of an aggregative adhesion fimbria (AAF) type III-encoding operon in enteroaggregative *Escherichia coli* as a sensitive probe for detecting the AAF-encoding operon family. Infect Immun 70, 4302-4311.

Brandts, J. F. and Lin, L. N. (1990) Study of strong to ultratight protein interactions using differential scanning calorimetry. Biochemistry 29, 6927-6940.

Chalton, D. A., Musson, J. A., Flick-Smith, H., Walker, N., McGregor, A., Lamb, H. K., Williamson, E. D., Miller, J., Robinson, J. H. and Lakey, J. H. (2006) Immunogenicity of a *Yersinia pestis* vaccine antigen monomerized by circular permutation. Infect Immun 74, 6624-6631.

Chen, T. H. and Elberg, S. S. (1977) Scanning electron microscopic study of virulent *Yersinia pestis* and *Yersinia pseudotuberculosis* type 1. Infect Immun 15, 972-977.

Choudhury, D., Thompson, A., Stojanoff, V., Langermann, S., Pinkner, J., Hultgren, S. J. and Knight, S. D. (1999) X-ray structure of the FimE-FimH chaperone-adhesin complex from uropathogenic *Escherichia coli* [see comments]. Science 285, 1061-1066.

Davis, K. J., Fritz, D. L., Pitt, M. L., Welkos, S. L., Worsham, P. L. and Friedlander, A. M. (1996) Pathology of experimental pneumonic plague produced by fraction 1-positive and fraction 1-negative *Yersinia pestis* in african green monkeys (Cercopithecus aethiops). Arch Pathol Lab Med 120, 156-163.

Drozdov, I. G., Anisimov, A. P., Samoilova, S. V., Yezhov, I. N., Yeremin, S. A., Karlyshev, A. V., Krasilnikova, V. M. and Kravchenko, V. I. (1995) Virulent non-capsulate *Yersinia pestis* variants constructed by insertion mutagenesis. J Med Microbiol 42, 264-268.

Du, Y., Galyov, E. and Forsberg, A. (1995) Genetic analysis of virulence determinants unique to *Yersinia pestis*. Contrib Microbiol Immunol 13, 321-324.

Du, Y., Rosqvist, R. and Forsberg, A. (2002) Role of fraction 1 antigen of *Yersinia pestis* in inhibition of phagocytosis. Infect Immun 70, 1453-1460.

Friedlander, A. M., Welkos, S. L., Worsham, P. L., Andrews, G. P., Heath, D. G., Anderson, G. W., Jr., Pitt, M. L., Estep, J. and Davis, K. (1995) Relationship between virulence and immunity as revealed in recent studies of the F1 capsule of *Yersinia pestis*. Clin Infect Dis 21 Suppl 2, S178-181.

Galyov, E. E., Smirnov, O., Karlishev, A. V., Volkovoy, K. I., Denesyuk, A. I., Nazimov, I. V., Rubtsov, K. S., Abramov, V. M., Dalvadyanz, S. M. and Zav'yalov, V. P. (1990) Nucleotide sequence of the *Yersinia pestis* gene encoding F1 antigen and the primary structure of the protein. Putative T and B cell epitopes. FEBS Lett 277, 230-232.

Henderson, N. S., So, S. S., Martin, C., Kulkarni, R. and Thanassi, D. G. (2004) Topology of the outer membrane usher PapE determined by site-directed fluorescence labeling. J Biol Chem 279, 53747-53754.

Hung, D. L. and Hultgren, S. J. (1998) Pilus biogenesis via the chaperone/usher pathway: An integration of structure and function. J Struct Biol 124, 201-220.

Hung, D. L., Knight, S. D., Woods, R. M., Pinkner, J. S. and Hultgren, S. J. (1996) Molecular basis of two subfamilies of immunoglobulin-like chaperones. EMBO J. 15, 3792-3805.

Jacob-Dubuisson, F., Striker, R. and Hultgren, S. J. (1994) Chaperone-assisted self-assembly of pili independent of cellular energy. J Biol Chem 269, 12447-12455.

Karlyshev, A. V., Galyov, E. E., Smirnov, O. Y., Abramov, V. M. and Zav'yalov, V. P. (1994) Structure and regulation of a gene cluster involved in capsule formation of *Yersinia pestis*.

In J. A. F. O. d. Kamp, (Ed.) *Biological membranes: Structure, biogenesis and dynamics.* Springer-Verlag, Berlin, pp. 321-330.

Knight, S. D., Berglund, J. and Choudhury, D. (2000) Bacterial adhesins: Structural studies reveal chaperone function and pilus biogenesis. Curr Opin Chem Biol 4, 653-660.

Li, H., Qian, L., Chen, Z., Thibault, D., Liu, G., Liu, T. and Thanassi, D. G. (2004) The outer membrane usher forms a twin-pore secretion complex. J Mol Biol 344, 1397-1407.

MacIntyre, S., Knight, S. D. and Fooks, L. J. (2004) Structure assembly and applications of the polymeric F1 antigen of *Yersinia pestis*. In E. Carniel and J. Hinnebusch, (Eds.), *Molecular and cellular biology of pathogenic yersinia*. Horizon Press, U.K.

MacIntyre, S., Zyrianova, I. M., Chernovskaya, T. V., Leonard, M., Rudenko, E. G., Zav'yalov, V. P. and Chapman, D. A. G. (2001) An extended hydrophobic interactive surface of *Yersinia pestis* caf1m chaperone is essential for subunit binding and F1 capsule assembly. Mol Microbiol 39, 12-25.

Makhatadze, G. I. and Privalov, P. L. (1995) Energetics of protein structure. Adv Protein Chem 47, 307-425.

Mulvey, M. A. (2002) Adhesion and entry of uropathogenic *Escherichia coli*. Cell Microbiol 4, 257-271.

Ng, T. W., Akman, L., Osisami, M. and Thanassi, D. G. (2004) The usher N terminus is the initial targeting site for chaperone-subunit complexes and participates in subsequent pilus biogenesis events. J Bacteriol 186, 5321-5331.

Nishiyama, M., Horst, R., Eidam, O., Herrmann, T., Ignatov, O., Vetsch, M., Bettendorff, P., Jelesarov, I., Grutter, M. G., Wuthrich, K., Glockshuber, R. and Capitani, G. (2005) Structural basis of chaperone-subunit complex recognition by the type 1 pilus assembly platform FimD. Embo J 24, 2075-2086.

Nishiyama, M., Vetsch, M., Puorger, C., Jelesarov, I. and Glockshuber, R. (2003) Identification and characterization of the chaperone-subunit complex-binding domain from the type 1 pilus assembly platform FimD. J Mol Biol 330, 513-525.

Perry, R. D. and Fetherston, J. D. (1997) *Yersinia pestis*--etiologic agent of plague. Clin. Microbiol. Rev. 10, 35-66.

Sauer, F. G., Barnhart, M., Choudhury, D., Knight, S. D., Waksman, G. and Hultgren, S. J. (2000a) Chaperone-assisted pilus assembly and bacterial attachment. Curr Opin Struc Biol 10, 548-556.

Sauer, F. G., Futterer, K., Pinkner, J. S., Dodson, K. W., Hultgren, S. J. and Waksman, G. (1999) Structural basis of chaperone function and pilus biogenesis. Science 285, 1058-1061.

Sauer, F. G., Knight, S. D., G., W. and Hultgren, S. J. (2000b) PapD-like chaperones and pilus biogenesis. Semin Cell Dev Biol 11, 27-34.

Sauer, F. G., Pinkner, J. S., Waksman, G. and Hultgren, S. J. (2002) Chaperone priming of pilus subunits facilitates a topological transition that drives fiber formation. Cell 111, 543-551.

So, S. S. and Thanassi, D. G. (2006) Analysis of the requirements for pilus biogenesis at the outer membrane usher and the function of the usher C-terminus. Mol Microbiol 60, 364-375.

Soto, G. E., Dodson, K. W., Ogg, D., Liu, C., Heuser, J., Knight, S., Kihlberg, J., Jones, C. H. and Hultgren, S. J. (1998) Periplasmic chaperone recognition motif of subunits mediates quaternary interactions in the pilus. Embo J 17, 6155-6167.

Soto, G. E. and Hultgren, S. J. (1999) Bacterial adhesins: Common themes and variations in architecture and assembly. J Bacteriol 181, 1059-1071.

Stathopoulos, C., Hendrixson, D. R., Thanassi, D. G., Hultgren, S. J., St Geme, J. W., 3rd and Curtiss, R., 3rd (2000) Secretion of virulence determinants by the general secretory pathway in gram-negative pathogens: An evolving story. Microbes Infect 2, 1061-1072.

Thanassi, D. G. and Hultgren, S. J. (2000) Assembly of complex organelles: Pilus biogenesis in gram-negative bacteria as a model system. Methods 20, 111-126.

Thanassi, D. G., Stathopoulos, C., Dodson, K., Geiger, D. and Hultgren, S. J. (2002) Bacterial outer membrane ushers contain distinct targeting and assembly domains for pilus biogenesis. J Bacteriol 184, 6260-6269.

Titball, R. W. and Williamson, E. D. (2001) Vaccination against bubonic and pneumonic plague. Vaccine 19, 4175-4184.

Titball, R. W. and Williamson, E. D. (2004) *Yersinia pestis* (plague) vaccines. Expert Opin Biol Ther 4, 965-973.

Vetsch, M., Erilov, D., Moliere, N., Nishiyama, M., Ignatov, O. and Glockshuber, R. (2006) Mechanism of fibre assembly through the chaperone-usher pathway. EMBO Rep 7, 734-738.

Vetsch, M., Puorger, C., Spirig, T., Grauschopf, U., Weber-Ban, E. U. and Glockshuber, R. (2004) Pilus chaperones represent a new type of protein-folding catalyst. Nature 431, 329-333.

Vetsch, M., Sebbel, P. and Glockshuber, R. (2002) Chaperone-independent folding of type 1 pilus domains. J Mol Biol 322, 827-840.

Worsham, P. L., Stein, M. P. and Welkos, S. L. (1995) Construction of defined F1 negative mutants of virulent *Yersinia pestis*. Contrib Microbiol Immunol 13, 325-328.

Zav'yalov, V. P., Chernovskaya, T. V., Chapman, D. A., Karlyshev, A. V., MacIntyre, S., Zavialov, A. V., Vasiliev, A. M., Denesyuk, A. I., Zav'yalova, G. A., Dudich, I. V., Korpela, T. and Abramov, V.M. (1997) Influence of the conserved disulphide bond, exposed to the putative binding pocket, on the structure and function of the immunoglobulin-like molecular chaperone Caf1M of *Yersinia pestis*. Biochem J 324 (Pt 2), 571-578.

Zavialov, A. V., Berglund, J., Pudney, A. F., Fooks, L. J., Ibrahim, T. M., MacIntyre, S. and Knight, S. D. (2003) Structure and biogenesis of the capsular F1 antigen from *Yersinia pestis*: Preserved folding energy drives fiber formation. Cell 113, 587-596.

Zavialov, A. V., Kersley, J., Korpela, T., Zav'yalov, V. P., MacIntyre, S. and Knight, S. D. (2002) Donor strand complementation mechanism in the biogenesis of non-pilus systems. Mol Microbiol 45, 983-995.

Zavialov, A. V., Tischenko, V. M., Fooks, L. J., Brandsdal, B. O., Aqvist, J., Zav'yalov, V. P., Macintyre, S. and Knight, S. D. (2005) Resolving the energy paradox of chaperone/usher-mediated fibre assembly. Biochem J 389, 685-694.

7
Relationship of the Lipopolysaccharide Structure of *Yersinia pestis* to Resistance to Antimicrobial Factors

Yuriy A. Knirel[1], Svetlana V. Dentovskaya[2], Olga V. Bystrova[1], Nina A. Kocharova[1], Sof'ya N. Senchenkova[1], Rima Z. Shaikhutdinova[2], Galina M. Titareva[2], Irina V. Bakhteeva[2], Buko Lindner[3], Gerald B. Pier[4], and Andrey P. Anisimov[2]

[1] N. D. Zelinsky Institute of Organic Chemistry, Russian Academy of Sciences, knirel@ioc.ac.ru
[2] State Research Center for Applied Microbiology and Biotechnology, Obolensk
[3] Research Center Borstel, Center for Medicine and Biosciences
[4] Channing Laboratory, Brigham and Women's Hospital, Harvard Medical School

Abstract. Disruption of lipopolysaccharide (LPS) biosynthesis genes in an epidemiologically significant *Yersinia pestis* strain showed that the ability to synthesize the full inner core of the LPS is crucial for resistances to the bactericidal action of antimicrobial peptides and to complement-mediated serum killing. Resistance to polymyxin B also requires a high content of the cationic sugar, 4-amino-4-deoxy-L-arabinose, in lipid A.

7.1 Introduction

For constant circulation in natural foci, the plague pathogen, *Yersinia pestis*, must penetrate into the host organism, counteract its protective bactericidal systems and reproduce to ensure bacteremia that is essential for further transmission of the infection by fleas to a new host. Each of these stages in the cyclic existence of *Y. pestis* is insured by numerous properties of the plague pathogen, including multi-functional pathogenicity factors such as lipopolysaccharide (LPS), which possesses a pleiotropic ability to counteract the defense mechanisms of insect vectors and mammalian hosts. In mammals, the LPS induces endotoxic shock, determines resistance to the bactericidal action of cationic peptides and normal sera, may partake in adhesive activity, and is necessary for the enzymatic activities of the plasminogen activator of *Y. pestis* (Anisimov et al. 2004; Bengoechea et al. 1998; Brubaker 1991; Butler 1983; Dmitrovskii 1994; Kukkonen et al. 2004; Porat et al. 1995; Straley 1993).

Owing to a frame-shift mutation in the O-antigen gene cluster inherited from *Yersinia pseudotuberculosis* (Skurnik et al. 2000), *Y. pestis* possesses a rough(R)-type LPS composed of an oligosaccharide core and lipid A, with no O-polysaccharide chain present. A single major LPS core glycoform is synthesized at 37°C (mammalian body temperature) whereas multiple glycoforms differing in terminal monosaccharides are produced at 25°C (flea temperature) (Knirel et al. 2005). The degree of acylation and the level of the cationic sugar, 4-amino-4-deoxy-L-arabinose (Ara4N), in lipid A increases with lower growth temperatures (Kawahara et al. 2002; Knirel et al. 2005; Rebeil et al. 2004). While a biological role for the structural variations in the LPS core remains to be determined, the modifications to the lipid A structure are evidently a part of the mechanism of optimal adaptation of *Y. pestis* to significantly different conditions in insect vectors and mammalian hosts. The less immunostimulatory LPS synthesized at mammalian temperature (37°C) compromises the host's ability to rapidly respond with a proper inflammatory response to infection (Kawahara et al. 2002; Rebeil et al. 2004). The LPS structure synthesized at flea temperatures confers resistance to antimicrobial peptides (Anisimov et al. 2005; Rebeil et al. 2004).

The aims of this work are to unravel the impact of particular LPS components on resistance of the bacteria to various antimicrobial factors and to understand better the biological significance of the temperature-dependent LPS structural variations. For these purposes, we identified genes that are involved with biosynthesis of the *Y. pestis* LPS, generated the corresponding nonpolar mutants and studied the resultant LPS structures and biological properties.

7.2 Methods and Results

7.2.1 Generation of Mutants

Parental strain *Y. pestis* 231 (ssp. pestis, bv. antiqua) was isolated in the Aksai focus, Kirghizia. When cultivated at 25°C, the pCD^- plasmidless attenuated derivative named KM260(11) was resistant to polymyxin B and normal human serum (NHS) (Anisimov et al. 2005).

The *Y. pestis* chromosome was found to harbor the *waa* gene cluster containing five genes (64469..69941, locus YPO0054-YPO0058) that are most likely involved in biosynthesis of the LPS inner core. *In silico* analysis of the *Y. pestis* strain CO92 genome sequence revealed several other putative genes for glycosyl transferases presumably involved in LPS biosynthesis that are located in different parts of the chromosome. Among them there are YPO0416 (434938..435708), YPO0417 (435822..437054), YPO0186 (203866..204846), YPO0187 (204939..205928), YPO3866 (4340131..4341228), and YPO2421 (2722822..2722328) genes. The YPO0654 (714210..715640) gene was suggested to encode ADP-L-*glycero*-D-*manno*-heptose synthase. Dispersion of *Y. pestis* LPS core biosynthesis genes among different regions of the chromosome indicates that they were acquired during different events of horizontal transfer.

Mutant strains with impaired LPS biosynthesis pathway were generated by one-step inactivation of the chromosomal genes mentioned above by λ Red recombination technology (Datsenko and Wanner 2000). Repeated electroporations of the PCR products into *Y. pestis* strains, KM260(11)/pKD46, gave each time single recombinant bacteria disrupted for the target LPS loci. The proper insertion of the *kan* cassette and subsequent deletion of the central parts of the target genes were verified by PCR.

The parental and mutant strains were grown in a New Brunswick Scientific fermentor in Brain Heart Infusion (BHI; Himedia Laboratories, Mumbai, India) at 25°C; kanamycin was used at final concentration of 40 μg·mL^{-1}.

7.2.2 Structure of Mutant Lipopolysaccharides

The LPSs were isolated by the phenol/chloroform/light petroleum procedure (Galanos, Lüderitz, and Westphal 1969) and purified by treatment with DNAse, RNAse, and Proteinase K followed by ultracentrifugation. Each LPS was degraded with dilute acetic acid to cleave the linkage between the core and lipid A moieties. The resultant water-soluble supernatant was fractionated by gel-permeation chromatography on Sephadex G-50 (S) to yield core oligosaccharides.

The whole LPSs and core oligosaccharides were analyzed by electrospray ionization Fourier transform ion-cyclotron resonance mass spectrometry in the negative ion mode using an Apex II instrument (Bruker Daltonics, MT) equipped with a 7 T actively shielded magnet. The data of the products from mutant strains were compared to those of the parental *Y. pestis* strain.

In the mass spectrum of the isolated core from the parental strain (Fig. 1A), there were several major series of ions due to a structural heterogeneity of the following types (Fig. 2). i) Alternation of 3-deoxy-D-*manno*-oct-2-ulosonic acid (Kdo) and D-*glycero*-D-*talo*-oct-2-ulosonic acid (Ko) at one of the terminal positions. These resulted in the appearance of two series with a 236-Da mass difference for ions that either contain or lack Ko. The latter came from compounds with terminal Kdo, which, in contrast to Ko, was cleaved upon mild acid hydrolysis of the LPS. The series for Ko-lacking ions was split into two for compounds with either Kdo or anhydro-Kdo at the reducing end (mass difference 18 Da). ii) Alternation of D-*glycero*-D-*manno*-heptose (DD-Hep) and D-Gal at another terminal position of the oligosaccharide. As a result, two series were observed with a mass difference of 30 Da. iii) Non-stoichiometric substitution with terminal D-GlcNAc giving rise to two series of ions with a 203-Da mass difference.

The mass spectrum of the core from YPO0186::*kan* mutant (Fig. 1B) showed no Gal-containing compounds and that from YPO0187::*kan* mutant (Fig. 1C) no compounds with DD-Hep. Therefore, the functions of the corresponding glycosyl transferases were assigned to the mutated genes. Mutation in either YPO0417 or YPO3866 resulted in the same product lacking GlcNAc (Fig. 1D). These and homology data enabled assignment of their functions as genes for ligase WaaL and UndP:GlcNAc transferase WecA, which transfer GlcNAc to LD-Hep II or undecaprenyl-phosphate, respectively. Similarly, based on the data of the corresponding mutants having a truncated inner core, it was concluded that YPO0416 (Fig. 1E), YPO0057 (Fig. 1F), and YPO0654 encode transferase WaaQ for distal L-*glycero*-D-*manno*-heptose (LD-Hep III), LD-Hep II transferase WaaF, and LD-Hep synthase, respectively. GlcNAc is present in the core of YPO0416::*kan* mutant but in a significantly lower content than in those with the inner core unaffected. Therefore, although not strictly necessary, LD-Hep III is important for effective incorporation of GlcNAc. No Glc is present in the LPS of YPO0057::*kan* mutant and, hence, addition

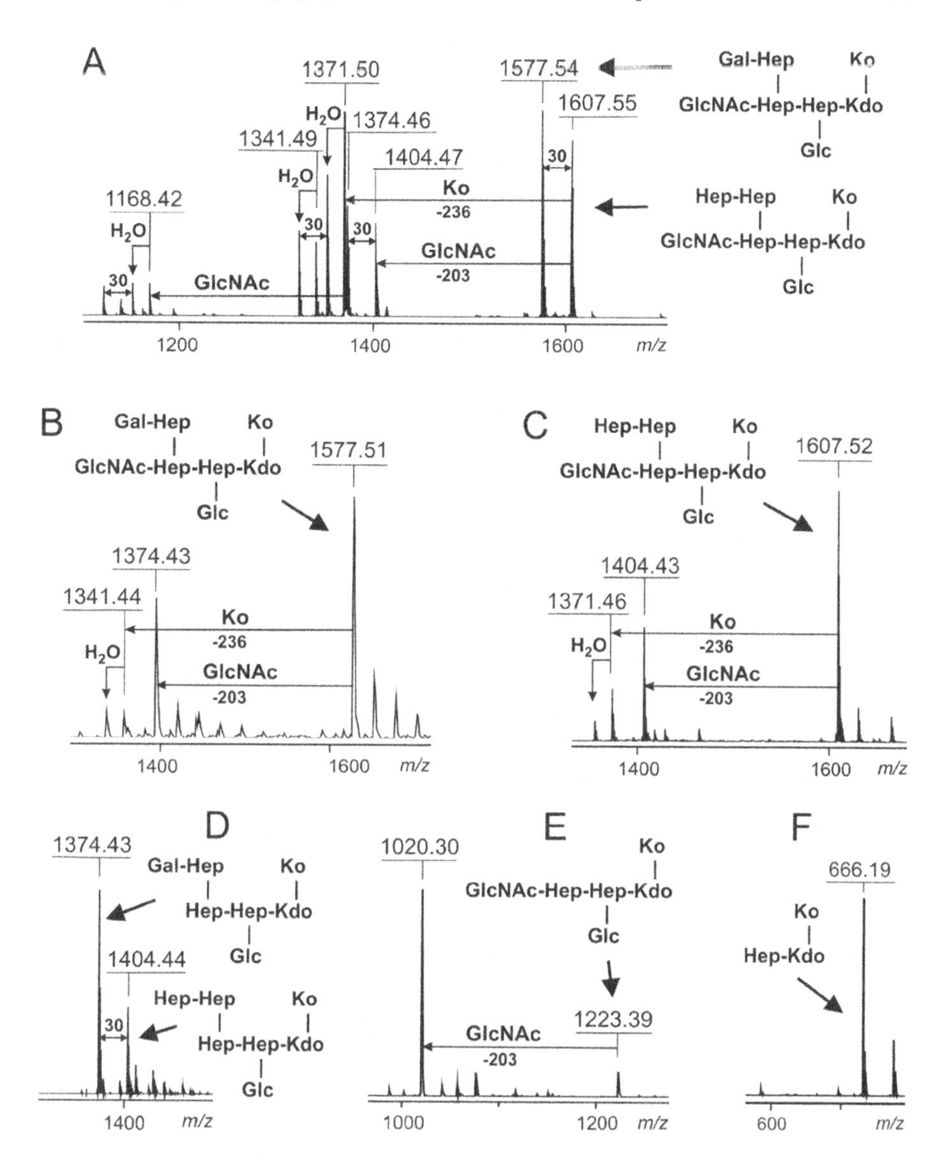

Fig. 1. High-resolution negative ion electrospray ionization mass spectra of the core oligosaccharides from the parental *Y. pestis* strain KM260(11) (A) and derived mutants with a mutation in the YPO0186 (B), YPO0187 (C), YPO0417 (D), YPO0416 (E), and YPO0057 (F) genes. The corresponding core structures are schematically shown in the insets.

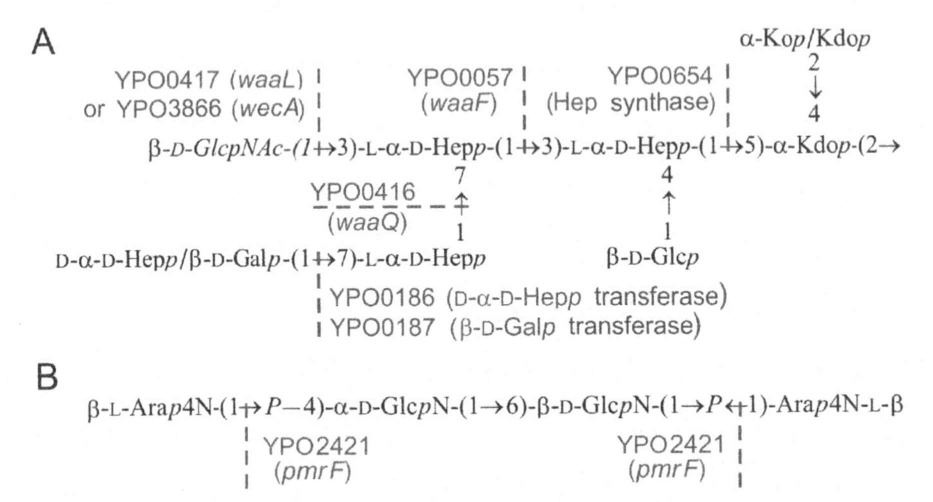

Fig. 2. Structure of the LPS core (A) and lipid A backbone (B) of *Y. pestis* (Knirel et al. 2005) and assignment of some gene functions involved in biosynthesis of the LPS. Abbreviations: L-α-D-Hep and D-α-D-Hep, L-*glycero*- and D-*glycero*-α-D-*manno*-heptose; Kdo, 3-deoxy-D-*manno*-oct-2-ulosonic acid; Ko, D-*glycero*-D-talo-oct-2-ulosonic acid; Ara4N, 4-amino-4-deoxy-L-arabinose. A non-stoichiometric content of GlcNAc is indicated by italics.

of Glc requires prior incorporation of LD-Hep II. The assigned core biosynthesis gene functions are summarized in Fig. 2.

MS studies of the whole LPSs showed that the mutant strains possess essentially the same lipid A moiety as the parental strain. The only exception was that the content of Ara4N in lipid A is lower in deeply truncated core mutants as compared with those having the complete inner core. While LPS molecules with two Ara4N residues prevail in the parental strain (Fig. 3A), most LPSs molecules in YPO0416::*kan*, YPO0057::*kan* (Fig. 3B), and YPO0654::*kan* mutants contain one or no Ara4N residue. The YPO2421::*kan* mutant is fully unable to incorporate Ara4N into the LPS (Fig. 3C), and it was concluded that YPO2421 encodes Ara4N transferase PmrF. The mutation in this gene had no influence on the LPS core structure.

7.2.3 Resistance of Mutants to Bactericidal Action of Polymyxin B

The sensitivity of *Y. pestis* strains to polymyxin B (PMB) was tested using different doses of this cationic antimicrobial peptide to calculate the minimum inhibitory concentration (MIC) as described previously (Anisimov et al. 2005). Table 1 shows that when grown at 25°C the YPO0416, YPO0057, YPO0654, and YPO2421 deficient derivatives were from 31 to 250 times less resistant to PMB (MIC <20 U·mL^{-1}) as compared with the wild-type parent strain and its derivatives carrying mutations in YPO0186, YPO0187, and YPO0417 loci (MIC >625 U·mL^{-1}).

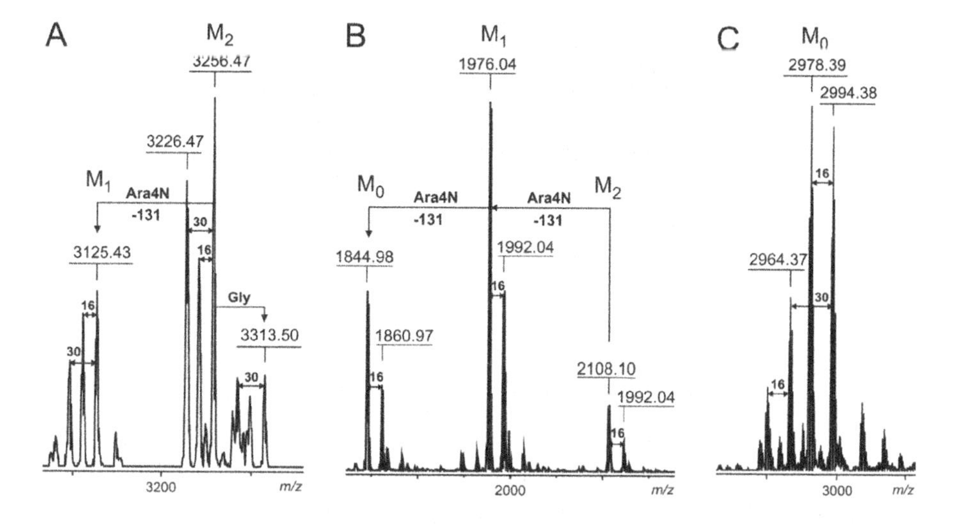

Fig. 3. Negative ion electrospray ionization mass spectra of the whole LPSs from *Y. pestis* strain KM260(11) (A), YPO0057::*kan* mutant (B), and YPO2421::kan mutant (C). Shown is the region of tetraacylated LPS ions with four 3-hydroxymyristoyl groups. Ions containing no, one, and two Ara4N residues are marked with M_0, M_1, and M_2, respectively. Mass differences of 30 and 16 Da correspond to Hep/Gal and Ko/Kdo alternations, respectively.

7.2.4 Resistance of Mutants to Bactericidal Action of Human Serum

A pool of normal human serum (NHS) was obtained from ten non-immunized healthy volunteers. The complement was inactivated by incubating NHS at 56°C for 30 min. Bactericidal properties of NHS were studied by incubation of bacteria with serum for 1 h as described earlier (Anisimov et al. 2005). When grown at 25°C, YPO0416, YPO0057, and YPO0654 deficient derivatives were highly sensitive to the bactericidal action of NHS but not heat-inactivated NHS (Table 1), whereas the other mutants were almost as resistant as the parent strain.

7.3 Conclusions

A single mutation in the gene for Ara4N transferase significantly reduces the polymyxin B resistance of *Y. pestis*, thus demonstrating a role of the cationic sugar in protection of the bacteria. At the same time, Ara4N has no influence on serum resistance. A mutation in either of YPO0186 and YPO0187, which are involved in biosynthesis of the variable outer core region of the *Y. pestis* LPS, does not affect polymyxin B and serum resistances of *Y. pestis*. In contrast, the susceptibility to both antimicrobial factors of the innate immune system is induced by mutation in either of the genes involved in synthesis of the inner heptose region of the LPS core, which is

Table 1. Susceptibility to antimicrobial factors of *Y. pestis* KM260(11) and derived mutants with defined LPS core structures. In LPS of all strains, there is also a core glycoform with the lateral Kdo instead of Ko. LA, wild-type lipid A; LA*, Ara4N-lacking lipid A. Dotted line indicates non-stoichiometric substitution with GlcNAc. MIC, minimum inhibitory concentration; CFU, colony-forming unit; PMB, polymyxin B; NHS, normal human serum; HIS, heat-inactivated serum. Data for susceptible strains are shown in bold face

Y. pestis strain	Core glycoform (schematic view)	MIC of PMB ($U{\cdot}mL^{-1}$)	Viable cells after 1-h incubation (lg $CFU{\cdot}mL^{-1}$)	
			NHS	HIS
Parental strain KM260(11)	Gal/DD-Hep–Hep Glc Ko | | | GlcNAc---Hep–Hep–Kdo–LA	1250	7.0 ± 0.81	7.2 ± 0.73
YPO0186	Gal–Hep Glc Ko | | | GlcNAc---Hep–Hep–Kdo–LA	2500	6.4 ± 0.65	6.8 ± 0.43
YPO0187	DD-Hep–Hep Glc Ko | | | GlcNAc---Hep–Hep–Kdo–LA	1250	6.3 ± 0.51	6.1 ± 0.72
YPO0417 (*waaL*)	Gal/DD-Hep–Hep Glc Ko | | | Hep–Hep–Kdo–LA	625	6.3 ± 0.71	6.8 ± 0.62
YPO0416 (*waaQ*)	Glc Ko | | GlcNAc---Hep–Hep–Kdo–LA	**20**	**3.6** ± 0.53	6.7 ± 0.70
YPO0057 (*waaF*)	Ko | Hep–Kdo–LA	**20**	**2.3** ± 0.19	6.1 ± 0.56
YPO0654 (*rfaE*)	Ko | Kdo–LA	**10**	**2.1** ± 0.23	6.7 ± 0.41
YPO2421 (*pmrF*)	Gal/DD-Hep–Hep Glc Ko | | | GlcNAc---Hep–Hep–Kdo–LA*	**20**	6.5 ± 0.55	6.7 ± 0.71

thus crucial for the bacterial resistance. The susceptibility to polymyxin B of mutants with a deeply truncated LPS core may be accounted for by a poorer incorporation of Ara4N into mutant lipid A.

7.4 Acknowledgments

This work was supported by the International Science and Technology Center/ Cooperative Threat Reduction Program of the US Department of Defense Partner

Project #1197p, the Russian Foundation for Basic Research (grant 06-04-49280), and the Council on Grants at the President of the Russian Federation for Support of Young Russian Scientists (grant MK-3998.2005.4 for O.V.B.).

7.5 References

Anisimov, A.P., Dentovskaya, S.V., Titareva, G.M., Bakhteeva, I.V., Shaikhutdinova, R.Z., Balakhonov, S.V., Lindner, B., Kocharova, N.A., Senchenkova, S.N., Holst, O., Pier, G.B., and Knirel, Y.A. (2005) Intraspecies and temperature-dependent variations in susceptibility of *Yersinia pestis* to bactericidal action of serum and polymyxin B. Infect. Immun. 73, 7324-7331.

Anisimov, A.P., Lindler, L.E., and Pier, G.B. (2004) Intraspecific diversity of *Yersinia pestis*. Clin. Microbiol. Rev. 17, 434-464.

Bengoechea, J.-A., Lindner, B., Seydel, U., Díaz, R., and Moriyón, I. (1998) *Yersinia pseudotuberculosis* and *Yersinia pestis* are more resistant to bactericidal cationic peptides than *Yersinia enterocolitica*. Microbiology 144, 1509-1515.

Brubaker, R.R. (1991) Factors promoting acute and chronic disease caused by yersiniae. Clin. Microbiol. Rev. 4, 309–324.

Butler, T. (1983) *Plague and Other Yersinia Infections*. Plenum Press, New York, NY.

Datsenko, K.A. and Wanner, B.L. (2000) One-step inactivation of chromosomal genes in *Escherichia coli* K-12 using PCR products. Proc. Natl. Acad. Sci. USA 97, 6640-6645.

Dmitrovskii, V.G. (1994) Toxic component of pathogenesis of plague infectious process: infective toxic shock. In: V.M. Stepanov (Ed.), *Prophylaxis and means of prevention of plague*. Scientific-Manufacturing Association of the Plague-Control Establishments, Almaty, Kazakhstan, pp. 15-16.

Domaradskii, I.V., Iaromiuk, G.A., Vasiukhina, L.V., and Korotaeva, A.V. (1963) On coagulation of blood plasma by plague and pseudotuberculosis microbes. Biull. Eksp. Biol. Med. 56, 79-82.

Galanos, C., Lüderitz, O., and Westphal, O. (1969) A new method for the extraction of R lipopolysaccharides. Eur. J. Biochem. 9, 245-249.

Kawahara, K., Tsukano, H., Watanabe, H., Lindner, B., and Matsuura, M. (2002) Modification of the structure and activity of lipid A in *Yersinia pestis* lipopolysaccharide by growth temperature. Infect. Immun. 70, 4092-4098.

Knirel, Y.A., Lindner, B., Vinogradov, E.V., Kocharova, N.A., Senchenkova, S.N., Shaikhutdinova, R.Z., Dentovskaya, S.V., Fursova, N.K., Bakhteeva, I.V., Titareva, G.M., Balakhonov, S.V., Holst, O., Gremyakova, T.A., Pier, G.B., and Anisimov, A.P. (2005) Temperature-dependent variations and intraspecies diversity of the structure of the lipopolysaccharide of *Yersinia pestis*. Biochemistry 45, 1731-1743.

Kukkonen, M., Suomalainen, M., Kyllönen, P., Lähteenmäki, K., Lång, H., Virkola, R., Helander, I.M., Holst, O., and Korhonen, T.K. (2004) Lack of O antigen is essential for plasminogen activation by *Yersinia pestis* and *Salmonella enterica*. Mol. Microbiol. 51, 215-225.

Porat, R., McCabe, W.R., and Brubaker, R.R. (1995) Lipopolysaccharide-associated resistance to killing of yersiniae by complement. J. Endotoxin Res. 2, 91-97.

Rebeil, R., Ernst, R.K., Gowen, B.B., Miller, S.I., and Hinnebusch, B.J. (2004) Variation in lipid A structure in the pathogenic yersiniae. Mol. Microbiol. 52, 1363-1373.

Skurnik, M., Peippo, A., and Ervelä, E. (2000) Characterization of the O-antigen gene clusters of *Yersinia pseudotuberculosis* and the cryptic O-antigen gene cluster of *Yersinia pestis*

shows that the plague bacillus is most closely related to and has evolved from *Y-pseudotuberculosis* serotype O:1b. Mol. Microbiol. 37, 316-330.
Sodeinde, O.A. and Goguen, J.D. (1988) Genetic analysis of the 9.5-kilobase virulence plasmid of *Yersinia pestis*. Infect. Immun. 56, 2743-2748.
Sodeinde, O.A. and Goguen, J.D. (1989) Nucleotide sequence of the plasminogen activator gene of *Yersinia pestis*: relationship to *ompT* of *Escherichia coli* and gene *E* of *Salmonella typhimurium*. Infect. Immun. 57, 1517-1523.
Sodeinde, O.A., Subrahmanyam, Y.V.B.K., Stark, K., Quan, T., Bao, Y., and Goguen, J.D. (1992) A surface protease and the invasive character of plague. Science 258, 1004-1007.
Straley, S.C. (1993) Adhesins in *Yersinia pestis*. Trends Microbiol. 1, 285-286.

Picture 11. Yuriy Knirel presents research on LPS structural variations and their biological effects. Photograph by A. Anisimov.

8
Characterization of Six Novel Chaperone/Usher Systems in *Yersinia pestis*

Suleyman Felek[1], Lisa M. Runco[2], David G. Thanassi[2], and Eric S. Krukonis[1,3]

[1] Department of Biologic and Materials Sciences, University of Michigan School of Dentistry
[2] Department of Molecular Genetics and Microbiology, State University of New York, Stony Brook
[3] Department of Microbiology and Immunology, University of Michigan School of Medicine, ekrukoni@umich.edu

8.1 Introduction

Yersinia pestis, the causative agent of plague, is a gram-negative pathogen that evolved from *Yersinia pseudotuberculosis* 1,500-20,000 years ago (Achtman et al. 1999). Plague has ravaged human populations for centuries and continues to occur in outbreaks throughout the world (Cantor 2001; Perry and Fetherston 1997). *Y. pestis* is usually transmitted by fleas that have fed on infected rodents prior to biting their human hosts. A number of virulence factors have been described in *Y. pestis* that contribute to disease including the 70-kb Yop-encoding plasmid pCD1 (Cornelis et al. 1998), plasminogen activator (Sodeinde et al. 1992), iron acquisition functions (Brubaker et al. 1965) and a surface-localized adhesin pH 6 antigen (Lindler et al. 1990).

Studies presented here focus on the function(s) of six annotated, but uncharacterized, chaperone/usher systems in *Y. pestis* virulence. Such systems are involved in secretion and assembly of surface structures that can play a variety of roles including: cell adhesion (Jones et al. 1995; Kuehn et al. 1992), biofilm formation (Tomaras et al. 2003; Vallet et al. 2001) and inhibition of phagocytosis (Du et al. 2002; Huang and Lindler 2004). There are two well-characterized chaperone/usher systems present in *Y. pestis*. The Caf1 system assembles an antiphagocytic capsule (Du et al. 2002) while the pH 6 antigen system assembles an anti-phagocytic adhesin (Huang and Lindler 2004; Makoveichuk et al. 2003; Yang et al. 1996). Analysis of the six novel chaperone/usher systems will likely advance our understanding of *Y. pestis* infections and may lead to additional vaccine targets.

Here we show that the newly characterized chaperon/usher system, encoded by *y0561-0563*, mediates adhesion of recombinant *E. coli* to human epithelial cells. Furthermore, five of six novel chaperon/usher loci resulted in increased biofilm-like formations in *E. coli* based on a crystal violet staining assay and two showed a dramatic (>10-fold) increase in crystal violet staining.

8.2 Experimental procedures

8.2.1 Bacterial Strains and Plasmids

Strains used in this study were *Y. pestis* KIM5-3001, *E. coli* DH5α, *E. coli* AAEC185 (*fim*⁻, Blomfield et al., 1991) and *Y. pseudotuberculosis* IP2666. Plasmid pMMB207 (moderate copy number, IPTG-inducible, (Morales et al. 1991)) was used in this study for cloning and expressing various *Yersinia* chaperone/usher loci. Gene deletions in *Y. pestis* KIM5 were performed by using PCR products as described by Drs. Don Court and Barry Wanner (Datsenko and Wanner 2000; Yu et al. 2000).

For cloning, the genes encoding chaperone-usher systems were amplified by PCR. PCR products and plasmid pMMB207 were cut by appropriate enzymes, gel purified and ligated. Because strain KIM5 is *pgm*⁻, it does not have the *y2388-y2392* locus (Buchrieser et al. 1999). Thus, we cloned this locus from *Y. pseudotuberculosis* strain IP2666. *Y. pestis* KIM and *Y. pseudotuberculosis* IP2666 have 100% nucleotide homology for effector proteins of this system. *E. coli* DH5α was transformed with the ligation products and chloramphenicol resistant colonies were selected. Clones were confirmed by DNA sequencing.

8.2.2 Transmission Electron Microscopy

An overnight culture (grown with aeration at 37°C) was diluted 1:50 into LB + 50 μM IPTG, and grown with aeration at 37°C until OD_{600} = ~0.4. Cultures were gently washed and resuspended in PBS, and allowed to adhere to formvar-carbon coated EM grids (Ernest F. Fullam, Latham, NY) for 2 minutes.

Bacteria were fixed on the grids with 1% glutaraldehyde in PBS for 1 minute, washed twice with PBS briefly, washed twice with water for 1 minute each, and stained for 20 seconds with 0.5% phosphotungstic acid. The grids were examined on an FEI BioTwinG2 transmission electron microscope at 80 kV accelerating voltage.

8.2.3 Adhesion Assay

E. coli AAEC185 strains containing pMMB207 and pMMB207-chaperone/usher systems were cultured overnight at 28°C with shaking in Luria Bertoni (LB) medium supplemented with 10 μg/ml chloramphenicol and 100 μM IPTG. Bacterial cultures were centrifuged and were resuspended in MEM. HEp-2 cells were infected with bacteria at an MOI of about 1 to 3 per cell after being washed twice with serum-free MEM. Plates were incubated at 37°C in 5% CO_2 for 2 h. The wells were washed twice with PBS and incubated for 10 min at room temperature with 500 μl double distilled water +0.1% Triton X-100 to remove cells from the plates. The wells were washed one more time with PBS and the PBS wash was combined with the Triton X-100 lysate. Ten fold dilutions in PBS were plated on LB agar plates containing 10 μg/ml chloramphenicol. The plates were incubated at 37°C overnight and the output CFUs were enumerated. Adhesion was presented as the number of cell associated CFUs divided by input CFUs after a 2 hr incubation x 100%.

8.2.4 Biofilm Formation Assay

Crystal violet staining was used to detect cells attached to polystyrene as described by O'Toole et. al. (O'Toole et al. 1999). Briefly, overnight cultures were diluted 1:100 into LB plus 10 μg/ml chloramphenicol and 100 μM IPTG in flat-bottom polystyrene culture plates (Costar Corning, NY) and incubated 24 h at 28°C or 37°C without shaking. The OD_{595} of cultures was read by a microplate reader (Perkin Elmer Lambda Reader, Norwalk, CT). Plates were washed three times with PBS and incubated for 15 min at RT with 0.01% crystal violet. The wells were washed three times with distilled water and the dye was solubilized with a 80% ethanol and 20% acetone mixture. Absorbance of the mixture was detected by a microplate reader. Crystal violet staining intensity was then normalized for bacterial density.

8.3 Results

8.3.1 Expression of *Y. pestis* Chaperone/Usher Systems in *E. coli*

The six novel chaperone usher systems present in *Y. pestis* (Fig. 1) along with the known adhesin locus encoding pH 6 antigen (*psaABC*) were cloned into the low copy-number inducible vector pMMB207 (Morales et al. 1991).

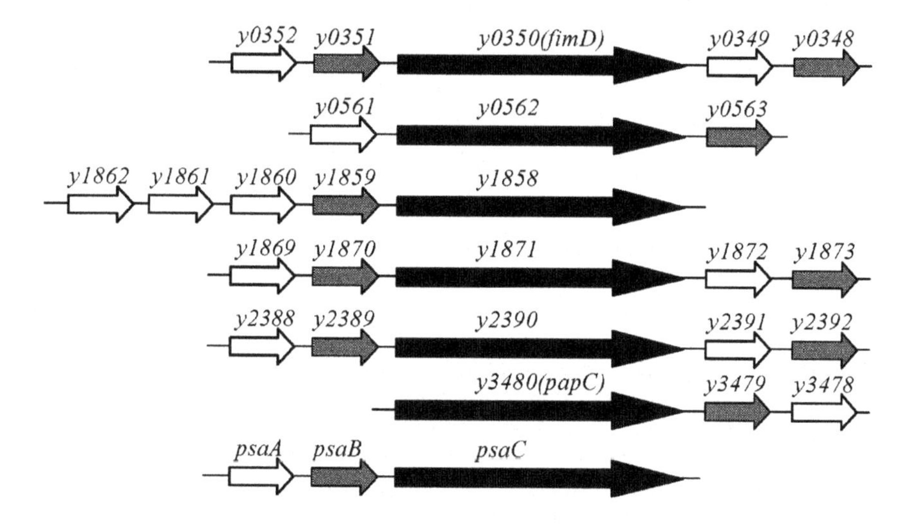

Fig. 1. Chaperone-usher systems in *Y. pestis* KIM genome investigated in this study (Deng et al. 2002). White arrows, putative target proteins; gray arrows, chaperones; black arrow, usher proteins.

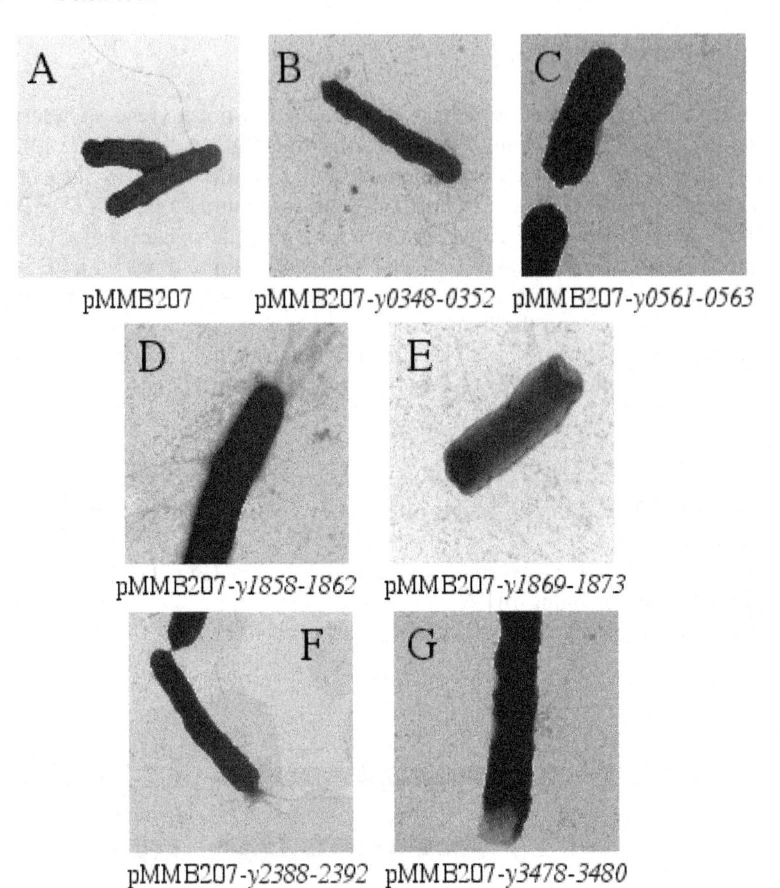

Fig. 2. Electron microscopy of the *fim⁻ E. coli* strain AAEC185 expressing each *Y. pestis* chaperone/usher system. Cells were induced at 37°C with 50 μM IPTG prior to processing for EM.

To confirm that the chaperone-usher systems are expressed in the outer membrane of *E. coli*, we cultured the strains overnight at 28°C under inducing conditions and outer membranes were prepared (Hantke 1984). We were able to detect secreted surface proteins of the predicted sizes for three loci; *y0348-0352* (17.5kD), *y0561-0563* (22kD) and *y1869-1873* (18.5 kD) (data not shown). The other three novel loci as well as pH 6 antigen could not be detected. However, in the case of pH 6 antigen we have shown the protein is sheared from the surface of the bacteria during processing for outer membrane preparations (S.F. and E.K. unpublished observations). This may also be the case for the other three chaperone usher systems.

In addition to assessing the expression of chaperone/usher-encoded proteins in outer membrane preparations, we also visualized novel surface appendages on *E. coli* AAEC185 using transmission electron microscopy (TEM). For these experiments, the strains were induced for chaperone/usher expression with 50 μM IPTG at 37°C. While expression of empty vector pMMB207 resulted in expression of flagella as represented by a thick appendage (Fig. 2A), expression of the *Y. pestis* chaperone/usher loci led to elaboration of various surface structures resembling fimbriae or pili in the *fim*⁻ strain of *E. coli*, AAEC185 (Fig. 2B-G). The fact that the structures each have unique features; straight versus kinked (Fig. 2C vs. B), polar versus centrally-located (Fig. 2F vs. C); suggests expression of the *Y. pestis* chaperone/usher loci do not simply induce some cryptic *E. coli* surface structure.

8.3.2 Chaperone/Usher-Mediated Adhesion to Cultured Cells

E. coli expressing various *Y. pestis* chaperone-usher systems was analyzed for binding to HEp-2 human epithelial cells. Of the novel chaperone/usher systems, only locus *y0561-0563* showed a significant (three-fold) increase in cell binding when expressed in *E. coli* AAEC185 (Fig. 3; $p<0.001$). As a positive control for adhesion, we tested the well-characterized chaperone/usher-dependent adhesin pH 6 antigen. pH 6 antigen mediated a ~25-fold increase in adhesion to HEp-2 cells when compared to *E. coli* with plasmid pMMB207 alone (Fig. 3).

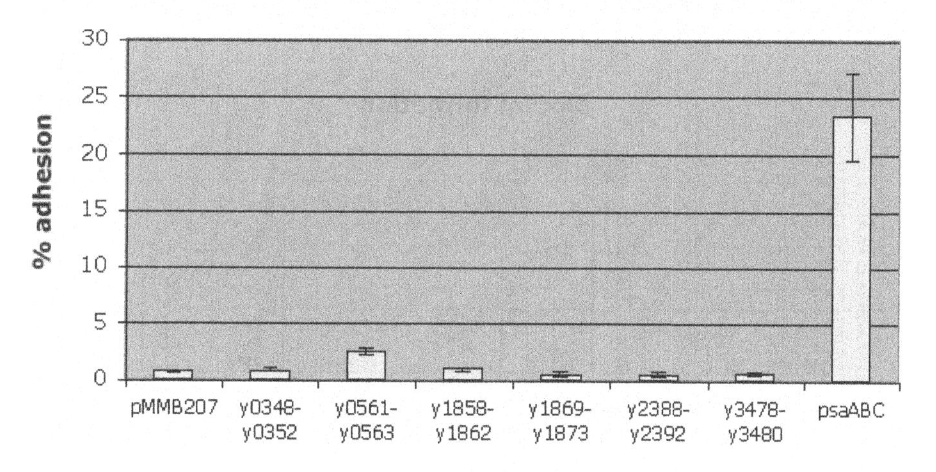

Fig. 3. Adhesion of *E. coli* AAEC185 (*fim*⁻) expression chaperone-usher systems to HEp-2 cells. Strains were cultured overnight at 28°C in LB medium supplemented with 10 μg/ml chloramphenicol and 100 μM IPTG and were added to cells cultures at an MOI of 1 to 3. Values are the averages and standard deviations of three independent experiments with duplicate assays from each experiment.

Although the *y0561-0563* locus could mediate attachment to HEp-2 cells when expressed in *E. coli*, there was no effect on *Y. pestis* KIM5-3001 adhesion when the *y0561-0563* locus was deleted (data not shown). We take this to suggest that other adhesins are likely redundant for the adhesive function of the surface appendage encoded by locus *y0561-0563*. When *Y. pestis* is grown at pH=6 37°, we do see a dramatic reduction in HEp-2 cell binding with a pH6 antigen (*ΔpsaA*) mutant, indicating we are able to reveal binding defects under appropriate conditions in *Y. pestis* (data not shown).

8.3.3 Induction of Biofilm-Like Structures by *Y. pestis* Chaperone/Usher Systems

Biofilms often play a role in establishment of infection or resistance to host clearance mechanisms (Fux et al. 2005). To analyze whether the chaperone/usher loci of *Y. pestis* can facilitate biofilm-like formations in *E. coli*, we performed a crystal-violet staining biofilm assay on polystyrene plates with *E. coli* DH5α expressing each of our seven cloned *Y. pestis* chaperone/usher loci. Of the novel chaperone/usher systems, only the *y0561-0563* locus conferred a dramatic (15-fold) increase in the ability of DH5α to make crystal violet-stainable deposits in microtiter wells at 28°C (Fig. 4). At 37°C five out of six loci conferred some increase in DH5α adherence to plastic and deposit formation, ranging from a 4 to 13-fold increase (Fig. 4). pH 6 antigen was also able to mediate biofilm formation in DH5α (Fig. 4), a previously unreported function for pH 6 antigen.

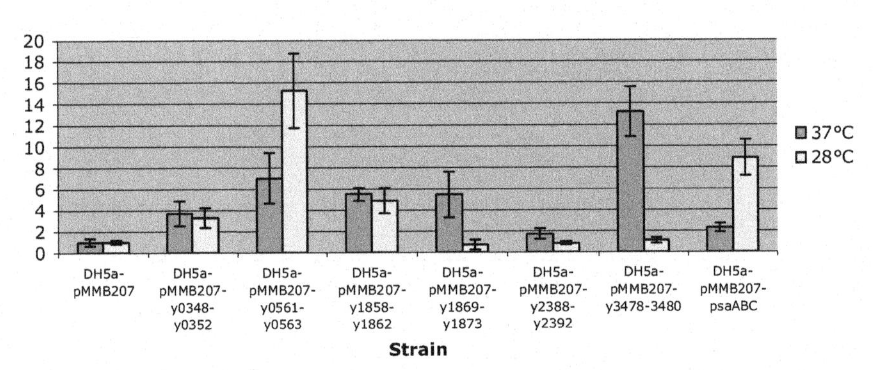

Fig. 4. Formation of biofilm-like deposits in *E. coli* DH5α. Overnight cultures of strains were diluted 1:100 into LB medium supplemented with 10 μg/ml chloramphenicol and 100 μM IPTG and incubated 24 h in polystyrene culture plates at 37°C or 28°C. Results are the averages and standard deviations of three independents experiments with at least duplicate assays from each experiment.

8.4 Discussion

Adhesins are important virulence factors for bacteria that provide successful colonization and determine the tissue tropism of the agent (Niemann et al. 2004; Ofek et al. 2003). Many bacterial pathogens maintain multiple adhesins to facilitate binding to different cell types. In some bacteria, more than 10 adhesins have been described (Ofek et al. 2003). In the present study, we found that a novel *Y. pestis* chaperone/usher locus, *y0561-0563*, encodes an adhesin for cultured human cells (Fig. 3). We also confirmed the strong adhesive activity of pH6 antigen from *Y. pestis*. Although a *Δ0561-0563* derivative of *Y. pestis* strain KIM5-3001 had no defect on cell binding, we hypothesize this may indicate that other adhesins function in this strain.

Biofilm formation is often an important virulence function in bacteria. It promotes colonization and protection of bacteria from host immune defenses. Attachment of the bacteria to a surface via an outer membrane protein or appendage is the first step of biofilm formation (Jefferson 2004). *Y. pestis* is known to produce *hms*-dependent biofilms at ambient temperatures, but not at mammalian temperatures, because HmsT is degraded at 37°C (Kirillina et al. 2004). Here we found that five novel chaperone/usher loci encode proteins capable of facilitating biofilm-like formations in *E. coli* (Fig. 4).

Based on these data we hypothesize that the surface appendages (Fig. 2) encoded by the novel *Y. pestis* chaperone/usher systems may contain particular activities that play an important role in colonization and virulence.

8.5 Acknowledgements

We would like to thank Dr. Greg Plano for providing us with the *Y. pestis* strain KIM5-3001 and Dr. James Bliska for providing *Y. pseudotuberculosis* strain IP2666.

8.6 References

Achtman, M., Zurth, K., Morelli, G., Torrea, G., Guiyoule, A., and Carniel, E. (1999) *Yersinia pestis*, the cause of plague, is a recently emerged clone of *Yersinia pseudotuberculosis*. PNAS 96, 14043-14048.

Blomfield, I.C., McClain, M.S., and Eisenstein, B.I. (1991) Type 1 fimbriae mutants of *Escherichia coli* K12: characterization of recognized afimbriate strains and construction of new *fim* deletion mutants. Mol. Microbiol. 5, 1439-1445.

Brubaker, R.R., Beesley, E.D., and Surgalla, M.J. (1965) *Pasteurella pestis*: role of pesticin I and iron in experimental plague. Science 149, 422-424.

Buchrieser, C., Rusniok, C., Frangeul, L., Couve, E., Billault, A., Kunst, F., Carniel, E., and Glaser, P. (1999) The 102-Kilobase pgm locus of *Yersinia pestis*: Sequence analysis and comparison of selected regions among different *Yersinia pestis* and *Yersinia pseudotuberculosis* strains. Infect. Immun. 67, 4851-4861.

Cantor, N. (2001) *In the Wake of the Plague*. New York: Perennial.

Cornelis, G.R., Boland, A., Boyd, A.P., Geuijen, C., Iriarte, M., Neyt, C., Sory, M.-P., and Stainier, I. (1998) The virulence plasmid of *Yersinia*, an antihost genome. Microbiol. Mol. Biol. Rev. 62, 1315-1352.

Datsenko, K.A., and Wanner, B.L. (2000) One-step inactivation of chromosomal genes in Escherichia coli K-12 using PCR products. PNAS 97, 6640-6645.

Deng, W., Burland, V., Plunkett III, G., Boutin, A., Mayhew, G.F., Liss, P., Perna, N.T., Rose, D.J., Mau, B., Zhou, S., Schwartz, D.C., Fetherston, J.D., Lindler, L.E., Brubaker, R.R., Plano, G.V., Straley, S.C., McDonough, K.A., Nilles, M.L., Matson, J.S., Blattner, F.R., and Perry, R.D. (2002) Genome sequence of *Yersinia pestis* KIM. J. Bacteriol. 184, 4601-4611.

Du, Y., Rosqvist, R., and Forsberg, A. (2002) Role of Fraction 1 antigen of *Yersinia pestis* in inhibition of phagocytosis. Infect. Immun. 70, 1453-1460.

Fux, C.A., Costerton, J.W., Stewart, P.S., and Stoodley, P. (2005) Survival strategies of infectious biofilms. Trends in Microbiology 13, 34-40.

Hantke, K. (1984) Cloning of the repressor protein gene of iron-regulated systems in *Escherichia coli* K12. Mol Gen Genet. 197, 337-341.

Huang, X.-Z., and Lindler, L.E. (2004) The pH 6 antigen is an antiphagocytic factor produced by *Yersinia pestis* independent of Yersinia outer proteins and capsule antigen. Infect. Immun. 72, 7212-7219.

Jefferson, K.K. (2004) What drives bacteria to produce a biofilm? FEMS Microbiol. Lett. 236, 163-173.

Jones, C., Pinkner, J., Roth, R., Heuser, J., Nicholes, A., Abraham, S., and Hultgren, S. (1995) FimH adhesin of Type 1 pili is assembled into a fibrillar tip structure in the Enterobacteriaceae. PNAS 92, 2081-2085.

Kirillina, O., Fetherston, J.D., Bobrov, A.G., Abney, J., and Perry, R.D. (2004) HmsP, a putative phosphodiesterase, and HmsT, a putative diguanylate cyclase, control Hms-dependent biofilm formation in *Yersinia pestis*. Mol. Microbiol 54, 75-88.

Kuehn, M.J., Heuser, J., Normark, S., and Hultgren, S.J. (1992) P pili in uropathogenic *E. coli* are composite fibres with distinct fibrillar adhesive tips. Nature 356, 252-255.

Lindler, L., Klempner, M., and Straley, S. (1990) *Yersinia pestis* pH 6 antigen: genetic, biochemical, and virulence characterization of a protein involved in the pathogenesis of bubonic plague. Infect. Immun. 58, 2569-2577.

Makoveichuk, E., Cherepanov, P., Lundberg, S., Forsberg, A., and Olivecrona, G. (2003) pH6 antigen of *Yersinia pestis* interacts with plasma lipoproteins and cell membranes. J. Lipid Res. 44, 320-330.

Morales, V.M., Backman, A., and Bagdasarian, M. (1991) A series of wide-host-range low-copy-number vectors that allow direct screening for recombinants. Gene 97, 39-47.

Niemann, H.H., Schubert, W.-D., and Heinz, D.W. (2004) Adhesins and invasins of pathogenic bacteria: a structural view. Microbes and Infection 6, 101-112.

O'Toole, G.A., Pratt, L.A., Watnick, P.I., Newman, D.K., Weaver, V.B., and Kolter, R. (1999) Genetic approaches to study of biofilms. Methods Enzymol. 310, 91-109.

Ofek, I., Hasty, D.L., and Doyle, R.J. (2003) *Bacterial Adhesion to Animal Cells and Tissues.* Washington, D.C.: ASM Press.

Perry, R.D., and Fetherston, J.D. (1997) *Yersinia pestis*-etiologic agent of plague. Clin. Micro. Rev. 10, 35-66.

Sodeinde, O., Subrahmanyam, Y., Stark, K., Quan, T., Bao, Y., and Goguen, J. (1992) A surface protease and the invasive character of plague. Science 258, 1004-1007.

Tomaras, A.P., Dorsey, C.W., Edelmann, R.E., and Actis, L.A. (2003) Attachment to and biofilm formation on abiotic surfaces by *Acinetobacter baumannii*: involvement of a novel chaperone-usher pili assembly system. Microbiology 149, 3473-3484.

Vallet, I., Olson, J.W., Lory, S., Lazdunski, A., and Filloux, A. (2001) The chaperone/usher pathways of *Pseudomonas aeruginosa*: Identification of fimbrial gene clusters (*cup*) and their involvement in biofilm formation. PNAS 98, 6911-6916.

Yang, Y., Merriam, J., Mueller, J., and Isberg, R. (1996) The *psa* locus is responsible for thermoinducible binding of *Yersinia pseudotuberculosis* to cultured cells. Infect. Immun. 64, 2483-2489.

Yu, D., Ellis, H.M., Lee, E.-C., Jenkins, N.A., Copeland, N.G., and Court, D.L. (2000) An efficient recombination system for chromosome engineering in *Escherichia coli*. PNAS 97, 5978-5983.

9
Polyamines in Bacteria: Pleiotropic Effects yet Specific Mechanisms

Brian W. Wortham[1], Chandra N. Patel[1,2], and Marcos A. Oliveira[1,3]

[1] Department of Pharmaceutical Sciences, University of Kentucky
[2] Present affiliation: Hematology/Cardiology, Bayer HealthCare LLC
[3] Present affiliation: Dept of Pharmaceutical Sciences, Feik School of Pharmacy, University of the Incarnate Word, oliveira@uiwtx.edu

Abstract. Extensive data in a wide range of organisms point to the importance of polyamine homeostasis for growth. The two most common polyamines found in bacteria are putrescine and spermidine. The investigation of polyamine function in bacteria has revealed that they are involved in a number of functions other than growth, which include incorporation into the cell wall and biosynthesis of siderophores. They are also important in acid resistance and can act as a free radical ion scavenger. More recently it has been suggested that polyamines play a potential role in signaling cellular differentiation in *Proteus mirabilis*. Polyamines have also been shown to be essential in biofilm formation in *Yersinia pestis*. The pleiotropic nature of polyamines has made their investigation difficult, particularly in discerning any specific effect from more global growth effects. Here we describe key developments in the investigation of the function of polyamines in bacteria that have revealed new roles for polyamines distinct from growth. We describe the bacterial genes necessary for biosynthesis and transport, with a focus on *Y. pestis*. Finally we review a novel role for polyamines in the regulation of biofilm development in *Y. pestis* and provide evidence that the investigation of polyamines in *Y. pestis* may provide a model for understanding the mechanism through which polyamines regulate biofilm formation.

9.1 Introduction

Polyamines are small MW polycationic natural compounds with positive charges tethered to a carbon skeleton. Due to their physical chemical nature they interact with biopolyanions, primarily nucleic acids. They are found in high concentrations (~1mM) in the cell, however they are mostly sequestered (Miyamoto et al. 1993). Bacteria synthesize primarily only two polyamines, putrescine and spermidine. Thermophilic bacteria, however, produce additional polyamines that provide stabilization to nucleic acids (Terui et al. 2005). One key question in the field has been the identification and characterization of specific mechanistic roles for polyamines. Generally polyamines have been broadly implicated in cell growth due to their ability to interact with nucleic acids and protein translation machinery (Tabor and Tabor 1985). This dominant characteristic has guided our understanding of their function however it has also over-shadowed the identification of functional roles that are not directly associated with growth. Furthermore, since a large body of work in bacteria has focused on *Escherichia coli*, there are potentially other functions for polyamines that have yet to be discovered.

9.2 Polyamine Biosynthesis

The steps in polyamine biosynthesis require a series of decarboxylases which utilize pyridoxal-5'-phosphate (PLP), a derivative of vitamin B6 (Sandmeier et al. 1994). The substrates for these decarboxylases are primarily the amino acids arginine, ornithine and in some cases lysine (Tabor and Tabor 1985). The decarboxylation of ornithine by ornithine decarboxylase (*speC*, locus tag y3347 in *Yersinia pestis* strain KIM) leads directly to the polyamine putrescine while decarboxylation of arginine by arginine decarboxylase (*speA*, locus tag y3313 in KIM) leads to agmatine. Agmatine can be converted to putrescine in two possible pathways. One pathway involves agmatinase (AG), a ureohydrolase, which produces urea and putrescine. AG is structurally and mechanistically similar to arginase, an enzyme that catalyzes the conversion of arginine to ornithine and urea. Neither agmatinase nor arginase can be detected in *Y. pestis* based on BLAST searches. This suggests that the two polyamine biosynthetic branches based on arginine and ornithine are uncoupled (Fig. 1). In *Y. pestis* agmatine is converted to putrescine in two steps via agmatine deiminase (*aguA*, locus tag *y3325* in KIM) and N-carbamoylputrescine amidohydrolase (*aguB*, locus tag *y3325* in KIM), as outlined in Fig. 1. The two branches of the polyamine pathway lead to the production of putrescine. Putrescine is then converted to spermidine by spermidine synthase (*speE*, locus tag *y0775* in KIM) an enzyme that utilizes decarboxylated S-adenosylmethionine (SAM), which is produced by S-adenosylmethionine decarboxylase (*speD*, locus tag *y0774* in KIM). Finally spermidine is degraded by spermidine N1-acetyltransferase (*speG*, locus tag *y1405* in KIM).

Another important characteristic of polyamine biosynthesis is the existence of isozymes for arginine, ornithine and lysine decarboxylase, which are involved in acid resistance mechanisms. All of these isozymes share a high degree of sequence similarity (>30%) which makes it difficult to assign function based solely on sequence comparisons (Hackert et al. 1994). These isozymes are often referred to as inducible enzymes. In *Y. pestis* there is only one isozyme that can be detected based on sequence analysis (locus tag *y2987* in KIM), however neither its activity nor its function has been investigated.

9.3 Polyamine Transport

Polyamine transporters are likely to exist in every bacterial cell. In fact a polyamine transporter has been identified and shown to be essential for survival of *Mycoplasma genitalium*, which contains no detectable polyamine biosynthetic enzymes (Mushegian and Koonin 1996). In general there are three major transport systems referred to as Pot. Polyamine uptake in *E. coli* follows the general order putrescine

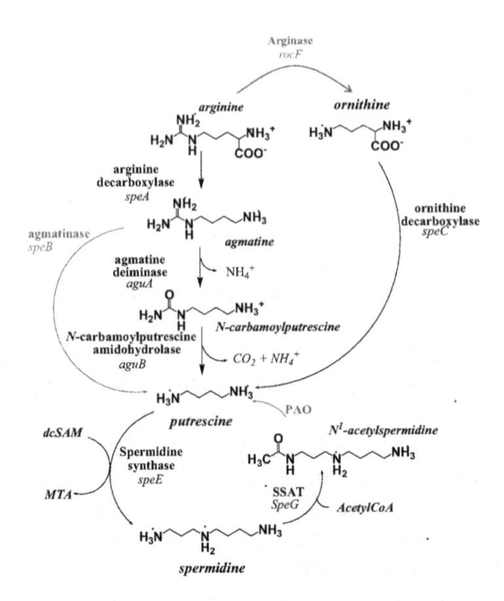

Fig. 1. A general polyamine biosynthetic pathway. With the exception of agmatinase, argininase and polyamine oxidase (PAO), all other indicated proteins have been identified by sequence comparison and some functionally characterized.

>spermidine (Kashiwagi et al. 1986). The best characterized polyamine uptake systems are the *E. coli* ATP binding cassette (ABC) transporters, *potABCD* a spermidine-preferential system and *potFGHI* a putrescine-specific system (Igarashi et al. 2001). The Pot transporters consist of four proteins, PotA an ATPase providing energy for transport, PotB with PotC form a channel though the bilayer and PotD is a periplasmic binding protein (PBP) with a polyamine recognition domain. This PotABCD system is considered spermidine preferential but it is also able to transport putrescine. The K_m for spermidine is 0.1 μM and 1.5 μM for putrescine (Igarashi and Kashiwagi 1999). The *potFGHI* system works the same way, with PotG as the ATPase, PotH and PotI as the channel forming proteins and PotF as the PBP. PotF is specific for putrescine with a K_m of 0.5 μM. The structures of the soluble periplasmic polyamine binding proteins in *E. coli* (PotD and PotF) have been determined and have a C_α backbone r.m.s.d. of 1.5 Å, with a sequence identity of 35% (Vassylyev et al. 1998). They also share a βαβ topology found in many periplasmic binding domains. The similarity in binding specificity for these two proteins has been narrowed to seven residues (Trp-37, Tyr-40, Ser-85, Glu-185, Trp-244, Asp-278, and Tyr-314) that are conserved in PotD and PotF (Vassylyev et al. 1998). These residues are critical to putrescine binding in PotF. PotD has six additional residues required for spermidine binding, which explains its dual binding capability. PotE is a single transmembrane transporter that can both import and excrete putrescine (Kashiwagi et al. 1997). In *E. coli*, *potE* is found in an operon with *speF*, encoding the inducible ornithine decarboxylase. The operon is induced at acidic pH and shows very low expression at neutral pH. The direction of transport is based on membrane potential and the excretion of putrescine is catalyzed by

ornithine. Thus PotE can act as an antiporter with a 1:1 exchange of putrescine for ornithine (Kashiwagi et al. 1997). Using PSI-BLAST on genomic sequences of *Y. pestis* CO92 and KIM we can identify putative proteins involved in the transport of putrescine, PotF, PotG, PotH and PotI, (*potFGHI*, locus tags *y2851*, *y2849*, *y2848*, *y2847*, respectively) with high sequence identify (>80%). The PotA, PotB, PotC, PotD proteins involved in specific transport of spermidine have a weaker (~30-45%) sequence identify (*potABCD,* locus tags *y1388*, *y1392*, *y1393*, *y1391*, respectively). The putative PotD of *Y. pestis* (*y1391*) has the weakest sequence identity (29%). A close inspection of residues necessary for polyamine binding of the putative PotD of *Y. pestis* reveals only half of the seven residues deemed necessary for polyamine binding. This suggests that *Y. pestis* PotD may either have a different polyamine binding mode or may be binding a different metabolite. PotE was not detected in *Y. pestis*.

9.4 Functions of Polyamines in Bacteria

9.4.1 Interaction of Polyamines and Nucleic Acids

Polyamines can be thought of as an organic cation that can act as a counterion to nucleic acids (Wallace et al. 2003). It has been estimated that 90% of spermidine in *E. coli* is bound to RNA, 5.1% to DNA and only 3.8% remains free. With putrescine 9.3%, 48%, and 39% was bound to DNA, RNA or is free, respectively (Miyamoto et al. 1993). Some of the longest and more complex polyamines have been found in thermophilic archaea and eubacteria. These polyamines can alter the melting temperature (T_m) of double stranded DNA (dsDNA), single stranded DNA (ssDNA) and RNA. The change in the T_m profile caused by polyamines showed no discernable mechanism but was thought to occur through a charge-charge or charge-dipole interaction with the DNA backbone (Terui et al. 2005).

Polyamines have also been shown to play a role in DNA condensation by stabilizing bends in DNA. This ability to bind and stabilize bent DNA is important since many DNA binding proteins, including transcription factors, recognize or induce bends in DNA upon binding (Pastre et al. 2005). One detailed analysis of polyamine interaction with DNA showed that polyamines bind specifically to bent adenine tracts in dsDNA from the *E. coli tyrT* promoter (Lindemose et al. 2005). Spermidine is able to reduce cleavage at bent adenine tracts by 50% at a concentration of 250 μM while putrescine needed to be at >1000 μM (Lindemose et al. 2005). Polyamines can also stabilize B to A or B to Z transitions in DNA (Balasundaram and Tyagi 1991; Pastre et al. 2005). The ability to stabilize secondary structure combined with evidence for sequence-specific binding gives polyamines the potential to play a role in transcription and translation of genes.

The role of polyamines as regulators of translation has lead to the description of a polyamine modulon. Using DNA array experiments comparing an *E. coli* strain with mutations in *speC* and *speB* grown with or without putrescine, Yoshida et al. (2004) were able to identify 309 mRNAs that were upregulated and 319 that were downregulated, out of 2,742 mRNAs. The addition of putrescine to the growth media of polyamine deletion mutants was shown to enhance the translation of key regulatory genes of which four were identified, *cya*, *rpoS*, *fecI*, and *fis*. All of these gene products function to regulate transcription. This set of four genes in turn is thought to upregulate 58 genes, many of which may be involved in growth.

9.4.2 Polyamines as Free Radical Scavengers

Part of the polyamine-induced growth defect is thought to occur as a result of the loss of the protective effects of polyamines in the cell resulting from oxidative stress-induced damage (Jung et al. 2003). There are least two different mechanisms that protect cells. First in *E. coli*, polyamines upregulate the *oxyR* and *katG* genes; both genes are involved in the cellular response to oxidative stress (Tkachenko et al. 2001). Second, polyamines may interact directly with free radical species and inactivate them. *In vitro* experiments with spermine have shown that it can react with a hydroxyl radical to form a dialdehyde; this product has a significantly reduced affinity for DNA. The result is that the DNA-bound polyamines exert a protective effect and when modified by reactive oxygen species, will be released from DNA and taken into a cellular detoxification system (Ha et al. 1998). In *E. coli*, spermidine, putrescine and cadaverine protected cells from oxidative stress (Chattopadhyay et al. 2003). Therefore, it is likely that genetic and chemical mechanisms work in concert to protect against oxidative stress.

9.4.3 Polyamines and Acid Resistance

Bacteria have developed strategies to survive in many environments. Bacterial survival at low pH is a requisite to survive gastric digestion and for enteropathogenic bacteria it is a virulence factor (Lin et al. 1996; Merrell and Camill 2000). The acid-inducible arginine decarboxylase (*adiA*), ornithine decarboxylase (*speF*) and lysine decarboxylase (*cadA*) are all polyamine biosynthetic enzymes that are part of the low pH response in bacteria (Foster 2004).

Three distinct systems have been identified for acid resistance in *E. coli* (Richard and Foster 2004). One of these, known as the ARG system, utilizes genes associated with the polyamine pathway and is dependent on arginine in the media and *adiA* expression. In *E. coli*, *adiA* is part of an operon with *adiC*, encoding an arginine-agmatine antiporter, and *adiY*, encoding a regulator that activates expression of *adiA* (Iyer et al. 2003). At pH 2.5 this system is able to raise the internal cellular pH by as much as two units (Foster 2004). Another polyamine-producing enzyme, lysine decarboxylase functions as a secondary, less effective system in acid response (Foster 2004). The *cadBA* operon encodes a lysine decarboxylase and a lysine-cadaverine antiporter. Recently it was demonstrated that the OmpC porin in *E. coli* could be closed by the addition of cadaverine (Samartzidou et al. 2003). OmpC is an outer membrane channel that allows free diffusion of small hydrophilic molecules (<600 Da) across the outer membrane (Nikaido and Vaara 1985). The cadaverine-induced inhibition enhanced survival in low pH media when compared to a cadaverine-insensitive porin (Samartzidou et al. 2003). This indicates that secretion of cadaverine and possibly other polyamines, may contribute to additional acid tolerance mechanisms.

9.4.4 Polyamines and the Cell Wall

Cationically charged polyamines have been shown to interact with cell envelopes. This is thought to occur by several mechanisms including a simple ionic interaction with anionic polysaccharides and proteins found on the cell surface (Field et al. 1989; Koski and Vaara 1991). In one set of experiments approximately 8% and 13% of the total polyamine content of *Salmonella typhimurium* and *E. coli*, respectively, was found to be noncovalently bound to the outer membrane of these bacteria. Polyamines have also been shown to stabilize bacterial spheroplasts and protoplasts from osmotic shock (Koski and Vaara 1991) and improve the survival rate of freeze-thawed *E. coli* cells (Souzu 1986).

Covalent interactions between polyamines and peptidoglycan have also been found. The evidence thus far has come almost exclusively in the form of anaerobic Gram-negative bacteria that have polyamines covalently linked to the peptidoglycan layer (Takatsuka and Kamio 2004). Murein lipoprotein that normally links the peptidoglycan layer to the outer membrane in Gram negative bacteria is not present in some bacteria. *Veillonella alcalescens*, *Veillonella parvula*, *Selenomonas ruminantiurn*, *Anaerovibrio lipolytica* and other members of the *Sporomusa* sub-branch of the *Firmicutes*, have all been shown to have specific polyamines covalently linked to the α-carboxyl group of D-glutamic acid residues found in peptidoglycan (Hamana et al. 2002; Hirao et al. 2000; Kamio and Nakamura 1987). In *S. ruminantiurn*, cadaverine is covalently linked to the peptidoglycan layer. More importantly, DL-α-Difluoromethyllysine (DFML), an inhibitor of lysine decarboxylase, has been shown to be bacteriocidal in *S. ruminantiurn* causing cell lysis. The addition of cadaverine prevented DFML activity (Kamio 1987; Kamio et al. 1986).

9.4.5 Polyamines, Iron Scavenging, Virulence and Signaling

The production of iron scavenging organic molecules called siderophores is known to be important for virulence of many pathogenic bacteria. The ability to chelate iron in the low-iron environment of a mammalian host is required for bacterial survival and pathogenesis (Litwin and Calderwood 1993). There is a class of bacterial siderophores that are built on polyamine backbones. *Bordetella pertussis*, the causative agent of whooping cough, produces a siderophore called alcaligin. Mutants that did not have functional ornithine decarboxylase (ODC) could not producc this siderophore. (Brickman and Armstrong 1996). Similarly, *Vibrio cholarae* makes a siderophore called vibriobactin. This siderophore uses norspermidine as the backbone structure (Griffiths et al. 1984).

In addition to biosynthesis, bacteria can acquire polyamines from the environment. In a genome wide screen, PotD was initially found to be important for the virulence of *Streptococcus pneumoniae* (Polissi et al. 1998). A deletion in *S. pneumoniae potD* did not interfere with *in vitro* growth however the mutant had significantly attenuated virulence in a murine animal model of both septicemia and pneumonia (Ware et al. 2006).

The data suggest that inhibitors of polyamine transporters may be an effective therapeutic approach against *S. pneumoniae*. Since there is a large body of evidence showing a link between polyamines and growth it is natural to associate levels of polyamines to growth status of a bacterial community. One example that suggests the concept of polyamines as a signaling molecule was demonstrated recently by Sturgill and Rather (2004). Using deletion mutants in polyamine biosynthetic genes, they demonstrated that polyamine deficiency disrupts the differentiation of *P. mirabilis* from a vegetative to a swarmer state. The swarmer-defective phenotype can be restored by exogenous putrescine. In addition, a mutation in a putative amino acid decarboxylase can restore swarming to a *P. mirabilis* putrescine-deficient mutant (Stevenson and Rather 2006).

9.4.6 A Novel Role for Polyamines in Biofilm Formation

Recently the first evidence for a link between polyamines and biofilm levels in *Y. pestis* has been published (Patel et al. 2006). Polyamine-deficient mutants of *Y. pestis* were generated with a single deletion in arginine (Δ*speA*) or ornithine (Δ*speC*) decarboxylase and a double deletion mutant (Δ*speA* Δ*speC*). The level of the polyamine putrescine compared to the parental strain $speA^+$ $speC^+$ (KIM6) was depleted progressively, with the highest levels found in *Y. pestis* Δ*speC* mutant (55% reduction), followed by the Δ*speA* mutant (95% reduction) and the Δ*speA* Δ*speC* mutant (>99% reduction). Spermidine on the other hand remained constant in the single mutants but was undetected in the double mutant. Biofilm levels were assayed by three independent measures: Congo red binding, crystal violet staining and confocal laser scanning microscopy. The levels of biofilm correlated to the level of putrescine as measured by HPLC-MS. The biofilm defect of the double deletion mutant was restored by genetic complementation with *speA* alone. The defect can also be restored with the addition of exogenous putrescine, in a dose dependent manner. The effects are distinct from growth since we observe little to no growth defects. The data show that there is a wide gap in putrescine concentration necessary for growth compared to that necessary for biofilm formation.

The effects of polyamines in biofilm have also recently been demonstrated in *Vibrio cholerae* (Karatan et al. 2005). In *V. cholerae*, deletion of a putative PotD homologue, *nspS*, disrupts biofilm formation. It is proposed that NspS is a polyamine sensor and that polyamines may be an environmental signal for biofilm through c-di-GMP.

9.5 Conclusions

The number of functions of polyamines distinct from growth has been increasing as we investigate polyamine deficiency in different bacteria; thus there may be more functions yet to be discovered. Studying polyamine deficiency in *Y. pestis* has led to the generation of a mutant that is dependent on exogenous putrescine for biofilm formation with little to no growth defects. This suggests that these *Y. pestis* mutants may offer a unique system to study the specific mechanism through which polyamines regulate biofilm in *Y. pestis*. In other bacterial systems this approach is not feasible because it has not possible to separate effects on biofilm from growth effects. Therefore, this system provides a tool that may allow for the selective screening of exogenous compounds that can specifically block biofilm formation. Because of the supporting evidence found with *V. cholerae*, compounds that target the polyamine pathway could have the potential for general use against other bacterial pathogens.

9.6 Acknowledgments

This work was supported by NIH Southeast Regional Center for Excellence in Biodefense (SERCEB) developmental grant UF4 AI057175 and UIW-Feik School of Pharmacy to M.A.O. BW was supported by UK-College of Pharmacy.

9.7 References

Balasundaram, D. and Tyagi, A.K. (1991) Polyamine--DNA nexus: structural ramifications and biological implications. Mol. Cell Biochem. 100, 129-140.

Brickman, T.J. and Armstrong, S.K. (1996) The ornithine decarboxylase gene *odc* is required for alcaligin siderophore biosynthesis in *Bordetella* spp.: putrescine is a precursor of alcaligin. J. Bacteriol. 178, 54-60.

Chattopadhyay, M.K., Tabor, C.W. and Tabor, H. (2003) Polyamines protect *Escherichia coli* cells from the toxic effect of oxygen. PNAS USA 100, 2261-2265.

Field, A.M., Rowatt, E. and Williams, R.J. (1989) The interaction of cations with lipopolysaccharide from *Escherichia coli* C as shown by measurement of binding constants and aggregation reactions. Biochem. J. 263, 695-702.

Foster, J.W. (2004) *Escherichia coli* acid resistance: tales of an amateur acidophile. Nat. Rev. Microbiol. 2, 898-907.

Griffiths, G.L., Sigel, S.P., Payne, S.M. and Neilands, J.B. (1984) Vibriobactin, a siderophore from *Vibrio cholerae*. J. Biol. Chem. 259, 383-385.

Ha, H.C., Sirisoma, N.S., Kuppusamy, P., Zweier, J.L., Woster, P.M. and Casero, R.A., Jr. (1998) The natural polyamine spermine functions directly as a free radical scavenger. PNAS USA 95, 11140-11145.

Hackert, M.L., Carroll, D.W., Davidson, L., Kim, S.O., Momany, C., Vaaler, G.L. and Zhang, L. (1994) Sequence of ornithine decarboxylase from *Lactobacillus* sp. strain 30a. J. Bacteriol. 176, 7391-7394.

Hamana, K., Saito, T., Okada, M., Sakamoto, A. and Hosoya, R. (2002) Covalently linked polyamines in the cell wall peptidoglycan of *Selenomonas*, *Anaeromusa*, *Dendrosporobacter*, *Acidaminococcus* and *Anaerovibrio* belonging to the *Sporomusa* subbranch. J. Gen. Appl. Microbiol. 48, 177-180.

Hirao, T., Sato, M., Shirahata, A. and Kamio, Y. (2000) Covalent linkage of polyamines to peptidoglycan in *Anaerovibrio lipolytica*. J. Bacteriol. 182, 1154-1157.

Igarashi, K., Ito, K. and Kashiwagi, K. (2001) Polyamine uptake systems in *Escherichia coli.* Res. Microbiol. 152, 271-278.
Igarashi, K. and Kashiwagi, K. (1999) Polyamine transport in bacteria and yeast. Biochem. J. 344, 633-642.
Iyer, R., Williams, C. and Miller, C. (2003) Arginine-agmatine antiporter in extreme acid resistance in *Escherichia coli.* J. Bacteriol. 185, 6556-6561.
Jung, I.L., Oh, T.J. and Kim, I.G. (2003) Abnormal growth of polyamine-deficient *Escherichia coli* mutant is partially caused by oxidative stress-induced damage. Arch. Biochem. Biophys. 418, 125-132.
Kamio, Y. (1987) Structural specificity of diamines covalently linked to peptidoglycan for cell growth of *Veillonella alcalescens* and *Selenomonas ruminantium.* J. Bacteriol. 169, 4837-4840.
Kamio, Y. and Nakamura, K. (1987) Putrescine and cadaverine are constituents of peptidoglycan in *Veillonella alcalescens* and *Veillonella parvula.* J. Bacteriol. 169, 2881-2884.
Kamio, Y., Pösö, H., Terawaki, Y. and Paulin, L. (1986) Cadaverine covalently linked to a peptidoglycan is an essential constituent of the peptidoglycan necessary for the normal growth in *Selenomonas ruminantium.* J. Biol. Chem. 261, 6585-6589.
Karatan, E., Duncan, T.R. and Watnick, P.I. (2005) NspS, a Predicted Polyamine Sensor, Mediates Activation of *Vibrio cholerae* Biofilm Formation by Norspermidine. J. Bacteriol. 187, 7434-7443.
Kashiwagi, K., Kobayashi, H. and Igarashi, K. (1986) Apparently unidirectional polyamine transport by proton motive force in polyamine-deficient *Escherichia coli.* J. Bacteriol. 165, 972-977.
Kashiwagi, K., Shibuya, S., Tomitori, H., Kuraishi, A. and Igarashi, K. (1997) Excretion and uptake of putrescine by the PotE protein in *Escherichia coli.* J. Biol. Chem. 272, 6318-6323.
Koski, P. and Vaara, M. (1991) Polyamines as constituents of the outer membranes of *Escherichia coli* and *Salmonella typhimurium.* J. Bacteriol. 173, 3695-3699.
Lin, J., Smith, M.P., Chapin, K.C., Baik, H.S., Bennett, G.N. and Foster, J.W. (1996) Mechanisms of acid resistance in enterohemorrhagic *Escherichia coli.* Appl. Environ. Microbiol. 62, 3094-3100.
Lindemose, S., Nielsen, P.E. and Mollegaard, N.E. (2005) Polyamines preferentially interact with bent adenine tracts in double-stranded DNA. Nucleic Acids Res. 33, 1790-1803.
Litwin, C.M. and Calderwood, S.B. (1993) Role of iron in regulation of virulence genes. Clin. Microbiol. Rev. 6, 137-149.
Merrell, D.S. and Camilli, A. (2000) Regulation of *Vibrio cholerae* genes required for acid tolerance by a member of the "ToxR-like" family of transcriptional regulators. J. Bacteriol. 182, 5342-5350.
Miyamoto, S., Kashiwagi, K., Ito, K., Watanabe, S. and Igarashi, K. (1993) Estimation of polyamine distribution and polyamine stimulation of protein synthesis in *Escherichia coli.* Arch. Biochem. Biophys. 300, 63-68.
Mushegian, A.R. and Koonin, E.V. (1996) A minimal gene set for cellular life derived by comparison of complete bacterial genomes. PNAS 93, 10268-10273.
Nikaido, H. and Vaara, M. (1985) Molecular basis of bacterial outer membrane permeability. Microbiol. Rev. 49, 1-32.
Pastre, D., Pietrement, O., Landousy, F., Hamon, L., Sorel, I., David, M.O., Delain, E., Zozime, A. and Le Cam, E. (2006) A new approach to DNA bending by polyamines and its implication in DNA condensation. Eur. Biophys. J. 35, 214-223.

Patel, C.N., Wortham, B.W., Lines, J.L., Fetherston, J.D., Perry, R.D. and Oliveira, M.A. (2006) Polyamines are essential for the formation of plague biofilm. J. Bacteriol. 188, 2355-2363.

Polissi, A., Pontiggia, A., Feger, G., Altieri, M., Mottl, H., Ferrari, L. and Simon, D. (1998) Large-scale identification of virulence genes from *Streptococcus pneumoniae*. Infect. Immun. 66, 5620-5629.

Richard, H. and Foster, J.W. (2004) *Escherichia coli* glutamate- and arginine-dependent acid resistance systems increase internal pH and reverse transmembrane potential. J. Bacteriol. 186, 6032-6041.

Samartzidou, H., Mehrazin, M., Xu, Z., Benedik, M.J. and Delcour, A.H. (2003) Cadaverine inhibition of porin plays a role in cell survival at acidic pH. J. Bacteriol. 185, 13-19.

Sandmeier, E., Hale, T.I. and Christen, P. (1994) Multiple evolutionary origin of pyridoxal-5'-phosphate-dependent amino acid decarboxylases. Eur. J. Biochem. 221, 997-1002.

Souzu, H. (1986) Fluorescence polarization studies on *Escherichia coli* membrane stability and its relation to the resistance of the cell to freeze-thawing. II. Stabilization of the membranes by polyamines. Biochim. Biophys. Acta 861, 361-367.

Stevenson, L.G. and Rather, P.N. (2006) A novel gene involved in regulating the flagellar gene cascade in *Proteus mirabilis*. J. Bacteriol. 188, 7830-7839.

Sturgill, G. and Rather, P.N. (2004) Evidence that putrescine acts as an extracellular signal required for swarming in *Proteus mirabilis*. Mol. Microbiol. 51, 437-446.

Tabor, C.W. and Tabor, H. (1985) Polyamines in microorganisms. Microbiol. Rev. 49, 81-99.

Takatsuka, Y. and Kamio, Y. (2004) Molecular dissection of the *Selenomonas ruminantium* cell envelope and lysine decarboxylase involved in the biosynthesis of a polyamine covalently linked to the cell wall peptidoglycan layer. Biosci. Biotechnol. Biochem. 68, 1-19.

Terui, Y., Ohnuma, M., Hiraga, K., Kawashima, E. and Oshima, T. (2005) Stabilization of nucleic acids by unusual polyamines produced by an extreme thermophile, *Thermus thermophilus*. Biochem. J. 388, 427-433.

Tkachenko, A., Nesterova, L. and Pshenichnov, M. (2001) The role of the natural polyamine putrescine in defense against oxidative stress in *Escherichia coli*. Arch. Microbiol. 176, 155-157.

Vassylyev, D.G., Tomitori, H., Kashiwagi, K., Morikawa, K. and Igarashi, K. (1998) Crystal structure and mutational analysis of the *Escherichia coli* putrescine receptor. Structural basis for substrate specificity. J. Biol. Chem. 273, 17604-17609.

Wallace, H.M., Fraser, A.V. and Hughes, A. (2003) A perspective of polyamine metabolism. Biochem. J. 376, 1-14.

Ware, D., Jiang, Y., Lin, W. and Swiatlo, E. (2006) Involvement of potD in *Streptococcus pneumoniae* polyamine transport and pathogenesis. Infect. Immun. 74, 352-361.

Yoshida, M., Kashiwagi, K., Shigemasa, A., Taniguchi, S., Yamamoto, K., Makinoshima, H., Ishihama, A. and Igarashi, K. (2004) A unifying model for the role of polyamines in bacterial cell growth, the polyamine modulon. J. Biol. Chem. 279, 46008-46013.

10
Intermediary Metabolism, Na^+, the Low Calcium-Response, and Acute Disease

Robert R. Brubaker

Department of Microbiology and Molecular Genetics, Michigan State University, brubake3@msu.edu

Abstract. The variables carriage of pCD, CO_2 tension, exogenous ATP, L-glutamate, Mg^{2+}, Na^+, pH, source of energy, and temperature are known to modulate the low calcium response of *Yersinia pestis in vitro*. The role of these effectors and the basis of their interactions are defined here with emphasis on known *Y. pestis*-specific missense mutations in glucose 6-phosphate dehydrogenase and aspartase, which preclude use of the hexose monophosphate pathway and prevent efficient catabolism of L-glutamic acid, respectively. A physiological Ca^{2+}-deficient rescue scenario is provided that permits essentially full-scale growth of virulent *Y. pestis* (<0.1 mM Na^+ and 25 mM L-glutamate at pH 6.5) with expression of pCD-encoded virulence effectors and their attendant type III secretion system. Multiplication in this environment indicates that Ca^{2+} prevents innate toxicity of Na^+. However, Na^+ actually promotes growth in Ca^{2+}-deficient medium at pH 9.0 due to the evident action of Na^+-translocating NADH-ubiquinone oxidoreductase. Another Ca^{2+}-deficient rescue scenario (100 mM Na^+ and 25 mM L-glutamate at pH 5.5) permitted growth while downregulating pCD-encoded functions. A consequence of the abrupt Na^+-mediated bacteriostasis typical of aspartase-deficient *Y. pestis* is conversion of L-glutamate to L-aspartate with release of the latter into culture supernatant fluids. Occurrence of this event *in vivo* would radically alter the equilibrium of host amino acid pools thereby contributing to enhanced lethality.

10.1 Introduction

Yersinia pseudotuberculosis typically causes chronic enteric disease and thus conforms to the epidemiological tenet of Theobald Smith, which states that an efficient parasite never kills its host outright because this act also eliminates the environment essential for survival of that parasite (Smith 1934). This restriction obviously does not apply to *Yersinia pestis*, the causative agent of bubonic plague, because this parasite must kill its host in order to assure subsequent dissemination via the flea vector. *Y. pestis* emerged from *Y. pseudotuberculosis* within the last 20,000 years (Achtman et al. 1999) and thus closely resembles its progenitor. This similarity suggests that a comparison of the two genomes would readily permit identification of factors responsible for acute disease. Obvious distinctions are the presence of two unique plasmids in *Y. pestis* (Ferber and Brubaker 1981) that were undoubtedly acquired recently by lateral transfer. Of these, ~10-kb pPCP encodes a plasminogen activator required for tissue invasion (Lähteenmäki et al. 1998) and ~100-kb pMT provides murine toxin (phospholipase D) necessary for survival in the flea (Hinnebusch et al. 2002) as well as capsular fraction 1, not required for lethality in mice (Burrows 1957) or man (Winter et al. 1960). However, transformation of pPCP into *Y. pseudotuberculosis* failed to enhance virulence in mice (Kutyrev et al. 1999) suggesting that chromosomally encoded functions are required for acute disease.

Comparison of the annotated chromosome of *Y. pestis* (Parkhill et al. 2001; Deng et al. 2002) with that of *Y. pseudotuberculosis* (Chain et al. 2004) revealed only 32 unique genes (primarily unknown ORFs) in plague bacilli. In contrast, the *Y. pestis* chromosome possessed inversions, transpositions, small deletions and numerous IS insertions not present in *Y. pseudotuberculosis* (Chain et al. 2004). These mutational events accounted for removal in *Y. pestis* of about 13% of all functional genes present in *Y. pseudotuberculosis*, and raised the possibility that acute disease reflects hereditary loss rather than gain. In addition, the study identified missense mutations in *Y. pestis* that account for radical differences in the metabolic flow of carbon and response to simulated mammalian intra- and extracellular environments. The purpose of this report is to explore the possibility that these missense mutations are associated with lethality.

10.1.1 The Low-Calcium Response

The modern era of research on plague commenced with the discovery that that level of Ca^{2+} contained in mammalian plasma (2.5 mM) is also necessary for *in vitro* growth of virulent *Yersinia pestis* at host but not room temperature (Kupferberg and Higuchi 1958; Higuchi et al. 1959). Knowledge of this unusual nutritional requirement facilitated discovery of a plasmid in *Yersinia enterocolitica* (Zink et al. 1980) that is required for tissue invasiveness. This ~70-kb plasmid is shared by all yersiniae pathogenic to man and is termed pYV in enteropathogenic *Y. enterocolitica* and *Yersinia pseudotuberculosis* and pCD in *Y. pestis*. pCD mediates an especially stringent (Brubaker 2005) "low calcium response" (LCR) defined as the ability to either grow in the presence of Ca^{2+} without producing certain pCD-encoded virulence effectors (Yops and LcrV) or to cease multiplication in Ca^{2+}-deficient medium while expressing these determinants (Brubaker and Surgalla 1964; Goguen et al. 1984). Loss of pCD results in avirulence and occurs spontaneously at the rate of 10^{-4} as judged by the ability of cured mutants to form colonies at 37°C on oxalated solid medium (Higuchi and Smith 1961) or to divide indefinitely in liquid Ca^{2+}-deficient media (Higuchi et al. 1959; Brubaker and Surgalla 1964).

The diverse roles of Yops and LcrV as virulence factors and their regulation have received intensive study (Brubaker 2003; Heesemann et al. 2006). Considerably less is known about the physiological basis of Ca^{2+}-dependency although starved cells exhibit a reduced adenylate energy charge suggesting a lesion in bioenergetics (Zahorchak et al. 1979). Alkaline pH and consequent increase in CO_2 tension also postponed the emergence of mutants lacking pCD (Delwiche et al. 1959) and the primary product of CO_2 fixation in *Y. pestis* was shown to be oxalacetate (Baugh et al. 1964). Results of parallel studies demonstrated that cells harboring pCD underwent lysis when grown at host temperature with D-glucose but not pen tose (Brownlow and Wessman 1960) and that both Mg^{2+} (Lawton et al. 1963;

Brubaker and Surgalla 1964) and Na^{+} (Brubaker 1967) favored a stringent LCR. However, studies of the dynamics of growth *in vitro* at 37°C were largely abandoned following the observation that Ca^{2+} alone was sufficient to prevent shift of bacterial populations to avirulence.

Two factors contributed to this change in emphasis. First, at least nine variables other than Ca^{2+} had been shown to modulate the LCR (carriage of pCD, CO_2 tension, exogenous ATP, L-glutamate, Mg^{2+}, Na^{+}, pH, source of energy, and temperature) thus complicating further attempts to isolate and characterize single effectors. Second, the development of recombinant technology and introduction of efficient methods of gene exchange offered the immediate promise of resolving the LCR at the molecular level.

10.1.2 Metabolic Lesions

The nutrition, bioenergetics, and flow of carbon in *Y. pestis* received considerable attention during the same era that defined variables of the LCR. These metabolic studies established that plague bacilli possessed a functional glycolytic (Embden-Meyerhof) pathway and tricarboxylic acid (Krebs) cycle (Brubaker 1972) but that the organisms lacked detectable glucose 6-phosphate dehydrogenase (Zwischenferment or Zwf), the initial enzyme of the hexose monophosphate (pentose phosphate) pathway (Mortlock 1962). In contrast, this pathway was functional in *Y. pseudotuberculosis* (Brubaker 1968) and a comparison of the two species demonstrated that significant Zwf is typically detectable in *Y. pseudotuberculosis* but not in *Y. pestis* (Mortlock and Brubaker 1962). This observation has now been extended to over 50 isolates of *Y. pestis* representing all Devignat biovars (Devignat 1951) from diverse geographic origin (R. Brubaker, unpublished observations).

A separate study compared the catabolic destruction of readily oxidizable L-amino acids by *Y. pestis* and *Y. pseudotuberculosis* (Dreyfus and Brubaker 1978). Results demonstrated rapid destruction of L-aspartate by *Y. pseudotuberculosis* whereas this amino acid was not catabolized by *Y. pestis* and actually accumulated in suspensions provided with L-glutamate, L-proline, or L-asparagine. This phenolmenon was favored by high pH and attributed to lack of detectable L-aspartate ammonia-lyase or aspartase (AspA) in all of 10 tested epidemic strains of *Y. pestis* although this nearly ubiquitous procaryote enzyme was present in all of 10 tested strains of *Y. pseudotuberculosis* (Dreyfus and Brubaker 1978). This deficiency prevents plague bacilli from efficiently catabolizing L-glutamate by transamination with oxalacetate as is accomplished by other Enterobacteriaceae, which cycle the resulting accumulated L-aspartate to fumarate via AspA (Halpern and Umbarger 1960). Over 50 diverse isolates of epidemic *Y. pestis* are now known to lack detectable AspA (R. Brubaker, unpublished observations) and, as shown below, this lesion profoundly influences the LCR.

10.1.3 Missense Mutations

As already noted, approximately 13% of the *Y. pseudotuberculosis* chromosome is no longer functional in *Y. pestis* as judged by disruption or outright loss of structural genes (Chain et al. 2004). These changes accounted for almost all of the defects previously noted in *Y. pestis* but not *Y. pseudotuberculosis* including lack of complete lipopolysaccharide and flagellar structure, inability to synthesize L-methionine, L-phenylalanine, and L-cysteine, and failure to ferment melibiose and rhamnose or hydrolyze urea (Brubaker 1991). Comparative annotation did not, however, provide an explanation for the absence of detectable Zwf or AspA in *Y. pestis* because these genes are of comparable size in both species (Parkhill et al. 2001; Deng et al. 2002; Chain et al. 2004). Furthermore, both defective enzymes of *Y. pestis* undergo evident normal transcription as judged by microarray analysis (Motin et al. 2004). Inspection of *zwf*, however, reveals a single base transition in the *Y. pestis* codon for amino acid position 158 causing translation of proline (encoded by CCC) rather than the serine residue (encoded by UCC) present in *Y. pseudotuberculosis*. It is well-established that proline causes bends in protein structure and disrupts α-helix thus misplacement of this amino acid is often sufficient to prevent enzymatic activity.

In the case of *aspA*, a single base transversion in the *Y. pestis* codon for amino acid 363 causes translation of leucine (encoded by UUG) rather than the valine residue (encoded by GUG) present in the active enzyme of *Y. pseudotuberculosis* (Parkhill et al. 2001; Deng et al. 2002; Chain et al. 2004). This substitution of one aliphatic amino acid for another at a putative silent region of the enzyme (Viola 1998) is the only change present in the two *aspA* genes (and attendant non-coding regions) thus its ability to cause loss of activity was initially viewed with skepticism. Further study, however, revealed production in *Y. pestis* of inactive cross-reacting immune protein to AspA (R. Brubaker, unpublished observations) and sole replacement of valine for leucine at amino acid position 363 in the cloned *aspA* of *Y. pestis* restored full enzymatic activity (R. Viola, personal communication).

10.2 Methods

Bacteria were preserved at –20°C in buffered glycerol (Beesley et al. 1967), directly inoculated onto the surface of slopes consisting of blood agar base (Difco, BD Diagnostics, Sparks, Md.), and incubated at 26°C for 24 h (*Y. pseudotuberculosis*) or 48 h (*Y. pestis*). Plague bacilli are often slow in adapting to new environments *in vitro* thus at least two transfers were made at 26°C in the same chemically-defined medium before inoculating and shifting a third transfer of the identical medium to 37°C. Cells of this third transfer received appropriate additions or adjustment of pH and were used to monitor growth, release of L-aspartate, and a source of crude extracts for enzymatic analysis.

10.2.1 Chemically-Defined Medium

The basic chemically defined medium used to identify the variables of Na^+, L-glutamic acid, and pH was a modification (Fowler and Brubaker 1994; Brubaker 2005) of that initially used to characterize the nutritional requirement for Ca^{2+} (Higuchi et al. 1959). First and second transfers incubated at 26°C contained *N*-(2-

hydroxyethyl)piperazine-*N'*-2-ethenesulfonic acid (HEPES) at a final concentration of 25 mM and pH of 7.0. HEPES was replaced in third transfers, typically incubated at 37°C, by another Good buffer (Good et al. 1965) depending on the desired initial pH. The latter, used at equimolar concentrations, were morpholinethanesulfonic acid (MES) at a final pH of 5.0, 5.5, or 6.0, morpholinopropanesulfonic acid (MOPS) at a final pH of 6.5, 7.0, and 7.5, tris(hydroxymethyl)aminomethane (TRIS) at a final pH of 8.0 and 8.5, and 2-(cyclohexylamino)ethane-2-sulfonic acid (CHES) at a final pH of 9.0. Stock solutions used to prepare the final medium were as follows:

1. Salt solution (x10, sterilized by filtration and stored at 4°C): Citric acid (0.1 M), MgO (200 mM), a Good buffer (250 mM), $FeSO_4$ (1.0 mM), and $MnSO_4$ (0.1 mM).
2. Amino acid solution (x2, sterilized by addition of a few drops of chloroform and stored at room temperature): DL-alanine (9.0 mM), L-isoleucine (7.6 mM), L-leucine (4.0 mM), L-methionine (3.2 mM), L-phenylalanine (4.8 mM), L-threonine (2.8 mM), L-valine (13.6 mM), L-arginine · HCl (2.0 mM), L-proline (14.0 mM), L-lysine · HCl (2.2 mM), glycine (5.2 mM), and L-tyrosine (2.2 mM).
3. Potassium phosphate solution (x100, and stored at 4°C): K_2HPO_4 (0.25 mM).
4. Vitamin solution (x10, sterilized by filtration and stored at 4°C): thiamin (0.03 mM), pantothenic acid (0.04 mM), and biotin (0.02 mM).
5. Energy source (x10, prepared by careful neutralization of D-gluconic acid with 10 N KOH, sterilized by filtration, and stored at 4°C): potassium D-gluconate (1.0 M).
6. Tryptophan solution (x100, prepared by addition of a few drops of 0.1 N KOH, sterilized by filtration, and stored at 4°C): L-tryptophan (10 mM).
7. Cysteine solution (x100, prepared immediately before use and sterilized by filtration): L-cysteine·HCl (100 mM).
8. pH indicator (x100, prepared in 95% ethanol and stored at room temperature): phenol red (1% w/v).

To prepare 1 liter of the basic medium, 500 ml of amino acid solution, 1 ml of pH indicator, and sufficient distilled water to achieve final volume were sterilized by autoclaving. After cooling, the medium aseptically received 100 ml of salts solution, 10 ml of potassium phosphate solution, 100 ml of vitamin solution, 40 ml of the energy source, 10 ml of tryptophan solution, and 10 ml of cysteine solution. When desired, calcium was added after autoclaving as a 1:100 dilution of a 0.4 M stock solution of $CaCl_2$ to yield a final concentration of 4.0 mM; sodium and glutamate were added in solid form as NaCl (100 mM) or L-glutamic acid (25.0 mM) before autoclaving. Final pH was adjusted within the constraints of the incorporated Good buffer by dropwise addition of 0.1 N KOH or HCl; if necessary, total volume was restored before use in experiments by addition of sterile distilled water.

Constituted media were placed aseptically into stoppered Erlenmeyer flasks (10% v/v), inoculated with cells removed from slopes of blood agar base at an optical density of 0.1 (620 nm), and aerated at 200 rpm in a model G76 water bath shaker (New Brunswick Scientific Co., Inc., New Brunswick, N. J.) set at 26°C. Upon entering the late logarithmic growth phase, the bacteria were transferred to a second culture of the same medium, which was similarly incubated. The latter was used to inoculate the third transfer at an optical density of 0.125 to 0.2; this culture was typically incubated at 37°C and used to determine kinetics of growth and prepare cells for enzymatic assays or analysis of accumulated L-aspartate.

10.2.2 Enzyme Assays

Zwf was determined by monitoring glucose 6-phosphate-dependent generation of NADPH by a minor modification of the assay used initially (Mortlock and Brubaker 1962). The reaction mixture consisted of 100 µmoles of Tris·Cl (pH 8.0), 2.0 µmoles of $MgCl_2$, 0.3 µmoles of $NADP^+$, and 100 µmoles of glucose 6-phosphate (Sigma) in a volume of 2.9 ml contained within a quartz cuvette. The reaction was started by addition of 0.1 ml of dialyzed cell-free extract and increase of absorbance at 340 nm was monitored with a Beckman DU spectrophotometer (Beckman Coulter, Fullerton, Calif.).

AspA was estimated by determining the L-aspartic acid-dependent release of NH_4^+ with Nessler's reagent as undertaken previously (Dreyfus and Brubaker 1978). The reaction mixture consisted of 250 µmoles of Tris·Cl (pH 7.0), 5.0 µmoles $MgCl_2$, and dialyzed cell-free extract in a total volume of 4.5 ml. The assay was started by addition of 250 µmoles of sodium L-aspartate in a volume of 0.5 ml and samples of 0.5 ml were removed at intervals and added to Eppendorf tubes containing 0.1 ml of 1.5 M trichloroacetic acid. The tubes were then centrifuged at highest speed for 1 min in a model II microfuge (Beckman Coulter) and then 0.5 ml of clear supernatant fluid was carefully removed and added to a tube containing 8.5 ml of distilled water. These samples received 1.0 ml of Nessler's reagent and, after incubation for 10 min, were assayed for absorbance at 480 nm as described for determination of asparaginase (Yellin and Wriston 1966). The resulting values were then evaluated against a standard curve prepared immediately before each determination comprised of known concentrations of NH_4Cl in samples of 10 ml containing the same concentrations of trichloroacetic acid and Nessler's reagent that were used to prepare samples for spectrophotometric analysis. In all cases, protein was determined by the method of Lowry et al. (1951).

10.2.3 Accumulation of L-Aspartate

Yersiniae were harvested by centrifugation (10,000 x g for 30 min), washed twice in cold 0.033 M potassium phosphate buffer (pH 7.0), and suspended at a concentration equivalent to 20 mg of protein per ml of distilled water. Portions (1.0 ml) of these suspensions was immediately added to 50 ml Erlenmeyer flasks containing 2.0 µmoles of $MgCl_2$ and 20 µmoles of sodium L-glutamate in 1.0 ml of 0.066 M potassium buffer (pH 7.0). The flask was aerated at 37°C as described above and samples of 0.1 ml were removed at intervals, sedimented for 1 min in a model 11 Microfuge (Beckman Coulter, Inc, Fullerton, Calif.), the supernatant fluid passed through a 4 mm Acrodisc HT Tuffryn low protein binding filter disc (Pall Life Sciences, Ann Arbor, Mich.), and frozen until analysis. After thawing, the samples were appropriately diluted in distilled water, applied (2.0 to 10.0 µl) to Whatman no.1 filter paper, and chromatographed with phenol:water (4:1 w/v) in a HCN atmosphere. The sheets were then dried, sprayed with 0.25% ninhydrin in water-saturated n-butanol (w/v) and, after full development of color in a water-saturated environment at 26°C, scanned and evaluated against a standard curve using the SigmaGel software program (Systat Software, Inc., Point Richmond, Calif.).

10.3 Results

In initial experiments, the nine variables noted above were analyzed to determine which combination was required for abrupt bacteriostasis following shift of yersiniae harboring pCD from 26°C to 37°C. The presence of Na^+ and L-glutamic acid were especially important in assuring prompt shutoff of growth. Provision of a seeming excess of D-gluconate (40 mM), an effective source of energy and metabolic CO_2, favored sustained growth with added Ca^{2+} but did not overcome the nutritional requirement for this cation. An evaluation of pH demonstrated that the onset of bacteriostasis in Ca^{2+}-deficient medium was most severe in neutral to slightly alkaline environments. Considered together, the most abrupt restriction of growth following shift to 37°C in chemically-defined medium lacking added Ca^{2+} occurred at pH 7.5 with 100 mM Na^+ and 25 mM L-glutamic acid. We use the term "worst-case scenario" (WCS) to define this variation (Fowler and Brubaker 1994; Brubaker 2005). As shown in Fig. 1A, WCS permitted full-scale growth of all three yersiniae provided Ca^{2+} (4.0 mM) was added. In contrast, Ca^{2+}-deficient WCS (Fig. 2B) caused immediate shutoff of *Y. pestis* as compared to *Y. pseudotuberculosis* or *Y. enterocolitica*, both of which underwent many residual doublings (Brubaker 2005).

Further manipulation of the basic Ca^{2+}-deficient synthetic medium provided three rescue scenarios (RS) that essentially negated the nutritional requirement of *Y. pestis* for Ca^{2+} at 37°C. These conditions were no added Na^+, 25.0 mM L-glutamate, and an initial pH of 6.5 (RS1), 100 mM Na^+, no added L-glutamate, and an initial pH of 9.0 (RS2), and 100 mM Na^+, 25.0 mM L-glutamate, and an initial pH of 5.5 (RS3). The abilities of *Y. pestis* harboring pCD to multiply within these three distinct Ca^{2+}-deficient environments is shown in Fig. 1C. RS1 closely mimics host cell cytoplasm and was the only physiological environment found to support Ca^{2+}-independent growth. The capability of RS1 to support sustained doubling demonstrates that, at pH 6.5 of cytoplasm, Ca^{2+} functions by eliminating a toxicity caused by Na^+. As shown for RS2, however, this effect is not ubiquitous in that Na^+, but not L-glutamate, was essential for Ca^{2+}-independent growth at pH 9.0. As noted below, this effect likely reflects the presence in *Y. pestis* (Parkhill et al. 2001; Deng et al. 2002) of Na^+-translocating NADH-ubiquinone oxidoreductase. Both RS1 and RS2 promoted full expression of Yops and LcrV comparable to WCS whereas little or no production of these effectors was observed in RS3 (Brubaker 2005). This result was interpreted as evidence suggesting that expression of one or more pCD-encoded function is necessary for the organisms to become vulnerable to the toxic effect of Na^+.

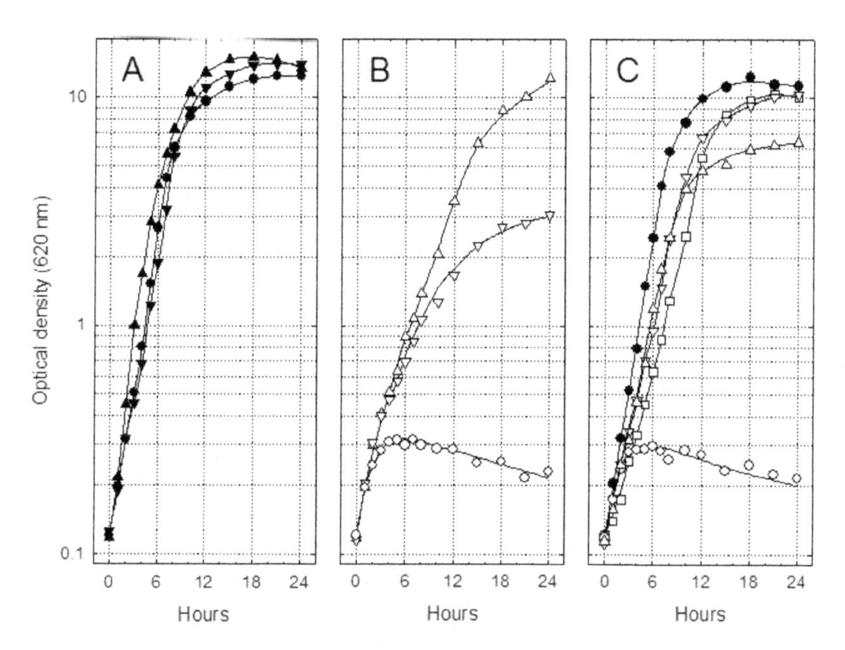

Fig. 1. Growth of (A) *Y. enterocolitica* WA (▲), *Y. pseudotuberculosis* PB1 (▼), and *Y. pestis* KIM (●) in WCS containing added Ca^{2+} (4.0 mM), (B) *Y. enterocolitica* WA (△), *Y. pseudotuberculosis* PB1 (▽), and *Y. pestis* KIM (○) in WCS lacking added Ca^{2+}, and (C) *Y. pestis* KIM in control WCS either with (●) or without added Ca^{2+} (○) or in RS1 (△), RS2 (□), and RS3 (▽).

L-aspartate is not a component of the chemically-defined media used in these experiments yet earlier study had detected accumulation of the amino acid in spent medium from cultures of *Y. pestis* (Dreyfus and Brubaker 1978). To further define this phenomenon, resting cells of *Y. enterocolitica*, *Y. pseudotuberculosis*, and *Y. pestis* were prepared for use in experiments concerned with the catabolic destruction of L-glutamate following cultivation for 7 h in WCS. In no case was released L-aspartate detected in suspensions of the enteropathogenic yersiniae supplied with L-glutamate. However, as shown in Fig. 2, significant L-aspartate arose after addition of L-glutamate to cultures of *Y. pestis*, provided that the organisms harbored pCD and had therefore ceased doubling prior to harvest (Brubaker 2005). Additional study showed that L-homoserine and L-proline also often accumulated with L-aspartate (data not shown) and that generation of L-aspartate was favored at alkaline pH (Dreyfus and Brubaker 1978) in the presence of Na^+ and minimized by added L-gluconate or phosphate ions (data not shown).

10.4 Conclusions

As already noted, discovery by Kupferberg and Higuchi (1958) of the essential role of Ca^{2+} in maintaining virulent populations of *Y. pestis* at 37°C resolved practical concerns of the day and permitted investigators to pursue other avenues of research.

Nevertheless, parallel studies in other laboratories had uncovered additional (often obscure) variables that directly related to the role of Ca^{2+}; this report is an attempt to integrate these findings into our current concept of the LCR. Evaluation of these variables proved worthwhile in that three environmental conditions were developed that permit near full-scale growth in the absence of Ca^{2+}. Since more than one variable was accountable for this accomplishment, we chose to term the permissive conditions rescue "scenarios", implying that the explanation for Ca^{2+}-independent growth is complex. Failure of RS media to achieve final optical densities entirely comparable to cultures containing Ca^{2+} was caused by unavoidable changes in critical variables during growth, especially metabolic increase in pH.

Perhaps the most dramatic result was the discovery that Ca^{2+} serves to inhibit pCD-dependent sensitivity to Na^+ at 37°C. It has been recognized for some time that Ca^{2+} downregulates virulence effectors now known to be encoded by pCD (Brubaker and Surgalla 1964). Accordingly, the possibility exists that one or more of these functions permit entrance of Na^+, an established cytotoxin, into the bacterial cell. Additional evidence supporting this notion was the finding that yersiniae were capable of excellent growth at pH 5.5 in RS3 but that this environment supported little or no expression of pCD-encoded functions. Another unexpected observation was that Na^+ was actually required for growth in RS2 at pH 9.0. This medium, like RS3, is not physiological but may nevertheless provide clues to the nature of Ca^{2+}-dependence. The fact that both Na^+ and highly alkaline pH are necessary for growth implicates Na^+-translocating NADH-ubiquinone oxidoreductase, an enzyme present in yersiniae that serves at alkaline pH as a primary electrogenic sodium pump (Zhou

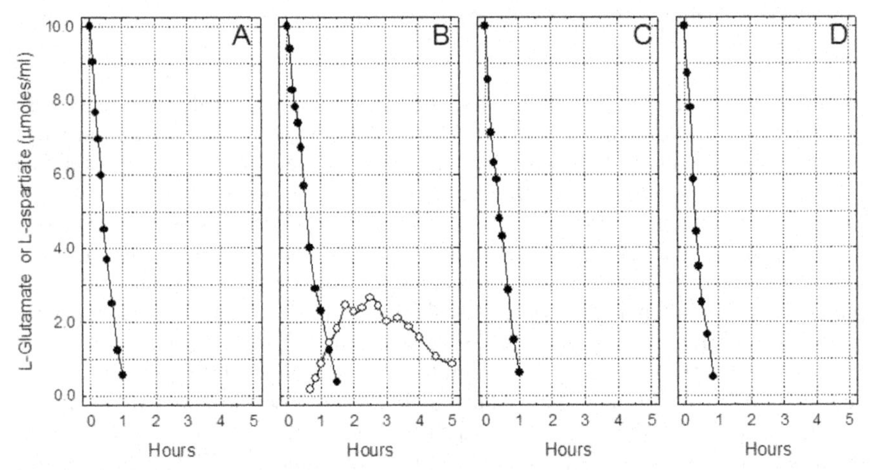

Fig. 2. Catabolic destruction of L-glutamate (●) with appearance of metabolic L-aspartate (○) by resting cells of *Yersinia pestis* strain KIM previously cultivated at 37°C in WCS (100 mM Na^+ and 25 mM L-glutamate); (A) pCD-positive cells previously grown with 4.0 mM Ca^{2+}, (B) pCD-positive cells previously grown without added Ca^{2+}, (C) pCD-negative cells previously grown with 4.0 mM Ca^{2+}, and (D) pCD-negative cells previously grown without added Ca^{2+}.

et al. 1999). The existence of this enzyme in *Y. pestis* assures that Na^+ cannot access cytoplasm at pH 9.0. If this is the only advantage that the oxidoreductase provides, then yersiniae must either lack or dismantle their sodium pump at physiological pH in Ca^{2+}-deficient environments such as cytoplasm.

The finding that only those yersiniae harboring pCD and starved for Ca^{2+} at 37°C release metabolic L-aspartate is of considerable significance. Accumulation of L-aspartate is stoichiometric but transient in the sense that it, in turn, undergoes further destruction. Appearance of detectable L-homoserine with L-aspartate suggests that L-aspartate may enter the anabolic aspartate amino acid family to form L-threonine. The latter is unable to generate L-methionine in *Y. pestis* (Englesberg and Ingraham 1957) but is instead hydrolyzed to glycine and acetaldehyde via threonine aldolase. Glycine is in equilibrium with L-serine and the latter undergoes unregulated degradation to pyruvate in *Y. pestis* (Dreyfus and Brubaker 1978). Further study will be required to determine if this essentially catabolic conversion defines the AspA-negative phenotype or if the release of L-aspartate *per se* is more important. In any event, it is now possible to at least speculate about the roles of the nine variables previously implicated as modulators of the LCR. The presence of pCD is obviously essential because this plasmid both encodes the major virulence factors of the species as well as some unknown function that facilitates sensitivity to Na^+. CO_2 tension likely favors generation of oxalacetate via phosphoenolpyruvate carboxykinase to restore the oxalacetate pool lost by transamination to L-aspartate. Exogenous ATP was initially thought to serve as a direct energy source but more likely functions to chelete free Mg^{2+} known to upregulate the LCR in the absence of Ca^{2+}. L-glutamate promotes upregulation of its major translocator (GltS) thereby favoring its uptake and subsequent transamination with oxalacetate to form L-aspartate. The latter is then removed from the bacterium, also via GltS, causing significant loss of potential carbon and energy. It is established that Na^+ functions as a porter for many other translocases and further work may demonstrate that internalized Na^+ exports other small molecules. The role of pH is critical but not fully understood; at the slightly acidic but physiological value of cytoplasm the bacteria become acutely sensitive to Na^+ if Ca^{2+} is absent. This observation is consistent with the possibility that at least one pCD-encoded activity is required for some critical bioenergetic function in cytoplasm that maintains an adenylate energy charge consistent with cell division. Finally, an effective source of energy is critical to maintain and assure the presence of metabolic CO_2.

Additional work will be required to determine if the missense mutations accountable for the absence of Zwf and AspA are associated with acute disease. Cells of *Y. pestis* (and possibly other yersiniae pathogenic to man) undergo prodigal catabolism at 37°C sparked by marked upregulation of respiratory cytochromes (Motin et al. 2004). These changes, which were monitored in RS1, included constitutive expression of ability to oxidize a variety of substrates not all of which were in the medium (e.g., pentoses, hexoses, and fatty acids). Loss of Zwf results in a quiet phenotype in *Escherichia coli* (Fraenkel 1968) and its removal from *Y. pestis* should not be detrimental to an organism that always has access to pentose. However, Zwf also functions to generate NADPH in procaryotes, thus its absence in *Y. pestis* may account for the "glucose effect" that lyses pCD-positive yersiniae in

Ca^{2+}-deficient medium (Brownlow and Wessman 1960). Loss of AspA may prove to be more important than Zwf in promoting acute disease. This mutation is associated with a stringent LCR characterized by accumulation of L-aspartate, which evidently undergoes subsequent catabolism via conversion to L-threonine (and then to glycine, which is promptly converted to L-serine and then to pyruvate). It is difficult to reconcile this unique pathway with lethality, although either L-threonine or glycine are required amino acids. However, the ability of *Y. pestis* to convert and release metabolic L-aspartate from L-glutamate would effectively disrupt the equilibrium of host metabolic pools. In this context, the concentration of L-aspartate is the lowest of all free host amino acids and that of the L-glutamine-L-glutamate complex is the highest (Soupart 1962).

At present, the investigator must decide if loss of these two important genes has occurred by evolutionary accident (chromosomal degeneration associated with permanent residence of *Y. pestis* within the protected environments of the host or flea vector) or design (selection for ability to cause acute disease). The argument favoring accident is that many parasites have undergone chromosomal degeneration following adaptation to a niche that no longer requires competition with saprophytes. In this case, loss of *zwf* and *aspA* would hold no more significance than loss of any other redundant function such as urease, ability to ferment rhamnose and melibiose, or synthesize L-phenylalanine, L-methionine, or glycine. The only evidence suggesting the alternative is that, despite the universal nature of AspA in procaryotes, the enzyme is not required for virulence in salmonellae (Yimga et al. 2006) and *aspA* is absent in the in the annotated genomes of *Mycobacterium tuberculosis* (Fleischmann et al. 2002), *Francisella tularensis* (Larsson et al. 2005), and rickettsiae (Andersson et al. 1998). The latter, like *Y. pestis*, also converts exogenous L-glutamate to L-aspartate (Bovarnick and Miller 1950). In addition, all of 10 tested enzootic Pestoides variants of *Y. pestis*, known to be avirulent in guinea pigs and man (Anisimov et al. 2004), expressed AspA activity (R. Brubaker, unpublished observations). The similarly attenuated 91001 strain of the enzootic microtus biovar also possesses the active *Y. pseudotuberculosis* codon at amino acid position 363 (GUG) encoding valine (Zhou et al. 2004). Hopefully, results of single exchange of *aspA* will determine if this association is fortuitous or actually accounts for lethality in man.

10.5 Acknowledgments

This work was sponsored by the NIH/NIAID Regional Center of Excellence for Biodefense and Emerging Infectious Diseases Research (RCE) Program. The author wishes to acknowledge membership within and support from the Region V 'Great Lakes' RCE (NIH award 1-U54-AI-057153). The technical assistance of Janet Fowler was indispensable.

10.6 References

Achtman, M., Zurth, K., Morelli, C., Torrea, G., Guiyoule, A. and Carniel, E. (1999) *Yersinia pestis*, the cause of plague, is a recently emerged clone of *Yersinia pseudotuberculosis*. PNAS 96, 14043-14048.

Andersson, S.G.E., Zomorodipour, A., Andersson, J.O., Sicheritz-Ponten, T., Alsmark, U.C.M., Podowski, R.M., Naslund, A.K., Eriksson, A.-S., Winkler, H.H. and Kurland, C.G. (1998) The genome sequence of *Rickettsia prowazekii* and the origin of mitochondria. Nature 396, 133-140.

Anisimov, A.P., Lindler, L.E. and Pier, G.B. (2004) Intraspecific diversity of *Yersinia pestis*. Clin. Microbiol. Rev. 17, 434-464.

Baugh, C.L., Lanham, J.W. and Surgalla, M.J. (1964) Eeffects of biocarbonate on growth of *Pasteurella pestis* II. Carbon dioxide fixation into oxalacetate by cell-free extracts. J. Bacteriol. 88, 553-558.

Beesley, E.D., Brubaker, R.R., Janssen, W.A. and Surgalla, M.J. (1967) Pesticins. III. Expression of coagulase and mechanism of fibrinolysis. J. Bacteriol. 94, 19-26.

Bovarnick, M.R. and Miller, J.C. (1950) Oxidation and transamination of glutamate by Typhus Rickettsiae. J. Biol. Chem. 184, 661-676.

Brownlow, W.J. and Wessman, G.E. (1960) Nutrition of *Pasteurella pestis* in chemically defined media at temperatures of 36 to 38 C. J. Bacteriol. 79, 299-304.

Brubaker, R.R. (1967) Growth of *Pasteurella pseudotuberculosis* in simulated intracellular and extracellular environments. J. Biol. Chem. 117, 403-417.

Brubaker, R.R. (1968) Metabolism of carbohydrates by *Pasteurella pseudotuberculosis*. J. Bacteriol. 95, 1698-1705.

Brubaker, R.R. (1972) The genus *Yersinia*: biochemistry and genetics of virulence. Curr. Top. Microbiol. Immunol. 57, 111-158.

Brubaker, R.R. (1991) Factors promoting acute and chronic diseases by yersiniae. Clin. Microbiol. Rev. 4, 309-324.

Brubaker, R.R. (2003) Interleukin-10 and Inhibition of Innate Immunity to Yersiniae: Roles of Yops and LcrV (V Antigen). Infect. Immun. 71, 3673-3681.

Brubaker, R.R. (2005) Influence of Na^+, dicarboxylic amino acids, and pH in modulating the low-calcium response of *Yersinia pestis*. Infect. Immun. 73, 4743-4752.

Brubaker, R.R. and Surgalla, M.J. (1964) The effect of Ca^{++} and Mg^{++} on lysis, growth, and production of virulence antigens by *Pasteurella pestis*. J. Infect. Dis. 114, 13-25.

Burrows, T.W. (1957) Virulence of *Pasteurella pestis*. Nature 179, 1246-1247.

Chain, P.S.G., Carniel, E., Larimer, F.W., Lamerdin, J., Stoutland, P.O., Regala, W.M., Georgescu, A.M., Vergez, L.M., Land, M.L., Motin, V.L., Brubaker, R.R., Fowler, J., Hinnebusch, J., Marceau, M., Medigue, C., Simonet, M., Chenal-Francisque, V., Souza, B., Dacheux, D., Elliott, J.M., Derbise, A., Hauser, L.J. and Garcia, E. (2004) Insights into the evolution of *Yersinia pestis* through whole-genome comparison with *Yersinia pseudotuberculosis*. PNAS 101, 13826-13831.

Delwiche, E.A., Fukui, G.M., Andrews, A.W. and Surgalla, M.J. (1959) Environmental conditions affecting the population dynamics and the retention of virulence of *Pasteurella pestis*: the role of carbon dioxide. J. Bacteriol. 77, 355-360.

Deng, W., Burland, V., Plunkett, G., III, Boutin, A., Mayhew, G.F., Liss, P., Perna, N.T., Rose, D.J., Mau, B., Zhou, S., Schwartz, D.C., Fetherston, J.D., Lindler, L.E., Brubaker, R.R., Plano, G.V., Straley, S.C., McDonough, K.A., Nilles, M.L., Matson, J.S., Blattner, F.R. and Perry, R.D. (2002) Genome Sequence of *Yersinia pestis* KIM. J. Bacteriol. 184, 4601-4611.

Devignat R (1951) Varietes de l'espece *Pasteurella pestis*. Nouvelle hypothese. Bull. W.H.O. 4, 247-263.

Dreyfus, L.A. and Brubaker, R.R. (1978) Consequences of aspartase deficiency in *Yersinia pestis*. J. Bacteriol. 136, 757-764.

Englesberg, E. and Ingraham, L. (1957) Meiotrophic mutatns of *Pasteurella pestis* and their use in the elucidation of nutritional requirements. PNAS 43, 369-372.

Ferber, D.M. and Brubaker, R.R. (1981) Plasmids in *Yersinia pestis*. Infect. Immun. 31, 839-841.

Fleischmann, R.D., Alland, D., Eisen, J.A., Carpenter, L., White, O., Peterson, J., DeBoy, R., Dodson, R., Gwinn, M., Haft, D., Hickey, E., Kolonay, J.F., Nelson, W.C., Umayam, L.A., Ermolaeva, M., Salzberg, S.L., Delcher, A., Utterback, T., Weidman, J., Khouri, H., Gill, J., Mikula, A., Bishai, W., Jacobs, W.R., Jr., Venter, J.C. and Fraser, C.M. (2002) Whole-genome comparison of *Mycobacterium tuberculosis* clinical and laboratory strains. J. Bacteriol. 184, 5479-5490.

Fowler, J.M. and Brubaker, R.R. (1994) Physiological basis of the low calcium response in *Yersinia pestis*. Infect. Immun. 62, 5234-5241.

Fraenkel DG (1968) Selection of *Escherichia coli* mutants lacking glucose-6-phosphate dehydrogenase or gluconate-6-phosphate dehydrogenase. J. Bacteriol. 95, 1267-1271.

Goguen, J.D., Yother, J. and Straley, S.C. (1984) Genetic analysis of the low calcium response in *Yersinia pestis* Mu dl (Ap *lac*) insertion mutants. J. Bacteriol. 160, 842-848.

Good N.E., Winget, G.D., Winter, W., Connolly, T.N., Izawa, S. and Singh, R.M. (1966) Hydrogen ion buffers for biological research. Biochemistry; 5, 467-477.

Halpern, Y.S. and Umbarger, H.E. (1960) Conversion of ammonia to amino groups in *Escherichia coli* K-12 mutants. J. Bacteriol. 80, 285-288

Heesemann J., Sing A. and Trülzsch, K. (2006) *Yersinia*'s stratagem: targeting innate and adaptive immune defense. Curr. Opin. Microbiol. 9, 1-7.

Higuchi, K., Kupferberg, L.L. and Smith, J.L. (1959) Studies on the nutrition and physiology of *Pasteurella pestis*. III. Effects of calcium ions on the growth of virulent and avirulent strains of *Pasteurella pestis*. J. Bacteriol. 77, 317-321.

Higuchi, K. and Smith, J.L. (1961) Studies on the nutrition and physiology of *Pasteurella pestis*. VI. A differential plating medium for the estimation of the mutation rate to avirulence. J. Bacteriol. 81, 605-608.

Hinnebusch, B.J., Rudolph, A.E., Cherepanov, P., Dixon, J.E., Schwan, T.G. and Forsberg, A. (2002) Role of *Yersinia* murine toxin in survival of *Yersinia pestis* in the midgut of the flea vector. Science 296, 733-735.

Kupferberg, L.L. and Higuchi, K. (1958) Role of calcium ions in the stimulation of growth of virulent strains of *Pasteurella pestis*. J. Bacteriol. 76,120-121.

Kutyrev, V., Mehigh, R.J., Motin, V.L., Pokrovskaya, M.S., Smirnov, G.B. and Brubaker, R.R. (1999) Expression of the plague plasminogen activator in *Yersinia pseudotuberculosis* and *Escherichia coli*. Infect. Immun. 67, 1359-1367.

Lähteenmäki, K., Virkola, R., Sarén, A., Emödy, L. and Korhonen, T.K. (1998) Expression of plasminogen activator Pla of *Yersinia pestis* enhances bacterial attachment to the mammalian extracellular matrix. Infect. Immun. 66, 5755-5762.

Larsson, P., Oyston, P.C.F., Chain, P., Chu, M.C., Duffield, M., Fuxelius, H.-H., Garcia, E., Halltorp, G., Johansson, D., Isherwood, K.E., Karp, P.D., Larsson, E., Liu, Y., Michell, S., Prior, J., Prior, R., Malfatti, S., Sjostedt, A., Svensson, K., Thompson, N., Vergez, L., Wagg, J.K., Wren, B.W., Lindler, L.E., Andersson, S.G.E., Forsman, M. and Titball, R.W. (2005) The complete genome sequence of *Francisella tularensis*, the causative agent of tularemia. Nat. Genet. 37, 153-159.

Lawton W.D., Erdman R.L. and Surgalla, M.J. (1963) Biosynthesis and purification of V and W antigen in *Pasteurella pestis*. J. Immunol. 91, 179-184.

Lowry, O.H., Rosebrough, N.J., Farr, A.L. and Randall, R.J. (1951) Protein measurement with the Folin phenol reagent. J. Biol. Chem. 193, 265-275.

Mortlock, R.P. (1962) Gluconate metabolism of *Pasteurella pestis*. J. Bacteriol. 84, 53-59.

Mortlock, R.P. and Brubaker, R.R. (1962) Glucose-6-phosphate dehydrogenase and 6-phosphogluconate dehydrogenase activities of *Pasteurella pestis* and *Pasteurella pseudotuberculosis*. J. Bacteriol. 84, 1122-1123.

Motin, V.L., Georgescu, A.M., Fitch, J.P., Gu, P.P., Nelson, D.O., Mabery, S.L., Garnham, J.B., Sokhansanj, B.A., Ott, L.L., Coleman, M.A., Elliott, J.M., Kegelmeyer, L.M., Wyrobek, A.J., Slezak, T.R., Brubaker, R.R. and Garcia, E. (2004) Temporal Global Changes in Gene Expression during Temperature Transition in *Yersinia pestis*. J. Bacteriol. 186, 6298-6305.

Parkhill, J., Wren, B.W., Thomson, N.R., Titball, R.W., Holden, M.T., Prentice, M.B., Sebaihia, M., James, K.D., Churcher, C., Mungall, K.L., Baker, S., Basham, D., Bentley, S.D., Brooks, K., Cerdeno-Tarraga, A.M., Chillingworth, T., Cronin, A., Davies, R.M., Davis, P., Dougan, G., Feltwell, T., Hamlin, N., Holroyd, S., Jagels, K., Karlyshev, A.V., Leather, S., Moule, S., Oyston, P.C., Quail, M., Rutherford, K., Simmonds, M., Skelton, J., Stevens, K., Whitehead, S. and Barrell, B.G. (2001) Genome sequence of *Yersinia pestis*, the causative agent of plague. Nature 413, 523-527.

Smith, T. (1934) *Parasitism and disease*. Princeton University Press, Princton, NJ.

Soupart, P. (1962) Free amino acids of blood and urine in the human. In: J.T. Holden (ed) *Amino acid pools*. Elsevier Publishing Co., Amsterdam, pp. 220-262.

Viola, R.E. (1998) L-aspartase: new tricks from an old enzyme. In: D.L. Purich (ed) *Advances in Enzymology and Related Areas of Molecular Biology*. John Wiley & Sons, Inc., New York, pp. 295-341.

Winter, C.C., Cherry, W.B. and Moody, M.D. (1960) An unusual strain of *Pasteurella pestis* isolated from a fatal human case of plague. Bull. Wld. Hlth. Org. 23, 408-409.

Yellin, T.O. and Wriston, J.C. (1966) Purification and properties of guinea pig serum asparaginase. Biochemistry 5, 1605-1612.

Yimga, M.T., Leatham, M.P., Allen, J.H., Laux, D.C., Conway, T. and Cohen, P.S. (2006) Role of gluconeogenesis and the tricarboxylic acid cycle in the virulence of *Salmonella enterica* serovar Typhimurium in BALB/c mice. J. Bacteriol. 74, 1130-1140.

Zahorchak, R.J., Charnetzky, W.T., Little, R.V. and Brubaker, R.R. (1979) Consequences of Ca^{2+} deficiency on macromolecular synthesis and adenylate energy charge in *Yersinia pestis*. J. Bacteriol. 39, 792-799.

Zhou, W., Bertsova, Y.V., Feng, B., Tsatsos, P., Verkhovskaya, M.L., Gennis, R.B., Bogachev, A.V. and Barquera, B. (1999) Sequencing and preliminary characterization of the Na$^+$-translocating NADH:ubiquinone oxidoreductase from *Vibrio harveyi*. Biochemistry 38, 16246-16252.

Zhou, D., Tong, Z., Song, Y., Han, Y., Pei, D., Pang, X., Zhai, J., Li, M., Cui, B., Qi, Z., Jin, L., Dai, R., Du, Z., Wang, J., Guo, Z., Wang, J., Huang, P. and Yang, R. (2004) Genetics of Metabolic Variations between *Yersinia pestis* Biovars and the Proposal of a New Biovar, microtus. J. Bacteriol. 186, 5147-5152.

Zink, D.L., Feeley, J.G., Wells, J.G., Vanderzant, C., Vickery, J.C., Roof, W.D. and O'Donovan, G.A. (1980) Plasmid-mediated tissue invasiveness in *Yersinia enterocolitica*. Nature (London) 283, 224-226.

Part III – Regulatory Mechanisms

Picture 12. Greg Plano presents research from his laboratory. Photograph by A. Anisimov.

Picture 13. Matthew Nilles presents data on control of secretion of Yops. Photograph by A. Anisimov.

11
Differential Gene Regulation in *Yersinia pestis* versus *Yersinia pseudotuberculosis*: Effects of Hypoxia and Potential Role of a Plasmid Regulator

Guangchun Bai, Eric Smith, Andrey Golubov, Janice Pata,
and Kathleen A. McDonough

Wadsworth Center, NYS Department of Health, Kathleen.McDonough@wadsworth.org

Abstract. The molecular basis of the biological differences between *Yersinia pestis* and *Yersinia pseudotuberculosis* remains largely unknown, and relatively little is known about environmental regulation of gene expression in these bacteria. We used a proteomic approach to explore the regulatory response of each bacterium to carbon dioxide-supplemented hypoxic conditions. Both organisms responded similarly and the magnitude of their responses was similar to what was observed in low iron conditions. We also identified proteins that were expressed at different levels in *Y. pestis* and *Y. pseudotuberculosis*, and found that SodB is expressed more strongly at both the protein and RNA levels in *Y. pseudotuberculosis* than in *Y. pestis*. Enzyme activity did not directly correlate with levels of protein expression, and we propose that an amino acid change difference between these orthologous proteins has the potential to affect catalytic activity. In addition, the upstream regulatory regions of several chromosomal genes were found to exhibit specific binding with a putative transcription factor, CDS4, from the *Y. pestis*-specific pPCP1 plasmid. The potential role of this protein in modulating *Y. pestis*- specific gene regulation warrants further investigation.

11.1 Introduction

Yersinia pestis and *Yersinia. pseudotuberculosis* share more than 90% DNA homology, yet the diseases these microbes cause and their modes of transmission are very different (Mollaret 1965; Bercovier et al. 1980; Ibrahim et al. 1993). *Y. pestis* causes bubonic plague and is transmitted from a rodent reservoir by a flea vector (Poland and Barnes 1979; Barnes 1982; Perry and Fetherston 1997). In contrast, *Y. pseudotuberculosis* is an enteric pathogen that is transmitted in contaminated food and water (Quan et al. 1981) but not by fleas (Une 1977; Hinnebusch et al. 1996; Erickson et al. 2006). Some of the differences between these organisms can be attributed to the acquisition by *Y. pestis* of two unique plasmids that confer upon it several new phenotypes (Brubaker et al. 1965; Ferber and Brubaker 1981; Ferber et al. 1981; Portnoy and Falkow 1981; Protsenko et al. 1983; Sodeinde and Goguen 1988; Sodeinde et al. 1988; McDonough and Falkow 1989; Perry and Fetherston 1997; Hinnebusch et al. 2002; Hinnebusch 2003). However, other virulence-associated differences are chromosomally encoded. These include the selective inactivation by one species of genes that are common to both organisms but required for virulence in only one (Rosqvist et al. 1988; Simonet et al. 1996; Han and Miller 1997; Hare and

McDonough 1999). These ‘niche’ genes, that is those that are consistently active in only one of these two closely related but distinct pathogens, are evidence of biological selection driving evolution of specialized pathogens.

Recent *Y. pestis* sequence analyses have revealed dynamic genomes with a large number of rearrangements and many inactivated genes, up to one third of which are interrupted by IS elements (Parkhill et al. 2001; Deng et al. 2002; Chain et al. 2004). Some of these *Y. pestis* pseudogenes are functional genes associated with enteric infection in *Y. pseudotuberculosis*, and it has become increasingly clear that reductive evolution has been a critical factor in the emergence of *Y. pestis* virulence (Hinchliffe et al. 2003; Stabler et al. 2003; Achtman et al. 2004; Chain et al. 2004; Zhou et al. 2004; Chain et al. 2006). These direct DNA – DNA sequence comparisons of *Y. pestis* and *Y. pseudotuberculosis* have been extremely useful for tracking inter-species deletions, insertions and mutations. However, this approach does not provide information about gene expression that may be critical to understanding plague pathogenesis.

Virulence is a complex phenotype, and bacteria orchestrate highly regulated series of interactions with their hosts. In many cases, simply having the right genes is not enough-- pathogens must express (or specifically not express) these genes at the right times and in the right combinations to cause infection. Gain or loss of even a small number of regulatory or functional genes can impact the expression of many other genes in an evolving genome, depending upon the environmental conditions to which the organism is exposed. Unlike *Y. pseudotuberculosis*, *Y. pestis* has had to adapt to survival within fleas as well as mammalian hosts. Recent studies have begun to explore the effects of these complex environments on *Y. pestis* gene expression (Lathem et al. 2005; Sebbane et al. 2006), but much remains to be learned about how specific gene regulatory pathways have evolved in *Y. pestis* versus *Y. pseudotuberculosis*.

Temperature and iron limitation are among the best-studied regulators of gene expression in *Y. pestis* (Perry 1993; Straley and Perry 1995; Perry and Fetherston 1997; Motin et al. 2004; Zhou et al. 2006). Iron availability and temperature vary depending on whether the bacterium is in the environment, within a flea, or infecting a mammalian host. For example, limited iron will be available within the host, which will also be maintained at a relatively constant temperature of 37°C. In contrast, flea temperatures are likely to be lower than 30°C, and iron availability in a digesting blood meal within the flea midgut may exceed that within the mammalian host. Bacteria will also be exposed to differing levels of oxygen and carbon dioxide as they move between their environmental source or flea vector and a mammalian host. Oxygen and carbon dioxide regulate virulence gene expression in several bacterial pathogens (Gronberg and Kihlstrom 1989; Stretton and Goodman 1998; Florczyk et al. 2003; Timm et al. 2003; Jydegaard-Axelsen et al. 2004; Sun et al. 2005), but the effect of these conditions has received little attention in Yersinia (Gronberg and Kihlstrom 1989). Hypoxic conditions are likely to be found in the subcutaneous tissue following a flea bite, within macrophages, and within the digestive tract, and most areas within a mammalian host are exposed to CO_2 levels (~5- 7%) that greatly exceed that of the outside environment (0.04%) (Guyton 1991; Lide 1997).

Table 1. The proteomes of *Y. pestis* (Pst) and *Y. pseudotuberculosis* (Ptb) were compared by 2D-GE in rich media with ambient air or hypoxic conditions, or in defined iron supplemented or iron limited PMH medium. Left panel shows that the total number of protein spots that differed between *Y. pestis* and *Y. pseudotuberculosis* were similar in BHI and iron supplemented control media. The numbers of spots that contained higher amounts of protein for each species are also shown. Right panel shows the response of each bacterium to hypoxic or low iron conditions, with the numbers of protein spots showing increases (up), or decreases (down), in response to each specified condition

	Pst/Ptb	
	BHI	PMH+
Total	338	335
Pst	155	145
Ptb	183	190

	Hypoxia		Low Iron	
	Pst	Ptb	Pst	Ptb
Total	236	210	245	239
up	111	124	109	156
down	125	86	136	143

11.2 Proteomic Comparison of *Y. pestis* and *Y. pseudotuberculosis*

We used two complementary proteomic approaches to explore differential gene expression in *Y. pestis* KIM6+ and *Y. pseudotuberculosis* PTB3 organisms exposed to either ambient air or hypoxia. These include two dimensional gel electrophoresis-mass spectrometry (2D-GEMS) (Florczyk et al. 2003) and multiplexed mass spectrometry-based protein quantitation using amine-reactive isobaric tagging reagents (iTraq) (Ross et al. 2004)). Iron limitation was also examined to provide a reference for the magnitude of the hypoxic response. Two dimensional gel electrophoresis was performed with *Y. pestis* and *Y. pseudotuberculosis* to evaluate the response of each proteome following exposure to ambient air (~20% O_2, ~0.04% CO_2) versus hypoxia (1.3% O_2, 5% CO_2) for 1, 10 or 24 hours at 28 or 37°C. Bacteria were pregrown overnight in BHI media at 28°C under aerated conditions. At the start of each experiment, cultures were diluted 1/10 into fresh media and transferred to ambient air or hypoxic (1.3% oxygen) conditions at 37°C in vented flasks with shaking. Hypoxic conditions were achieved in a Forma Scientific Series II incubator that depletes and maintains oxygen at specified levels by injecting a regulated flow of nitrogen gas, as described previously (Purkayastha et al. 2002; Florczyk et al. 2003). Protein preparation and 2D gel electrophoresis were performed using standard techniques as described (McDonough et al. 2000; Florczyk et al. 2001).

11.2.1 Response to Hypoxia is Comparable to That of Iron Limitation

The magnitudes of *Y. pestis* and *Y. pseudotuberculosis* responses to CO_2- supplemented hypoxic conditions were similar to those obtained in response to iron limiting conditons (Table 1). 2D gels were analyzed visually and using PDQuest

Table 2. Proteins that were detected at higher levels in either *Y. pseudotuberculosis* or *Y. pestis*, as measured by both 2D-GEMS and iTraq. Fold increases are shown for iTraq data

Y. pseudotuberculosis dominant	Fold Increase
• SodB detoxification	9.0
• DsbA translation/modification	3.3
• OppA transport	3.5
Y. pestis dominant	
• RpsF protein synthesis	3.7
• AhpC detoxification	2.8
• GroES chaperone	2.0
• ClpB(YQ) protease	3.0
• Udp nucleoside salvage	5.0

(BioRad) to detect protein spots that differed at least two fold in intensity between conditions/strains. Relatively few differences between the ambient air and hypoxic cultures were observed visually with either *Y. pestis* or *Y. pseudotuberculosis* at 1h, but a significant number occurred by 10 h (Table 1), and continued through 24 h (not shown). The individual contributions of hypoxia versus carbon dioxide are currently being investigated, although it is expected that most of these changes are due to hypoxia.

More iron-responsive changes were detected in our protein-based analysis than in a recent microarray-based study (245 versus 88) (Zhou et al. 2006), although the general agreement between studies is quite good considering the different systems used to measure differential gene expression. The number of changes observed at the protein level will include the additional detection of posttranscriptionally regulated gene products and posttranslational protein modifications such as specific cleavage, degradation, methylation, phosphorylation and/or glycosylation. In some cases, individual proteins will show up as multiple spots due to these effects. Proteome analysis can also detect proteins expressed from orfs that are not annotated in genome databases, and would likely be missed by orf-based array studies (Jungblut et al. 2001). Together, these features make proteomic comparisons an excellent complement to RNA-based expression analyses (Cash 2003).

Y. pestis versus *Y. pseudotuberculosis* gels were also compared to identify species-specific effects on protein expression (Table 2). Some of these differentially expressed proteins were identified by LC/MS/MS, as described previously (Gazdik and McDonough 2005). iTraq analyses were done in parallel with the 2D-GEMS experiments to more rapidly identify differentially expressed proteins on a global scale, and several genes associated with these proteins were selected for further analysis.

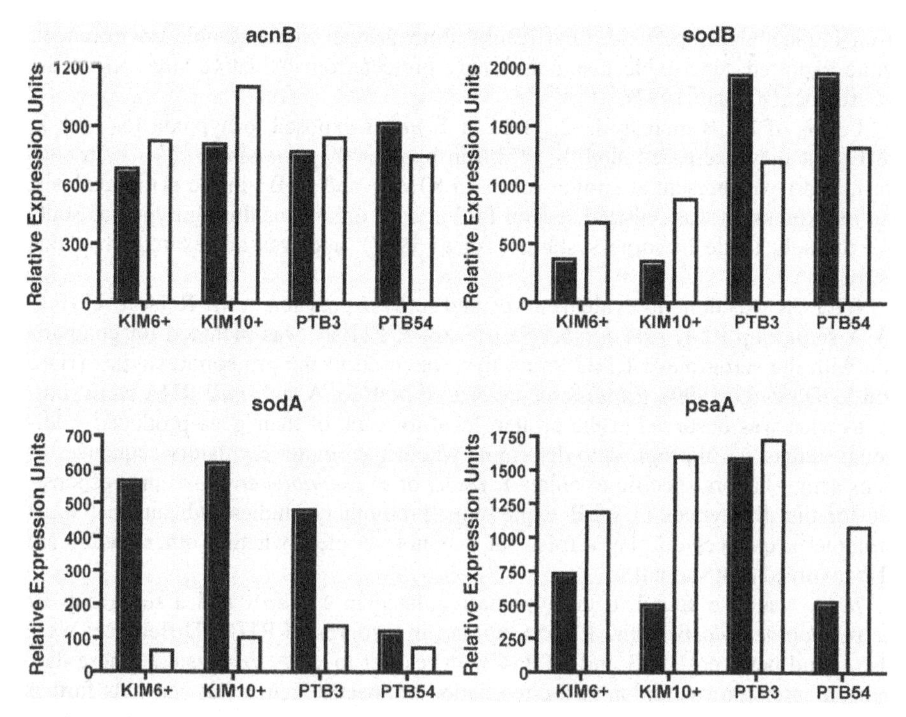

Fig. 1. qRT PCR for selected mRNA's from *Y. pestis* (KIM6+, KIM10+), and *Y. pseudotuberculosis* (PTB3, PTB54) bacteria grown at 28°C in ambient air (solid bars) or hypoxic (open bars) conditions for 10 hours. Results shown are representative of at least two experiments. cDNA samples were diluted to concentrations required to achieve linear amplification for each template.

11.3 Differential Expression of SodB

SodB was one of the proteins that was most strongly expressed in *Y. pseudotuberculosis* relative to *Y. pestis* in both the the iTraq and 2D-GEMs analyses. Superoxide dismutases are involved in oxygen detoxification, and are essential for the virulence of several pathogens (Roggenkamp et al. 1997; Narasipura et al. 2003; Ammendola et al. 2005; Gee et al. 2005; Bakshi et al. 2006). Three types have been identified based on their metal cofactors, and pathogenic Yersinia carry genes for all three (Parkhill et al. 2001; Deng et al. 2002). SodA, which requires a manganese cofactor, is essential for full virulence of *Yersinia enterocolitica* (Roggenkamp et al. 1997; Najdenski et al. 2004), but not *Y. pseudotuberculosis* or *Y. pestis* (Sebbane et al. 2006). Neither the iron-containing SodB nor the copper/zinc-associated SodC have been investigated in Yersinia. In *Escherichia coli*, SodA and SodB are located in the cytoplasm, and SodC is periplasmic. The cytoplasmic enzymes are thought to protect from intracellular superoxide that results from growth in an aerobic environment, while exported superoxide dismutases may protect from oxidative stressors in the environment (Touati 1997).

Levels of SodB increased ~2.5 fold in *Y. pestis* exposed to hypoxia for 10 h at 28°C, but they decreased slightly (25%) in hypoxic *Y. pseudotuberculosis*. In contrast, SodA was present at similar levels in KIM6+ and PTB3 in the iTraq analysis, but its expression was reduced several fold in both organisms during hypoxia. SodC is extremely labile in some systems (Touati 1997), and was not examined in this assay.

RT PCR was done to evaluate *sod*A and *sod*B expression at the RNA level (Fig. 1). A serogroup III *Y. pseudotuberculosis* strain, PTB54, was included for comparison with the serogroup I PTB3 strain that was used in the proteomic studies (Hare and McDonough 1999). Expression patterns of both *sod*A and *sod*B RNA were similar to what was observed at the protein level for each of their gene products. Additional studies are in progress to determine whether promoter regulatory sequences or trans acting factors specific to either *Y. pestis* or *Y. pseudotuberculosis* are responsible for the differences in *sod*B expression. Preliminary studies indicate that *sod*B promoter sequences do play a role, but it is not yet clear whether other factors are also involved (not shown).

*psa*A was also found to be hypoxia-regulated in *Y. pestis* and a serogroup III *Y. pseudotuberculosis* strain, PTB54, but not in serogroup I PTB3. Differences were also found between PTB3 and PTB54 with respect to *sod*A expression. These data indicate interstrain variation in the regulation of specific genes that warrants further investigation in both *Y. pseudotuberculosis* and *Y. pestis*. Expression of *acn*B, which encodes Aconitase B, was similar in all conditions and served as a constitutive control for these experiments (Fig. 1).

The differing levels of SodB expression in *Y. pestis* and *Y. pseudotuberculosis* were further examined in a native polyacrylamide gel with nitroblue tetrazolium to measure superoxide dismutase activity, as described (Beauchamp and Fridovich 1971). Surprisingly, superoxide dismutase activity was not reduced in *Y. pestis* relative to any of five *Y. pseudotuberculosis* strains tested, including PTB3, which expressed higher levels of both RNA and protein (Fig. 2). SodB activity may be masked by the activity of co-migrating SodA or SodC due the similar sizes of these proteins, and this possibility is currently being investigated. However, levels of *sod*B RNA were much higher than those of *sod*A or *sod*C (not shown), so we think it unlikely that co-migration of other superoxide dismutases is the sole explanation for these results. An alternative possibility is that the *Y. pseudotuberculosis* SodB is less active than the *Y. pestis* SodB.

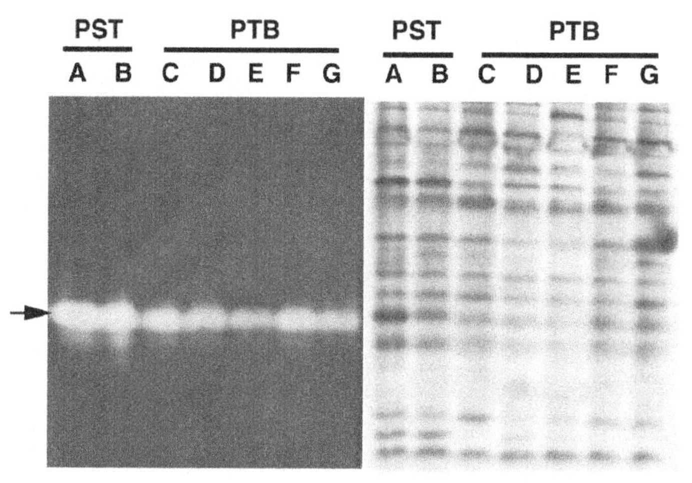

Fig. 2. Zymogram (left) showing superoxide dismutase activity (arrow), and coomassie-stained (right) polyacrylamide gels. PST refers to *Y. pestis*, and PTB denotes Yptb strains. Lanes contain: A, KIM6+; B, KIM10+; C, PTB3; D, PTB54; E, YPIII; F, PTB52; G, PTB4.

11.4 Structural Comparison of *Y. pestis* vs *Y. pseudotuberculosis* SodB

The *Y. pestis* and *Y. pseudotuberculosis* orthologs differ by a single amino acid, and the potential effects of this amino acid substitution were explored with a modeling approach. Homology models of the SodB proteins from *Y. pestis* and *Y. pseudotuberculosis* were constructed using the program MODELLER (Eswar et al. 2003) and were based on the 2.1 Å resolution crystal structure of iron superoxide dismutase from *Pseudomonas ovalis* (PDB code 1DT0, chain A; (Bond et al. 2000)). The *Yersinia* proteins share 70% sequence identity with the *P. ovalis* protein. The *Y. pestis* and *Y. pseudotuberculosis* models align with the *P. ovalis* structure with root mean squared deviations of 0.16 and 0.14 Å, respectively, over 191 Cα atoms and were based on sequences alignments having E-values of 7e-67 and 3e-67.

In the homology models, the single amino acid difference between the two proteins (Ile 37 in *Y. pestis*, Asn 37 in *Y. pseudotuberculosis*; equivalent to Val 36 in *P. ovalis*) is located on the surface of the protein, in a long α-helix (Fig. 3A). Because this position is more than 11Å away from the predicted dimerization interface, differences in enzyme activity are unlikely to be due to differences in the oligomeric state. Amino acid 37 is, however, located on the opposite side of the helix from amino acid 35, which is located in the active site and is strictly conserved as a tyrosine. Even small changes in the structure of the active site could significantly alter activity. The modeling suggests that the positions of the Tyr 35 hydroxyl group in the two enzymes differ by about 1 Å (Fig. 3B) and is the largest difference between

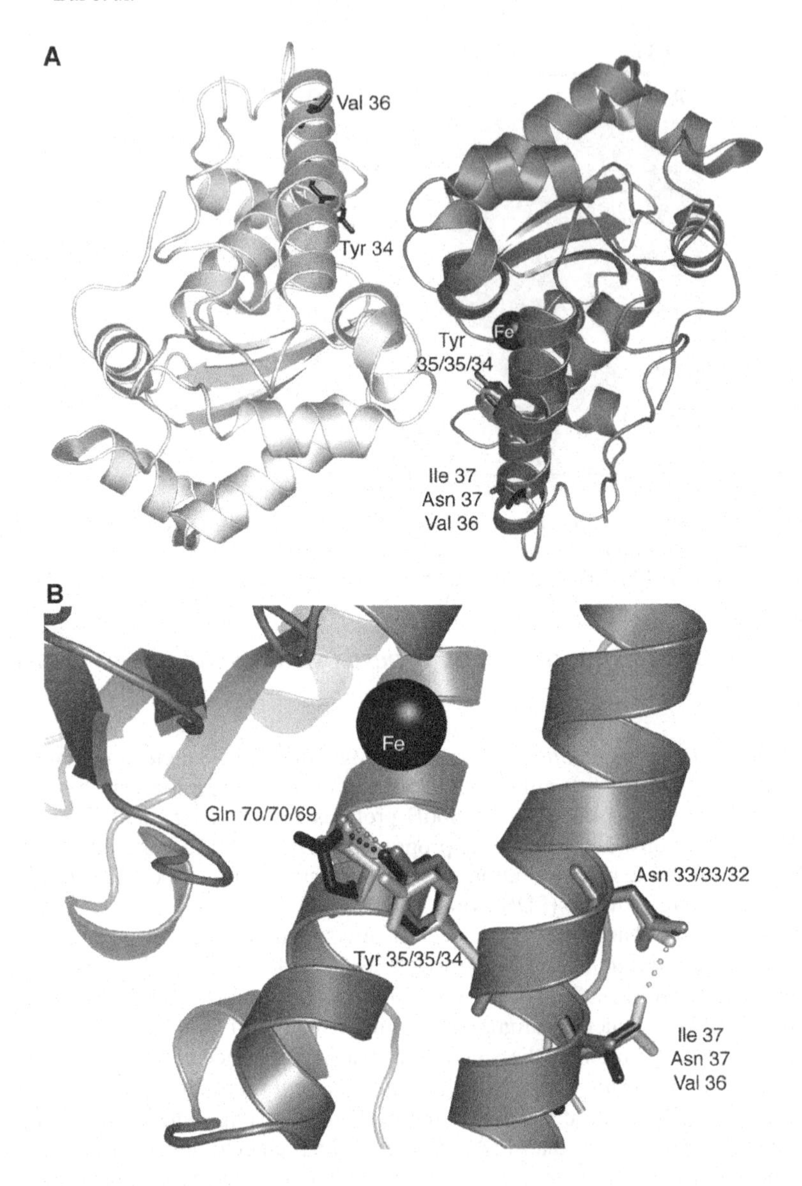

Fig. 3. Homology modeling of SodB proteins from *Y. pestis* and *Y. pseudotuberculosis* based on the crystal structure of iron superoxide dismutase from *P. ovalis* (Bond et al. 2000). (A) Overview. (B) Active site region. Side chains are shown superimposed on the ribbon diagram of the *P. ovalis* structure: *Y. pestis* (dark gray), *Y. pseudotuberculosis* (light gray) and *P. ovalis* (black), with residue numbers listed in that order. The iron atom bound to one subunit in the crystal structure is shown as a black sphere. Dotted lines represent hydrogen bonds observed in the *P. ovalis* structure or predicted in the homology models.

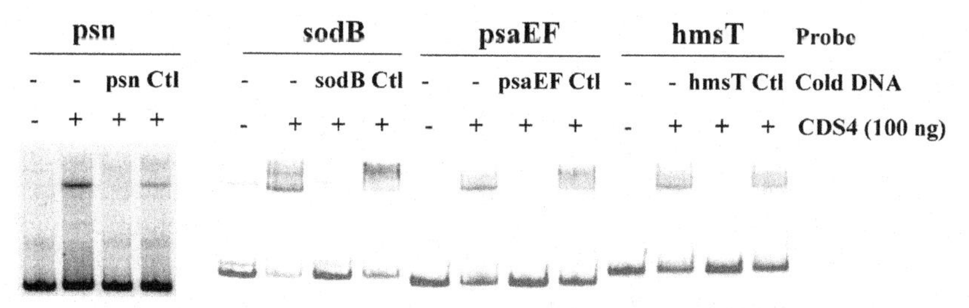

Fig. 4. EMSA results showing specific binding of CDS4 with upstream regulatory regions of several chromosomal genes, used as probes. Competition assays were done using 500 fold excess of cold (unlabelled) probe DNA where specified, or a non-specific DNA control fragment (Ctl). + denotes addition of purified CDS4 protein; - marks negative control lanes in which no CDS4 was added.

the two enzymes in the active site region. In the *P. ovalis* structure, the equivalent tyrosine, Tyr 34, hydrogen bonds to Gln 69, which in turn is hydrogen bonded to a water molecule coordinated to the iron at the active site. The homology models suggest that the altered position of Tyr 35 in the *Y. pseudotuberculosis* protein could be induced by hydrogen bonding of Asn 37 to Asn 33 (Fig. 3B), which could not occur with Ile 37 in the *Y. pestis* protein. Further studies are needed to determine whether this amino acid change alters SodB catalytic activity in *Y. pseudotuberculosis*, perhaps as a balance for its higher level of expression.

11.5 Plasmid-based Transcription Factor Binds Chromosomal Genes

As mentioned earlier, addition or loss of genes encoding regulatory factors could have effects on gene expression that differentiate *Y. pestis* from *Y. pseudotuberculosis* in ways that would not be obvious by DNA-DNA comparisons. We noted differences in the proteomic profiles of KIM6+ and KIM10+, that differ only by the absence of the *Y. pestis*-specific plasmid pPCP1 from KIM10+ (not shown). CDS4 is an uncharacterized orf on pPCP1 that is annotated as a potential transciption factor, and we reasoned that it could affect gene expression in *Y. pestis*. We evaluated CDS4 protein interactions with selected promoter sequences by eletrophoretic mobility shift assays (EMSA) as a preliminary measure of its potential for regulation of chromosomal genes in *Y. pestis*.

CDS4 was amplified by PCR, cloned into pET28a+ and expressed with an N-terminal His-tag in *E. coli*. Soluble CDS4 protein was then purified by affinity chromatography as previously described (Bai et al. 2005), and the purity was evaluated by SDS-PAGE. EMSA was performed using 100 ng of purified CDS4 with 0.05 pmol ^{33}P-end-labeled DNA probe, as described (Seoh and Tai 1999; Bai et al.

2005). Approximately 500-fold excess of unlabeled DNA fragments was used for competition experiments.

The mobility of probes generated from DNA regions upstream of *psn*, *sodB*, *psaEF* and *hmsT* were obviously retarded in the presence of CDS4 protein (Fig. 4). The reappearance of the free probes was detected in addition of excess amount of unlabeled probe DNA, but not an unrelated control DNA of similar size (Fig. 4), indicating specific binding of CDS4 with these probes. A number of additional probes were tested that did not bind with CDS4 (not shown). We are currently defining the CDS4 binding motif and investigating the possibility that *psn*, *sodB*, *psaEF* and *hmsT* are regulated by CDS4 protein in *Y. pestis*. Levels of *sod*B RNA were similar in log phase KIM6+ and KIM10+ grown at 28°C (Fig. 1), indicating that CDS4-mediated regulation of sodB expression, if it occurs, is likely to be restricted to selected conditions. This is not surprising as regulation of *sod*B expression in other bacteria is complex and controlled by numerous factors (Masse and Gottesman 2002). We expect that continued study of environmental gene regulation in *Y. pestis* versus *Y. pseudotuberculosis* will provide new clues about the biology of these two important pathogens and their path of biological divergence.

11.6 Acknowledgements

We gratefully acknowledge Q. Lin, T. Ryan, K. Chave, H. Vasudeva-Rao, the UAlbany Proteomics Facility and the Wadsworth Center Biochemistry Core for advice and technical assistance. This work was supported in part by National Institutes of Health grant AI06160602.

11.7 References

Achtman, M., Morelli, G., Zhu, P., Wirth, T., Diehl, I., Kusecek, B., Vogler, A.J., Wagner, D.M., Allender, C.J., Easterday, W.R., Chenal-Francisque, V., Worsham, P., Thomson, N.R., Parkhill, J., Lindler, L.E., Carniel, E. and Keim, P. (2004). Microevolution and history of the plague bacillus, *Yersinia pestis*. PNAS. U S A 101, 17837-17842.

Ammendola, S., Ajello, M., Pasquali, P., Kroll, J.S., Langford, P.R., Rotilio, G., Valenti, P. and Battistoni, A. (2005). Differential contribution of sodC1 and sodC2 to intracellular survival and pathogenicity of *Salmonella enterica* serovar Choleraesuis. Microbes Infect. 7, 698-707.

Bai, G., McCue, L.A. and McDonough, K.A. (2005). Characterization of *Mycobacterium tuberculosis* Rv3676 (CRPMt), a cyclic AMP receptor protein-like DNA binding protein. J. Bacteriol. 187, 7795-7804.

Bakshi, C.S., Malik, M., Regan, K., Melendez, J.A., Metzger, D.W., Pavlov, V.M. and Sellati, T.J. (2006). Superoxide dismutase B gene (*sod*B)-deficient mutants of *Francisella tularensis* demonstrate hypersensitivity to oxidative stress and attenuated virulence. J. Bacteriol. 188, 6443-6448.

Barnes, A.M. (1982). Surveillance and control of bubonic plague in the United States. Symp. Zool. Soc. London 50, 237-270.

Beauchamp, C. and Fridovich, I. (1971). Superoxide dismutase improved assays and an assay applicable to acrylamide gels. Anal. Biochem. 44, 276-287.

Bercovier, H., Mollaret, H.H., Alonso, J.M., Brault, J., Fanning, G.R., Steigerwalt, A.G. and Brenner, D.J. (1980). Intra- and interspecies relatedness of *Yersinia pestis* by DNA hybridization and its relationship to *Yersinia pseudotuberculosis*. Curr. Microbiol. 4, 225-229.

Bond, C.J., Huang, J., Hajduk, R., Flick, K.E., Heath, P.J. and Stoddard, B.L. (2000). Cloning, sequence and crystallographic structure of recombinant iron superoxide dismutase from *Pseudomonas ovalis*. Acta Crystallogr. D Biol. Crystallogr. 56, 1359-1366.

Brubaker, R.R., Beesley, E.D. and Surgalla, M.J. (1965). *Pasteurella pestis*: role of pesticin I and iron in experimental plague. Science 149, 422-424.

Cash, P. (2003). Proteomics of bacterial pathogens. Adv. Biochem. Eng. Biotechnol. 83, 93-115.

Chain, P.S.G., Carniel, E., Larimer, F.W., Lamerdin, J., Stoutland, P.O., Regala, W.M., Georgescu, A.M., Vergez, L.M., Land, M.L., Motin, V.L., Brubaker, R.R., Fowler, J., Hinnebusch, J., Marceau, M., Medigue, C., Simonet, M., Chenal-Francisque, V., Souza, B., Dacheux, D., Elliott, J.M., Derbise, A., Hauser, L.J. and Garcia, E. (2004). Insights into the evolution of *Yersinia pestis* through whole-genome comparison with *Yersinia pseudotuberculosis*. PNAS 101, 13826-13831.

Chain, P.S.G., Hu, P., Malfatti, S.A., Radnedge, L., Larimer, F., Vergez, L.M., Worsham, P., "Chu, M.C. and Andersen, G.L. (2006). Complete genome sequence of *Yersinia pestis* strains Antiqua and Nepal516: evidence of gene reduction in an emerging pathogen. J. Bacteriol. 188, 4453-4463.

Deng, W., Burland, V., Plunkett, G., III, Boutin, A., Mayhew, G.F., Liss, P., Perna, N.T., Rose, D.J., Mau, B., Zhou, S., Schwartz, D.C., Fetherston, J.D., Lindler, L.E., Brubaker, R.R., Plano, G.V., Straley, S.C., McDonough, K.A., Nilles, M.L., Matson, J.S., Blattner, F.R. and Perry, R.D. (2002). Genome Sequence of *Yersinia pestis* KIM. J. Bacteriol. 184, 4601-4611.

Erickson, D.L., Jarrett, C.O., Wren, B.W. and Hinnebusch, B.J. (2006). Serotype differences and lack of biofilm formation characterize *Yersinia pseudotuberculosis* infection of the *Xenopsylla cheopis* flea vector of *Yersinia pestis*. J. Bacteriol. 188, 1113-1119.

Eswar, N., John, B., Mirkovic, N., Fiser, A., Ilyin, V.A., Pieper, U., Stuart, A.C., Marti-Renom, M.A., Madhusudhan, M.S., Yerkovich, B. and Sali, A. (2003). Tools for comparative protein structure modeling and analysis. Nucleic Acids Res. 31, 3375-3380.

Ferber, D.M. and Brubaker, R.R. (1981) Plasmids in *Yersinia pestis*. Infect. Immun. 31, 839-841.

Ferber, D.M., Fowler, J.M. and Brubaker, R.R. (1981). Mutations to tolerance and resistance to pesticin and colicins in *Escherichia coli*. J. Bacteriol. 146, 506-511.

Florczyk, M.A., McCue, L.A., Purkayastha, A., Currenti, E., Wolin, M.J. and McDonough, K.A. (2003). A family of *acr*-coregulated *Mycobacterium tuberculosis* genes shares a common DNA motif and requires Rv3133c (*dos*R or *dev*R) for expression. Infect. Immun. 71, 5332-5343.

Florczyk, M.A., McCue, L.A., Stack, R.F., Hauer, C.R. and McDonough, K.A. (2001). Identification and characterization of mycobacterial proteins differentially expressed under standing and shaking culture conditions, including Rv2623 from a novel class of putative ATP-binding proteins. Infect. Immun. 69, 5777-5785.

Gazdik, M. A. and McDonough, K.A. (2005). Identification of cyclic AMP-regulated genes in *Mycobacterium tuberculosis* complex bacteria under low-oxygen conditions. J. Bacteriol. 187, 2681-2692.

Gee, J.M., Valderas, M.W., Kovach, M.E., Grippe, V.K., Robertson, G.T., Ng, W.L., Richardson, J.M., Winkler, M.E. and Roop, R.M. 2nd. (2005). The *Brucella abortus* Cu, Zn superoxide dismutase is required for optimal resistance to oxidative killing by murine macrophages and wild-type virulence in experimentally infected mice. Infect. Immun. 73, 2873-2880.

Gronberg, A. and Kihlstrom, E. (1989). Structural variations and growth potential of *Yersinia enterocolitica* under different culture conditions. Apmis 97, 227-235.

Guyton, A.C., Ed. (1991). *Textbook of medical physiology, 8th ed.* Philadelphia, PA, W.B. Saunders Company.

Han, Y.W. and Miller, V.L. (1997). Reevaluation of the virulence phenotype of the *inv yad*A double mutants of *Yersinia pseudotuberculosis*. Infect. Immun. 65, 327-330.

Hare, J. M. and McDonough, K.A. (1999). High-frequency RecA-dependent and -independent mechanisms of congo red binding mutations in *Yersinia pestis*. J. Bacteriol. 181, 4896-4904.

Hinchliffe, S.J., Isherwood, K.E., Stabler, R.A., Prentice, M.B., Rakin, A., Nichols, R.A., Oyston, P.C.F., Hinds, J., Titball, R.W. and Wren, B.W. (2003) Application of DNA microarrays to study the evolutionary genomics of *Yersinia pestis* and *Yersinia pseudotuberculosis*. Genome Res. 13, 2018-2029.

Hinnebusch, B.J. (2003). Transmission factors: *Yersinia pestis* genes required to infect the flea vector of plague. Adv. Exp. Med. Biol. 529, 55-62.

Hinnebusch, B.J., Perry, R.D. and Schwan, T.G. (1996). Role of the *Yersinia pestis* hemin storage (hms) locus in the transmission of plague by fleas. Science 273, 367-370.

Hinnebusch, B.J., Rudolph, A.E., Cherepanov, P., Dixon, J.E., Schwan, T.G. and Forsberg, A. (2002). Role of *Yersinia* murine toxin in survival of *Yersinia pestis* in the midgut of the flea vector. Science 296, 733-735.

Ibrahim, A., Goebel, B.M., Liesack, W., Griffiths, M. and Stackebrandt, E. (1993). The phylogeny of the genus *Yersinia* based on 16S rDNA sequences. FEMS Microb. Letters 114, 173-178.

Jungblut, P.R., Muller, E.C., Mattow, J. and Kaufmann, S.H. (2001). Proteomics reveals open reading frames in *Mycobacterium tuberculosis* H37Rv not predicted by genomics. Infect. Immun. 69, 5905-5907.

Jydegaard-Axelsen, A.M., Hoiby, P.E., Holmstrom, K., Russell, N. and Knochel, S. (2004). CO_2- and anaerobiosis-induced changes in physiology and gene expression of different *Listeria monocytogenes* strains. Appl. Environ. Microbiol. 70, 4111-4117.

Lathem, W.W., Crosby, S.D., Miller, V.L. and Goldman, W.E. (2005). Progression of primary pneumonic plague: A mouse model of infection, pathology, and bacterial transcriptional activity. PNAS 102, 17786-17791.

Lide, D.R., Ed. (1997). *CRC handbook of chemistry and physics*. Cleveland, Ohio, CRC Press.

Masse, E. and Gottesman, S. (2002). A small RNA regulates the expression of genes involved in iron metabolism in *Escherichia coli*. PNAS U S A 99, 4620-4625.

McDonough, K. A. and Falkow, S. (1989). A *Yersinia pestis*-specific DNA fragment encodes temperature-dependent coagulase and fibrinolysin-associated phenotypes. Mol. Microbiol. 3, 767-775.

McDonough, K.A., Florczyk, M.A. and Kress, Y. (2000). Intracellular passage within macrophages affects the trafficking of virulent tubercle bacilli upon reinfection of other macrophages in a serum-dependent manner. Tuber Lung Dis. 80, 259-271.

Mollaret, H. H. (1965). Sur la nomenclature et la taxonomic du bacille de malassez et vignal. Int. Bull. Bacteriol. Nomencl. Taxon. 15, 97-106.

Motin, V.L., Georgescu, A.M., Fitch, J.P., Gu, P.P., Nelson, D.O., Mabery, S.L., Garnham, J.B., Sokhansanj, B.A., Ott, L.L., Coleman, M.A., Elliott, J.M., Kegelmeyer, L.M., Wyrobek, A.J., Slezak, T.R., Brubaker, R.R. and Garcia, E. (2004) Temporal global changes in gene expression during temperature transition in *Yersinia pestis*. J. Bacteriol. 186, 6298-6305.

Najdenski, H.M., Golkocheva, E.N., Vesselinova, A.M. and Russmann, H. (2004). Comparison of the course of infection of virulent *Yersinia enterocolitica* serotype O:8 with an isogenic *sod*A mutant in the peroral rabbit model. Int. J. Med. Microbiol. 294, 383-393.

Narasipura, S.D., Ault, J.G., Behr, M.J., Chaturvedi, V. and Chaturvedi, S. (2003). Characterization of Cu, Zn superoxide dismutase (SOD1) gene knock-out mutant of *Cryptococcus neoformans* var. gattii: role in biology and virulence. Mol. Microbiol. 47, 1681-1694.

Parkhill, J., Wren, B.W., Thomson, N.R., Titball, R.W., Holden, M.T., Prentice, M.B., Sebaihia, M., James, K.D., Churcher, C., Mungall, K.L., Baker, S., Basham, D., Bentley, S.D., Brooks, K., Cerdeno-Tarraga, A.M., Chillingworth, T., Cronin, A., Davies, R.M., Davis, P., Dougan, G., Feltwell, T., Hamlin, N., Holroyd, S., Jagels, K., Karlyshev, A.V., Leather, S., Moule, S., Oyston, P.C., Quail, M., Rutherford, K., Simmonds, M., Skelton, J., Stevens, K., Whitehead, S. and Barrell, B.G. (2001). Genome sequence of *Yersinia pestis*, the causative agent of plague. Nature 413, 523-527.

Perry, R.D. (1993) Acquisition and storage of inorganic iron and hemin by the yersiniae. Trends Microbiol. 1, 142-147.

Perry, R.D. and Fetherston, J.D. (1997). *Yersinia pestis* - etiologic agent of plague. Clin. Microbiol. Rev. 10, 35-66.

Poland, J.D. and Barnes, A.M. (1979). Plague. in J.F. Steele (ed.), *CRC Handbook Series in Zoonoses*, Vol. 1, Section A. Bacterial, rickettsial and mycotic diseases-1979. CRC Press, Inc., Boca Raton, Fla., pp. 515-516.

Portnoy, D.A. and Falkow, S. (1981). Virulence-associated plasmids from *Yersinia enterocolitica* and *Yersinia pestis*. J. Bacteriol. 148, 877-883.

Protsenko, O.A., Anisimov, P.I., Mosharov, O.T., Konnov, N.P., Popov, Y.A. and Kokushkin, A.M. (1983) Detection and characterization of *Yersinia pestis* plasmids determining pesticin I, fraction I antigen, and "mouse" toxin synthesis. Soviet Genet. 19, 838-846.

Purkayastha, A., McCue, L.A. and McDonough, K.A., et al. (2002). Identification of a *Mycobacterium tuberculosis* putative classical nitroreductase gene whose expression is coregulated with that of the acr aene within macrophages, in standing versus shaking cultures, and under low oxygen conditions. Infect. Immun. 70, 1518-1529.

Quan, T. J., Barnes, A.M. and Polland, J.D. (1981). Yersinioses, p. 723-745. In A. Balows and W. J. Hausler, (ed.). *Diagnostic procedures for bacterial, mycotic and parasitic infections,* 6th edition. American Public Health Association, Washington, D. C.

Roggenkamp, A., Bittner, T., Leitritz, L., Sing, A. and Heesemann, J. (1997). Contribution of the Mn-cofactored superoxide dismutase (SodA) to the virulence of *Yersinia enterocolitica* serotype O8. Infect. Immun. 65, 4705-4710.

Rosqvist, R., Skurnik, M. and Wolf-Watz, H. (1988). Increased virulence of *Yersinia pseudotuberculosis* by two independent mutations. Nature 334, 522-525.

Ross, P.L., Huang, Y.N., Marchese, J.N., Williamson, B., Parker, K., Hattan, S., Khainovski, N., Pillai, S., Dey, S., Daniels, S., Purkayastha, S., Juhasz, P., Martin, S., Bartlet-Jones, M., He, F., Jacobson, A. and Pappin, D.J. (2004). Multiplexed protein quantitation in Saccharomyces cerevisiae using amine-reactive isobaric tagging reagents. Mol. Cell. Proteomics 3, 1154-1169.

Sebbane, F., Lemaîre, N., Sturdevant, D.E., Rebeil, R., Virtaneva, K., Porcella, S.F. and Hinnebusch, B.J. (2006). Adaptive response of *Yersinia pestis* to extracellular effectors of innate immunity during bubonic plague. PNAS 103, 11766-11771.

Seoh, H.K. and Tai, P.C. (1999). Catabolic repression of *sec*B expression is positively controlled by cyclic AMP (cAMP) receptor protein-cAMP complexes at the transcriptional level. J. Bacteriol. 181, 1892-1899.

Simonet, M., Riot, B., Fortineau, M. and Berche, P. (1996). Invasin production by *Yersinia pestis* is abolished by insertion of an IS*200*-like element within the *inv* gene. Infect. Immun. 64, 375-379.

Sodeinde, O.A. and Goguen, J.D. (1988). Genetic analysis of the 9.5-kilobase virulence plasmid of *Yersinia pestis*. Infect. Immun. 56, 2743-2748.

Sodeinde, O.A., Sample, A.K., Brubaker, R.R. and Goguen, J.D. (1988). Plasminogen activator/coagulase gene of *Yersinia pestis* is responsible for degradation of plasmid-encoded outer membrane proteins. Infect. Immun. 56, 2749-2752.

Stabler, R.A., Hinds, J., Witney, A.A., Isherwood, K., Oyston, P., Titball, R., Wren, B., Hinchliffe, S., Prentice, M., Mangan, J.A. and Butcher, P.D. (2003). Construction of a *Yersinia pestis* microarray. Adv. Exp. Med. Biol. 529, 47-49.

Straley, S.C. and Perry, R.D. (1995). Environmental modulation of gene expression and pathogenesis in *Yersinia*. Trends Microbiol. 3, 310-317.

Stretton, S. and Goodman, A.E. (1998). Carbon dioxide as a regulator of gene expression in microorganisms. Antonie Van Leeuwenhoek 73, 79-85.

Sun, L., Fukamachi, T., Saito, H. and Kobayashi, H. (2005). Carbon dioxide increases acid resistance in *Escherichia coli*. Lett. Appl. Microbiol. 40, 397-400.

Timm, J., Post, F.A., Bekker, L.G., Walther, G.B., Wainwright, H.C., Manganelli, R., Chan, W.T., Tsenova, L., Gold, B., Smith, I., Kaplan, G. and McKinney, J.D. (2003). Differential expression of iron-, carbon-, and oxygen-responsive mycobacterial genes in the lungs of chronically infected mice and tuberculosis patients. PNAS U S A 100, 14321-14326.

Touati, D. (1997). Superoxide dismutases in bacteria and pathogen protists. Oxidative stress and the molecular biology of antioxidant defenses. J.G. Scandlios. New York, Cold Spring Harbor Laboratory Press: 447-493.

Une, T. (1977). Studies on the pathogenicity of *Yersinia enterocolitica*. III. Comparative studies between *Y. enterocolitica* and Y. *pseudotuberculosis*. Microbiol. Immunol. 21, 505-516.

Zhou, D., Han, Y., Song, Y., Tong, Z., Wang, J., Guo, Z., Pei, D., Pang, X., Zhai, J., Li, M., Cui, B., Qi, Z., Jin, L., Dai, R., Du, Z., Bao, J., Zhang, X., Yu, J., Wang, J., Huang, P. and Yang, R. (2004). DNA microarray analysis of genome dynamics in *Yersinia pestis*: Insights into bacterial genome microevolution and niche adaptation. J. Bacteriol. 186, 5138-5146.

Zhou, D., Qin, L., Han, Y., Qiu, J., Chen, Z., Li, B., Song, Y., Wang, J., Guo, Z., Zhai, J., Du, Z., Wang, X. and Yang, R. (2006). Global analysis of iron assimilation and fur regulation in *Yersinia pestis*. FEMS Microbiol. Lett. 258, 9-17.

Picture 14. Researchers at a poster session. Photograph by R. Perry.

12

Two-Component System Regulon Plasticity in Bacteria: A Concept Emerging from Phenotypic Analysis of *Yersinia pseudotuberculosis* Response Regulator Mutants

Claire Flamez, Isabelle Ricard, Sonia Arafah, Michel Simonet, and Michaël Marceau

Faculté de Médecine Henri Warembourg, Institut Pasteur de Lille, Université Lille II,
michael.marceau@ibl.fr

Abstract. In bacteria, the most rapid and efficient means of adapting gene transcription to extracellular stresses often involves sophisticated systems referred to as two-component systems (2CSs). Although highly conserved throughout the bacterial world, some of these systems may control distinct cell events and have differing contributions to virulence, depending on the species considered. This chapter summarizes the work performed by our group - from the initial PhoP-PhoQ and PmrA-PmrB studies to the most recent genome-scale preliminary analyses - in an attempt to highlight the contribution of 2CS regulon plasticity to the acquisition of some of *Yersinia pseudotuberculosis*' specific features.

12.1 Two-Component Systems, What They are and What They do

By their very nature, two component systems are efficient means of performing transcriptional adaptation in response to sometimes life-threatening environmental changes; one component (the sensor) directly detects the signal outside the cell and transduces it through the inner membrane to the second (the regulator). These signal transduction systems are based on phosphate transfers between histidine and aspartate residues (Fig. 1). After detection of the cognate stimulus (in the periplasm, in most cases), sensors from the same system will phosphorylate each other (explaining why these elements are also referred to as "sensor-kinases"). The phosphate is transferred to the partner response regulator, which then becomes active as a transcriptional regulator, provided it has a DNA-binding domain - this is not always the case, since some response regulators may act as enzymes or exhibit other activities (for review, see West and Stock 2001). The genes coding for the sensor and its cognate response regulator are usually arranged in operons (the transcription of which is generally controlled by the response regulators themselves), along with the effector genes.

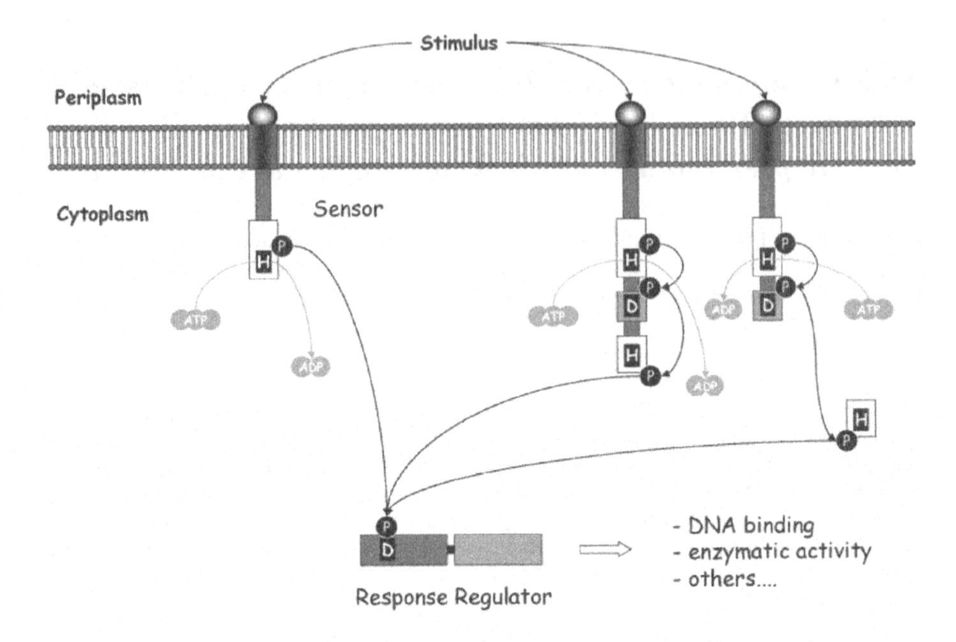

Fig. 1. Models for signal transduction via 2CSs. The prototype two-component system is shown on the left-hand part of the Figure. In some systems (referred to as multicomponent phosphorelays (on the right)), the situation appears to be somewhat more complex, with an intermediate round of phosphotransfers involving a second, conserved aspartate residue harbored by the sensor itself (which is then referred to as a hybrid sensor-kinase). The ultimate phosphate transfer from this residue to the response regulator involves a histidine-containing phosphotransfer (HPt) domain or protein.

12.2 Transcriptional Control of the *pmrF* Operon by the PhoP-PhoQ and PmrA-PmrB Systems: The First Example of 2CS Regulon Plasticity in *Yersinia*

12.2.1 Early Studies of Transcriptional Regulation of the *pmrF* Operon

In vitro, Yersinia pseudotuberculosis is highly resistant to polymyxin B and cationic alpha-helical peptides (Bengoechea et al. 1998). As in other Gram-negative bacteria (including *Salmonella enterica*), this phenotype requires the polypeptide products of the *pmrHFIGKLM* seven-coding-sequence operon (the *pmrF* operon) in order to minimize the initial interactions between the above-mentioned microbicidal compounds and the bacterium's outer membrane (Gunn et al. 2000). In *Salmonella*, transcriptional activation of the *pmrF* operon is induced *in vitro* by high iron concentrations (in experimental practice, $FeCl_3$ or $FeSO_4$ are the salts used). This type of upregulation requires the PmrA-PmrB two-component system, where PmrB is the integral membrane sensor-kinase and PmrA is the cognate regulator which binds to the *pmrF* operon's promoter region (Wosten et al. 2000). A similar induction may also be observed at low pH or following a drop in the extracellular Mg^{2+} concentration. In addition to PmrA-PmrB, this process also requires the PhoP-PhoQ 2CS, which appears to be a major virulence regulator in many pathogenic species (Groisman 2001; Soncini et al. 1996). In *Salmonella*, PhoP (the response regulator) modulates (rather than directly promotes) transcription of the *pmrF* operon by controlling the biosynthesis of PmrD, a post-translational activator of PmrA (Fig. 2a) (Kox et al. 2000; Roland et al. 1994). The results obtained by our group with the *Salmonella* LT2 strain are consistent with this model: resistance to polymyxin is induced by Ca^{2+} and Mg^{2+} ion limitation but decreases dramatically either at low Fe^{3+} concentrations or in *pmrA* or *phoP* mutants.

Both PmrA-PmrB and PhoP-PhoQ systems are found in *Y. pseudotuberculosis*. As expected, resistance to polymyxin was enhanced by low Mg^{2+} and low pH environments in this species but, in contrast to *Salmonella,* it was not reduced when bacterial cells were grown in iron-chelated media. Furthermore, inactivation of PmrA or PmrB did not have any impact on polymyxin resistance, suggesting that the *pmrF* operon plays an identical role in the two species but is differently regulated. In line with this observation, we found that *pmrF* transcription in *Y. pseudotuberculosis*

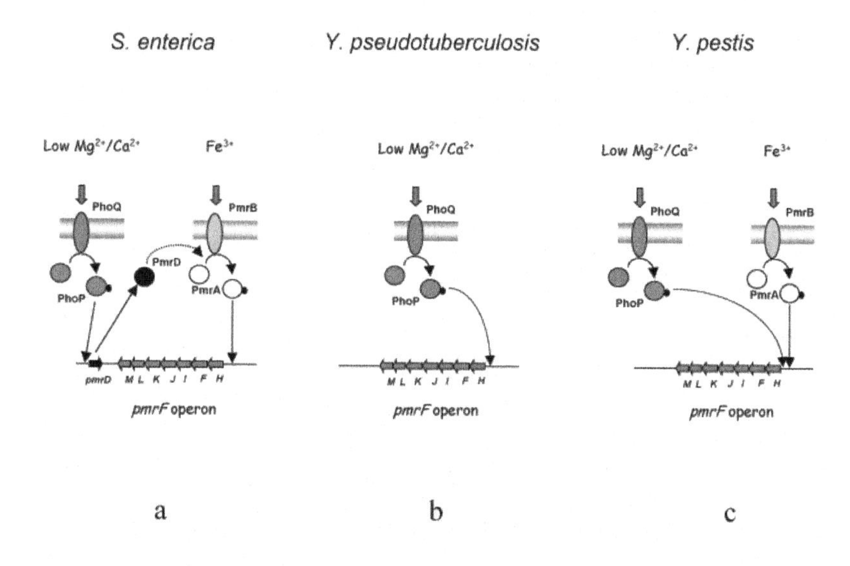

Fig. 2. Comparison of the *Salmonella enterica*, *Y. pseudotuberculosis* and *Y. pestis* models for transcriptional regulation of the *pmrF* operon.

was PmrA-PmrB independent, with PhoP interacting directly with the *pmrF* operon promoter region (Fig. 2b) (Marceau et al. 2004). As reported by another group, this process appears to be similar but not identical in *Yersinia pestis* (Fig. 2c). In this species, both PhoP and PmrA were shown to independently activate transcription of the *pmrF* operon (Winfield et al. 2005). Confirming the differences between these two models, a *phoP* mutant was found to be as susceptible to polymyxin as a *Yersinia pmrF* mutant in *Y. pseudotuberculosis*, whereas the same phenotype in *Y. pestis* requires the inactivation of both 2CSs. However, the contrast between these two new models and the *Salmonella* paradigm illustrates the possibility that bacterial gene regulation networks may rapidly evolve - probably as the consequence of the different needs prompted by dissimilar lifestyles.

12.2.2 PhoP-PhoQ and PmrA-PmrB Regulon Plasticity: Broader Than Expected?

As in many other bacterial species, *phoP*- or *pmrA*-defective *Y. pseudotuberculosis* mutants were found to be less resistant to low pH (Bearson et al. 1998; Foster et al. 1994; Hyytiainen et al. 2003; Llama-Palacios et al. 2003), which may suggest partial conservation of the respective PhoP-PhoQ and PmrA-PmrB regulons between species. In contrast, *phoP* disruption enhanced *Y. pseudotuberculosis'* tolerance to bile salts, whereas a *pmrA* knock-out reduced it. Surprisingly, these two response regulators were found to have opposite contributions to this phenotype in *Salmonella* (Froelich et al. 2006; van Velkinburgh and Gunn 1999). These observations may suggest further regulatory network remodeling and show that in a given situation, even bacteria that are taxonomically very close may behave in a completely opposite fashion via use of the same, highly conserved systems. However, the results of these phenotypic analyses are preliminary and cannot be extrapolated to answer the question of whether genes shared by these two closely-related species are regulated in an opposite manner by their 2CSs.

12.3 Do the Other *Y. pseudotuberculosis* 2CS Regulons Undergo Plasticity? Preliminary Studies

The results described above raised the possibility that *Y. pseudotuberculosis* 2CSs other than PmrA-PmrB and PhoP-PhoQ might also control distinct cell events, when compared with their counterparts in other bacteria. To test this hypothesis, we decided to compare the contributions to stress resistance of all the 2CSs that could be detected in *Y. pseudotuberculosis*, relative to their counterparts in other bacteria.

12.3.1 An *in silico* Inventory of 2CSs in *Y. pseudotuberculosis*

Based on the presence of specific functional domains, we identified 25 putative sensors and 28 potential response-regulators in the *Y. pseudotuberculosis* IP32953 genome sequence (Chain et al. 2004). Most of these proteins are highly conserved, with amino acid sequences more than 60% identical to their respective counterparts in other bacteria. On this basis, the first 13 2CSs (listed in Table 1) could easily be reconstructed *in silico*, whereas the remaining 11 could only be assembled and/or assigned after additional computational analyses, essentially using the detection of peptide signatures at consistent positions (for example, the PmrB sensor-kinase harbors a conserved ExxxE motif reported as binding Fe^{3+} (Wosten et al. 2000)) and the fact that two partner subunit-encoding genes are often found within the same polycistronic unit.

HydG and HydH (alternatively referred to as ZraS and ZraR, respectively) were not found in CO92 and KIM but were detected in the other three *Y. pestis* genomes. Four additional systems (UhpA-UhpB, BaeR-BaeS, RstA-RstB and EvgA-EvgS) are supposedly inactive (due to frameshifts, IS insertions in their coding sequences or larger deletions) in some of the sequenced strains only. Interestingly, inactivation of EvgS may result from distinct genetic events with regard to the *Y. pestis* species in question. Of the five orphan subunits, only YPTB3350 has a predicted function.

12.3.2 Comparative Analysis of *Y. pseudotuberculosis* Response Regulator Mutant Phenotypes Reveals Further Discordances with Other Enterobacterial Species

Based on the above inventory, we inactivated the twenty-two remaining response regulators belonging to complete 2CSs (in addition to PmrA and PhoP) using standard, site-directed mutagenesis techniques. After using the wild-type *Yersinia* strain to determine the optimal conditions for survival assays, we assessed each mutant's *in vitro* resistance to the various types of stress encountered by *Yersinia* cells in the digestive tract: high osmolarity (1 M sodium chloride), oxidative stress (2.5 mM hydrogen peroxide), the presence of bile (15% sodium choleate), inorganic (pH 3) and organic (pH 4.3) acidity and the presence of the antimicrobial peptide polymyxin B (4 $\mu g\ ml^{-1}$). Of these 24 mutants, only 8 (*phoP*, *ompR*, *pmrA* and *ntrC*-, *arcA*-, *rstA*-, *rcsB*- and *yfhA*-like) displayed significantly altered ($p < 0.05$) resistance phenotypes towards at least one of these five environmental stresses (summarized in Table 2). Complementation with the corresponding intact regulatory gene restored the wild-type phenotype, indicating that the engineered mutations did not have a polar effect. It is noteworthy that all the mutants grew as well in LB medium as the wild type strains from which they were derived.

Some mutant phenotypes were expected and consistent with those (when known) of the corresponding putative ortholog mutants in other enteropathogenic species: for example, *Y. pseudotuberculosis ompR* mutants are less resistant to low pH, bile salts and osmotic upshifts (Bang et al. 2000; Hirakawa et al. 2003; Mizuno and Mizushima 1987), whereas ArcA-deficient strains are highly susceptible to hydrogen peroxide (Lu et al. 2002). However, some of the mutants revealed new or unexpected 2CS contributions: tolerance towards bile salts seems to require NtrC, which is supposedly specific for the cellular response to nitrogen limitation (Ninfa et al. 1995). Mutation of the *rstA* gene was associated with a greater susceptibility to oxidative stress, and mutations in *yfhA* and *ompR* were both responsible for higher resistance to polymyxin B - two phenotypes that have never been reported in the literature with these three response regulator mutants. Moreover, abrogated PhoP production

locus	component	relevant modules	*Y. pseudotub.* IP32953	*Y. pestis* CO92	*Y. pestis* antiqua	*Y. pestis* medievalis	*Y. pestis* Nepal 516	*Y. pestis* KIM	function
1	NtrB	HK	YPTB0023	YPO0023	YPA_3519	YP_0024	YPN_0250	y3805	Nitrogen assimilation
	NtrC	RR	YPTB0022	YPO0022	YPA_3520	YP_0023	YPN_0249	y3806	
2	CpxA	HK	YPTB0069	YPO0073	YPA_3470	YP_0073	YPN_3778	y0069	Protein misfolding
	CpxR	RR	YPTB0070	YPO0074	YPA_3469	YP_0074	YPN_3777	y0067	
3	EnvZ	HK	YPTB3763	YPO0137	YPA_3332	YP_0138	YPN_3928	y3917	Osmotic regulation
	OmpR	RR	YPTB3764	YPO0136	YPA_3333	YP_0137	YPN_3929	y3916	
4a	ArcB	HK, Hpt	YPTB3500	YPO3555	YPA_3748	YP_3809	YPN_3437	y0126	Aerobic respiration control
4b	ArcA	RR	YPTB0601	YPO0458	YPA_4049	YP_3725	YPN_0329	y3721	
5	CreC	HK	YPTB3170	YPO0895	YPA_0374	YP_3591	YPN_3094	y3279	Catabolic regulation.
	CreB	RR	YPTB3171	YPO0896	YPA_0373	YP_3592	YPN_3095	y3280	
6	PhoQ	HK	YPTB2435	YPO1633	YPA_1891	YP_1763	YPN_1998	y1793	Ca^{2+}/Mg^{2+} metabolism, virulence
	PhoP	RR	YPTB2434	YPO1634	YPA_1890	YP_1764	YPN_1997	y1794	
7a	CheA	Hpt, HK	YPTB2405	YPO1666	YPA_1855	YP_1796	YPN_1694	y1827	Chemotaxis regulation
7b	CheY	RR	YPTB2397	YPO1680	YPA_1842	YP_1810	YPN_1951	y1842	
8a	BarA	HK, RR, Hpt	YPTB0750	YPO3381	YPA_2880	YP_0304	YPN_0710	y0808	Hydrogen peroxide sensitivity
8b	UvrY	RR	YPTB1735	YPO1865	YPA_1243	YP_1528	YPN_2259	y2443	
9	RstB	HK	YPTB2231	YPO2309	YPA_1657	IS 266	YPN_1769	y2140	Possibly involved in stress response.
	RstA	RR	YPTB2230	YPO2308	YPA_1656	YP_2093	YPN_1768	y2139	
10	KdpD	HK	YPTB2919	YPO2689	YPA_2419	YP_2492	YPN_1176	y1262	Resistance to low turgor pressure
	KdpE	RR	YPTB2920	YPO2688	YPA_2418	YP_2491	YPN_1175	y1261	
11	BaeS	HK	YPTB2817	F/S 370	F/S 370	YP_2718	F/S 370	F/S 370	Resistance to extracellular stresses
	BaeR	RR	YPTB2818	YPO2853	YPA_2292	YP_2719	YPN_1285	y1381	
12	YfhK	HK	YPTB2876	YPO2916	YPA_2355	YP_2541	YPN_1222	y1313	Sigma-54 dependent regulation
	YfhA	RR	YPTB2874	YPO2914	YPA_2353	YP_2543	YPN_1224	y1315	
13	PhoR	HK	YPTB0918	YPO3204	YPA_2697	YP_0728	YPN_0885	y0979	Control of the phosphate regulon
	PhoB	RR	YPTB0917	YPO3205	YPA_2698	YP_0727	YPN_0884	y0978	
14a	RcsC	HK, RR	YPTB1257	YPO1217	YPA_0931	YP_0920	YPN_2760	y2971	Control of exopolysaccharide biosynthesis ?
14b	YojN	HPt	YPTB1259	F/S 642	F/S 642	F/S 642	F/S 642	F/S 642	
	RcsB	RR	YPTB1258	YPO1218	YPA_0932	YP_0919	YPN_2759	y2970	
15	UhpB	HK	YPTB3846	IS 266	IS 266	YP_3370	IS 266	IS 266	Sugar phosphate transport
	UhpA	RR	YPTB3847	YPO4012	YPA_4100	YP_3371	YPN_3660	y4033	
16	PmrB	HK	YPTB0468	YPO3508	YPA_0058	YP_0575	YPN_3252	y0676	Fe^{3+}-induced regulation ?
	PmrA	RR	YPTB0469	YPO3507	YPA_0057	YP_0576	YPN_3251	y0677	
17	ZraS	HK	YPTB1204	NF	YPA_1084	YP_0984	YPN_2825	NF	Response to zinc and lead.
	ZraR	RR	YPTB1205	NF	YPA_1085	YP_0983	YPN_2824	NF	
18	CopS / CusS	HK	YPTB1992	YPO2000	YPA_1382	YP_1847	YPN_1480	y2308	Copper resistance
	CopR / CusR	RR	YPTB1993	YPO2001	YPA_1383	YP_1848	YPN_1481	y2307	
19a	EvgS	HK, RR	YPTB1922	F/S 419	YPA_1302	YP_1666	NF	y2387	Possible regulation of virulence
19b	EvgA/ FimZ	RR	YPTB1923	YPO1925	YPA_1303	YP_1667	NF	y2386	
20a	YehU / LytS	HK	YPTB3788	YPO3943	YPA_3772	YP_3305	YPN_3591	y3885	Unknown function.
20b	YehT / LytR	RR	YPTB0840	YPO3287	YPA_2858	YP_0397	YPN_0812	y0902	
21a	NarX	HK	YPTB1957	YPO1959	YPA_1342	YP_1704	NF	y2351	Nitrate-nitrite metabolism
21b	NarP	RR	YPTB2763	YPO3041	YPA_2232	YP_2664	YPN_1344	y1441	
22	-	HK	YPTB2728	YPO3008	YPA_2197	YP_2632	YPN_1375	y1473	Unknown function.
	-	RR	YPTB2729	YPO3009	YPA_2198	YP_2633	YPN_1374	y1472	
23	-	HK	YPTB2718	YPO2997	YPA_2187	YP_2622	YPN_1385	y1483	Unknown function
	-	RR	YPTB2719	YPO2998	YPA_2188	YP_2623	YPN_1384	y1482	
24	-	HK, RR	YPTB0311	YPO0256	YPA_4028	YP_0409	YPN_3413	y0513	Possible PI regulating system
	-	RR	YPTB0310	YPO0255	YPA_4029	YP_0408	YPN_3414	y0512	
Orphans									
25	-	HK, RR	YPTB3808	YPO3965	YPA_3793	YP_3328	YPN_3614	y3864	Unknown function
26	-	RR	YPTB3801	YPO3958	YPA_3786	YP_3321	YPN_3607	y3871	Unknown function
27	-	RR	YPTB2099	YPO2173	YPA_1531	YP_1972	YPN_1640	y2148	RpoS regulator
28	FleR	RR	YPTB3350	YPO0712	YPA_3080	YP_3024	YPN_0570	y3466	Flagellar reg.
29	-	RR	YPTB1603	YPO1576	YPA_0872	YP_1464	YPN_2404	y2588	Unknown function

Table 1. The twenty-four complete *Y. pseudotuberculosis* 2CS (including the seven complex phosphotranfer systems) and five orphans (with no obvious partners) and their possible counterparts found in the five *Y. pestis* genome sequences (Chain et al. 2006; Deng et al. 2002; Parkhill et al. 2001; Song et al. 2004) currently available in GenBank (ftp://ftp.ncbi.nih.gov/genomes/Bacteria/). With the exception of PhoP-PhoQ and PmrA-PmrB, all the 2CS functions listed here are putative and – even though highly probable– have not been verified by experimental evidence. Unlike the situation in *Escherichia coli*, *yehT* and *yehU* are found at separate locations on the *Yersinia* chromosome. The NarX sensor and NarP regulator are part of distinct 2CSs (NarX-NarL and NarQ-NarP), both of which reportedly contribute to regulation of nitrate/nitrite metabolism (Stewart 1994) in other enterobacteria. The fact that NarQ and NarL are absent from the *Yersinia* genomes and that NarX can cross-activate NarP in *E. coli* strongly suggests that these apparent orphans may work as cognate partners in *Yersinia*. Abbreviations: PI: pathogenicity island; IS: insertion sequence after amino acid no.; F/S: frameshift after amino acid no.; NF: not found.

in *Y. pseudotuberculosis* was linked to increased bacterial resistance to hydrogen peroxide - a phenotypic modification never previously described in *phoP* mutants in other prokaryotes. In addition to confirming the contrasting results reported above with PmrA and PhoP, our experiments showed that ArcA was necessary for survival of *Y. pseudotuberculosis* in the presence of sodium choleate, whereas it seems not to be essential in *Salmonella* (Lu et al. 2002). Moreover, the contribution of PhoP was found to be essential for resistance to high osmolarity (Oyston et al. 2000) in *Y. pseudotuberculosis* - again, as in *Y. pestis* but in contrast to the situation in *Salmonella* (Flamez, unpublished results).

Of the selected mutants, only *phoP* and *ompR* mutants displayed the expected, attenuated virulence in the mouse model. However, RstA and YfhA were also identified as virulence regulators in *Y. pseudotuberculosis*, despite the fact that they are not considered as such in other bacterial pathogens. In contrast, PmrA, RcsB and ArcA are known to mediate virulence in other bacterial species (Gunn et al. 2000; Humphreys et al. 2004; Sengupta et al. 2003) but were found to be non-essential in *Y. pseudotuberculosis* when administered orally or intravenously. This type of situation was also observed with CpxR, KdpE, UvrY and EvgA - all known to control virulence in other pathogens (Altier 2005; Cotter and Jones 2003; Gal-Mor and Segal 2003; Nakayama and Watanabe 1998; Tomenius et al. 2006). However, these numerous discordances may also result from distinct stresses encountered by the various pathogens during their respective courses of infection.

12.3.3 What Can We Conclude From These Preliminary Analyses?

These numerous unexpected phenotypes of *Y. pseudotuberculosis* response regulator mutants (relative to their counterparts in other bacterial species) are consistent with the emerging concept whereby regulons are constantly being remodeled (Lozada-Chavez et al. 2006). However, biological phenomena other than regulon plasticity (such as alternative/redundant regulation pathways or cellular functions) could perhaps also account for some of the phenotypic discordances reported here. Furthermore, these analyses may well have been complicated by the existence of interconnections between these two systems (i.e. cross-regulation) and the fact that strain-to-strain

Table 2. Comparison in *Y. pseudotuberculosis* strain IP32777 *vs* other pathogens of the contribution of response regulators to resistance to five stresses likely to be encountered following an oral challenge. Elements that are reportedly not required for these resistance phenotypes are not listed here. New or unexpected phenotypes are indicated in bold characters.

Response regulator	Stress *in vitro*	*Y. pseudotuberculosis*		Other bacteria		
		Mutant resistance phenotype	Role in virulence (in mice) ?	Mutant resistance phenotype	experimental evidence	Role in virulence (in mice) ?
RcsB	**bile salts**	**increased (1)**	no	**unknown**	-	Yes
NtrC	**bile salts**	**reduced**	no	**unknown**	-	Yes
RstA	**oxidative**	**reduced**	yes	**unknown**	-	not shown
YfhA	**polymyxin**	**increased**	yes	**unknown**	-	not shown
OmpR	low pH	reduced	yes	reduced	Bang et al. 2000	yes
	high osmolarity	reduced		reduced	Mizuno et al. 1987	
	bile salts	reduced		reduced	Brzostek et al. 2003	
	polymyxin	**increased**		**unknown**	-	
ArcA	**bile salts**	**reduced**	no	**unchanged**	Lu et al. 2002	yes
	oxidative	reduced		reduced	Turlin et al. 2005	
PhoP	low pH	reduced	yes	reduced	Bearson et al. 1998	yes
	high osmolarity	reduced		reduced	Oyston et al. 2000	
	bile salts	**increased**		**reduced**	van Velkinburgh and Gunn 1999	
	oxidative	**increased**		**reduced**	Tu et al. 2006	
	polymyxin	reduced		reduced	Gunn et al. 1996	
PmrA	low pH	reduced	not significant	reduced	Hagiwara et al. 2004	yes
	bile salts	**reduced**		**unchanged**	Froelich et al. 2006	
	polymyxin	unchanged		reduced	Roland et al. 1993	

(1) Possibly via an RcsA-dependent mechanism

differences may exist. Given that *Y. pseudotuberculosis* arrays are now available, transcriptome comparisons of wild-type and mutant response regulators should provide further information on ecological adaptation and, therefore, the divergence of bacterial species through 2CS network remodeling.

12.4 Acknowledgments

C. Flamez received a postgraduate fellowship from the Région Nord-Pas-de-Calais and from the Institut Pasteur de Lille. We thank C. Mullet for technical assistance, P. Vincent for assistance with statistical analysis and Dr Robert Perry and the scientific committee of the Ninth International Symposium on *Yersinia* for selecting our work for oral presentation.

12.5 References

Altier, C. (2005) Genetic and environmental control of *Salmonella* invasion. J Microbiol. 43, Spec No, 85-92.

Bang, I.S., Kim, B.H., Foster, J.W. and Park, Y.K. (2000) OmpR regulates the stationary-phase acid tolerance response of *Salmonella enterica* serovar Typhimurium. J. Bacteriol. 182, 2245-2252.

Bearson, B.L., Wilson, L. and Foster, J.W. (1998) A low pH-inducible, PhoPQ-dependent acid tolerance response protects *Salmonella typhimurium* against inorganic acid stress. J. Bacteriol. 180, 2409-2417.

Bengoechea, J.A., Lindner, B., Seydel, U., Diaz, R. and Moriyon, I. (1998) *Yersinia pseudotuberculosis* and *Yersinia pestis* are more resistant to bactericidal cationic peptides than *Yersinia enterocolitica*. Microbiology. 144, 1509-1515.

Brzostek, K., Raczkowska, A. and Zasada, A. (2003) The osmotic regulator OmpR is involved in the response of *Yersinia enterocolitica* O:9 to environmental stresses and survival within macrophages. FEMS Microbiol. Lett. 228, 265-271.

Chain, P.S., Carniel, E., Larimer, F.W., Lamerdin, J., Stoutland, P.O., Regala, W.M., Georgescu, A.M., Vergez, L.M., Land, M.L., Motin, V.L., Brubaker, R.R., Fowler, J., Hinnebusch, J., Marceau, M., Medigue, C., Simonet, M., Chenal-Francisque, V., Souza, B., Dacheux, D., Elliott, J.M., Derbise, A., Hauser, L.J. and Garcia, E. (2004) Insights into the evolution of *Yersinia pestis* through whole-genome comparison with *Yersinia pseudotuberculosis*. Proc. Natl. Acad. Sci. U. S. A. 101, 13826-13831.

Chain, P.S., Hu, P., Malfatti, S.A., Radnedge, L., Larimer, F., Vergez, L.M., Worsham, P., Chu, M.C. and Andersen, G.L. (2006) Complete genome sequence of *Yersinia pestis* strains Antiqua and Nepal516: evidence of gene reduction in an emerging pathogen. J. Bacteriol. 188, 4453-4463.

Cotter, P.A. and Jones, A.M. (2003) Phosphorelay control of virulence gene expression in *Bordetella*. Trends Microbiol. 11, 367-373.

Deng, W., Burland, V., Plunkett, G., 3rd, Boutin, A., Mayhew, G.F., Liss, P., Perna, N.T., Rose, D.J., Mau, B., Zhou, S., Schwartz, D.C., Fetherston, J.D., Lindler, L.E., Brubaker, R.R., Plano, G.V., Straley, S.C., McDonough, K.A., Nilles, M.L., Matson, J.S., Blattner, F.R. and Perry, R.D. (2002) Genome sequence of *Yersinia pestis* KIM. J. Bacteriol. 184, 4601-4611.

Foster, J.W., Park, Y.K., Bang, I.S., Karem, K., Betts, H., Hall, H.K. and Shaw, E. (1994) Regulatory circuits involved with pH-regulated gene expression in *Salmonella typhimurium*. Microbiology. 140, 341-352.

Froelich, J.M., Tran, K. and Wall, D. (2006) A *pmrA* constitutive mutant sensitizes *Escherichia coli* to deoxycholic acid. J. Bacteriol. 188, 1180-1183.

Gal-Mor, O. and Segal, G. (2003) Identification of CpxR as a positive regulator of *icm* and *dot* virulence genes of *Legionella pneumophila*. J. Bacteriol. 185, 4908-4919.

Groisman, E.A. (2001) The pleiotropic two-component regulatory system PhoP-PhoQ. J. Bacteriol. 183, 1835-1842.

Gunn, J.S. and Miller, S.I. (1996) PhoP-PhoQ activates transcription of *pmrAB*, encoding a two-component regulatory system involved in *Salmonella typhimurium* antimicrobial peptide resistance. J. Bacteriol. 178, 6857-6864.

Gunn, J.S., Ryan, S.S., Van Velkinburgh, J.C., Ernst, R.K. and Miller, S.I. (2000) Genetic and functional analysis of a PmrA-PmrB-regulated locus necessary for lipopolysaccharide modification, antimicrobial peptide resistance, and oral virulence of *Salmonella enterica* serovar typhimurium. Infect. Immun. 68, 6139-6146.

Hagiwara, D., Yamashino, T. and Mizuno, T. (2004) A genome-wide view of the *Escherichia coli* BasS-BasR two-component system implicated in iron-responses. Biosci. Biotechnol. Biochem. 68, 1758-1767.

Hirakawa, H., Nishino, K., Hirata, T. and Yamaguchi, A. (2003) Comprehensive studies of drug resistance mediated by overexpression of response regulators of two-component signal transduction systems in *Escherichia coli.* J. Bacteriol. 185, 1851-1856.

Humphreys, S., Rowley, G., Stevenson, A., Anjum, M.F., Woodward, M.J., Gilbert, S., Kormanec, J. and Roberts, M. (2004) Role of the two-component regulator CpxAR in the virulence of Salmonella enterica serotype Typhimurium. Infect. Immun. 72, 4654-4661.

Hyytiainen, H., Sjoblom, S., Palomaki, T., Tuikkala, A. and Tapio Palva, E. (2003) The PmrA-PmrB two-component system responding to acidic pH and iron controls virulence in the plant pathogen *Erwinia carotovora* ssp. carotovora. Mol. Microbiol. 50, 795-807.

Kox, L.F., Wosten, M.M. and Groisman, E.A. (2000) A small protein that mediates the activation of a two-component system by another two-component system. Embo J. 19, 1861-1872.

Llama-Palacios, A., Lopez-Solanilla, E., Poza-Carrion, C., Garcia-Olmedo, F. and Rodriguez-Palenzuela, P. (2003) The *Erwinia chrysanthemi phoP-phoQ* operon plays an important role in growth at low pH, virulence and bacterial survival in plant tissue. Mol. Microbiol. 49, 347-357.

Lozada-Chavez, I., Janga, S.C. and Collado-Vides, J. (2006) Bacterial regulatory networks are extremely flexible in evolution. Nucleic Acids Res. 34, 3434-3445.

Lu, S., Killoran, P.B., Fang, F.C. and Riley, L.W. (2002) The global regulator ArcA controls resistance to reactive nitrogen and oxygen intermediates in *Salmonella enterica* serovar Enteritidis. Infect. Immun. 70, 451-461.

Marceau, M., Sebbane, F., Ewann, F., Collyn, F., Lindner, B., Campos, M.A., Bengoechea, J.A. and Simonet, M. (2004) The *pmrF* polymyxin-resistance operon of *Yersinia pseudotuberculosis* is upregulated by the PhoP-PhoQ two-component system but not by PmrA-PmrB, and is not required for virulence. Microbiology 150, 3947-3957.

Mizuno, T. and Mizushima, S. (1987) Isolation and characterization of deletion mutants of *ompR* and *envZ*, regulatory genes for expression of the outer membrane proteins OmpC and OmpF in *Escherichia coli.* J. Biochem. (Tokyo) 101, 387-396.

Nakayama, S. and Watanabe, H. (1998) Identification of *cpxR* as a positive regulator essential for expression of the *Shigella sonnei virF* gene. J. Bacteriol. 180, 3522-3528.

Ninfa, A.J., Atkinson, M.R., Kamberov, E.S., Feng, J. and Ninfa, E.G. (1995). Control of nitrogen assimilation by the NRI-NRII two-component system of enteric bacteria, In: J.A. Hoch and T.J. Silhavy (ed.), *Two-component signal transduction.* American Society for Microbiology, Washington, D.C. pp. 67–88.

Oyston, P.C., Dorrell, N., Williams, K., Li, S.R., Green, M., Titball, R.W. and Wren, B.W. (2000) The response regulator PhoP is important for survival under conditions of macrophage-induced stress and virulence in *Yersinia pestis.* Infect. Immun. 68, 3419-3425.

Parkhill, J., Wren, B.W., Thomson, N.R., Titball, R.W., Holden, M.T., Prentice, M.B., Sebaihia, M., James, K.D., Churcher, C., Mungall, K.L., Baker, S., Basham, D., Bentley, S.D., Brooks, K., Cerdeno-Tarraga, A.M., Chillingworth, T., Cronin, A., Davies, R.M., Davis, P., Dougan, G., Feltwell, T., Hamlin, N., Holroyd, S., Jagels, K., Karlyshev, A.V., Leather, S., Moule, S., Oyston, P.C., Quail, M., Rutherford, K., Simmonds, M., Skelton, J., Stevens, K., Whitehead, S. and Barrell, B.G. (2001) Genome sequence of *Yersinia pestis*, the causative agent of plague. Nature. 413, 523-527.

Roland, K.L., Martin, L.E., Esther, C.R. and Spitznagel, J.K. (1993) Spontaneous *pmrA* mutants of *Salmonella typhimurium* LT2 define a new two-component regulatory system with a possible role in virulence. J. Bacteriol. 175, 4154-4164.

Roland, K.L., Esther, C.R. and Spitznagel, J.K. (1994) Isolation and characterization of a gene, pmrD, from *Salmonella typhimurium* that confers resistance to polymyxin when expressed in multiple copies. J. Bacteriol. 176, 3589-3597.

Sengupta, N., Paul, K. and Chowdhury, R. (2003) The global regulator ArcA modulates expression of virulence factors in *Vibrio cholerae*. Infect. Immun. 71, 5583-5589.

Soncini, F.C., Garcia Vescovi, E., Solomon, F. and Groisman, E.A. (1996) Molecular basis of the magnesium deprivation response in *Salmonella typhimurium*: identification of PhoP-regulated genes. J. Bacteriol. 178, 5092-5099.

Song, Y., Tong, Z., Wang, J., Wang, L., Guo, Z., Han, Y., Zhang, J., Pei, D., Zhou, D., Qin, H., Pang, X., Zhai, J., Li, M., Cui, B., Qi, Z., Jin, L., Dai, R., Chen, F., Li, S., Ye, C., Du, Z., Lin, W., Yu, J., Yang, H., Huang, P. and Yang, R. (2004) Complete genome sequence of *Yersinia pestis* strain 91001, an isolate avirulent to humans. DNA Res. 11, 179-197.

Stewart, V. (1994) Dual interacting two-component regulatory systems mediate nitrate- and nitrite-regulated gene expression in *Escherichia coli*. Res. Microbiol. 145, 450-454.

Tomenius, H., Pernestig, A.K., Jonas, K., Georgellis, D., Mollby, R., Normark, S. and Melefors, O. (2006) The *Escherichia coli* BarA-UvrY two-component system is a virulence determinant in the urinary tract. BMC Microbiol. 6, 27.

Tu, X., Latifi, T., Bougdour, A., Gottesman, S. and Groisman, E.A. (2006) The PhoP/PhoQ two-component system stabilizes the alternative sigma factor RpoS in *Salmonella enterica*. Proc. Natl. Acad. Sci. U. S. A. 103, 13503-13508.

Turlin, E., Sismeiro, O., Le Caer, J.P., Labas, V., Danchin, A. and Biville, F. (2005) 3-phenylpropionate catabolism and the *Escherichia coli* oxidative stress response. Res. Microbiol. 156, 312-321.

van Velkinburgh, J.C. and Gunn, J.S. (1999) PhoP-PhoQ-regulated loci are required for enhanced bile resistance in *Salmonella* spp. Infect. Immun. 67, 1614-1622.

West, A.H. and Stock, A.M. (2001) Histidine kinases and response regulator proteins in two-component signaling systems. Trends Biochem. Sci. 26, 369-376.

Winfield, M.D., Latifi, T. and Groisman, E.A. (2005) Transcriptional regulation of the 4-amino-4-deoxy-L-arabinose biosynthetic genes in *Yersinia pestis*. J. Biol. Chem. 280, 14765-14772.

Wosten, M.M., Kox, L.F., Chamnongpol, S., Soncini, F.C. and Groisman, E.A. (2000) A signal transduction system that responds to extracellular iron. Cell. 103, 113-125.

Picture 15. Susan Straley introduces Michaël Marceau. Photograph by A. Anisimov.

13
Regulatory Elements Implicated in the Environmental Control of Invasin Expression in Enteropathogenic *Yersinia*

Ann Kathrin Heroven, Katja Böhme, Hien Tran-Winkler, and Petra Dersch

Institut für Mikrobiologie, Technische Universität Braunschweig,
p.dersch@tu-bs.de

Abstract. During infections of the intestinal tract *Yersinia pseudotuberculosis* penetrates the epithelial cell layer through M-cells into the Peyer's patches. This early step in the infection process is primarily mediated by the outer membrane protein invasin. Expression of the invasin gene is activated by the MarR-type regulatory protein RovA in response to environmental conditions, including temperature and growth phase. In order to gain insight into the nature of the underlying control systems, mutagenesis and gene bank screens were used to identify regula components modulating the levels of invasin and RovA. We found that the *inv* and *rovA* genes were both subjected to silencing by the nucleoid-associated protein H-NS. Under inducing conditions, RovA appears to disrupt the silencer complex, through displacement of H-NS from an extended AT-rich region located upstream of the *inv* and *rovA* promoters. Furthermore, a LysR-type regulatory protein, RovM with homology to HexA/PecT of phytopathogenic *Erwinia* species was shown to interact specifically with the *rovA* regulatory region and represses *rovA* transcription in addition to H-NS. Disruption of the *rovM* gene significantly enhanced internalization of *Y. pseudotuberculosis* into host cells and higher numbers of the mutant bacteria were detectable in gut-associated lymphatic tissues and organs in infected mice. In addition, the histone-like protein YmoA, which has a global effect on the bacterial physiology, was found to activate *rovA* expression through RovM. Together, our studies showed, that H-NS, RovM and YmoA are key regulators implicated in the environmental control of virulence factors, which are important for the initiation of a *Yersinia* infection.

13.1 Introduction

During the course of infection the enteropathogenic *Yersinia* species, *Yersinia enterocolitica* and *Yersinia pseudotuberculosis* encounter the host mucosal surfaces overlaying the intestinal cells. Subsequently, the bacteria invade and translocate the epithelial cell layer through M-cells and spread into underlying lymphoid tissues (Peyer's patches). This initial step in the infection process is primarily mediated by the 103 kDa outer membrane protein invasin, which mediates tight binding to host β_1-integrin receptors on the apical surface of the M-cells, resulting in the internalization of the bacteria and translocation across the intestinal mucosa (Clark et al. 1998; Isberg and Leong 1990). Mutation of the *inv* gene revealed that this adhesion factor is not essential for the infection of mice and does not affect the outcome or alter the lethal dose at which 50% of the mice succumb to infection. However, loss of the *inv* gene delayed the infection process and resulted in a reduced rate of colonization of the Peyer's patches, implying that invasin functions predominantly during the initial steps of an *Y. pseudotuberculosis* infection (Marra and Isberg 1996; Pepe and Miller 1993).

13.1.1 Environmental Control of Invasin Expression

First, *in vitro* expression studies showed that expression of invasin is subject to tight regulation by multiple environmental signals, including temperature, growth phase, but also nutrients, osmolarity and pH. In both enteropathogenic *Yersinia* species, invasin is maximally expressed at moderate growth temperatures in bacteria grown to late stationary phase (Nagel et al. 2001; Pepe et al. 1994). These conditions are usually found outside hosts but it has been assumed that increased synthesis of invasin under these conditions could prime the bacteria to guarantee a rapid and efficient penetration of the intestinal tracts, shortly after ingestion. This is supported by a recent *in vivo* expression study in our laboratory in which we used *inv-gfp* expressing *Y. pseudotuberculosis* for oral infections of mice. We found that invasin is highly expressed in the Peyer's patches and organs during the first hours after oral uptake but cannot be detected at later time points during the infection. As invasin is of little use after passing through M-cells but renders the bacterium more susceptible to host immune responses, repression of its expression at 37°C may be advantageous for the persistence of *Yersinia* at later stages of the infection.

13.1.2 Identification of RovA, an Activator of *inv* Expression

The first regulator of *inv* expression was found by a transposon mutagenesis strategy in *Y. enterocolitica*. Two independent transposon mutants which showed negligible invasin levels harbored an insertion in the same gene named *rovA* for regulator of virulence A (Revell and Miller 2000). RovA was also identified in *Y. pseudotuberculosis* by a different genetic complementation procedure screening for induction of a non-expressed *inv-phoA* fusion in *Escherichia coli* (Nagel et al. 2001). The *rovA* gene encodes a 143 amino acid protein, which is also present in *Yersinia pestis* and shows considerable homology to members of the SlyA/Hor/Rap family of global transcriptional regulators. These regulatory proteins were shown to both positively and negatively modulate the expression of multiple genes and are implicated in the control of a wide range of physiological processes that are involved in environmental and host-associated stress adaptation and virulence. For instance, SlyA of *Salmonella typhimurium* is important for environmental adaptation, survival of the bacteria in macrophages and for virulence in mice (Libby et al. 1994; Linehan et al. 2005; Navarre et al. 2005; Stapleton et al. 2002). SlyA of *E. coli* activates the expression of the cryptic hemolysin ClyA and controls genes crucial for surviving heat and acidic stress conditions (Oscarsson et al. 1996; Spory et al. 2002). Other RovA homologues, such as PecS/Hor of phytopathogenic *Erwinia* species and Rap in *Serratia marcescens*, regulate the synthesis of secondary metabolites, including pigments, antibiotics and extracellular macerating enzymes, such as cellulase and pectate lyases, implicated in plant pathogenicity (Thomson et al. 1997). Phylogenetic analysis indicated that the RovA/SlyA/Hor proteins belong to the large family of MarR-type transcriptional regulators, which are widely distributed in various species of bacteria and archaea, and represent a subgroup of the family placed in a separate phylogenetic tree between other MarR-like subfamilies.

13.2 The Transcriptional Regulator RovA

To date, a three-dimensional structure of a RovA/SlyA/Hor protein is not available. However, in an attempt to define the structural organization and functional domains of RovA, we identified various missense and deletion mutants of RovA, which were impaired in their ability to activate *inv* expression. The characterization of the mutants and *in silico* modeling indicated that the overall structure of the RovA protein appears to resemble that of other MarR-type proteins (Tran et al. 2005).

13.2.1 Structure of the RovA Regulator Protein

Crosslinking and gel filtration experiments revealed that RovA forms stable homodimers in solution. The analysis of terminal deletion mutants of RovA further showed that the dimerization domain seems to include the first N-terminal and the last two C-terminal helices, which all contribute to the formation of an extensive and well-packed dimer interface. The RovA protein further employs an internal winged-helix DNA binding motif, consisting of a Helix-Turn-Helix motif (HTH) followed by two β-sheet wings. This internal region exhibits its highest homology to all the other MarR-type proteins, and most of our isolated non-functional RovA mutant proteins, which were strongly impaired in DNA binding, harbored amino acid substitutions within this central region. It is still unknown how RovA interacts with DNA, but the DNA-bound structure of the MarR-type protein OhrR of *Bacillus subtilis* suggests that DNA binding is achieved by interaction of the HTH recognition helix with the major groove of the DNA (Hong et al. 2005).

13.2.2 The DNA-Binding Sites of RovA

In our first attempt to explore the underlying mechanism of RovA-mediated activation of gene transcription, we examined the interaction of RovA with the *inv* regulatory region, defined the requirements of RovA for *inv* promoter recognition and characterized the RovA DNA-binding sequence. DNA-binding studies to define RovA binding sites revealed that the RovA protein protects two separate regions in the *inv* promoter. Interaction with both binding sequences is absolutely required for full activation of the *inv* promoter, as upstream deletions removing these RovA binding sites abolished environmentally controlled RovA-mediated transcription (Heroven et al. 2004). The inspection of the RovA target sites did not lead to the identification of a typical palindromic consensus sequence as found for most DNA-binding proteins, but revealed high AT abundance and the occurrence of poly(AT) stretches in the protected binding regions. This is a typical characteristic of curved DNA sequences, but the *inv* regulatory DNA fragments did not migrate aberrantly in polyacrylamide gels at 4°C, indicating that DNA bending is not an important feature of RovA binding.

However, it is possible that a special local topology or high flexibility of the DNA is required for RovA-DNA complex formation. To determine the nucleotides that are essential for RovA binding, we introduced modifications within the sequence spanning the protected RovA-binding regions and tested RovA interaction (Heroven et al. 2004). We found that a sequence with a short non-palindromic consensus $^{A}/_{T}$ATTAT$^{A}/_{T}$ which is found in each target site is important for RovA-DNA interaction.

13.2.3 Influence of RovA on *Yersinia* Physiology and Virulence

Besides DNA-binding mechanism and dimerization, RovA possesses additional functions and properties that apparently allow RovA to specifically activate virulence gene expression. We found that the last four C-terminal amino acids are important for gene activation. Deletions of these final residues result in a complete loss of *inv* and *rovA* promoter activation, although the mutant protein forms dimers and binding to its DNA target sites is still intact. This suggests that the C-terminal tail of the protein enhances transcription through interaction with RNA polymerase (Tran et al. 2005).

Similar to other members of the SlyA/Hor family, RovA was shown to be a global regulator that influences the expression of multiple other genes implicated in bacterial metabolism and physiology, stress adaptation and virulence. First, it was observed that a *Y. enterocolitica rovA* mutant has a much more attenuated phenotype than wild-type *Yersinia* and the *inv* mutant (Dube et al. 2003; Revell and Miller 2000). Most significantly, the LD_{50} of the *rovA* mutant is about 70-fold higher than that of the wild-type and the progression of the infection, in particular the dissemination of the bacteria into deeper tissues is considerably reduced. Mice orally infected with a *Y. enterocolitica rovA* mutant also show fundamental differences in their inflammatory response. They fail to induce IL-1α expression and have a greatly reduced inflammation in the Peyer′s patches (Dube et al. 2001). Mouse infection experiments in our laboratory demonstrated that a *rovA* mutant strain of *Y. pseudotuberculosis* is also significantly attenuated in virulence. After oral infection, considerable less bacteria of the *rovA* mutant strain were recovered from the Peyer′s patches and the mesenterial lymph nodes of infected BALB/c mice, and no or very low numbers of bacteria were found in the liver, kidney and spleen (Heroven and Dersch 2006; Nagel et al. 2003). Strikingly, the RovA regulatory protein of *Y. pestis*, which lacks a functional *inv* gene, is also required for bubonic plague. Recently, it was shown that a *Y. pestis rovA* mutant is strongly attenuated (LD_{50} is about 80-fold increased) and shows a significantly delayed dissemination of the bacteria to the lungs and the spleen after subcutaneous inoculation (Cathelyn et al. 2006).

All these data suggest that RovA regulates numerous factors in addition to invasin that are required for full virulence of all pathogenic *Yersinia* species. In fact, two-dimensional gel electrophoresis showed that the expression of at least 50 proteins is either up- or downregulated in a *Y. pseudotuberculosis rovA* mutant (Nagel et al. 2003). In order to define the RovA regulon and to analyze RovA influence (activation/repression) on the expression of these genes, we developed a screening system to find additional RovA-dependent promoters in the genome of *Y. pseudotuberculosis.*

To do so, DNA fragments of the *Y. pseudotuberculosis* genome and virulence plasmid were inserted into the multiple cloning site of a promoter probe vector, encoding a promoterless *phoA* gene. The activity of the cloned promoter fragment was then tested in the presence and absence of RovA. Up to date, 30 different RovA-dependent promoters of *Y. pseudotuberculosis* have been isolated and a direct interaction of purified RovA with the promoter fragment was demonstrated. The selected clones include the regulatory sequences of transcriptional regulators, proteins involved in metabolism and stress adaptation, and cell envelope associated components, such as transporters, iron sequestration systems, flagella and adhesins. The latter includes the pH6 antigen (H. Tran, unpublished results), which was also identified as a RovA-dependent virulence gene of *Y. pestis* (Cathelyn et al. 2006). Recent comparisons of the transcriptional profiles from the wild-type strain YPIII and the isogenic *rovA* mutant by using whole genome microarrays complemented our previous analysis and led to the identification of additional loci that are indirectly regulated by the RovA regulatory protein.

13.3 Regulation of *inv* and *rovA* Expression

Expression studies revealed that *inv* and *rovA* of *Y. pseudotuberculosis* follow the same temperature- and growth phase-dependent expression pattern, indicating that environmental control of *inv* in response to these parameters occurs through alteration of *rovA* expression. Primer extension and Northern blot analysis showed that *rovA* transcription is promoted by two separate σ^{70}-dependent and temperature-controlled promoters $P1_{rovA}$ and $P2_{rovA}$ mapped 76 nt and 343 nt upstream of the translational start unit (Heroven et al. 2004). The underlying molecular mechanisms that mediate *rovA* expression in response to temperature and other environmental signals are largely unknown. However, our first attempt to identify the components of the signaling pathways controlling *rovA* expression showed that multiple regulatory factors, which may act on the transcriptional and the post-transcriptional level are implicated in the regulatory network of RovA.

13.3.1 Autoregulation of *rovA* Expression

The first regulator, which was shown to influence *rovA* expression was RovA itself (Heroven et al. 2004; Nagel et al. 2001). We found that the expression of a *rovA-lacZ* fusion is repressed in a *rovA* mutant background. Autoactivation is direct and involves specific interaction of two or more RovA dimers with an extended AT-rich sequence upstream of the *rovA* promoters, leading to the formation of a high-molecular-weight RovA-DNA complex (Heroven et al. 2004). Our analysis of the *rovA* promoter region also revealed an additional, low affinity RovA target site downstream of the *rovA* promoter P1, which is not required for *rovA* activation (Heroven et al. 2004). This raised the question whether RovA can activate, but also block, *rovA* expression under specific circumstances. One important factor in *rovA* promoter activity is the concentration of RovA in the cell. We found that the level of RovA-mediated activation of *rovA* transcription is reduced when higher amounts of

RovA are produced. Reduction in *rovA* promoter activity under high RovA concentrations is still seen when the regulatory sequences upstream of P1 are deleted, but not when the downstream binding site of RovA is mutated (Heroven et al. 2004). This suggests a negative regulatory function for this binding site in *rovA* expression. It is likely that this target site acts as a threshold valve, which is occupied after a certain level of the RovA protein has been reached upon activation of *rovA* transcription. Such a concentration-dependent regulatory system allows *Yersinia* to quickly induce and finely adjust RovA synthesis to permit the most favorable production of RovA-dependent virulence and stress responsive factors.

13.3.2 The Nucleoid-Associated Protein H-NS Represses *inv* and *rovA*

During the characterization of the regulatory sequences it became evident that continuous deletions of the AT-rich upstream regions of the *inv* between positions -207 to -31 and *rovA* promoters between positions -547 and -429 resulted in a progressive increase in the basal transcription of both promoters without RovA (Heroven et al. 2004). This suggested a second regulatory protein which interacts with the RovA-dependent promoters of *inv* and *rovA,* and negatively affects their activity. Evaluation of potential candidates revealed participation of the nucleoid-associated protein H-NS, which is involved in the packaging of chromosomal DNA in the bacterial nucleoid (Dorman, 2004). The binding sites of the H-NS protein and RovA in the *inv* and *rovA* regulatory regions are superimposed, and the presence of RovA alleviates H-NS mediated repression of both promoters. This indicates that RovA, in addition to direct stimulation of the RNA polymerase, acts as an antirepressor of H-NS mediated silencing (Heroven et al. 2004; Tran et al. 2005). This was the first case showing an involvement of H-NS in the modulation of *Yersinia* gene expression. Most notably, affected genes are predominantly expressed at moderate temperature and not at 37°C, making their regulation distinct from other H-NS-dependent genes. In agreement with our results, a similar repressive effect of H-NS on *inv* transcription has also recently been reported for *Y. enterocolitica* (Ellison and Miller 2006).

13.3.3 Identification of Additional Genes Regulating *rovA*

In order to further decipher the regulatory network of environmentally controlled *rovA* expression, a genetic screen for molecular components of the *rovA* signaling cascade was performed. A plasmid-born gene library from *Y. pseudotuberculosis* YPIII was introduced into a merodiploid *Y. pseudotuberculosis rovA-lacZ* fusion strain. About $2 \cdot 10^4$ transformants were screened on LB broth agar plates containing the indicator X-Gal. Darker and lighter blue colonies, which exhibited a significantly increased or decreased *rovA-lacZ* expression level at 25°C during stationary phase growth were isolated (Heroven and Dersch 2006). In total, 13 independent gene bank plasmids were obtained, encoding regulatory factors which led to a >2-fold difference in the *rovA-lacZ* expression levels. One of the plasmids encoded the nucleoid-associated protein H-NS, demonstrating that our genetic screen was effective. Two other independent clones encoded a LysR-type transcriptional regulator protein, that we named RovM. RovM exhibits significant homology (65-66% identity) to PecT of phytopathogenic *Erwinia* species and HexA of *Photorhabdus temperata*, a bacterium that is mutuallistically associated with nematodes and pathogenic to insect larvae (Harris et al. 1998; Joyce and Clarke 2003; Mukherjee et al. 2000; Surgey et al. 1996). These proteins have pleiotropic effects on gene expression and were shown to regulate motility, biofilm formation and exoenzyme production, implicated in host interaction and pathogenesis. In addition, two clones carrying the gene for the nucleoid-associated protein YmoA were found to affect *rovA* expression.

13.3.4 The LysR-regulator RovM Controls *rovA*, Virulence and Motility

A detailed characterization of RovM showed that this novel LysR-type transcriptional regulator represses *rovA* expression of *Y. pseudotuberculosis* directly and affects host cell invasion, motility and virulence of this pathogen. Removal of the *rovM* gene induces expression of *rovA* (Fig. 1A) and the RovA-dependent internalization factor invasin, leading to a 2-fold increase of *Y. pseudotuberculosis* uptake into cultured human epithelial cells (Heroven and Dersch 2006). Oral infection studies using the BALB/c mouse model system further demonstrated that RovM also seems to affect *rovA* expression *in vivo* and alters pathogenesis of a *Y. pseudotuberculosis* infection. Considerably higher numbers of a *rovM* mutant were isolated from the Peyer′s patches, the mesenterial lymph nodes, liver and spleen compared with the wild type. In the opposite, significantly lower numbers of a *rovM* overexpressing strain were detected in all dissected lymphatic tissues and no bacteria were isolated from the organs (Heroven and Dersch 2006).

In contrast to other LysR-type regulators, which generally activate genes that are divergently transcribed from the LysR regulator gene, RovM represses the global virulence regulator gene *rovA*, which is not closely linked on the *Y. pseudotuberculosis* genome. DNA-binding studies with purified His-tagged RovM demonstrated that this regulatory protein interacts directly with the *rovA* regulatory region close to the upstream promoter P1 (-80 to -47) and leads to the formation of a single RovM-DNA complex. The identified binding site includes palindromic sequences, which contain the conserved T-N_{11}-A core motif, and shows similarity to other LysR-type binding sites. The RovM binding site in the *rovA* promoter is distinct and does not overlap with the binding site of H-NS. RovM and H-NS can interact independently with these binding sequences, but they are both required for effective silencing of the *rovA* gene (Heroven and Dersch 2006). It is also known from our binding studies that both regulatory proteins are able to bind simultaneously to the *rovA* regulatory region, which may lead to structural alterations of the *rovA* promoter and/or the formation of a higher order RovM-H-NS-DNA nucleoprotein silencing complex.

Additional regulatory studies using different *rovM-lacZ* fusion constructs and a RovM-specific polyclonal antibody further showed that the synthesis of the RovM regulatory protein in *Y. pseudotuberculosis* is under positive autoregulatory control. Furthermore, expression of *rovM* is affected by the composition of the bacterial growth medium, suggesting that the regulation of *rovA* in response to nutrient availability is mediated through RovM (Heroven and Dersch 2006).

13.3.5 The Histone-like Protein YmoA Activates *rovA* Through RovM

In addition to H-NS and RovM, the nucleoid-associated YmoA protein, a member of a growing Hha-YmoA family of low-molecular-mass proteins, was also shown to affect *rovA* and *inv* expression in *Y. pseudotuberculosis*. The effect of this global regulatory protein on virulence gene expression in enteropathogenic *Yersinia* is less clear. Overexpression of the *ymoA* gene was found to slightly induce the expression of a chromosomal *rovA-lacZ* fusion, whereas no RovA protein and significantly less invasin protein were detected in a *ymoA* knock-out mutant strain (Fig. 1A). This strongly suggests that YmoA activates *rovA* expression in *Y. pseudotuberculosis*. This finding was surprising as it has been reported that YmoA has no effect on *rovA*, but a negative effect on the expression of the RovA-dependent invasin gene in *Y. enterocolitica* (Ellison et al. 2003). This suggests that *rovA* regulation by YmoA is different in the two related enteropathogenic *Yersinia* species. However, it is also possible that the distinct phenotype observed with *Y. enterocolitica* is specific for the *ymoB::*Tn*5* insertion mutation used in the study, which was assumed to have a polar affect on the *ymoA* gene downstream (Ellison et al. 2003).

Recent studies on the mechanism of action of the Hha-YmoA family proteins further showed that they interact with H-NS or H-NS homologous proteins such as StpA (Nieto et al. 2002; Paytubi et al. 2004) and modulate the expression of virulence genes (Madrid et al. 2006). Up to date, it is unclear, whether the Hha/YmoA proteins bind DNA themselves or if their regulatory activity is dependent upon the interaction with H-NS-family members. Purified YmoA and H-NS in the μmolar range bind alone and simultaneously to the *inv* promoter *in vitro,* and expression of *ymoA* from a multi-copy plasmid leads to a 50% repression of an *inv-phoA* fusion in *Y. enterocolitica* and *Y. pseudotuberculosis*. This suggests a functional partnership in invasin regulation when higher amounts of the proteins are present (Ellison et al. 2003; Ellison and Miller 2006)(K. Böhme, unpublished results). However, expression studies in *Y. pseudotuberculosis* revealed that YmoA is apparently far less abundant in the bacterial cell than H-NS, and only very low amounts of the protein are detectable in whole cell extracts under usual growth conditions (e.g. 25°C, LB medium). Therefore, it is questionable whether effects observed under conditions in which YmoA is highly overexpressed are really relevant for virulence gene expression during infection.

To gain a deeper insight into the role of YmoA in *rovA* and *inv* expression we compared independent *ymoA* knock-out mutants with the wild type strain and found that absence of YmoA caused a significant increase in the negative regulator RovM (Fig. 1A). This effect does not seem to be direct but appears to require additional regulatory factors. Most strikingly, stability of *Y. pseudotuberculosis* YmoA seems to be reduced at 37°C compared to 20-25°C, similar to what has been observed for YmoA of *Y. pestis* (Jackson et al. 2004). This suggests that temperature-mediated control of *rovA* expression might occur through YmoA. The molecular mechanism of this signal transduction pathway and regulation of the components in response to different environmental cues are subjects of our current investigations.

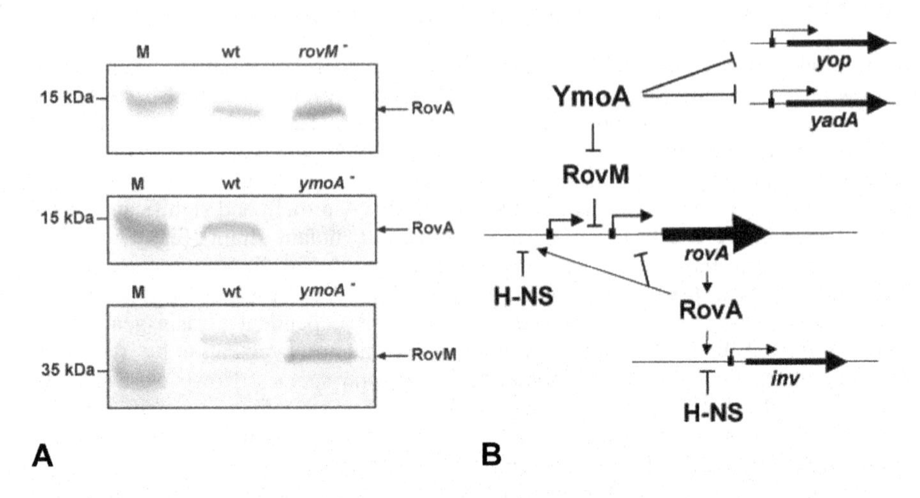

Fig. 1. The RovA regulatory network of *Y. pseudotuberculosis*. (A) Whole-cell extracts from overnight cultures of *Y. pseudotuberculosis* wild-type (wt), a *rovM* and a *ymoA* mutant strain were analyzed by Western blotting with a polyclonal antibody directed against RovA or RovM. (B) Expression of the *rovA* gene is subject to silencing by the nucleoid-associated protein H-NS and the LysR-type regulator RovM. RovA can alleviate *rovA* repression by disruption of the H-NS/RovM repression complex and direct activation of RNA polymerase. Activation of *rovA* expression leads to an accumulation of RovA and allows binding to an additional lower affinity site downstream of the *rovA* promoters. RovM synthesis itself is repressed by the small histone-like protein YmoA, which also inhibits expression of the Yop proteins and the adhesin YadA.

13.4 Conclusions

In the past few years it became evident that multiple *Yersinia* virulence genes, including the internalization factor invasion, are regulated by the global MarR-type regulator protein RovA. RovA itself is controlled by a complex regulatory network of multiple regulatory factors in response to different environmental parameters, such as temperature, nutrient availability and growth phase. Based on our current knowledge, we postulate the following model to describe the *rovA* regulatory network (Fig. 1B), in which H-NS, RovM and the YmoA proteins play a key role in adjusting the expression of different *Yersinia* virulence factors.

13.5 Acknowledgements

The work was supported by the Deutsche Forschungsgemeinschaft grant DE616/3.

13.6 References

Cathelyn, J.S., Crosby, S.D., Lathem, W.W., Goldman, W.E. and Miller, V.L. (2006) RovA, a global regulator of *Yersinia pestis*, specifically required for bubonic plague. Proc. Natl. Acad. Sci. USA 103, 13514-13519.

Clark, M.A., Hirst, B.H. and Jepson, M.A. (1998) M-cell surface beta1 integrin expression and invasin-mediated targeting of *Yersinia pseudotuberculosis* to mouse Peyer's patch M cells. Infect. Immun. 66, 1237-1243.

Dorman, C.J. (2004) H-NS, a universal regulator for a dynamic genome. Nat. Rev. Microbiol. 2, 391-400.

Dube, P.H., Revell, P.A., Chaplin, D.D., Lorenz, R.G., and Miller, V.L. (2001) A role for IL-1 alpha in inducing pathologic inflammation during bacterial infection. Proc. Natl. Acad. Sci. USA 98, 10880-10885.

Dube, P.H., Handley, S.A., Revell, P.A. and Miller, V.L. (2003) The *rovA* mutant of *Yersinia enterocolitica* displays differential degrees of virulence depending on the route of infection. Infect. Immun. 71, 3512-3520.

Ellison, D.W., Young, B., Nelson, K. and Miller, V.L. (2003) YmoA negatively regulates expression of invasin from *Yersinia enterocolitica*. J. Bacteriol. 185, 7153-7159.

Ellison, D.W. and Miller, V.L. (2006) H-NS represses *inv* transcription in *Yersinia enterocolitica* through competition with RovA and interaction with YmoA. J. Bacteriol. 188, 5101-5112.

Harris, S.J., Shih, Y.L., Bentley, S.D. and Salmond, G.P. (1998) The *hexA* gene of *Erwinia carotovora* encodes a LysR homologue and regulates motility and the expression of multiple virulence determinants. Mol. Microbiol. 28, 705-717.

Heroven, A. and Dersch, P. (2006) RovM, a novel LysR-type regulator of the virulence activator gene *rovA*, controls cell invasion, virulence and motility of *Yersinia pseudotuberculosis* Mol. Microbiol. 62, 1469-1483.

Heroven, A.K., Nagel, G., Tran, H.J., Parr, S. and Dersch, P. (2004) RovA is autoregulated and antagonizes H-NS-mediated silencing of invasin and *rovA* expression in *Yersinia pseudotuberculosis*. Mol. Microbiol. 53, 871-888.

Hong, M., Fuangthong, M., Helmann, J.D. and Brennan, R.G. (2005) Structure of an OhrR-ohrA operator complex reveals the DNA binding mechanism of the MarR family. Mol. Cell 20, 131-141.

Isberg, R.R. and Leong, J.M. (1990) Multiple beta 1 chain integrins are receptors for invasin, a protein that promotes bacterial penetration into mammalian cells. Cell 60, 861-871.

Jackson, M.W., Silva-Herzog, E. and Plano, G.V. (2004) The ATP-dependent ClpXP and Lon proteases regulate expression of the *Yersinia pestis* type III secretion system via regulated proteolysis of YmoA, a small histone-like protein. Mol. Microbiol. 54, 1364-1378.

Joyce, S.A. and Clarke, D.J. (2003) A *hexA* homologue from *Photorhabdus* regulates pathogenicity, symbiosis and phenotypic variation. Mol. Microbiol. 47, 1445-1457.

Libby, S.J., Goebel, W., Ludwig, A., Buchmeier, N., Bowe, F., Fang, F.C., Guiney, D.G., Songer, J.G. and Heffron, F. (1994) A cytolysin encoded by *Salmonella* is required for survival within macrophages. Proc. Natl. Acad. Sci. USA 91, 489-493.

Linehan, S.A., Rytkonen, A., Yu, X.J., Liu, M. and Holden, D.W. (2005) SlyA regulates function of *Salmonella* pathogenicity island 2 (SPI-2) and expression of SPI-2-associated genes. Infect. Immun. 73, 4354-4362.

Madrid, C., Basalobre, C., Garcia, J. and Juarez, A. (2007) The novel Hha/YmoA family of nucleoid-associated proteins, use of structural mimicry to modulate the activity of the H-NS family of proteins. Mol. Microbiol. 63, 7-14.

Marra, A. and Isberg, R.R. (1996) Analysis of the role of invasin during *Yersinia pseudotuberculosis* infection of mice. Ann. N. Y. Acad. Sci. 797, 290-292.

Mukherjee, A., Cui, Y., Ma, W., Liu, Y. and Chatterjee, A.K. (2000) *hexA* of *Erwinia carotovora* ssp. *carotovora* strain Ecc71 negatively regulates production of RpoS and *rsmB* RNA, a global regulator of extracellular proteins, plant virulence and the quorum-sensing signal, N-(3-oxohexanoyl)-L-homoserine lactone. Environ. Microbiol. 2, 203-215.

Nagel, G., Lahrz, A. and Dersch, P. (2001) Environmental control of invasin expression in *Yersinia pseudotuberculosis* is mediated by regulation of RovA, a transcriptional activator of the SlyA/Hor family. Mol. Microbiol. 41, 1249-1269.

Nagel, G., Heroven, A.K., Eitel, J. and Dersch, P. (2003) Function and regulation of the transcriptional activator RovA of Yersinia pseudotuberculosis. Adv. Exp. Med. Biol. 529, 285-287.

Navarre, W.W., Halsey, T.A., Walthers, D., Frye, J., McClelland, M., Potter, J.L., Kenney, L.J., Gunn, J.S., Fang, F.C. and Libby, S.J. (2005) Co-regulation of *Salmonella enterica* genes required for virulence and resistance to antimicrobial peptides by SlyA and PhoP/PhoQ. Mol. Microbiol. 56, 492-508.

Nieto, J.M., Madrid, C., Miquelay, E., Parra, J.L., Rodriguez, S. and Juarez, A. (2002) Evidence for direct protein-protein interaction between members of the enterobacterial Hha/YmoA and H-NS families of proteins. J. Bacteriol. 184, 629-635.

Oscarsson, J., Mizunoe, Y., Uhlin, B.E. and Haydon, D.J. (1996) Induction of haemolytic activity in *Escherichia coli* by the *slyA* gene product. Mol. Microbiol. 20, 191-199.

Paytubi, S., Madrid, C., Forns, N., Nieto, J.M., Balsalobre, C., Uhlin, B.E. and Juarez, A. (2004) YdgT, the Hha paralogue in *Escherichia coli*, forms heteromeric complexes with H-NS and StpA. Mol. Microbiol. 54, 251-263.

Pepe, J.C. and Miller, V.L. (1993) The biological role of invasin during a *Yersinia enterocolitica* infection. Infect. Agents Dis. 2, 236-241.

Pepe, J.C., Badger, J.L. and Miller, V.L. (1994) Growth phase and low pH affect the thermal regulation of the *Yersinia enterocolitica inv* gene. Mol. Microbiol. 11, 123-135.

Revell, P.A. and Miller, V.L. (2000) A chromosomally encoded regulator is required for expression of the *Yersinia enterocolitica inv* gene and for virulence. Mol. Microbiol. 35, 677-685.

Spory, A., Bosserhoff, A., von Rhein, C., Goebel, W. and Ludwig, A. (2002) Differential regulation of multiple proteins of *Escherichia coli* and *Salmonella enterica* serovar Typhimurium by the transcriptional regulator SlyA. J. Bacteriol. 184, 3549-3559.

Stapleton, M.R., Norte, V.A., Read, R.C. and Green, J. (2002) Interaction of the *Salmonella typhimurium* transcription and virulence factor SlyA with target DNA and identification of members of the SlyA regulon. J. Biol. Chem. 277, 17630-17637.

Surgey, N., Robert-Baudouy, J. and Condemine, G. (1996) The *Erwinia chrysanthemi pecT* gene regulates pectinase gene expression. J. Bacteriol. 178, 1593-1599.

Thomson, N.R., Cox, A., Bycroft, B.W., Stewart, G.S., Williams, P. and Salmond, G.P. (1997) The Rap and Hor proteins of *Erwinia, Serratia* and *Yersinia*, a novel subgroup in a growing superfamily of proteins regulating diverse physiological processes in bacterial pathogens. Mol. Microbiol. 26, 531-544.

Tran, H.J., Heroven, A.K., Winkler, L., Spreter, T., Beatrix, B. and Dersch, P. (2005) Analysis of RovA, a transcriptional regulator of *Yersinia pseudotuberculosis* virulence that acts through antirepression and direct transcriptional activation. J. Biol. Chem. 280, 42423-42432.

14 Regulation of the Phage-Shock-Protein Stress Response in *Yersinia enterocolitica*

Andrew J. Darwin

Department of Microbiology, New York University School of Medicine,
darwia01@med.nyu.edu

Abstract. The phage-shock-protein (Psp) system of *Yersinia enterocolitica* encodes a stress response that is essential for viability when the secretin component of its Ysc type III secretion system is produced. Therefore, *Y. enterocolitica psp* null mutants are completely avirulent in a mouse model of infection. This article summarizes what is known about the regulation of the *Y. enterocolitica* Psp system. *psp* gene expression is induced by the overproduction of secretins, some cytoplasmic membrane proteins, or disruption of the F_0F_1-ATPase. All of these may deplete the proton-motive force, which could be the inducing signal for the Psp system. None of these Psp triggers induce two other extracytoplasmic stress responses (RpoE and Cpx), which suggests that the inducing signal of the Psp system is specific. The induction of *psp* gene expression requires the cytoplasmic membrane proteins PspB and PspC, which interact and presumably work together to achieve their regulatory function. However, the regulatory role of PspBC does not completely explain why they are essential for survival during secretin-stress, suggesting that they have a second unrelated role. Finally, current ideas about how PspB/C might sense the inducing trigger(s) are briefly discussed, including a consideration of whether there might be any unidentified signal transduction components that communicate with the Psp system.

14.1 The Discovery of the Psp System

The Psp system was originally identified in *Escherichia coli* K-12 by Peter Model's laboratory (Brissette et al. 1990). They described an *E. coli* protein that was produced at a high level during infection of the bacterial cell by a filamentous bacteriophage. This *E. coli* protein was named phage-shock-protein-A (PspA, which is encoded by the first gene of the *pspABCDE* operon). Induction of PspA synthesis was caused by the mislocalization of a single phage encoded protein in the bacterial cell envelope (Brissette et al. 1990). This protein, pIV, is a member of the secretin family of proteins, which normally form oligomeric pores in the outer membrane of bacteria (Genin and Boucher 1994). Secretins are not specific to phage. They are critical components of all type 2 and type 3 secretion systems and play a role in the assembly of type IV pili. Besides secretin mislocalization, many other stimuli were shown to induce PspA synthesis in *E. coli*, including the mislocalization of some other cell envelope proteins and extremes of temperature, osmolarity or ethanol concentration (reviewed by Model et al. 1997). All of these have the potential to damage the cell envelope and dissipate the proton-motive force (PMF), which may be the inducing signal for the Psp system. In support of this, the proton ionophore carbonylcyanide *m*-chlorophenylhydrazone (CCCP) also induces *E. coli* PspA synthesis (Weiner and Model 1994).

14.2 The Psp system in *Yersinia*

For several years the Psp system was only known and studied in *E. coli* K-12, but *E. coli psp* null strains had only subtle phenotypes (Model et al. 1997). However, a *Yersinia. enterocolitica* signature-tagged mutagenesis screen led to the identification of a *pspC* transposon insertion mutant that was completely avirulent in a mouse model of infection (Darwin and Miller 1999). Compared to the wild type strain, the *pspC* mutant had an exaggerated growth defect when the Ysc type 3 secretion system was synthesized (Darwin and Miller 2001). This presumably explains the attenuation phenotype. The *pspC* null mutant also had a growth defect *in vitro* when only the YscC secretin component of the type 3 secretion system was overproduced and mis-localized in the cell envelope (Darwin et al. 2001).

14.2.1 Genetic Components of the *Y. enterocolitica* Psp System

The *Y. enterocolitica* Psp system is encoded by two unlinked loci (Fig. 1). The first of these is comprised of the *pspABCDycjXF* operon and the divergently transcribed *pspF* gene. Interestingly, the first four genes of the *Y. enterocolitica* and *E. coli pspA* operons are homologous, whereas the genes at the end of their *pspA* operons differ. The fifth and final gene of the *E. coli pspA* operon is *pspE*, encoding a rhodanese enzyme (Adams et al. 2002), for which there is no equivalent in *Y. enterocolitica.* The final two genes of the *Y. enterocolitica pspA* operon are *ycjXF*. These genes are also present in *E. coli*, but are several kilobases downstream of its *pspA* operon and not coordinately regulated with it.

The *pspF-pspABCDycjXF* locus encodes all of the essential components for normal regulation of the Psp system and for a successful response to secretin-induced stress (see below). However, there is one more component of the Psp system, which is encoded by the unlinked *pspG* gene (Fig. 1). PspG is not essential for normal regulation or functioning of the Psp response, but its overexpression can partially suppress the phenotype of a complete *pspA* operon deletion mutant (Green and Darwin 2004).

14.2.2 The *psp* Genes of Other *Yersinia* Species

Genome sequence information reveals that all of the *Y. enterocolitica psp* genes are conserved in the two other *Yersinia* species that are pathogenic for humans, *Yersinia. pestis* and *Yersinia. pseudotuberculosis* (Fig. 1). The *psp* loci are arranged identically and, as expected, the predicted Psp protein sequences are very similar between the different species (data not shown). Therefore, although *Y. pestis* and *Y. pseudotuberculosis psp* null mutants have not been reported in the literature, it seems likely that the Psp system functions similarly in all three species. However, it is not yet known if the Psp system is essential for the virulence of *Y. pestis* and *Y. pseudotuberculosis.*

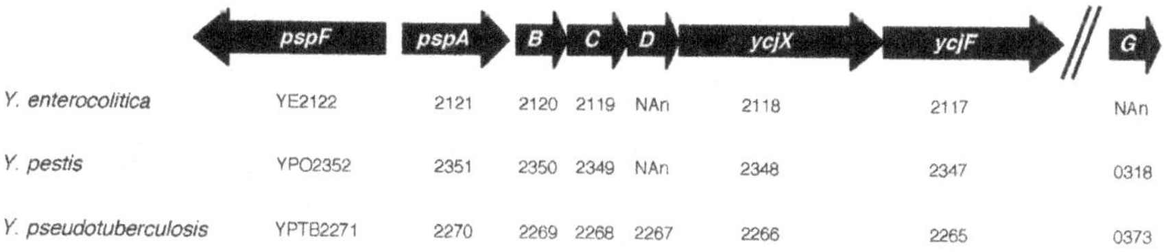

	pspF	*pspA*	*B*	*C*	*D*	*ycjX*	*ycjF*	*G*
Y. enterocolitica	YE2122	2121	2120	2119	NAn	2118	2117	NAn
Y. pestis	YPO2352	2351	2350	2349	NAn	2348	2347	0318
Y. pseudotuberculosis	YPTB2271	2270	2269	2268	2267	2266	2265	0373

Fig. 1. The *psp* loci of human pathogenic *Yersinia* species. Relative size and orientation of each gene is represented by an arrow. Genome annotation identifiers are shown below each open reading frame. The genomes used for this analysis were from *Y. enterocolitica* strain 8081, *Y. pestis* strain CO92 and *Y. pseudotuberculosis* strain IP 32953. "NAn" indicates that the gene has not been annotated in the genome sequence. However, the *pspD* gene is present between *pspC* and *ycjX* in both *Y. enterocolitica* 8081 and *Y. pestis* CO92. The *pspG* gene of *Y. enterocolitica* is also present and located between genes YE3851 and YE3852.

14.2.3 Brief Description of the Psp Proteins

In addition to *Yersinia* species and *E. coli*, the Psp system is conserved in many other bacterial species, although there are differences between which *psp* genes are present (Darwin 2005). However, in Psp$^+$ species, the *pspF-pspABC* genes are always conserved, which suggests that they encode the essential components of the system. This is supported by the observation that in frame deletion mutations in any of the *pspFABC* genes cause robust phenotypes in *Y. enterocolitica*, whereas deletions of *pspD-ycjF* or of *pspG* do not. We do not yet know the full functions of most of the critical Psp proteins, but we are beginning to get insight in some cases, especially in terms of their roles in regulating *psp* gene expression. Indeed, PspF, PspA, PspB and PspC all appear to play important roles in regulating the Psp response itself (Fig. 2).

PspF is a DNA-binding transcriptional regulator of the enhancer-binding protein family that activates the RpoN- (σ^{54}) dependent promoters located upstream of *pspA* and *pspG* (Green et al. 2004; Maxson and Darwin 2006a). PspF activity is apparently controlled by PspA, which has been shown to bind to PspF in *E. coli* and prevent it from activating transcription (Dworkin et al. 2000; Elderkin et al. 2005). This presumably also occurs in *Y. enterocolitica* because a Δ*pspA* mutation causes constitutive high-level expression of the *pspA* and *pspG* promoters (Darwin et al. 2001; Green et al. 2004). The cytoplasmic membrane proteins PspB and PspC are required for induction of *psp* gene expression in response to various stimuli (see below). An attractive model is that upon activation PspB and/or PspC interact with PspA and prevent it from interfering with PspF activity (Fig. 2). Support for interactions involving PspABC has been published for *E. coli* (Adams et al. 2003).

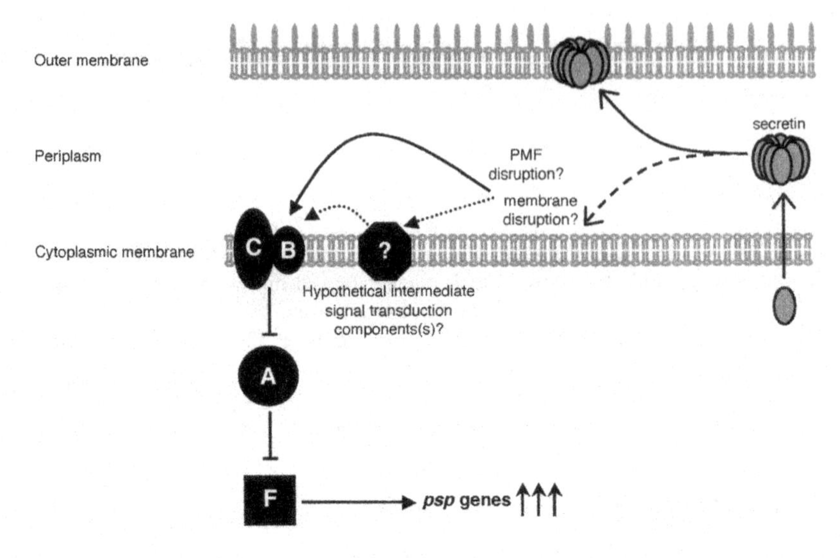

Fig. 2. Proposed regulatory organization of the Psp system. A mislocalized secretin (e.g. YscC) probably disrupts the cytoplasmic membrane and reduces the PMF (Guilvout et al. 2006). Other inducing conditions (not shown) probably have a similar effect. PspB and PspC either sense the resulting inducing signal directly, or possibly by communicating with an intermediary protein. The activated PspBC complex may then interact with PspA and prevent it from inhibiting the transcriptional activator PspF. This allows PspF to induce expression of the *psp* genes (*pspA* operon and *pspG*).

Little is known about the functions of PspD, YcjX, YcjF and PspG. Simultaneous deletion of the genes encoding all of these proteins does not affect the regulation of *psp* gene expression or cause a secretin-sensitive phenotype (data not shown). However, overexpression of the *pspG* gene partially alleviates the secretin sensitive phenotype of a complete *pspABCDycjXF* deletion mutant (Green et al. 2004). This suggests that PspG, at least when present at a high concentration, has a similar function to one or more of the proteins encoded in the *pspA* operon.

14.3 Physiological Function of the Psp System

Although this article largely focuses on regulation of the *Y. enterocolitica* Psp system, it is important to briefly consider the proposed physiological role of the Psp response itself. However, the quick answer to the question of "what is the physiological role of the Psp response?" is "we don't know". Nevertheless, mounting evidence suggests that the Psp system allows the cell to deal with specific perturbations of its cytoplasmic membrane or, as seems more likely, a decrease in PMF that results from these perturbations. Much attention for explaining this function has focused specifically on the PspA protein. For example, the *E. coli* PspA protein has been associated with survival in stationary phase at pH 9 (Weiner et al. 1994), efficient protein export by the *sec* and *tat* systems (Kleerebezem and Tommassen 1993; DeLisa et al. 2004) and maintenance of membrane potential during the mislocalization of an outer membrane porin (Kleerebezem et al. 1996). In *Salmonella enterica* serovar Typhimurium a *pspA* null mutation causes reduced resistance to the proton ionophore CCCP (Becker et al. 2005). Furthermore, a mislocalized secretin (PulD), which is a potent inducer of *psp* gene expression, can insert into the cytoplasmic membrane of *E. coli*, where it may form a pore allowing dissipation of the PMF (Guilvout et al. 2006). The authors hypothesized that PspA might somehow prevent this PMF dissipation because mislocalization of PulD was lethal in a *pspA* null strain. Taken together, all of these observations have led to speculation that the peripheral cytoplasmic membrane protein PspA is an important effector that somehow helps the cell to maintain its PMF.

The assignment of PspA as a critical effector of the Psp system is supported by the data described above and by the fact that PspA is by far the most abundant Psp protein when the system is induced. However, the situation is complicated by our data from *Y. enterocolitica*. Mislocalization of the YscC secretin is lethal in some *psp* null mutants, such as *pspC*, but not in a *pspA* in frame deletion mutant (Darwin et al. 2001). Therefore, whilst PspA may well be an important effector of the Psp system, it is dispensable for the ability of *Y. enterocolitica* to survive secretin-induced stress. This means that if mislocalized YscC dissipates the PMF then at least one other protein, besides PspA, must be able to counteract this in *Y. enterocolitica*. The most likely proteins responsible for this function are PspB and PspC, as discussed below.

14.4 Identification of Inducers of the *Y. enterocolitica* Psp System

14.4.1 The Psp System is an Extracytoplasmic Stress Response

When our laboratory began to study regulation of the *Y. enterocolitica* Psp system, the first question we addressed was whether or not the Psp system should be classified as an extracytoplasmic stress response system (ESR). Other ESRs, such as the RpoE and Cpx systems, studied extensively in *E. coli*, are only activated by stimuli that affect the cell envelope (Duguay and Silhavy 2004). Known proteinaceous or environmental inducers of the Psp system in *E. coli* suggested that it might also be an ESR. However, a non-biased random approach to identify inducers of the Psp system had not been attempted. Therefore, we designed a screen to randomly identify two classes of genes that would induce *psp* gene expression: those that caused *psp* induction when they were overexpressed; and those that caused *psp* induction when they were disrupted (Maxson and Darwin 2004). We hypothesized that, if the Psp system is an ESR, the overexpressed genes would encode cell envelope components or proteins that affected cell envelope function. Similarly, the inactivated genes would encode proteins important for cell envelope integrity and/or the production of cell envelope components.

Table 1. Inducers of the *Y. enterocolitica* Psp Response

Gene	Comments	Predicted Subcellular Location [a]
Overexpression inducers		
YE3548	YsaC secretin (T3SS) [b]	OM
YE3565	Yts1D secretin (T2SS)	OM
YE3636	Secretin, unknown function	OM
YE3976	Secretin, unknown function	OM
YE0566	Unknown function	IM
YE0693	AmpE, unknown function	IM
YE3435	YggT, unknown function	IM
Inactivation inducers		
YE4203	GlmS, cell envelope biogenesis	CY
YE4208	AtpA, F_0F_1-ATPase subunit	IM
YE4212	AtpB, F_0F_1-ATPase subunit	IM

[a]OM = outer membrane, IM = inner membrane, CY = cytoplasm
[b]T3SS = type 3 secretion system, T2SS = type 2 secretion system

A *Y. enterocolitica* strain with a single-copy Φ(*pspA-lacZ*) operon fusion was randomly mutagenized with a transposon encoding an outward facing *tac* promoter, which could be induced with isopropyl-β-D-thiogalactopyranoside (IPTG). Mutants with increased Φ(*pspA-lacZ*) expression were identified and selected for further analysis. The mutants were divided into two classes. The first class were those with an IPTG-dependent increase in Φ(*pspA-lacZ*) expression, suggesting that it was due to overexpression of a downstream gene from the *tac* promoter of the transposon. The second class were those where the increase in Φ(*pspA-lacZ*) expression was IPTG-independent, suggesting that it resulted from gene disruption by the transposon.

Only seven genes were found to induce the Psp response when they were overexpressed (Table 1 and Maxson et al. 2004). Four of these encode secretins, the identification of which served to validate the screen (the *yscC* gene was not identified because the screen was done in a strain lacking the virulence plasmid). Besides the secretins, the three other overexpression inducers are all predicted to encode inner membrane proteins of unknown function. The identification of these predicted cell envelope proteins, together with the failure to identify any cytoplasmic proteins, supported our hypothesis that the Psp system responds exclusively to extracytoplasmic stress. Furthermore, the fact that only seven overexpression inducers were identified suggests that the inducing signal(s) of the Psp system is specific and induction cannot be caused by the overexpression of many different envelope proteins. This assertion was further supported by our observation that none of these overexpression inducers was able to significantly activate the RpoE or Cpx systems (Maxson et al. 2004). Similarly, known overexpression inducers of the RpoE and Cpx systems did not induce the Psp response.

The disruption of three genes was found to induce Φ(*pspA-lacZ*) expression (Table 1). One of these, *glmS*, encodes glucosamine-6-phosphate synthetase. In the absence of exogenous glucosamine, GlmS is required to synthesize precursors of several cell envelope components. Therefore, without glucosamine in the growth medium we found that the *Y. enterocolitica glmS* null mutant grew to a low cell density before apparently undergoing cellular lysis. Not surprisingly, in addition to inducing the Psp response, the *glmS* mutation also caused robust induction of the RpoE and Cpx systems (Maxson et al. 2004). Nevertheless, the identification of the *glmS* mutation further served to characterize Psp as an ESR.

14.4.2 Inactivation of the F_0F_1-ATPase Induces the Psp Response

Our screen for inducers of the Psp response also identified disruption mutations of the *atpA* and *atpB* genes, which encode components of the F_0F_1-ATPase (Maxson et al. 2004). This enzyme complex interconverts PMF and ATP, with the direction of the reaction depending on the needs of the cell. The identification of these mutations was particularly exciting because it provided further evidence of a link between cellular energy status and induction of the Psp response. Interestingly, these *atp* mutations only caused significant induction of the Psp response when the bacteria were grown in media buffered to alkaline pH (Maxson et al. 2004). This may be because the external pH would be higher than that of the cytoplasm and the ΔpH gradient across the cytoplasmic membrane would make a negative contribution to the PMF. The activity of the F_0F_1-ATPase might be required to compensate for this and, without it, the PMF may drop sufficiently to activate the Psp system. The *atp* null mutations represented the only inducers identified in our screen that are not predicted to directly affect the integrity of the cell envelope. This suggests that it is the drop in PMF, rather than any preceding change in cytoplasmic membrane integrity, which provides the inducing signal for the Psp response.

14.5 PspB and PspC are Essential Signal Transduction Components

14.5.1 Positive Regulation by PspB and PspC

Another of our regulatory interests in the *Y. enterocolitica* Psp system is to identify proteins required for detecting the inducing signal and transducing it to other Psp components in order to activate *psp* gene expression. Work done in *E. coli* had suggested that the cytoplasmic membrane proteins PspB and PspC were required for positive regulation of the Psp system in response to some inducing signals (e.g. Weiner et al. 1991). Therefore, the starting point for our work has been to characterize the roles of PspB and PspC in *Y. enterocolitica*.

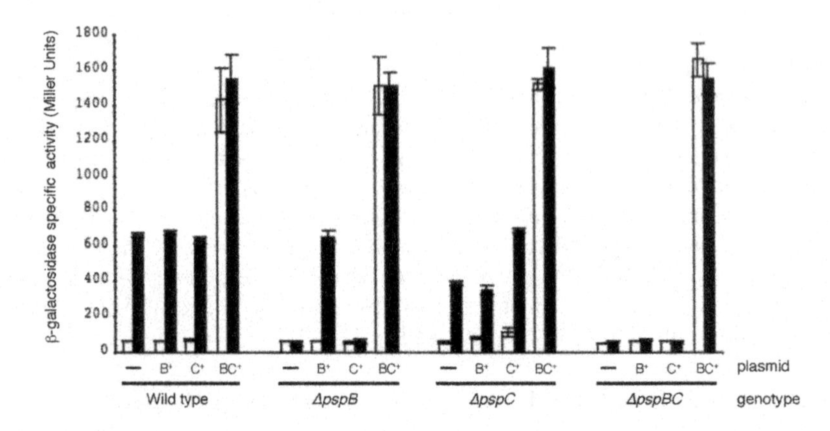

Fig. 3. PspB and PspC must both be present for normal regulation of the *Y. enterocolitica* Psp response. Strains had a single copy Φ(*pspA-lacZ*) operon fusion and either intact *pspB* and *pspC* genes or in frame deletions as indicated (genotype). Strains had either an empty arabinose-inducible expression plasmid vector (–) or the same plasmid expressing *pspB*, *pspC* or *pspBC* as indicated (plasmid). Each strain also contained either an empty IPTG-inducible expression plasmid vector (white bars) or the same plasmid expressing the *ysaC* secretin gene (black bars). Data used to compile this figure were taken from a table in an article produced by Blackwell Publishing (Maxson and Darwin 2006b).

Our work revealed that in frame deletion mutations in either *pspB* or *pspC* abolished induction of Φ(*pspA-lacZ*) expression in response to overproduction of the three inner membrane inducers described above or disruption of the F_0F_1-ATPase (Maxson et al. 2006b). Induction by the overproduced secretin YsaC was also abolished by a Δ*pspB* mutation. However, whilst a Δ*pspC* mutation significantly reduced YsaC-dependent induction, it did not completely abolish it. The significance of the latter observation is not yet clear.

These data suggested that both PspB and PspC are required for induction of the Psp system in response to various inducing triggers. However, this proposed regulatory role for PspB and C predicts that removal of one will reduce the level of the other (and of all other Psp proteins). Therefore, to solidify our conclusion that both PspB and PspC are required for positive regulation we complemented the regulatory defects of Δ*pspB/C* mutants with multicopy plasmids encoding PspB and/or PspC expressed from an arabinose-inducible promoter. The data clearly showed that PspB and PspC must both be present for normal induction of the Psp system in response to secretin overexpression (Fig. 3 and Maxson et al. 2006b). Using two-hybrid and cross-linking analysis we also discovered that the PspB and PspC proteins interact (Maxson et al. 2006b). We hypothesize that this interaction is important for the regulatory function of these proteins.

Taken together, all of these data suggest that PspBC form a transmembrane complex that detects the inducing signal and communicates with PspA to control its ability to negatively regulate *psp* gene expression (Fig. 2). However, much of this model, especially the dynamics of the protein-protein interactions, awaits direct experimental evidence.

14.5.2 PspB and PspC are Dual Function Proteins

Published analysis of PspBC in *E. coli* had suggested that the only role of these proteins was to positively regulate *psp* gene expression. However, our work unexpectedly revealed that this is not their only role in *Y. enterocolitica.* Both PspB and PspC play a physiological role in supporting cell viability when a secretin is mislocalized (Maxson et al. 2006b). Furthermore, they can carry out this function independently of each other, unlike their regulatory role where they must work together. How this function is achieved is not known. However, we speculate that PspB and PspC might counteract a decrease in the PMF or something that results from it. Interestingly, overproduction of only PspG can also facilitate growth during secretin mislocalization (Green et al. 2004). Therefore, PspG may share a common physiological role with PspB and PspC, whereas it does not share their regulatory function.

14.6 Are There Other Signal Transduction Components?

Finally, my laboratory has been considering the question of how PspBC might sense an inducing signal, such as a decrease in the PMF. A simple hypothesis is that one or both of them sense the signal directly. Their cytoplasmic membrane location is well suited for this. Another intriguing possibility is that they monitor the activity of one or more other PMF-responsive proteins. In support of this second idea, it was recently reported that the redox-responsive sensory protein ArcB is essential for induction of the *E. coli* Psp response by the pIV secretin and some other stimuli (Jovanovic et al. 2006). This suggests that PspBC might somehow monitor the activation status of the ArcB sensor. However, ongoing work in our laboratory indicates that a *Y. enterocolitica arcB* null mutation does not affect induction or function of the Psp response (D. Savitzky and A. Darwin, unpublished data). Furthermore, using random genetic approaches we have so far failed to identify any other proteins that are essential for induction of the Psp response in *Y. enterocolitica* (J. Seo and A. Darwin, unpublished data). Therefore, we currently favor the hypothesis that PspB and/or PspC sense the inducing signal directly.

14.7 Acknowledgements

I thank the past and present members of my laboratory whose work is discussed in this article: Rebecca Green, Michelle Maxson, Diana Savitzky and Jin Seo. Psp research in my laboratory is supported by grant AI052148 from the National Institutes of Health.

14.8 References

Adams, H., Teertstra, W., Demmers, J., Boesten, R. and Tommassen, J. (2003) Interactions between phage-shock proteins in *Escherichia coli*. J. Bacteriol. 185, 1174-1180.

Adams, H., Teertstra, W., Koster, M. and Tommassen, J. (2002) *pspE* (phage-shock protein E) of *Escherichia coli* is a rhodanese. FEBS Lett. 518, 173-176.

Becker, L.A., Bang, I., Crouch, M. and Fang, F.C. (2005) Compensatory role of PspA, a member of the phage shock protein operon, in *rpoE* mutant *Salmonella enterica* serovar Typhimurium. Mol. Microbiol. 56, 1004-1016.

Brissette, J.L., Russel, M., Weiner, L. and Model, P. (1990) Phage shock protein, a stress protein of *Escherichia coli*. Proc. Natl. Acad. Sci. USA 87, 862-866.

Darwin, A.J. (2005) The phage-shock-protein response. Mol. Microbiol. 57, 621-628.

Darwin, A.J. and Miller, V.L. (1999) Identification of *Yersinia enterocolitica* genes affecting survival in an animal host using signature-tagged transposon mutagenesis. Mol. Microbiol. 32, 51-62.

Darwin, A.J. and Miller, V.L. (2001) The *psp* locus of *Yersinia enterocolitica* is required for virulence and for growth *in vitro* when the Ysc type III secretion system is produced. Mol. Microbiol. 39, 429-444.

DeLisa, M.P., Lee, P., Palmer, T. and Georgiou, G. (2004) Phage shock protein PspA of *Escherichia coli* relieves saturation of protein export via the Tat pathway. J. Bacteriol. 186, 366-373.

Duguay, A.R. and Silhavy, T.J. (2004) Quality control in the bacterial periplasm. Biochim. Biophys. Acta 1694, 121-134.

Dworkin, J., Jovanovic, G. and Model, P. (2000) The PspA protein of *Escherichia coli* is a negative regulator of σ^{54}-dependent transcription. J. Bacteriol. 182, 311-319.

Elderkin, S., Bordes, P., Jones, S., Rappas, M. and Buck, M. (2005) Molecular determinants for PspA-mediated repression of the AAA transcriptional activator PspF. J. Bacteriol. 187, 3238-3248.

Genin, S. and Boucher, C.A. (1994) A superfamily of proteins involved in different secretion pathways in gram-negative bacteria: modular structure and specificity of the N-terminal domain. Mol. Gen. Genet. 243, 112-118.

Green, R.C. and Darwin, A.J. (2004) PspG, a new member of the *Yersinia enterocolitica* phage shock protein regulon. J. Bacteriol. 186, 4910-4920.

Guilvout, I., Chami, M., Engel, A., Pugsley, A.P. and Bayan, N. (2006) Bacterial outer membrane secretin PulD assembles and inserts into the inner membrane in the absence of its pilotin. EMBO J. 25, 5241-5249.

Jovanovic, G., Lloyd, L.J., Stumpf, M.P., Mayhew, A.J. and Buck, M. (2006) Induction and function of the phage shock protein extracytoplasmic stress response in *Escherichia coli*. J. Biol. Chem. 281, 21147-21161.

Kleerebezem, M., Crielaard, W. and Tommassen, J. (1996) Involvement of stress protein PspA (phage shock protein A) of *Escherichia coli* in maintenance of the protonmotive force under stress conditions. EMBO J. 15, 162-171.

Kleerebezem, M. and Tommassen, J. (1993) Expression of the *pspA* gene stimulates efficient protein export in *Escherichia coli*. Mol. Microbiol. 7, 947-956.

Maxson, M.E. and Darwin, A.J. (2004) Identification of inducers of the *Yersinia enterocolitica* phage shock protein system and comparison to the regulation of the RpoE and Cpx extracytoplasmic stress responses. J. Bacteriol. 186, 4199-4208.

Maxson, M.E. and Darwin, A.J. (2006a) Multiple promoters control expression of the *Yersinia enterocolitica* phage-shock-protein A (*pspA*) operon. Microbiology 152, 1001-1010.

Maxson, M.E. and Darwin, A.J. (2006b) PspB and PspC of *Yersinia enterocolitica* are dual function proteins: regulators and effectors of the phage-shock-protein response. Mol. Microbiol. 59, 1610-1623.
Model, P., Jovanovic, G. and Dworkin, J. (1997) The *Escherichia coli* phage-shock-protein (*psp*) operon. Mol. Microbiol. 24, 255-261.
Weiner, L., Brissette, J.L. and Model, P. (1991) Stress-induced expression of the *Escherichia coli* phage shock protein operon is dependent on σ^{54} and modulated by positive and negative feedback mechanisms. Genes Dev. 5, 1912-1923.
Weiner, W. and Model, P. (1994) Role of an *Escherichia coli* stress-response operon in stationary-phase survival. Proc. Natl. Acad. Sci. USA 91, 2191-2195.

Picture 16. Attendees at the banquet. Photograph by R. Perry.

15
Functional Quorum Sensing Systems Affect Biofilm Formation and Protein Expression in *Yersinia pestis*

Alexander G. Bobrov[1], Scott W. Bearden[2], Jacqueline D. Fetherston[1], Arwa Abu Khweek[1], Kenneth D. Parrish[3], and Robert D. Perry[1]

[1] Department of Microbiology, Immunology and Molecular Genetics, University of Kentucky, bobrov@uky.edu
[2] Present affiliation: Bacterial Diseases Branch, Division of Vector-Borne Infectious Diseases, The National Center for Zoonotic, Vector-Borne, and Enteric Diseases, Centers for Disease Control and Prevention
[3] Present affiliation: Advanced Implant & Perio Center

Abstract. Gram-negative bacteria predominantly use two types of quorum sensing (QS) systems - LuxI-LuxR, responsible for synthesis of N-acylhomoserine lactones (AHL or AI-1 signal molecule), and LuxS, which makes furanones (AI-2 signal molecule). We showed that LuxS and two LuxI-LuxR (YtbIR and YpsIR) systems are functional in *Y. pestis*. Four different AHL molecules were detected in *Y. pestis* extracts using TLC bioassays. Our data suggest that YtbIR is responsible for the production of long chain AHLs. Confocal laser scanning microscopy showed that biofilm formation is decreased in an *ytbIR ypsIR luxS* mutant. Two-dimensional gel electrophoresis revealed altered levels of protein expression in a *Y. pestis* triple QS mutant at 26°C and 37°C.

15.1 Introduction

In the phenomenon known as quorum sensing (QS), bacterial cells communicate with each other via small extracellular molecules that indicate population density. The N-acylhomoserine lactones (AHLs) of Gram-negative bacteria have been the most extensively studied signaling molecules in QS systems and are also referred to as AI-1. AHLs vary in length, side chain substitutions, and the degree of acyl saturation, making them relatively species specific - the AHL produced by one bacterial species generally does not serve as a density signal for broad range of bacteria. AHLs are synthesized by the LuxI family of synthase proteins. At sufficiently high concentrations, AHLs diffuse back into the bacterial cell and bind to a LuxR-type regulatory protein. The LuxR-AHL complex activates the expression of a variety of genes (Fuqua et al. 2001).

In animal and plant pathogens, such as *Agrobacterium tumefaciens, Aeromonas hydrophila, Erwinia chrysanthemi, Pseudomonas aeruginosa*, and *Vibrio anguillarum*, AHL systems control the expression of a number of exported products that are proven or putative virulence factors. One rationale for QS regulation is that the production of virulence factors by isolated bacteria may cause a neutralizing host response. However, coordinated expression by a large population may overwhelm host defenses and lead to invasion of cells, tissues, and blood vessels, followed by dissemination.

An independent, non-AHL, QS autoinducer (termed AI-2) has been identified. In this system, the signaling molecules are furanones produced by LuxS (Surette et al. 1999; Chen et al. 2002). Unlike AI-1, the AI-2 autoinducer is not species specific; AI-2 from *Vibrio harveyi* serves as a signal for *Salmonella* and *Escherichia*. Homologues of *luxS* are also widespread among bacteria - examples include *Borrelia burgdorferi, Clostridium perfringens,* enterohemorrhagic and enteropathogenic *E. coli, Haemophilus influenzae, Klebsiella pneumoniae, Mycobacterium tuberculosis, Neisseria gonorrhoeae, Neisseria meningitidis, Salmonella typhi, Salmonella typhimurium, Staphylococcus aureus, Streptococcus pneumoniae, Vibrio cholerae, V. anguillarum, Yersinia enterocolitica,* and *Y. pestis* (Schauder and Bassler 2001). There is evidence that the AI-2 system regulates expression of virulence factors in *S. typhimurium* and *V. cholerae*. In enterohemorrhagic and enteropathogenic *E. coli*, LuxS regulates the expression of intimin, the intimin receptor, and the type III secretion system encoded within the locus of enterocyte effacement (LEE) pathogenicity island (Sperandio et al. 1999).

We have identified three QS loci in the genome of *Y. pestis*, two being LuxIR-like and one LuxS-like, and have investigated the role of these systems in the production of signal molecules as well as biofilm formation.

15.2 Materials and Methods

15.2.1 Bacterial Strains, Plasmids, Culture Conditions and Recombinant DNA Techniques

Y. pestis strains and newly constructed plasmids used in this study are listed in Table 1. *E. coli* cells were grown on Luria broth (LB). *Y. pestis* cells were grown on TBA plates and in Heart Infusion Broth and LB. Cell growth was monitored on a Spectronic Genesys5 spectrophotometer at 620 nm. Where required, ampicillin (Ap; 100 μg/ml), kanamycin (Km; 50 μg/ml), or streptomycin (Sm; 50 μg/ml) was added to cultures. Standard recombinant DNA methods (Ausubel et al. 1987) were used to isolate and construct the various plasmids. Genomic DNA was isolated by the CTAB method. The plasmids were transformed into *E. coli* and *Y. pestis* cells using a standard $CaCl_2$ procedure and electroporation, respectively. When necessary, plasmid DNA or PCR products were sequenced by Retrogen, Inc. Synthetic oligonucleotide primers were purchased from Integrated DNA Technologies. To construct a deletion in *ytb*IR, two fragments of 646 and 643 bp, respectively, were amplified by PCR from either end of the *ytb*IR locus using pWSKYtbIR and primer pairs, luxIR1/2 (GACTAGTTTTGGCATACATTTGTTCAGC and TTCACCTCCCAGTTAAGA CT) and luxIR3/4 (TGTCGATATAATCACGCAGTG and GCTCTAGAGGCGT GTTAGCATTATTTGTC). The PCR products were ligated together and the ligation reaction was used as a template for a second PCR reaction with primers luxIR1 and

Table 1. Bacterial strains and plasmids used in this study

Strains or plasmids	Relevant characteristics	Reference or source
Y. pestis strains		
KIM6+	Pgm^+ Pla^+ Yps^+ Ytb^+ $LuxS^+$	Fetherston et al. 1992
KIM6	Pgm^-(Δpgm^-) Pla^+ Yps^+ Ytb^+ $LuxS^+$	Fetherston et al. 1992
KIM10+	Pgm^+ Pla^- Yps^+ Ytb^+ $LuxS^+$	Fetherston et al. 1992
KIM6-2083+	Pgm^+ Pla^+ Yps^+ Ytb^+ $LuxS^-$ (*luxS::kan2083)*	This study
KIM6-2106+	Pgm^+ Pla^+ Yps^- (Δ*ypsIR2106)* Ytb^+ $LuxS^+$	This study
KIM6-2107+	Pgm^+ Pla^+ Yps^+ Ytb^- (Δ*ytbIR2107)* $LuxS^+$	This study
KIM6-2108+	Pgm^+ Pla^+ Yps^- Ytb^- $LuxS^+$	This study
KIM6-2109+	Pgm^+ Pla^+ Yps^- Ytb^- $LuxS^-$	This study
KIM5(pCD1Ap)+	Pgm^+ Pla^+ Yps^+ Ytb^+ $LuxS^+$ Lcr^+	
KIM5-2083 (pCD1Ap)+	Pgm^+ Pla^+ Yps^+ Ytb^+ $LuxS^-$ Lcr+	This study
Plasmids		
pCD1Ap	71.7 kb, pCD1 with *bla* cassette inserted into *'yadA* , Lcr^+; Ap^r	Gong et al. 2001
pWSKYps1	4.1-Kb *Cla*I-*Sac*I fragment from KIM10+ cloned into pWSK29 (Wang and Kushner, 1991); Km^r *ypsIR*$^+$	This study
pWSKYps2	3.7-Kb *Xho*I-*Eco*RV fragment from pLuxYsp-1 cloned into pWSK29; Ap^r	This study
pKNGYps3	715-bp *Sal*I-*Dra*I and 1550-bp *Dra*I-*Xba*I fragments from pLuxYps2 ligated into pKNG101; Sm^r *ΔypsIR2106*	This study
pWSKYtbIR	8.065-Kb *Bgl*II fragment from KIM10+ ligated into the *Bam*HI site of pWSK29; Km^r *ytbIR*$^+$	This study
pBSYtbIR	1.289-Kb PCR fragment cloned into the *Eco*RV site of pBluescript; Ap^r	This study
pKNGYtbIR	1.2-Kb *Bam*HI-*Xba*I fragment from pBSYtbIR ligated into the same sites of pKNG101; Sm^r *ΔytbIR2107*	This study
pLuxS1	0.82-Kb *Hind*III fragment from KIM10+ cloned into pBluescript II KS (Stratagene); Ap^r *luxS*$^+$	This study
pLuxS2	0.85-kb *Xma*I-*Sal*I from pLuxS1 cloned into pKNG101 (Kaniga et al., 1991); Sm^r	This study
pLuxS3	1.27-kb *Bam*HI fragment consisting of Km^r gene from pUC4K (Yanisch-Perron et al.1985) cloned in *Bam*HI of *luxS* gene in pLUXS2 ; *luxS::kan2083*	This study

luxIR4. The 1289 bp fragment, which removes 1203 bp from the *ytb*IR locus, was cloned into the suicide vector pKNG101, generating pKNGYbtIR. To construct a

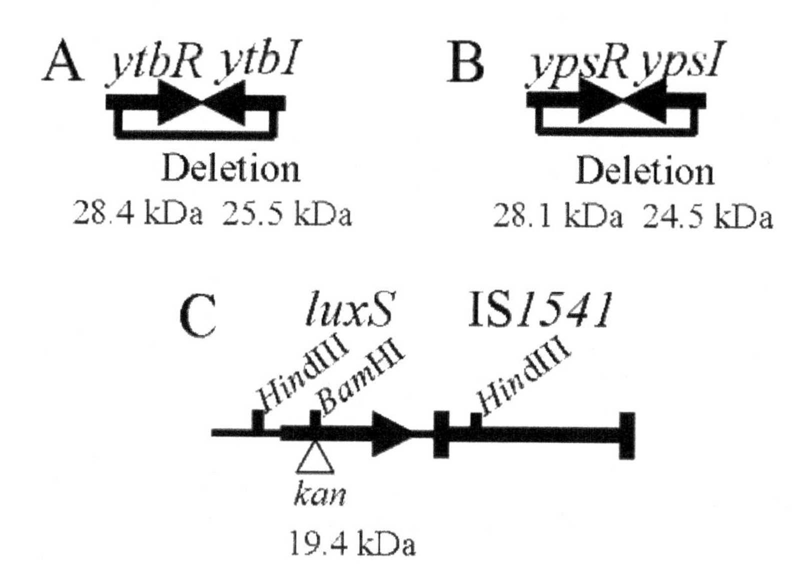

Fig. 1. Genetic organization of *Y. pestis ytbIR*, *ypsIR,* and *luxS*. The construction of mutations in *ytbIR* (A), *ypsIR*(B) and *luxS* (C) is shown.

deletion in *ypsIR*, a 1.5-kb *Dra*I fragment was eliminated from the *ypsIR* locus and cloned into pKNG101 yielding pKNGYps3. To construct a mutation in *luxS*, a 1.27-kb *Bam*HI fragment containing a Km^r gene from pUC4K was ligated into the *Bam*HI of a cloned *luxS* gene, generating the pLuxS3 suicide plasmid. Suicide plasmids were electroporated into KIM6+. Cells from Sm^r colonies were grown overnight without Sm and plated on TBA sucrose plates to isolate sucrose-resistant mutants that had completed allelic exchange. The triple QS mutant was constructed sequentially with the three suicide plasmids. All mutations were confirmed by PCR. Fig. 1 depicts the three QS loci and constructed mutations.

15.2.2 Assays

V. harveyi BB170, which does not respond to AI-1, was used in AI-2 autoinducer bioassays of cell-free culture media as previously described (Surette and Bassler 1998; Surette et al. 1999). Briefly, an overnight culture of *V. harveyi* BB170 was diluted 1:5,000 into fresh AB medium. Supernatants from late-exponential phase cultures grown in glucose-free LB were added to a final concentration of 10% (v/v) to early exponential phase BB170 cultures and light production was monitored for 5 hours using a TopCount luminescence counter with 96-well format (Packard).

E. coli MG4 (pKDT17) (Pearson et al. 1997) was used in liquid assays to measure AI-1 autoinducer levels in culture supernatants from early-stationary phase cultures of *Y. pestis* strains grown in LB with or without 50mM MOPS (pH 7.1) at 26°C or 37°C. The pKDT17 reporter plasmid contains a *lasB-lacZ* fusion that is induced by certain exogenous autoinducer molecules. *E. coli* (pKDT17) was grown in LB overnight at 30°C, back diluted to an OD620 of 0.1 and incubated with supernatants for 5 hours at 30°C. The β-galactosidase activity was measured as described previously (Miller 1992).

15.2.3 Thin Layer Chromatography

Two different strains were used to detect AHL signal molecules in *Y. pestis*: *Agrobacterium tumefaciens* NTL4 (pCF218 + pCF372) which carries *traR* and a *traI-lacZ* fusion, and *Chromobacterium violaceum* strain CV026. The *Agrobacterium* strain produces a blue color in the presence of 5-bromo-4-chloro-3-indolyl-β-D-galactopyranoside (X-Gal) in response to AHLs (Fuqua and Winans 1996; Luo et al. 2001) while a purple pigment is produced by CV026 in response to AHLs (Throup et al. 1995).

The extraction of AHLs and thin layer chromatography was described by Shaw et al. (1997). *Y. pestis* strains were grown in LB with or without 50mM MOPS (pH 7.1) at 26°C or 37°C and AHLs were extracted three times with ethyl acetate. The extracts were concentrated using rotary evaporation. AHL extracts and synthetic standards were applied to a C_{18} TLC plate (Macherey-Nagel) and developed in 60:40 (v/v) methanol/water until the front reached the top. The TLC plate was dried and overlaid with 1% LB or 1% ABT agar containing *C. violaceum* CV026 or *A. tumefaciens* NTL4 (pCF218 + pCF372), respectively, and incubated at room temperature until AHL spots were detected.

15.2.4 Confocal Laser Scanning Microscopy

Samples were processed for confocal microscopy as previously described (Lawrence and Neu 1999). Briefly *Y. pestis* strains, grown overnight on TBA slants, were used to inoculate the defined medium, TMH, to an optical density at 620 nm of 0.1. Glass coverslips were placed in the cultures which were incubated overnight at 30°C. The coverslips were washed twice, incubated with FITC-conjugated wheat germ agglutinin (WGA) for 20 minutes, rinsed with distilled water to remove unbound WGA, and examined by confocal laser scanning microscopy (CLSM).

15.2.5 Two-Dimensional Gel Electrophoresis

Y. pestis strains were cultured in PMH2 with 10mM $FeCl_3$ at either 26 or 37°C to early stationary phase. Proteins were radiolabeled for 2 hours with ^{35}S-amino acids (DuPont NEN Research Products). Samples were prepared as described in the Bio-Rad manual "2-D electrophoresis for proteomics". Supernatants were subjected to isoelectric focusing at between pH 3 to 10 by using precast Immobiline^{TM} DryStrips (Amersham Biosciences), followed by reducing SDS-PAGE in precast Excel 8-18% acrylamide gradient gels (Amersham Biosciences) or 10% polyacrylamide gels. Labeled proteins were detected by autoradiography.

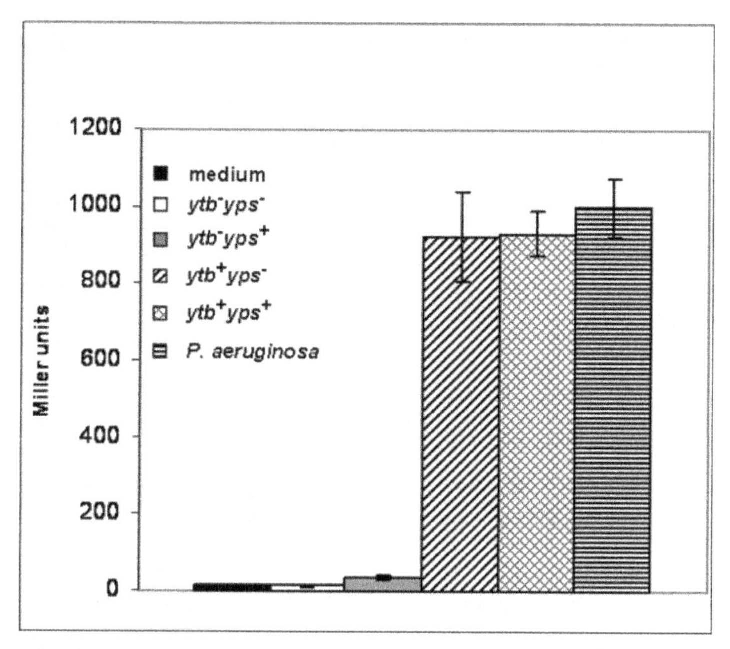

Fig. 2. Activation of *lasB::lacZ* reporter by *Y. pestis* supernatants. *E. coli* cells carrying pKDT17 (encoding the *lasB::lacZ* reporter) were exposed to cell-free culture supernatants from *Y. pestis* strains, *E. coli* (negative control), or *P. aeruginosa* (positive control) and samples were collected after 5 h of incubation.

15.3 Results and Discussion

15.3.1 QS Loci in *Y. pestis*

The *Y. pestis* KIM genome contains two *luxIR* systems and a *luxS* homolog. The two *luxIR* homologs are nearly identical (99%) to the *Yersinia pseudotuberculosis* genes and were designated *ytbIR* and *ypsIR* in keeping with the *Y. pseudotuberculosis* nomenclature (Deng et al. 2002; Atkinson et al. 1999). The deduced amino acid sequence of *Y. pestis* LuxS shows 87-89% identity to LuxS proteins from *E. coli, S. typhimurium, Shigella flexneri* and several other bacteria (Blattner et al. 1997; McClelland et al. 2001; Wei et al., 2003). The *ytbIR, ypsIR,* and *luxS* loci were cloned from *Y. pestis* KIM6+ genomic DNA. *Y. pestis* strains with single, double, and triple mutations in the QS systems were constructed.

Table 2. Activation of *lasB::lacZ* reporter depends on pH and temperature

conditions	OD_{620} ~ 1.0		OD_{620} ~ 1.7		OD_{620} ~ 2.6-2.8		OD_{620} ~ 2.1-2.3	
	pH	MU[a]	pH	MU	pH	MU	pH	MU
37°C	6.92	543	7.37	683	-	-	7.85	459
37°C MOPS pH 7.1	6.94	673	7.06	824	-	-	7.15	937
26°C	6.85	1315	6.92	1297	7.54	1567	-	-
26°C MOPS pH 7.1	6.9	1146	6.92	1450	7.09	1582	-	-

[a] Miller units of β-galactosidase activity from *lasB::lacZ* reporter

15.3.2 Detection and Characterization of AHLs Produced by *Y. pestis*

We used a *P. aeruginosa lasB::lacZ* reporter to detect AI-1 signal molecules in *Y. pestis* supernatants. This reporter responds to autoinducers with a 12-carbon acyl group but will recognize autoinducers with 8-14 carbon acyl groups. *Y. pestis* produced AI-1 molecules that activated the reporter. Supernatants from cells grown at 26°C gave higher values than those from 37°C grown cells (Fig. 2). In *Y. pseudotuberculosis*, the AHL levels were almost undetectable when the cells were grown at 37°C. Yates et al. (2002) showed that during the growth of *Y. pseudotuberculosis* at 37°C in LB, the medium becomes alkaline when the cells reach stationary phase. The combination of temperature and pH leads to lactonolysis, i.e. ring opening of the AHLs. However, lactonolysis could be reduced using LB buffered with MOPS (pH 7.1) (Yates et al. 2002). Using LB and buffered LB at 26°C, we found no difference between the activating ability of supernatants obtained from stationary phase cultures despite some differences in pH (Table 2). The activity of the reporter upon addition of supernatant from stationary phase cultures grown at 37°C was significantly lower. When we added supernatants from cultures grown in buffered medium at 37°C, a two-fold increase in *lasB::lacZ* reporter activity was observed. Therefore our results are in agreement with the previous data for *Y. pseudotuberculosis* (Yates et al. 2002), suggesting that pH and temperature dependent lactonolysis may also occur in *Y. pestis*. Comparison of the various *luxIR* mutants showed that the Ytb system plays a critical role in the activation of the *lasB* reporter. While a mutation in the *yps* operon gave a slight reduction in the activating ability of *Y. pestis* supernatants towards *lasB::lacZ*, deletion of *ybtRI* almost completely abolished the stimulatory activity of supernatants from this mutant (Fig. 2).

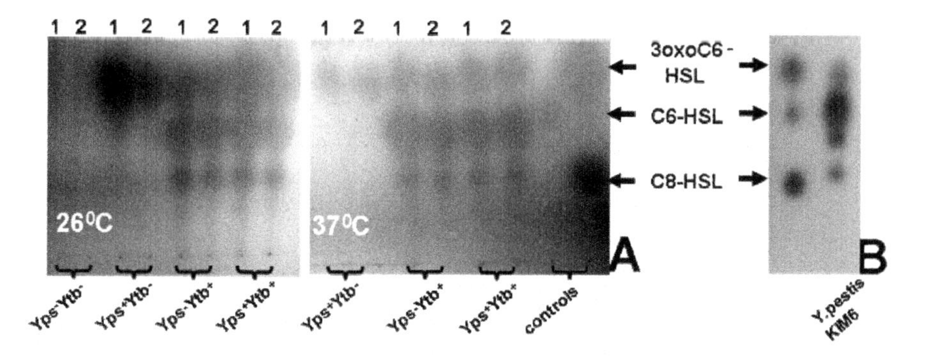

Fig. 3. TLC chromatograms of AHLs extracted from *Y. pestis* strains. A and B AHLs detected using *A. tumefaciens* NTL4 (pCF218 + pCF372) and *C. violaceum* CV026, respectively. 1 and 2 HSLs extracted from strains grown in LB medium and LB medium buffered with MOPS, respectively.

We used TLC to examine the spectrum of AHLs produced by *Y. pestis*. Three compounds migrating similar to 3-oxo-C6-HSL, C6-HSL and C8-HSL (Fig. 3A) were detected using *A. tumefaciens* NTL4 (pCF218 + pCF372). The same compounds were identified in *Y. pseudoberculosis* (Atkinson *et al.* 1999). A fourth compound migrating between C6-HSL and C8-HSL, similar to N-3-oxo-octanoyl homoserine lactone (3-oxo-C8-HSL) produced by *P. aeruginosa* PAO1 (data not shown), was detected using *C. violaceum* CV026 (Fig. 3B). Yates et al. (2002) identified a similarly migrating compound from *Y. pseudoberculosis* supernatant as C7-HSL. The amount of all AHLs, especially 3-oxo-C6-HSL, was reduced at 37°C, likely due to lactonolysis. The AHL profile of the *yps* mutant did not differ from that of the wild type strain in the levels of C6-HSL and C8-HSL but did have a 2-3 fold decrease in the amount of 3-oxo-C6-HSL. In supernatants from the *ytb* mutant, only 3-oxo-C6-HSL and traces of C6-HSL were detected. Our data suggests that the Ytb system is responsible for the synthesis of 3-oxo-C6-HSL, C6-HSL and C8-HSL and the Yps system produces 3-oxo-C6-HSL. These results are in agreement with recently published data on the AHL profiles in *Y. pseudotuberculosis ytbI* and *ypsI* mutants (Ortori et al. 2006) and those of Kirwan et al. (2006) where expression of YtbI (termed YspI in strain CO92 nomenclature) resulted in production of 3-oxo-C6-HSL and 3-oxo-C8-HSL.

15.3.3 Functionality of the *Y. pestis* LuxS System

We used a bioassay to demonstrate that the *Y. pestis* produces a functional AI-2 signal molecule. Supernatants obtained from KIM6+, a *luxS* mutant and a complemented *luxS* mutant were tested in luminescence bioassays using *V. harveyi* strain BB170 that recognizes only the AI-2 molecule. Activity of the culture supernatants

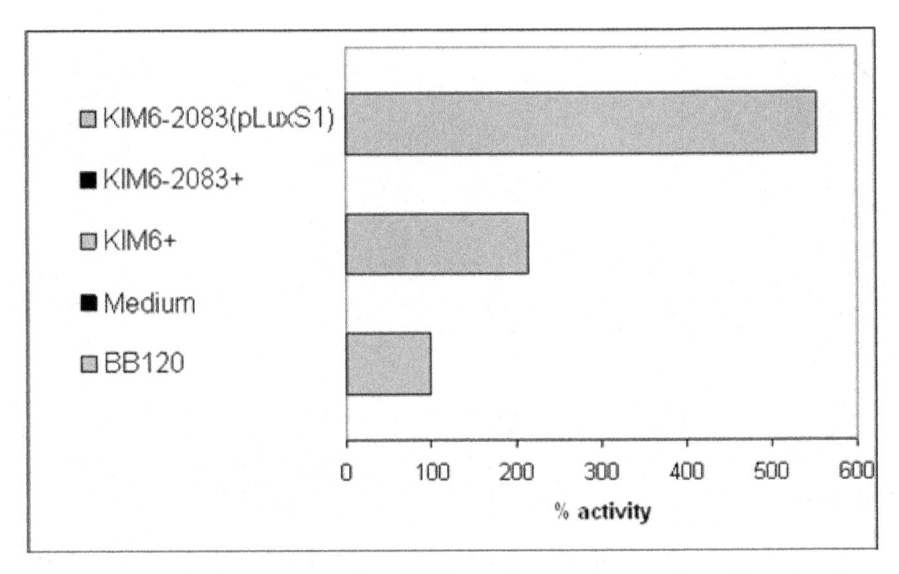

Fig. 4. LuxS activity in *Y. pestis* strains. Cell-free culture supernatants from *Y. pestis* cultures were tested, in a bioassay, for the presence of AI-2 signal molecule produced by LuxS. Activity present was compared to that produced by *V. harveyi* BB120 which was normalized to 100%.

from *Y. pestis* strains at 5 h of incubation was compared to the AI-2- producing *V. harveyi* BB120 strain which was normalized to 100% (Fig. 4). *Y. pestis* KIM6+ showed a high level of activity while the *luxS::kan* mutant, KIM6-2083+, lost all activity. Complementing this mutant with the $luxS^+$ recombinant plasmid, pLuxS2, restored high-level production of the AI-2 molecule. Uninoculated medium had no stimulatory activity (Fig. 4). 2-D gel electrophoresis of *Y. pestis* proteins identified only modest differences in protein expression between the parental strain and the *luxS::kan* mutant (data not shown).

To determine the role of LuxS in the pathogenesis of plague, mice were infected subcutaneously with KIM5(pCD1Ap)+ or KIM5-2083(pCD1Ap)+ cells grown at 26°C to approximate conditions of a flea bite. LD_{50} analysis did not show a loss of virulence in the *luxS::kan* mutant strain. Further studies will be necessary to assess whether this mutation has more modest affects on the disease course and/or organ specificities.

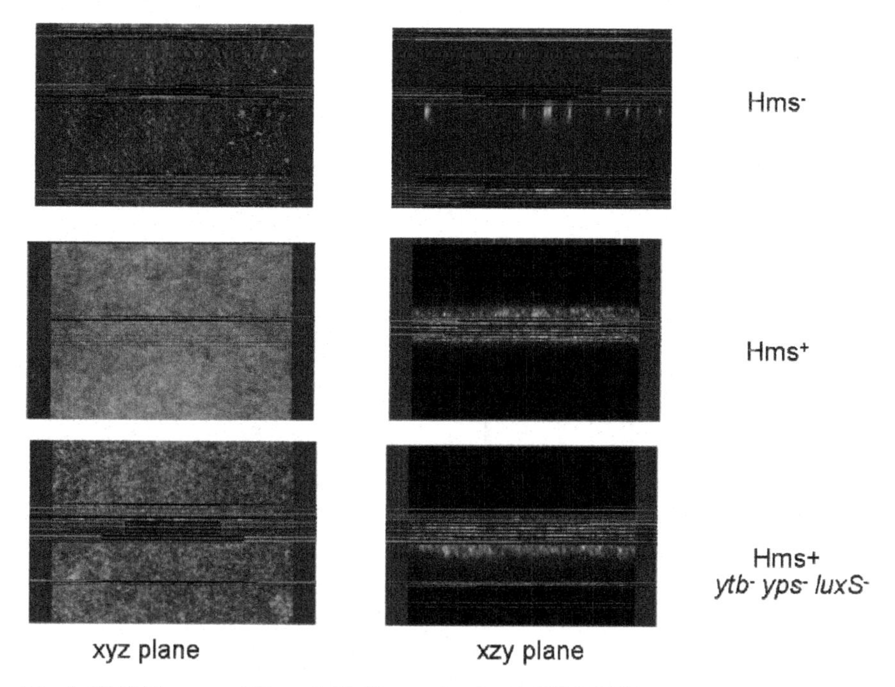

Fig. 5. CLSM images of *Y. pestis* biofilms stained with FITC-WGA.

15.3.4. Effect of QS Systems on Biofilm Formation and Protein Synthesis in *Y. pestis*

In a number of other bacteria, regulation of biofilm formation by quorum sensing systems has been shown (Parsek and Greenberg 2005). In many cases, quorum sensing systems are required for the early stages of biofilm development. Biofilm formation has been shown to be critical for transmission of plague to mammals by some fleas (Hinnebusch et al. 1996; Jarrett et al. 2004). In vitro biofilm formation in *Y. pestis* has been demonstrated using crystal violet staining and Congo red binding assays as well as CLSM (Kirillina et al. 2004). We could not detect any difference between the parent strain and the triple QS mutant by Congo red binding and crystal violet staining assays. However, using CLSM we consistently observed a mild defect in biofilm development in the triple QS mutant (Fig. 5). Similar results were obtained using cells expressing the green fluorescent protein (data not shown). However, the two AHL systems did not appear to play any role in the blockage of *Xenopsylla cheopis* and transmission of plague to mammals (Jarrett et al. 2004). Whether quorum sensing systems are important for transmission of plague by some other flea species or unblocked fleas (Eisen et al. 2006) remains to be elucidated.

The total protein profile of KIM6+ and the triple QS mutant (KIM6-2109) was analyzed by 2-D gel electrophoresis. For *Y. pestis* grown at 26°C, at least seven proteins that were highly expressed in KIM6+ were missing or reduced in KIM6-

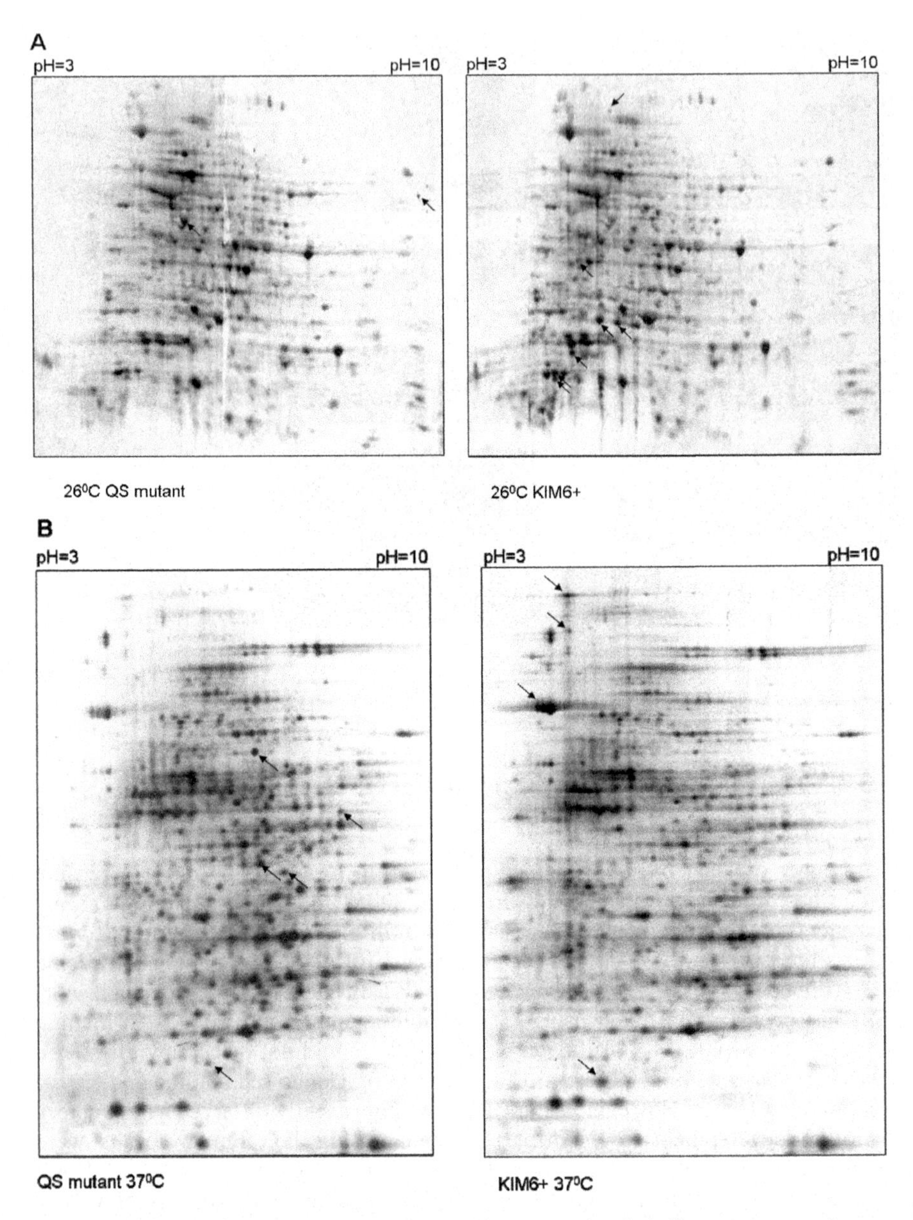

Fig. 6. Effect of QS on *Y. pestis* protein expression. Arrowheads indicate representative proteins whose expression was altered.

2109 while there were two proteins in the QS mutant that appear to be missing in KIM6+ (Fig. 6A). This differential expression might explain the defect in biofilm formation of the *Y. pestis* triple QS mutant at ambient temperature. At least eleven proteins were differentially expressed in KIM6+ compared to the triple QS mutant when the bacteria were grown at 37°C (Fig. 6B). Moreover, a different set of the proteins was affected at 37°C compared to 26°C indicating that QS regulation in *Y. pestis* is temperature-dependent. Using protein microarrays, Chen et al. (2006) have suggested that expression of a number of virulence-associated proteins including F1, LcrV, KatY and pH6 were decreased in an *ypeRI/yspRI* mutant [*ytb* and *yps* were annotated as *ysp* and *ype*, respectively in CO-92 genome (Parkhill et al. 2001)]. Swift et al. (1999) noted a slight delay in time of death from infection with a *ypeR* mutant compared to the wild-type strain of *Y. pestis*. This suggests that quorum sensing may play a role in the pathogenesis of plague. During a mammalian infection *Y. pestis* exists in acidic conditions as a result of inflammation, thus breakdown of the AHLs as a result of lactonolysis at 37°C might be reduced.

15.4 Conclusions

Using several bioassays we demonstrated that *Y. pestis* has functional LuxI-LuxR and LuxS systems. Four compounds which are likely 3-oxo-C6-HSL, C6-HSL, C8-HSL and 3-oxo-C8-HSL were detected in *Y. pestis* supernatants. Our data suggest that YtbIR system was responsible for synthesis 3-oxo-C6-HSL, C6-HSL and C8-HSL. In contrast YpsIR system failed to produce long chain molecules. CLSM demonstrated that biofilm formation in vitro was reduced in a triple QS mutant suggesting that development of plague biofilm in fleas and transmission of bubonic plague to mammals can be affected by QS. Differential expression of proteins in the QS mutant and wild type strain at different temperatures suggests that QS can regulate protein expression in *Y. pestis* during infection of fleas, at ambient temperature, as well as during infection of mammals at 37°C.

15.5 Acknowledgements

We thank Clay Fuqua for the *Agrobacterium* AHL reporter and control strains as well as protocols. We thank Dave Erickson and Kelly Babb for helpful discussions. This project was supported by a University of Kentucky Research Professorship Award to RDP and by Public Health Service grant AI25098 from the US National Institutes of Health.

15.6 References

Ausubel, F.M., Brent, R., Kingston, R.E., Moore, D.D., Seidman, J.G., Smith, J.A. and Struhl, K. (1987) *Current Protocols in Molecular Biology*. New York, N. Y.: John Wiley & Sons.

Atkinson, S., Throup, J.P., Stewart, G.S.A.B. and Williams, P. (1999) A hierarchical quorum sensing system in *Yersinia pseudotuberculosis* is involved in the regulation of motility and clumping. Mol. Microbiol. 33, 1267-1277.

Blattner, F.R., Plunkett, G. III, Bloch, C.A., Perna, N.T., Burland, V.,Riley, M., Collado-Vides, J., Glasner, J. D., Rode, C. K., Mayhew, G. F., Gregor, J., Davis, N.W.,

Kirkpatrick, H.A., Goeden, M.A., Rose, D.J., Mau, B. and Shao, J. (1997) The complete genome sequence of *Escherichia coli* K-12 Science 277, 1453-1474.

Chen, X., Schauder, S., Potier, N., Van Dorsselaer, A., Pelczer, I. Bassler, B.L. and Hughson, F. M. (2002) Structural identification of a bacterial quorum-sensing signal containing boron. Nature 415:545-549.

Deng, W., Burland, V., Plunkett G. III, Boutin, A., Mayhew, G.F. Liss, P., Perna, N.T., Rose, D.J., Mau, B., Zhou, S. Schwartz, D.C., Fetherston, J.D., Lindler, L.E., Brubaker, R.R., Plano, G.V., Straley, S.C., McDonough, K.A., Nilles, M.L., Matson, J.S., Blattner, F.R. and Perry, R.D. (2002) Genome sequence of *Yersinia pestis* KIM. J. Bacteriol. 184, 4601-4611.

Eisen, R.J., Bearden, S.W., Wilder, A.P., Montenieri, J.A., Antolin, M.F., and Gage, K.L. (2006) Early-phase transmission of *Yersinia pestis* by unblocked fleas as a mechanism explaining rapidly spreading plague epizootics. Proc. Natl. Acad. Sci. USA 103, 15380-15385.

Fetherston, J.D., Schuetze, P. and Perry, R.D. (1992) Loss of the pigmentation phenotype in *Yersinia pestis* is due to the spontaneous deletion of 102 kb of chromosomal DNA which is flanked by a repetitive element. Mol. Microbiol. 6, 2693-2704.

Fuqua, W.C. and Winans, S. C. (1996) Conserved *cis*-acting promoter elements are required for density-dependent transcription of *Agrobacterium tumefaciens* conjugal transfer genes. J. Bacteriol. 178, 435-440.

Fuqua, C., Parsek, M.R. and Greenberg. E. P. (2001) Regulation of gene expression by cell-to-cell communication: acyl-homoserine lactone quorum sensing. Annu. Rev. Genet. 35, 439-468.

Gong, S., Bearden, S.W., Geoffroy, V.A., Fetherston, J.D. and Perry, R.D. (2001) Characterization of the *Yersinia pestis* Yfu ABC inorganic iron transport system. Infect. Immun. 69, 2829-2837.

Hinnebusch, B.J., Perry, R.D. and Schwan, T.G. (1996) Role of the *Yersinia pestis* hemin storage (*hms*) locus in the transmission of plague by fleas. Science 273, 367-370.

Jarrett, C.O., Deak, E., Isherwood, K.E., Oyston, P.C., Fischer, E.R., Whitney, A.R., Kobayashi, S.D., DeLeo, F.R. and Hinnebusch, B.J. (2004) Transmission of *Yersinia pestis* from an infectious biofilm in the flea vector. J. Infect. Dis. 190, 783-792.

Kaniga, K., Delor, I. & Cornelis, G. R. (1991) A wide-host-range suicide vector for improving reverse genetics in Gram-negative bacteria: inactivation of the *blaA* gene of *Yersinia enterocolitica*. Gene 109, 137-141.

Kirillina, O., Fetherston, J.D., Bobrov, A.G., Abney, J. and Perry, R.D. (2004) HmsP, a putative phosphodiesterase, and HmsT, a putative diguanylate cyclase, control Hms-dependent biofilm formation in *Yersinia pestis*. Mol. Microbiol 54, 75-88.

Kirwan, J.P., Gould, T.A., Schweizer, H.P., Bearden, S.W., Murphy, R.C., and Churchill, M.E.A. (2006) Quorum-sensing signal synthesis by the *Yersinia pestis* acyl-homoserine lactone synthase YspI. J. Bacteriol. 188, 784-788.

McClelland M., Sanderson, K.E., Spieth J., Clifton S.W., Latreille P., Courtney L., Porwollik S., Ali J., Dante M., Du F., Hou S., Layman D., Leonard S., Nguyen C., Scott K., Holmes A., Grewal N., Mulvaney E., Ryan, E., Sun H., Florea L., Miller W., Stoneking T., Nhan M., Waterston R. and Wilson R.K. (2001) Complete genome sequence of *Salmonella enterica* serovar Typhimurium LT2. Nature 413, 852-856.

Ortori, C.A., Atkinson, S., Chhabra, S.R., Camara, M., Williams, P. and Barrett, D.A. (2006) Comprehensive profiling of *N*-acylhomoserine lactones produced by *Yersinia psedotuberculosis* using liquid chromatography coupled to hybrid quadrupole-linear ion trap mass spectrometry. Anal. Bioanal. Chem. In press.

Parkhill, J., Wren, B.W., Thomson, N.R., Titball, R.W., Holden, M.T., Prentice, M.B., Sebaihia, M., James, K.D., Churcher, C., Mungall, K.L., Baker, S., Basham, D., Bentley, S.D.,

Brooks, K., Cerdeno-Tarraga, A.M., Chillingworth, T., Cronin, A., Davies, R.M., Davis, P., Dougan, G., Feltwell, T., Hamlin, N., Holroyd, S., Jagels, K., Karlyshev, A.V., Leather, S., Moule, S., Oyston, P.C., Quail, M., Rutherford, K., Simmonds, M., Skelton, J., Stevens, K., Whitehead, S. and Barrell. B.G. (2001) Genome sequence of *Yersinia pestis*, the causative agent of plague. Nature 413, 523-527.

Parsek, M.R. and Greenberg, E. P. (2005) Sociomicrobiology: the connections between quorum sensing and biofilms. Trends Microbiol.13, 27-33.

Pearson, J.P., Pesci, E.C. and Iglewski B.H. (1997) Roles of *Pseudomonas aeruginosa las* and *rhl* quorum-sensing systems in control of elastase and rhamnolipid biosynthesis genes. J. Bacteriol. 179, 5756-5767.

Perry, R.D., Pendrak, M.L. and Schuetze, P. (1990) Identification and cloning of a hemin storage locus involved in the pigmentation phenotype of *Yersinia pestis*. *J Bacteriol* 172, 5929-5937.

Schauder, S. and Bassler, B.L. (2001) The language of bacteria. Genes Dev. 15, 1468-1480.

Shaw, P.D., Ping, G., Daly, S.L., Cha, C., Cronan, J.E.J., Rinehart, K.L. and Farrand S.K. (1997) Detecting and characterizing *N*-acyl-homoserine lactone signal molecules by thin-layer chromatography. Proc. Natl. Acad. Sci. USA 94, 6036-6041.

Sperandio, V., Mellies, J.L., Nguyen, W., Shin, S. and Kaper, J.B. (1999) Quorum sensing controls expression of the type III secretion gene transcription and protein secretion in enterohemorrhagic and enteropathogenic *Escherichia coli*. *Proc Natl Acad Sci USA* 96, 15196-1520

Surette, M.G., and Bassler, B.L. (1998) Quorum sensing in *Escherichia coli* and *Salmonella typhimurium*. Proc. Natl. Acad. Sci. USA 95:7046-7050.

Surette, M.G., Miller, M.B. and Bassler, B.L. (1999) Quorum sensing in *Escherichia coli*, *Salmonella typhimurium*, and *Vibrio harveyi*: a new family of genes responsible for autoinducer production. Proc. Natl. Acad. Sci. USA 96, 1639-1644.

Swift, S., Isherwood, K.E., Atkinson, S., Oyston. P.C.F. and Stewart, G.S.A.B. (1999) Quorum Sensing in Aeromonas and Yersinia In: England, R., Hobbs, G., Bainton, N. and Roberts, D. McL. (Eds.), *Microbial Signalling and Communication*. Cambridge University Press, Cambridge, pp. 85-104.

Throup, J.P., Camara, M., Bainton, N.J., Briggs, G.S., Chhabra, S.R., Bycroft, B.W., Williams, P. and Stewart, G.S.A.B. (1995) Characterisation of the *yenI/yenR* locus from *Yersinia enterocolitica* mediating the synthesis of two *N*-acyl homoserine lactone signal molecules. Mol. Microbiol. 17, 345-356.

Yanisch-Perron, C., Vieira, J. and Messing. J. (1985) Improved M13 phage cloning vectors and host strains: nucleotide sequences of the M13mp18 and pUC19 vectors. Gene 33, 103-119.

Yates, E. A., Philipp, B., Buckley, C., Atkinson, S., Chhabra, S.R., Sockett, R.E., Goldner, M., Dessaux, Y., Camara, M., Smith, H. and P. Williams, P. (2002) *N*-acylhomoserine Lactones undergo lactonolysis in a pH-, temperature-, and acyl chain length-dependent manner during growth of *Yersinia pseudotuberculosis* and *Pseudomonas aeruginosa*. Infect. Immun. 70, 5635-5646.

Wang, R.F. and Kushner, S.R. (1991) Construction of versatile low-copy-number vectors for cloning, sequencing and gene expression in *Escherichia coli*. Gene 100, 195-199.

Wei, J., Goldberg, M.B., Burland, V., Venkatesan, M.M., Deng, W., Fournier, G., Mayhew, G.F., Plunkett, G. III, Rose, D.J., Darling, A., Mau, B., Perna, N.T., Payne, S.M., Runyen-Janecky, L.J., Zhou, S., Schwartz, D.C. and Blattner, F.R. (2003) Complete genome sequence and comparative genomics of *Shigella flexneri* serotype 2a strain 2457T. Infect. Immun. 71, 2775-2786.

16
Analysis of *Yersinia pestis* Gene Expression in the Flea Vector

Viveka Vadyvaloo[1], Clayton Jarrett[1], Daniel Sturdevant[2], Florent Sebbane[3], and B. Joseph Hinnebusch[1]

[1] Laboratory of Zoonotic Pathogens, and [2] Research Technology Section, Rocky Mountain Laboratories, National Institutes of Health, vadyvaloov@niaid.nih.gov
[3] Inserm, U801, Institut Pasteur de Lille, Université de Lille 2

Abstract. *Yersinia pestis* is the causative agent of plague. Unlike the other pathogenic *Yersinia* species, *Y. pestis* has evolved an arthropod-borne route of transmission, alternately infecting flea and mammalian hosts. Distinct subsets of genes are hypothesized to be differentially expressed during infection of the arthropod vector and mammalian host. Genes crucial for mammalian infection are referred to as virulence factors whilst genes playing a role in the flea vector are termed transmission factors. This article serves as a review of known factors involved in flea-borne transmission and introduces an '*in vivo*' microarray approach to elucidating the genetic basis of *Y. pestis* infection of- and transmission by the flea.

16.1 Biological Mechanism of Transmission

Yersinia pestis infections persist in wild rodent species throughout the world, and are transmitted largely by flea bite. *Y. pestis* is able to undergo a development process in the flea midgut which involves its multiplication to eventually form a thick cohesive aggregate of cells in between the spines of the proventricular valve. This thick mass of cells has been identified as a biofilm (Darby et al. 2002; Jarrett et al. 2004).

The flea digestive tract is where the storage, digestion and absorption of the blood meal occur. Anatomically, the digestive tract is made up of a hindgut, esophagus and midgut lumen, with the latter 2 organs being separated by a valve-like structure, the proventriculus. The proventriculus is central to the mechanism of transmission. It is made up of layers of muscles that surround densely packed cuticle-covered spines. The proventriculus is usually closed but during feeding it opens and closes in a synchronized manner to help propel the blood meal into the midgut and prevent its leakage out of the gut. Transmission can occur when the thick mass of *Y. pestis* bacterial cells block the proventriculus, preventing passage of the blood meal into the midgut lumen (Bacot and Martin 1914). The blocked flea will then attempt repeatedly to feed and the blood meal will be regurgitated along with some dislodged bacteria back into the bite site, resulting in transmission.

16.1.1 *Y. pestis in vivo* and *in vitro* Biofilm

A bacterial biofilm is characterized by a dense multicellular association of bacterial cells that are embedded by an extracellular matrix of bacterial origin, and that adhere to an interface (Davey and O'Toole 2000). The *Y. pestis* extracellular matrix is formed by the gene products of the *hmsHFRS* genes belonging to the hemin storage (*hms*) locus (Darby et al. 2002; Jarrett et al. 2004; Perry et al. 1990). Additionally, the gene products coded by the hemin storage gene locus are responsible for colonization and blockage of the flea proventriculus (Hinnebusch et al. 1996; Jarrett et al. 2004). Studies carried out *in vitro* show that in *Y. pestis,* biofilm formation is confined to the lower temperature range of 21 to 28°C, which matches the flea body temperature, and biofilm is absent at the mammalian body temperature of 37°C (Jones et al. 1999; Pendrak and Perry 1993; Perry et al. 1990).

In general, biofilms constitute a physiologically heterogeneous population of cells even within a single species biofilm (Sternberg et al. 1999; Werner et al. 2004). The heterogeneity lies in the different growth rates exhibited by cells due to different nutrient accessibilities to the layers of cells within the biofilm, *e.g.*, deeper layers of cells have lower oxygen concentrations compared to the oxygenated liquid media-biofilm interface (Yu et al. 2004). An *in vitro Y. pestis* biofilm can be studied by using a glass flowcell apparatus (Jarrett et al. 2004). Flowcells can be described as flow-through biofilm culture devices because they are composed of narrow glass chambers that have a port of entry and exit which allows media to be continuously pumped through, thereby removing non-adherent cells and providing constant nutrition to the developing biofilm. The flowcell is designed to enable visualization of this dynamic biofilm using confocal microscopy after staining the intact biofilm with a fluorescent dye. Erickson et al. (Erickson et al. 2006) used this method effectively to study the ability of *Y. pseudotuberculosis* and *Y. pestis* to form biofilms *in vitro*. Since a dynamic biofilm can be maintained in a flowcell system, this may provide a useful method when studying potential *Y. pestis* biofilm-specific factors.

16.1.2 Known Transmission Factors of *Y. pestis*

Since most studies focus on understanding *Y. pestis* virulence mechanisms in the mammal, relatively few factors that act to produce a transmissible infection in the flea vector have been discovered. The *hms* genes introduced above represents one of these factors. The name, hemin storage, was derived from the observation that the *hms* genes are involved in the pigmentation phenotype produced by whole *Y. pestis* cells binding to hemin or its structural analogue, Congo Red (Burrows and Jackson 1956; Surgalla and Beesley 1969). Another *Y. pestis* factor that influences biofilm formation, a phosphoheptoisomerase involved in LPS synthesis, is also required for normal proventricular blockage in fleas (Darby et al. 2005). The Yersinia murine toxin or Ymt is a third important factor shown to affect flea-borne transmission of *Y. pestis*. The *ymt* gene codes for an intracellular phospholipase which enables *Y. pestis* to survive in the digestive environment of the flea gut (Hinnebusch et al. 2002). The actual function of Ymt is not known but deletion of the *ymt* gene results in spheroplasting and lysis of *Y. pestis* cells with inability of the bacteria to establish an initial infection (Hinnebusch et al. 2002).

16.2 Methods

To identify bacterial factors that are involved in the ability of *Y. pestis* to infect and block fleas, we compared the gene expression profiles of *Y. pestis* KIM6+ in blocked fleas and in flowcells by microarray.

16.2.1 Isolation of RNA from Blocked Flea Midguts

In our laboratory the rat flea *Xenopsylla cheopis* is used to study transmission of *Y. pestis*. *X. cheopis* is the most efficient vector of the plague organism (Burroughs 1947). *X. cheopis* fleas are infected by allowing them to feed on freshly prepared heparinized mouse blood containing approximately 10^9 cells/ml in an artificial feeding system (Hinnebusch et al. 1998; Hinnebusch et al. 1996). The infection by *Y. pestis* is left to develop and fleas are fed every 3 days and blockage is monitored over a period of 1-4 weeks. Blockage tends to occur at a rate of 25-45% in the *Y. pestis* KIM6+ strain (Hinnebusch et al. 1996). Blocked fleas were dissected in a droplet of phosphate buffered saline under a dissecting microscope, after being anesthetized with a non-lethal dose of isofluorane. The midguts were removed from the flea and placed immediately in 500 µl of RNAprotect™ bacterial reagent (Qiagen) in which they were macerated with the rounded edge of the piston of a 1 ml plastic syringe. Approximately 30 blocked midguts were pooled and stored at –80°C until RNA extraction and processing.

16.2.2 Isolation of RNA from Flowcell Biofilms

Y. pestis KIM6+ was grown for 48 h at 21°C in LB/100mM MOPS pH 7 medium and diluted to a concentration of 1 x 10^7 cells/ml. A single channel of the borosilicate glass flowcell (Stovall; Greensboro, N.C.) was injected with the diluted bacterial suspension (0.4 ml). The flowcell was connected to a reservoir of sterile LB /100mM MOPS pH 7 medium via a peristaltic pump at the influent end and to a discard reservoir at the effluent end. After a 20-min period to allow bacteria to attach to the glass surface, sterile medium was pumped through the flowcell at 0.3 ml/min. The flow was stopped after 48 h when a thick, adherent biofilm was observed to be attached to the borosilicate glass surface of the flowcell. This biofilm was washed out through one port of the flowcell chamber using RNAprotect™, into an eppendorf tube containing more RNAprotect™ bacterial reagent, via the opposite port of the flowcell chamber. The samples were then stored at –80°C until RNA extraction and processing. Once all the samples were collected, RNA was isolated and processed together for hybridization to the custom Rocky Mountain laboratories *Y. pestis* chip (Sebbane et al. 2006).

16.3 Results

The microarray data were statistically analyzed using software from Partek (Partek Incorporated, St Louis, MO), Significance Analysis of Microarrays or SAM (Tusher et al. 2001) that is incorporated into a Microsoft Excel (Microsoft Corporation, Redmond, WA) template designed for microarray data analysis. The analysis software determines whether the difference in expression of a gene is significantly different in the 2 growth environments. All statistical analyses must be in agreement before a gene is declared to be differentially expressed. The microarray data for the 6 replicates in the flea and flowcell biofilm conditions were found to be reproducible and discrete by these statistical analyses used in the experiment.

16.3.1 The Polyamine Transport (*pot*) Locus is Highly Up-Regulated in the Flea

Two genes were found to be most highly upregulated in the flea biofilm relative to *in vitro* flowcell conditions. These genes, *y1392*, with a 21-fold increase in gene expression; and *y1391*, with a 34-fold increase in gene expression are linked in a 4 gene locus (Fig. 1). The gene sequences were individually subjected to a BLAST search, revealing that the sequences were highly similar to the polyamine transport (uptake) genes of several different bacterial species. Gene expression signals for the 2 other linked genes, *y1390* and *y1393*, were detected only in the flea samples. The *y1390* and *y1393* genes have homology to a γ-aminobutyrate aminotransferase (*goaG*) involved in polyamine catabolism, and another polyamine transport gene, respectively.

What are polyamines? The polyamines putrescine, spermidine and spermine are organic cations that are required for cell growth and differentiation (Wallace et al. 2003) and that modulate DNA, RNA and protein synthesis (Igarashi and Kashiwagi 2000; Tabor and Tabor 1985; Yoshida et al. 2004).

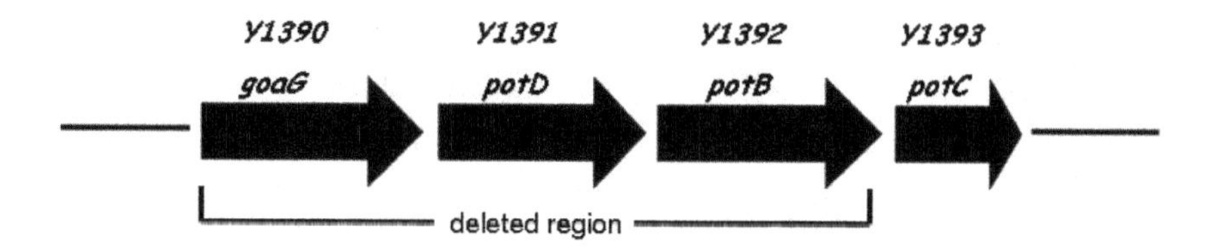

Fig. 1. Organization of the *Y. pestis* polyamine transport (*pot*) locus involved in uptake of exogenous polyamines. The deleted region indicates the genes that were deleted to create the *Y. pestis* KIM6+ *Δpot* mutant.

There have also been many other specific functional roles ascribed to polyamines, including protection against oxidative stress in *Escherichia coli* (Tkachenko et al. 2001), repression of polyamine production in *E. coli* (Moore and Boyle 1991), increasing polyamine uptake in *Pseudomonas aeruginosa* (Lu et al. 2002), and extracellular signalling for swarming behaviour in *Proteus mirabilis* (Sturgill and Rather 2004). Of most interest to us, however, was the recent report that polyamines play a role in the formation of a *Y. pestis* biofilm *in vitro* (Patel et al. 2006).

In the report on polyamines in *Y. pestis* biofilm (Patel et al. 2006), the polyamine biosynthesis genes *speA* and *speC* that encode arginine decarboxylase and ornithine decarboxylase were deleted, resulting in polyamine-deficient *Y. pestis* mutants. The single mutants (*ΔspeA* and *ΔspeC*) and a double mutant (*ΔspeA ΔspeC*) had reduced biofilm formation and inability to form a biofilm, respectively. Moreover, genetic complementation of the *ΔspeA ΔspeC* double mutant with *speA*, or the addition of exogenous putrescine resulted in restoration of biofilm formation.

16.3.2 Deletion of the *pot* Locus in *Y. pestis* KIM6+

The polyamine uptake genes were highly upregulated in the flea, suggesting to us that polyamines may be important in an *in vivo* biofilm. We hypothesized that if this was the case then a deletion in the polyamine transport locus may result in altered biofilm formation and inability to form a proventricular biofilm in the flea vector. Therefore, we constructed a mutant in *Y. pestis* KIM6+ by deleting the first 3 genes of the *pot* locus *y1390*, *y1391* and *y1392* (Fig. 1) and this mutant was referred to as *Y. pestis* KIM6+ *Δpot*.

A growth curve experiment of the *Y. pestis* KIM6+ *Δpot* mutant relative to wild-type *Y. pestis* KIM6+ (Fig. 2) revealed that the mutant had a normal growth rate, indicating that the mutation did not result in any general fitness cost to the bacteria.

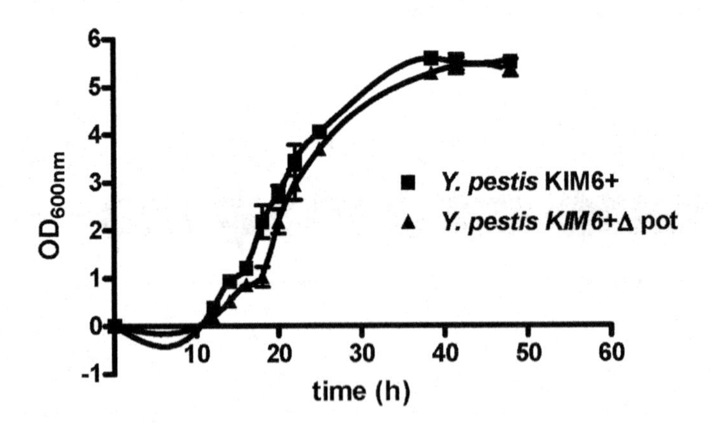

Fig. 2. Growth curve of *Y. pestis* KIM6+ and *Y. pestis* KIM6+ *Δpot* in LB medium at 21°C.

16.3.2.1 *In vitro* Biofilm Formation of *Y. pestis* KIM6+*Δpot*

In vitro studies to quantify biofilm formation of *Y. pestis* KIM6+ *Δpot* and the wild-type *Y. pestis* KIM6+ were done using a modification of the method of Christensen et al. (Christensen et al. 1985). Bacteria were grown in TMH medium (Straley and Bowmer 1986) in 96-well microtitre plates. Once the biofilm formed after 48 hours at room temperature, the medium was removed and the wells were washed gently with distilled water. The biofilm was then stained with 0.05% safranin O for 10 min after which the stain was removed and the wells washed thoroughly and allowed to air-dry. A solution of 30% acetic acid was used to dissolve the bound safranin and the absorbance was read at 450 nm. The wildtype and mutant were observed to make an equal amount of biofilm (Fig. 3).

16.3.2.2 Flea Infections with *Y. pestis* KIM6+*Δpot*

We infected *X. cheopis* fleas with the *Y. pestis* KIM6+ *Δpot* mutant (Hinnebusch et al. 1996) and found that the rate of infection and blockage was comparable to the wildtype *Y. pestis* KIM6+ (data not shown). This indicates that the deletion of the *pot* locus has no effect on the ability of *Y. pestis* KIM6+ to form a biofilm on the proventriculus or to cause blockage. One can speculate that the high induction of the *pot* transport genes is because the bacteria preferentially use an exogenous source of polyamine rather than synthesizing it. Fleas are allowed to feed on mammalian blood, which contains approximately 0.21 μmol/L putrescine, 6.56 μmol/L spermidine and 3.92 μmol/L spermine (Cooper et al. 1978) and is probably the source of the exogenous polyamine. In addition, it has been previously reported that polyamine uptake and utilization is upregulated by exogenous polyamine (Lu et al. 2002). Moreover, we observed that the constitutive ornithine decarboxylase gene (*y2987*), and the *aguA* (*y3324*) and *aguB* (*y3324*) genes involved in the conversion of agmatine to putrescine (Patel et al. 2006) are only induced in the flowcell biofilm. No

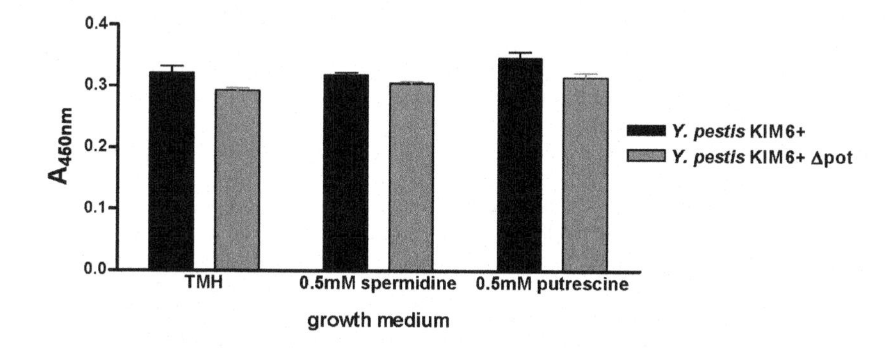

Fig. 3. Quantitative *in vitro* biofilm assay in TMH medium indicating that there is no difference in amount of biofilm formed by wildtype *Y. pestis* KIM6+ and *Y. pestis* KIM6+ *Δpot* and that this is independent of supplementing the TMH medium with exogenous polyamines, spermidine or putrescine.

signal for these genes was detected in the flea, indicating a requirement for polyamine biosynthesis *in vitro* only.

16.4 Conclusions

We can conclude from our findings that deletion of the polyamine transport system that is highly upregulated in the flea has no effect on biofilm formation either *in vitro* or in the flea and that the *pot* genes are not required to produce a transmissible infection in the flea. However, one must consider that the *pot* deletion mutant we used has functional polyamine biosynthesis genes, which would allow the mutant to synthesize polyamines *de novo* if it was unable to transport them into the cell from the flea gut. It may be interesting to infect fleas with *Y. pestis* strains in which both polyamine biosynthetic and polyamine transport genes were deleted. It is possible that this *Y. pestis* mutant would be unable to synthesize or take up polyamines and this may give a clearer indication whether polyamines indeed play a major role in flea biofilm formation.

16.5 References

Bacot, A.W. and Martin, C.J. (1914) Observations on the mechanism on the transmission of plague by fleas. J. Hyg. 13, 423-439.

Burroughs, A.L. (1947) Sylvatic plague studies. the vector efficiency of nine species of fleas compared with *Xenopsylla cheopis*. J. Hyg. 45, 371-396.

Burrows, T.W. and Jackson, S. (1956) The pigmentation of *Pasteurella pestis* on a defined medium containing haemin. Br. J. Exp. Pathol. 37, 570-576.

Christensen, G.D., Simpson, W.A., Younger, J.M., Baddour, L.M., Barrett, F.F., Melton, D.M. and Beachey, E.H. (1985) Adherence of coagulase-negative staphylococci to plastic tissue culture plates: a quantitative model for the adherence of staphylococci to medical devices. J. Clin. Microbiol. 22, 996-1006.

Cooper, K.D., Shukla, J.B. and Rennert, O.M. (1978) Polyamine compartmentalization in various human disease states. Clin. Chim. Acta 82, 1-7.

Darby, C., Ananth, S.L., Tan, L. and Hinnebusch, B.J. (2005) Identification of *gmhA*, a *Yersinia pestis* gene required for flea blockage, by using a *Caenorhabditis elegans* biofilm system. Infect. Immun. 73, 7236-7242.

Darby, C., Hsu, J.W., Ghori, N. and Falkow, S. (2002) *Caenorhabditis elegans*: plague bacteria biofilm blocks food intake. Nature 417, 243-244.

Davey, M.E. and O'Toole, G.A. (2000) Microbial biofilms: from ecology to molecular genetics. Microbiol. Mol. Biol. Rev. 64, 847-867.

Erickson, D.L., Jarrett, C.O., Wren, B.W. and Hinnebusch, B.J. (2006) Serotype differences and lack of biofilm formation characterize *Yersinia pseudotuberculosis* infection of the *Xenopsylla cheopis* flea vector of *Yersinia pestis*. J. Bacteriol. 188, 1113-1119.

Hinnebusch, B.J., Fischer, E.R. and Schwan, T.G. (1998) Evaluation of the role of the *Yersinia pestis* plasminogen activator and other plasmid-encoded factors in temperature-dependent blockage of the flea. J. Infect. Dis. 178, 1406-1415.

Hinnebusch, B.J., Perry, R.D. and Schwan, T.G. (1996) Role of the *Yersinia pestis* hemin storage (*hms*) locus in the transmission of plague by fleas. Science 273, 367-370.

Hinnebusch, B.J., Rudolph, A.E., Cherepanov, P., Dixon, J.E., Schwan, T.G. and Forsberg, A. (2002) Role of Yersinia murine toxin in survival of *Yersinia pestis* in the midgut of the flea vector. Science 296, 733-735.

Igarashi, K. and Kashiwagi, K. (2000) Polyamines: mysterious modulators of cellular functions. Biochem. Biophys. Res. Commun. 271, 559-564.

Jarrett, C.O., Deak, E., Isherwood, K.E., Oyston, P.C., Fischer, E.R., Whitney, A.R., Kobayashi, S.D., DeLeo, F.R. and Hinnebusch, B.J. (2004) Transmission of *Yersinia pestis* from an infectious biofilm in the flea vector. J. Infect. Dis. 190, 783-792.

Jones, H.A., Lillard, Jr., J.W. and Perry, R.D. (1999) HmsT, a protein essential for expression of the haemin storage (Hms+) phenotype of *Yersinia pestis*. Microbiology 145, 2117-2128.

Lu, C.D., Itoh, Y., Nakada, Y. and Jiang, Y. (2002) Functional analysis and regulation of the divergent *spuABCDEFGH-spuI* operons for polyamine uptake and utilization in *Pseudomonas aeruginosa* PAO1. J. Bacteriol. 184, 3765-3773.

Moore, R.C. and Boyle, S.M. (1991) Cyclic AMP inhibits and putrescine represses expression of the *speA* gene encoding biosynthetic arginine decarboxylase in *Escherichia coli*. J. Bacteriol. 173, 3615-3621.

Patel, C.N., Wortham, B.W., Lines, J.L., Fetherston, J.D., Perry, R.D. and Oliveira, M.A. (2006) Polyamines are essential for the formation of plague biofilm. J. Bacteriol. 188, 2355-2363.

Pendrak, M.L. and Perry, R.D. (1993) Proteins essential for expression of the Hms+ phenotype of *Yersinia pestis*. Mol. Microbiol. 8, 857-864.

Perry, R.D., Pendrak, M.L. and Schuetze, P. (1990) Identification and cloning of a hemin storage locus involved in the pigmentation phenotype of *Yersinia pestis*. J. Bacteriol. 172, 5929-5937.

Sebbane, F., Lemaitre, N., Sturdevant, D.E., Rebeil, R., Virtaneva, K., Porcella, S.F. and Hinnebusch, B.J. (2006) Adaptive response of *Yersinia pestis* to extracellular effectors of innate immunity during bubonic plague. Proc. Natl. Acad. Sci. USA 103, 11766-11771.

Sternberg, C., Christensen, B.B., Johansen, T., Toftgaard Nielsen, A., Andersen, J.B., Givskov, M. and Molin, S. (1999) Distribution of bacterial growth activity in flow-chamber biofilms. Appl. Environ. Microbiol. 65, 4108-4117.

Straley, S.C. and Bowmer, W.S. (1986) Virulence genes regulated at the transcriptional level by Ca2+ in *Yersinia pestis* include structural genes for outer membrane proteins. Infect. Immun. 51, 445-454.

Sturgill, G. and Rather, P.N. (2004) Evidence that putrescine acts as an extracellular signal required for swarming in *Proteus mirabilis*. Mol. Microbiol. 51, 437-446.

Surgalla, M.J. and Beesley, E.D. (1969) Congo red-agar plating medium for detecting pigmentation in *Pasteurella pestis*. Appl. Microbiol. 18, 834-837.

Tabor, C.W. and Tabor, H. (1985) Polyamines in microorganisms. Microbiol. Rev. 49, 81-99

Tkachenko, A., Nesterova, L. and Pshenichnov, M. (2001) The role of the natural polyamine putrescine in defense against oxidative stress in *Escherichia coli*. Arch. Microbiol. 176, 155-157.

Wallace, H.M., Fraser, A.V. and Hughes, A. (2003) A perspective of polyamine metabolism. Biochem. J. 376, 1-14.

Werner, E., Roe, F., Bugnicourt, A., Franklin, M.J., Heydorn, A., Molin, S., Pitts, B. and Stewart, P.S. (2004) Stratified growth in *Pseudomonas aeruginosa* biofilms. Appl. Environ. Microbiol. 70, 6188-6196.

Yoshida, M., Kashiwagi, K., Shigemasa, A., Taniguchi, S., Yamamoto, K., Makinoshima, H., Ishihama, A. and Igarashi, K. (2004) A unifying model for the role of polyamines in bacterial cell growth, the polyamine modulon. J. Biol. Chem. 279, 46008-46013.

Yu, T., de la Rosa, C. and Lu, R. (2004) Microsensor measurement of oxygen concentration in biofilms: from one dimension to three dimensions. Water Sci. Technol. 49, 353-358.

Picture 17. Irina Zudina and Andrey Filippov at a poster session. Photograph by A. Anisimov.

17
Regulation of Biofilm Formation in *Yersinia pestis*

Alexander G. Bobrov[1], Olga Kirillina, and Robert D. Perry

[1] Department of Microbiology, Immunology and Molecular Genetics, University of Kentucky, bobrov@uky.edu

Abstract. Plague biofilm development is controlled by positive (HmsT) and negative (HmsP) regulators. The GGDEF-domain protein HmsT appears to have diguanylate cyclase activity to synthesize bis-(3'-5')-cyclic dimeric guanosine monophosphate (c-di-GMP) from 2 GTP molecules. The EAL domain of HmsP has phosphodiesterase activity and likely degrades c-di-GMP. This second messenger molecule probably influences biofilm development by activating the glycosyl transferase activity of HmsR. Here we demonstrate the in vitro pH optimum for phosphodiesterase activity of HmsP and that an alanine substitution in residue L508, D626, or E686 within the EAL domain affects this enzymatic activity and the biological function of the protein. Finally, protein-protein interactions and the cytoplasmic location of the enzymatic domains of HmsT and HmsP are evaluated.

17.1 Biofilm and Transmission of Bubonic Plague

Bacteria exist in two distinct states: as planktonic or single, free-living cells or as a biofilm, a multicellular community of bacteria attached to a solid surface and surrounded by an extracellular matrix, where exopolysaccharide is a major component (Costerton et al. 1995; Sutherland 2001). In oriental rat fleas (*Xenopsylla cheopis*), and a number of other fleas, transmission of plague from fleas to mammals depends upon the blockage of the proventriculus of flea. This blockage is formed by a biofilm of dense aggregates of plague bacilli attached to spines of the flea proventriculus (Bacot and Martin 1914; Bibikova and Klassovskii 1974; Hinnebusch et al. 1996; Jarrett et al. 2004). Blockage prevents passage of the blood meal into the mid-gut causing increased feeding attempts. These feeding attempts dislodge bacteria from the biofilm and backflow of the bloodmeal contaminate the mammalian bite wound. In contrast, *Oropsylla montana* (which infests ground and rock squirrels) is immediately infectious for mammals without the macro-phenomenon of blockage (Eisen et al. 2006). The transmission mechanism and any role for biofilm formation in this unblocked transmission method remains to be elucidated. Once injected into mammals, plague bacilli likely revert to the planktonic state to invade and disseminate within the host. This is suggested by the fact that, in vitro, a biofilm is not formed at 37°C (Jackson and Burrows 1956; Perry et al. 2004) and that biofilm formation plays no significant role in the progression of bubonic plague in mice (Kutyrev et al. 1992; Lillard et al. 1999).

In this report we describe the molecular mechanisms which may control life-style switch in *Y. pestis* between a biofilm and planktonic state.

17.2 Genes Essential for Biofilm Development in *Y. pestis*

The blockage of fleas is caused by Pgm$^+$ cells of *Y. pestis* which are able to bind Congo red (CR) or hemin and clump in liquid medium at 26°C but not at 37°C. The *hmsHFRS* operon, encoded within the *pgm* locus, is required for CR binding, blockage of fleas, and biofilm formation in fleas and the nematode *Caenorhabditis elegans* (Jackson and Burrows 1956; Bibikova and Klassovskii 1974; Perry et al. 1990; Hinnebusch et al. 1996; Darby et al. 2002; Jarrett et al. 2004; Kirillina et al. 2004). A fifth essential gene, *hmsT*, lies elsewhere in the *Y. pestis* genome - its gene product is a positive regulator of biofilm development (Hare and McDonough 1999; Jones et al. 1999; Kirillina et al. 2004). The gene encoding a negative regulator of the CR binding phenotype, *hmsP*, was recently identified (Kirillina et al. 2004). Mutations in *speA* and *speC*, which encode polyamine biosynthetic enzymes, also completely disrupt in vitro plague biofilm formation (Patel et al. 2006).

An unidentified enzymatic activity of YrbH, in addition to synthesis of arabinose 5-phosphate, is critical for biofilm formation of *Y. pestis* in *C. elegans*. Mutations in *yrbH* and *waa* or *gmhA* led to LPS truncation and disruption of plague biofilm formation in *C. elegans* and fleas, respectively. It was proposed that these alterations in LPS could interfere with transport of the polysaccharide required for biofilm formation (Darby et al. 2005; Tan and Darby 2006).

Crystal violet (CV) staining is a measure of cell attachment and thus biofilm formation in bacteria (O'Toole et al. 1999). CR stains certain polysaccharides and has been used to detect the extracellular matrix of some bacterial biofilms (Alison and Sutherland 1984). The Hms$^+$ strain of *Y.* pestis KIM shows high levels of CV staining, from strong attachment to glass test tubes, and CR binding while individual mutations in the four genes of the *hmsHFRS* operon or in *hmsT* cause a loss of CR binding and CV staining (Kirillina et al. 2004; Forman et al. 2006). Confocal laser scanning microscopy (CLSM) confirms these results, showing a biofilm structure for the Hms$^+$ strain while *hmsHFRS* or *hmsT* mutants failed to adhere to glass and form a biofilm (Kirillina et al. 2004).

Conserved polysaccharide deacetylase and group 2 glycosyltransferase motifs in HmsF and HmsR, respectively, suggests their involvement in the biosynthesis of exopolysaccharide. HmsR is 58% similar to IcaA, a poly-*N*-acetyl-D-glucosamine transferase from *Staphylococcus epidermidis*. Moreover, all HmsHFRS proteins have a high degree of similarity (50-83%) to PgaABCD proteins from *Escherichia coli.* The *pga* operon is essential for production of poly-β-1,6-*N*-acetyl-D-glucosamine (PGA) required for biofilm development in *E. coli* (Mack et al. 1996; Jones et al. 1999; Wang et al. 2004). In addition, dispersin B, a glycoside hydrolase shown to specifically cleave PGA, prevented formation of *Y. pestis* biofilm (Itoh et al. 2005). Consequently, the HmsHFRS proteins are likely responsible for production of β-1,6-*N*-acetyl-D-glucosamine essential for plague biofilm development.

17.3 Phosphodiesterase HmsP and Diguanylate Cyclase HmsT Control Biofilm Formation in *Y. pestis*

HmsT, a positive regulator of plague biofilm formation, has a highly conserved GGDEF domain named after a five conserved residue signature motif (Jones et al. 1999). The GGDEF domain is predicted to synthesize an unusual molecule of bis-(3'-5')-cyclic dimeric guanosine monophosphate (c-di-GMP). Proteins containing the GGDEF domain have been termed diguanylate cyclases. HmsP encodes a highly conserved EAL domain named after a three amino acid motif which degrades c-di-GMP via its phosphodiesterase activity (Kirillina et al. 2004; Bobrov et al. 2005; Römling et al. 2005). An *hmsP* mutant of *Y. pestis* KIM exhibits increased CV staining compared to the Hms$^+$ parent and CLSM showed an extremely thick mat-like biofilm formed by this mutant. Consequently, HmsP is a negative regulator of biofilm formation (Kirillina et al. 2004; Bobrov et al. 2005).

C-di-GMP is a novel bacterial second messenger which was discovered by Moishe Benziman's research group during studies on regulation of cellulose biosynthesis in *Gluconobacter xylinum.* They demonstrated that a diguanylate cyclase synthesizes one molecule of c-di-GMP from two molecules GTP and that c-di-GMP allosterically activates a cellulose synthase enzyme leading to a 1000-fold increase in cellulose production. C-di-GMP phosphodiesterase inhibits cellulose production in this bacterium through degradation of c-di-GMP (Ross et al. 1987). Since cellulose is a major component of biofilm in many bacteria this model suggests a wide-spread mechanism of regulation of bacterial biofilm formation. Recent studies have confirmed a role for c-di-GMP in controlling biofilm formation in a number of bacteria including *Pseudomonas aeruginosa, Vibrio cholerae, Salmonella enterica* and *E. coli.* These bacteria produce cellulose, PGA, or other exopolysaccharides that are all regulated by c-di-GMP levels. While GGDEF-domain proteins increased cellular levels of c-di-GMP and promote biofilm formation, inhibition of biofilm development by EAL domain proteins correlated with decreased cellular levels of c-di-GMP. Moreover, diguanylate cyclase activity with purified GGDEF-domain proteins and phosphodiesterase activity with EAL-domain proteins has been demonstrated (Römling et al. 2005).

Since the GGEE residues of HmsT are essential for CR binding and biofilm formation in *Y. pestis,* the GGDEF domain probably is responsible for diaguanylate cyclase activity of HmsT which is essential for the control of biofilm formation by this positive regulator (Kirillina et al. 2004). HPLC analysis showed that over-expression of HmsT elevates cellular levels of c-di-GMP approximately three-fold in a *Y. pestis hmsT* mutant, again demonstrating that positive regulation by HmsT likely occurs through c-di-GMP synthesis (Simm et al. 2005).

Complementation of an *hmsP* mutant by over-expressing HmsP causes a drastic reduction in biofilm development and CR binding. A similar complementation was achieved with just the EAL domain of HmsP suggesting that the EAL domain alone is sufficient for negative regulation. Using a purified his-tag fusion of the EAL domain of HmsP, we demonstrated that this protein has phosphodiesterase (PDE)

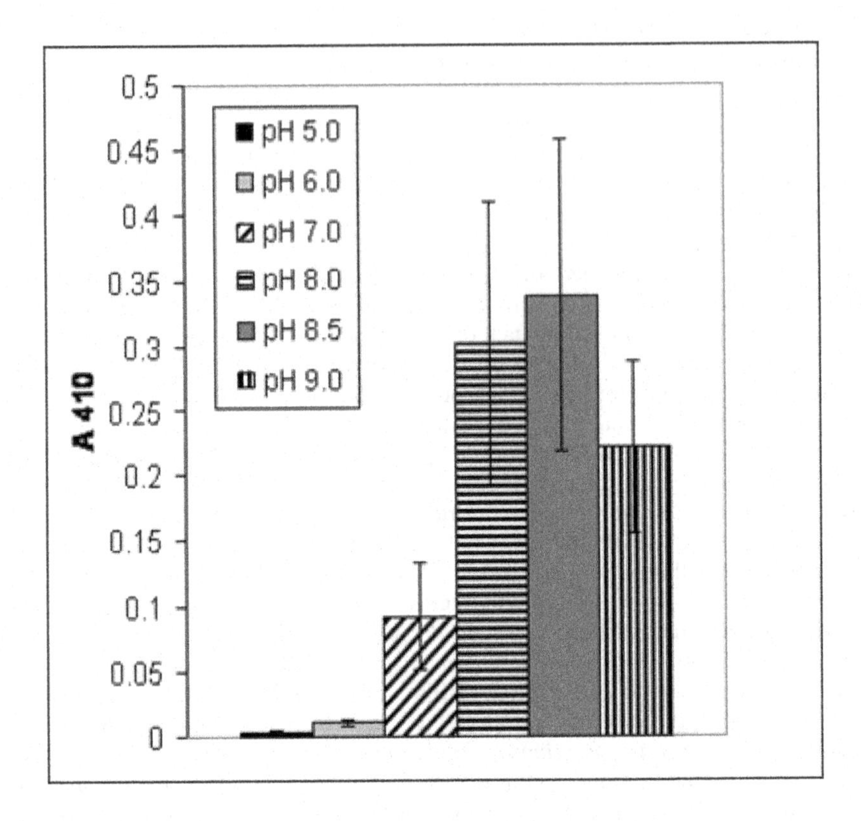

Fig. 1. Determination of optimum pH for phosphodiesterase activity of EAL-HmsP. 10 µg of affinity purified, dialyzed His-tag-EAL-HmsP was tested for PDE activity in a reaction buffer (0.05 M Tris, 0.05 M Bis-Tris, 0.1M Na acetate, 1mM $MnCl_2$, 2.5 mM of bis-*p*NPP) at the indicated pH's.

activity in the presence of Mn^{2+} against the artificial substrate bis(*p*-nitrophenyl) phosphate (bis-*p*NPP). This in vitro activity was considerably higher at 37°C than at room temperature (Bobrov et al. 2005). In this study we have determined that the in vitro pH optimum for the PDE activity of EAL-HmsP is between 8 and 9 (Fig.1).

Previously, we demonstrated that the conserved glutamate 506 residue was critical for PDE activity and the biological activity of EAL-HmsP (Bobrov et al. 2005). Here we report on the effects of site-directed mutagenesis of additional selected conserved residues within the EAL domain of HmsP. We substituted alanine for leucine 508, aspartate 626 or glutamate 686. No enzymatic activity was apparent for the EAL-HmsP-D626A mutant as well as for the negative control EAL-HmsP-E506A variant. In contrast, the L508A and E686A mutants showed only an ~50% decrease in PDE activity against bis-*p*NPP (Fig. 2C). To test the importance of these residues for biological function we expressed these variants in an *hmsP* mutant. The

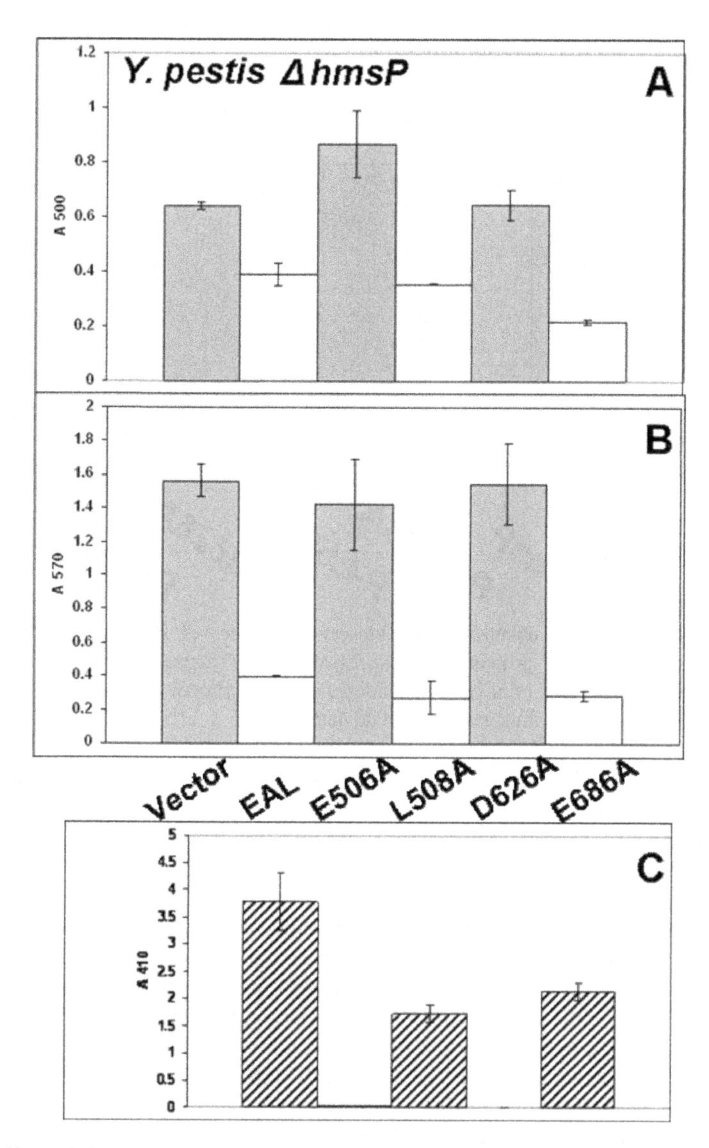

Fig. 2. Effect of alanine substitutions in conserved residues of the EAL domain of HmsP on PDE activity and biofilm formation in *Y. pestis*. The variant EAL-HmsP proteins were purified and tested for PDE activity (A). Conditions for PDE activity, Crystal violet staining (B) and CR binding (C) assays have been described (Bobrov et al. 2005). Values are the averages of at least two independent experiments with duplicate assays for each experiment. Error bars indicate standard deviations.

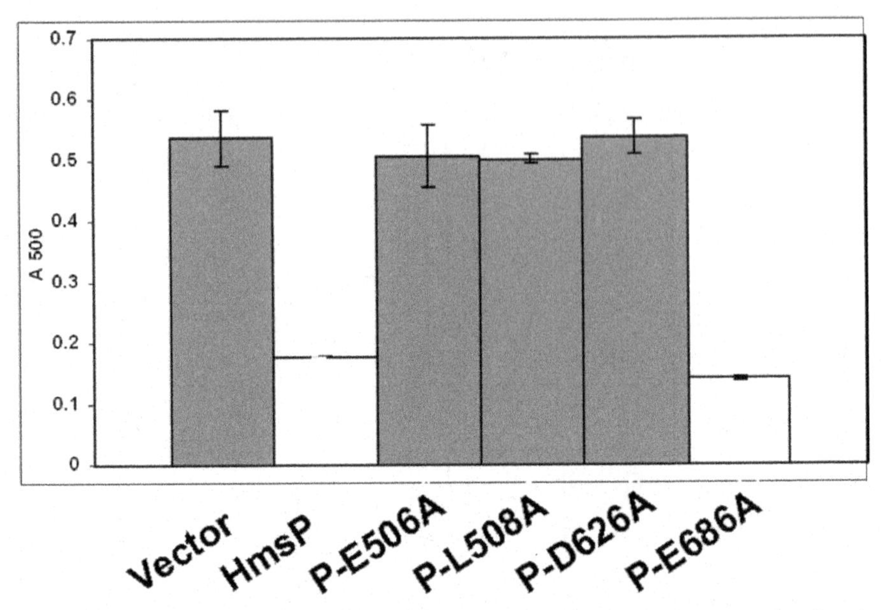

Fig. 3. Effect of alanine substitutions in conserved residues of HmsP on CR binding in *Y. pestis*. The variant HmsP proteins were expressed in *hmsP* mutant and tested for CR binding. Values are the averages of at least two independent experiments with duplicate assays for each experiment. Error bars indicate standard deviations.

expression of the wild-type EAL-HmsP domain and those with partial phosphodiesterase activity (EAL-HmsP-L508A and EAL-HmsP-E686A) inhibited CV staining and CR binding. In contrast, variant EAL-domain proteins lacking PDE activity did not complement the *hmsP* mutation demonstrating no reduction in CV staining or CR binding (Fig. 2A and 2B). Our data show that the 50% reduction in PDE activity against the artificial substrate is sufficient for functional complementation when the variant proteins are over-expressed and indicate that phosphodiesterase activity of the HmsP EAL domain is required for negative regulation of biofilm formation in *Y. pestis*. When a full length HmsP protein carrying the L508A substitution was overexpressed in the *hmsP* mutant it failed to inhibit CR binding (Fig. 3). There are several possible explanations for the intriguing phenotypic difference between the EAL-HmsP-L508A and the HmsP-L508A proteins. Since they are encoded by different plasmid expression systems, the total levels of protein may differ. If the level of inner membrane (IM)-bound HmsP-L508A is lower than the soluble variant, the lower enzymatic activity may not be sufficient to complement the *hmsP* mutation. The full length HmsP has HAMP and (likely inactive) GGDEF domains that may affect enzymatic activity of the protein. Nevertheless, our results suggest that HmsP and HmsT control the level of c-di-GMP available to HmsR thus modulating polysaccharide synthesis. In this model, c-di-GMP synthesized by hmsT serves as an allosteric activator for hmsR. Conversely, HmsP prevents HmsR activation and biofilm development through degradation of this second messenger.

17.4 Topology and Protein-Protein Interactions of Hms Proteins

While two Hms proteins, the putative β-barrel protein HmsH and HmsF are localized to the outer membrane, HmsR, HmsS, HmsT and HmsP are found in the inner membrane (Perry et al. 2004; unpublished observations). It has been shown in other bacteria that proteins with GGDEF and EAL domains have these enzymatic domains in the cytoplasm (Römling et al. 2005). Assembly of the polysaccharide repeating unit occurs in cytoplasm as well. Several software programs predict that HmsP has two transmembrane domains with the N- and C-termini of the protein located in the periplasm. HmsT is predicted to have five transmembrane domains with the N-terminus in the cytoplasm and the C-terminus in the periplasm. Both predictions place the enzymatic GGDEF and EAL domains in the bacterial periplasm.

In order to determine whether the enzymatic domains of HmsP and HmsT lie in the cytoplasm or periplasm, we constructed full length and truncated C-terminal translational fusions with β-galactosidase and alkaline phosphatase. Alkaline phosphatase exhibits enzymatic activity only if the PhoA domain is secreted beyond the inner membrane while β-galactosidase shows activity only if it remains in the cytoplasm. For HmsP, our results indicate the opposite of the predicted topology where two transmembrane domains separate the central periplasmic segment from cytosolic N- and C-terminal portions. Thus the C-terminal HAMP, likely defective GGDEF, and EAL domains all reside in the cytoplasm (unpublished results). Likewise our results for HmsT suggested a cytoplasmic location of the C-terminus containing the GGDEF domain. A series of translational fusions indicate that HmsT has four transmembrane domains, not the five predicted by computational analyses. The change in the number of transmembrane domains shifts the C-terminal GGDEF domain to a cytoplasmic location (unpublished observations). Thus the enzymatic domains of *Y. pestis* Hms proteins are cytoplasmic indicating that synthesis and degradation of c-di-GMP occurs in the cytoplasm as demonstrated for similar systems in other bacteria (Römling et al. 2005).

It has been shown that membrane complexes are required for the production of cellulose (Kimura et al. 2001). We also propose that Hms IM proteins interact with each other forming an enzymatic complex for regulation of polysaccharide production. A bacterial two-hybrid system based on interaction-mediated reconstitution of adenylate cyclase activity (Karimova et al. 1998) was used to study the interactions of Hms proteins. Based on topology studies, the cytosolic parts of the Hms proteins were fused to T25 and T18 fragments of *Bordetella pertussis* adenylate cyclase and tested for catabolic activation of *lac* operons in an *E. coli cya* strain. If two proteins interact, cAMP will be produced and transcription of the *lac* operon will be induced. By this criterion, we detected interaction of HmsP with itself, HmsT and HmsR but not with HmsS (unpublished data). HmsT also appears to form a dimer and interact with HmsP but not with HmsR or HmsS. Western blot analysis showed the presence of monomer and dimer forms of HmsP in *Y. pestis* cells partially confirming our two-hybrid results (unpublished observations).

17.5 Conclusions

The interactions of HmsP and HmsT with themselves and other Hms proteins suggests a local mechanism of regulation of plague biofilm development mediated through protein-protein interactions. We propose that an increase in levels of c-di-GMP by HmsT nearby HmsR causes activation of this enzyme leading to polysaccharide production and development of plague biofilm. Degradation of HmsT at 37°C by specific proteases (Lon, ClpP, and/or ClpX), as well as a higher phopshodiesterase activity of HmsP causes lower levels of c-di-GMP within the Hms protein complex (Perry et al. 2004; Bobrov et al. 2005). This reduces the enzymatic activity of HmsR and thus inhibits biofilm formation at higher, mammalian temperatures.

We propose that c-di-GMP is responsible for the life style switch of *Y. pestis*. In a flea, at ambient temperatures, increases in the levels of c-di-GMP mediated by HmsT cause a transition to a biofilm state and blockage of the flea. In fleas that become blocked this leads to a short period of efficient transmission of plague to mammals. At 37°C, levels of c-di-GMP decrease mediated by increased HmsP activity and the degradation of HmsT; this results in inhibition of biofilm biosynthesis. Thus, in mammals, the plague bacilli likely switch to a planktonic state for dissemination throughout the host and for expression of surface virulence factors that would not be masked by an exopolysaccharide layer.

17.6 Acknowledgments

We are grateful to Renato Morona for plasmids pRMCD28 and pRMCD70, Colin Manoil for strain *E. coli* CC118, Gouzel Karimova and Daniel Ladant for bacterial two hybrid system strains and plasmids and Standa Forman for HmsT, HmsR and HmsS constructs to test protein-protein interactions. This study was supported by Public Health Service grant AI25098 from the US National Institutes of Health.

17.7 References

Allison, D.G. and Sutherland, I.W. (1984) A staining technique for attached bacteria and its correlation to extracellular carbohydrate production. J. Microbiol. Methods 2, 93-99.

Bacot, A.W., and Martin, C.J. (1914) LXVII. Observations on the mechanism of the transmission of plague by fleas. J. Hyg. 13, 423-439.

Bibikova, V.A., and Klassovskii, L.N. (1974) *The Transmission of Plague by Fleas (in Russian).* Moscow: Meditsina.

Bobrov, A.G., Kirillina, O. and Perry, R.D. (2005) The phosphodiesterase activity of the HmsP EAL domain is required for negative regulation of biofilm formation in *Yersinia pestis*. FEMS Microbiol. Lett. 247, 123-130.

Costerton, J.W., Lewandowski, Z., Cladwell, D.E., Korber, D.R. and Lappin-Scott, H.M. (1995) Microbial biofilms. Annu. Rev. Microbiol. 49, 711-745.

Darby, C., Hsu, J.W., Ghori, N. and Falkow, S. (2002). *Caenorhabditis elegans*: plague bacteria biofilm blocks food intake. Nature 417, 243-244.

Darby, C., Ananth, S.L., Tan, L., and Hinnebusch, B.J. (2005) Identification of *gmhA*, a *Yersinia pestis* gene required for flea blockage, using a *Caenorhabditis elegans* biofilm system. Infect. Immun. 73, 7236-7242.

Eisen, R.J., Bearden, S.W., Wilder, A.P., Montenieri, J.A., Antolin, M.F., and Gage, K.L. (2006) Early-phase transmission of *Yersinia pestis* by unblocked fleas as a mechanism explaining rapidly spreading plague epizootics. Proc. Natl. Acad. Sci. USA 103, 15380-15385.

Forman, S., Bobrov, A.G., Kirillina, O., Craig, S.K., Abney, J., Fetherston J.D. and Perry, R.D. (2006) Identification of critical amino acid residues in the plague biofilm Hms proteins. Microbiology 152, 3399-3410.

Hare, J.M. and McDonough, K.A. (1999) High-frequency RecA-dependent and -independent mechanisms of Congo red binding mutations in *Yersinia pestis*. J. Bacteriol. 181, 4896-4904.

Hinnebusch, B.J., Perry, R.D. and Schwan, T.G. (1996) Role of the *Yersinia pestis* hemin storage (*hms*) locus in the transmission of plague by fleas. Science 273, 367-370.

Itoh, Y., Wang, X., Hinnebusch, B.J., Preston, J.F., III and Romeo, T. (2005) Depolymerization of β-1,6-*N*-acetyl-D-glucosamine disrupts the integrity of diverse bacterial biofilms. J. Bacteriol. 187, 382-387.

Jackson, S. and Burrows, T.W. (1956) The pigmentation of *Pasteurella pestis* on a defined medium containing haemin. Br. J. Exp. Pathol. 37, 570-576.

Jarrett, C.O., Deak, E., Isherwood, K.E., Oyston, P.C., Fischer, E.R., Whitney, A.R., Kobayashi, S.D., DeLeo, F.R. and Hinnebusch, B.J. (2004) Transmission of *Yersinia pestis* from an infectious biofilm in the flea vector. J. Infect. Dis. 190, 783-792.

Jones, H.A., Lillard, J.W., Jr. and Perry, R.D. (1999) HmsT, a protein essential for expression of the haemin storage (Hms^+) phenotype of *Yersinia pestis*. Microbiology 145, 2117-2128.

Karimova, G., Pidoux, J., Ullmann, A. and Ladant D. (1998) A bacterial two-hybrid system based on a reconstituted signal transduction pathway. Proc. Natl. Acad. Sci. USA 95, 5752-5756.

Kimura, S., Chen, H.P., Saxena, I.M., Brown, R.M., Jr. and Itoh, T. (2001) Localization of c-di-GMP-binding protein with the linear terminal complexes of *Acetobacter xylinum*. J. Bacteriol. 183, 5668–5674.

Kirillina, O., Fetherston, J.D., Bobrov, A.G., Abney, J. and Perry, R.D. (2004) HmsP, a putative phosphodiesterase, and HmsT, a putative diguanylate cyclase, control Hms-dependent biofilm formation in *Yersinia pestis*. Mol. Microbiol. 54, 75-88.

Kutyrev, V.V., Filippov, A.A., Oparina, O.S. and Protsenko, O.A. (1992) Analysis of *Yersinia pestis* chromosomal determinants Pgm^+ and Pst^s associated with virulence. Microb. Pathog. 12, 177-186.

Lillard, J.W., Jr., Bearden, S.W., Fetherston, J.D., and Perry, R.D. (1999) The haemin storage (Hms^+) phenotype of *Yersinia pestis* is not essential for the pathogenesis of bubonic plague in mammals. Microbiology 145, 197-209.

Mack, D., Fischer, W., Krokotsch, A., Leopold, K., Hartmann, R., Egge, H. and Laufs, R. (1996) The intercellular adhesin involved in biofilm accumulation of *Staphylococcus epidermidis* is a linear β-1,6-linked glucosaminoglycan: purification and structural analysis. J. Bacteriol. 178, 175–183.

O'Toole, G.A., Pratt, L.A., Watnick, P.I., Newman, D.K., Weaver, V.B. and Kolter, R. (1999) Genetic approaches to study of biofilms. Meth. Enzymol. 310, 91-109.

Patel, C. N., Wortham, B.W., Lines, J.L., Fetherston, J.D., Perry, R.D. and Oliveira, M.A. (2006) Polyamines are essential for the formation of plague biofilm. J. Bacteriol. 188, 2355-2363.

Perry, R.D., Pendrak, M.L. and Schuetze, P. (1990) Identification and cloning of a hemin storage locus involved in the pigmentation phenotype of *Yersinia pestis*. J. Bacteriol. 172, 5929-5937.

Perry, R.D., Bobrov, A.G., Kirillina, O., Jones, H.A., Pedersen, L.L., Abney, J. and Fetherston, J.D. (2004) Temperature regulation of the hemin storage (Hms^+) phenotype of *Yersinia pestis* is posttranscriptional. J. Bacteriol. 186, 1638-1647.

Römling, U., Gomelsky, M. and Galperin, M.Y. (2005) C-di-GMP: the dawning of a novel bacterial signalling system. Mol. Microbiol. 57, 629-639.

Ross, P., Weinhouse, H., Aloni, Y., Michaeli, D., Weinberger-Ohana, P., Mayer, R., Braun, S., de Vroom, E., van der Marel, G.A., van Boom, J.H. and Benziman, M. (1987) Regulation of cellulose synthesis in *Acetobacter xylinum* by cyclic diguanylic acid. Nature 325, 279–281

Simm, R., Fetherston, J.D., Kader, A., Römling, U. and Perry, R.D. (2005) Phenotypic convergence mediated by GGDEF-domain-containing proteins. J. Bacteriol. 187, 6816-6823.

Sutherland, I.W. (2001) The biofilm matrix - an immobilized but dynamic microbial environment. Trends Microbiol. 9, 222-227.

Tan, L. and Darby, C. (2004) A movable surface: formation of *Yersinia* sp. biofilms on motile *Caenorhabditis elegans*. J. Bacteriol. 186, 5087-5092.

Tan, L. and Darby, C. (2006)*Yersinia pestis* YrbH is a multifunctional protein required for both 3-deoxy-D-manno-oct-2-ulosonic acid biosynthesis and biofilm formation. Mol. Microbiol. 61, 861-70.

Wang, X., Preston, J.F., III and Romeo, T. (2004) The *pgaABCD* locus of *Escherichia coli* promotes the synthesis of a polysaccharide adhesin required for biofilm formation. J. Bacteriol. 186, 2724-2734.

Picture 18. Robert Perry opens the 9th International Symposium on *Yersinia*. Photograph by A. Anisimov.

18
Environmental Stimuli Affecting Expression of the Ysa Type Three Secretion Locus

Shirly Mildiner-Earley[1], Kimberly A. Walker[1], and Virginia L. Miller[2]

[1] Molecular Microbiology ,Washington University
[2] Molecular Microbiology and Pediatrics, Washington University, virginia@borcim.wustl.edu

Abstract. *Yersinia enterocolitica* has two type III secretion systems (TTSS): The well characterized Ysc-Yop system and the relatively uncharacterized Ysa-Ysp system. Detection of Ysps in culture supernatants has only been observed *in vitro* when cultures are grown at low temperature (26°C) and in high salt (290mM NaCl). Previous reports demonstrated that expression from the *ysaE* promoter was activated by high salt. In this study, we report a new environmental stimulus for *ysa* gene expression; in the presence of high salt, growth on solid surface stimulates expression 7-fold compared to growth in high salt broth. These new data indicate that, in the presence of salt, solid surface is an extremely robust signal for the Ysa system.

18.1 Introduction

Gram-negative bacteria can deliver proteins into the cytosol of host cells via contact dependent type III secretion systems (TTSS). *Yersinia enterocolitica* has two such systems. The Ysc TTSS is encoded on the virulence plasmid and has been thoroughly studied (Cornelis 2002). The Ysc effectors, designated Yops, have been shown to be involved in different stages of pathogenesis, including the prevention of phagocytosis and disruption of host mitogen-activated protein kinase pathways (MAP) (Orth et al. 1999). Induction of Yop secretion can be accomplished *in vitro* at 37°C in the absence of Ca^{2+} ions (Michiels et al. 1990). The second TTSS is encoded on the chromosome and is designated the Ysa TTSS (Haller et al. 2000). This TTSS secretes at least 15 Ysps, or Yersinia secreted proteins, most of which are unrelated to other known TTSS effectors. *In vitro*, secretion of Ysps into culture supernatants can be induced at 26°C in the presence of high NaCl concentrations (290mM). In addition, translocation of several of the Ysps in a Ysa dependent manner into tissue culture cells has been demonstrated (Matsumoto and Young 2006).

We previously showed that *ysaE*, a gene homologous to AraC-like regulators in *Salmonella* and *Shigella* (InvF and MxiE, respectively) is regulated by a putative two component system, YsrRS, in the presence of high concentrations of NaCl (Walker and Miller 2004). *ysaE* is the first gene of a putative operon encoding the apparatus genes as well as a chaperone, *sycB*, and genes encoding the translocon, *yspBC*. YsrRS appears to be required for the transcription of the *ysa* genes. A Δ*ysrS* strain shows reduced *ysaE* and *sycB* expression, and Δ*ysrS* and Δ*ysrR* strains are defective for secretion *in vitro*. A recent report suggests that RcsB is also required for *ysa* gene expression (Venecia and Young 2005). The Rcs two component system is composed of RcsC, the sensor kinase, RcsB, the response regulator, and RcsD, the protein containing the histidine phosphotransfer domain necessary for the phosphorelay from RcsC to RcsB (Takeda et al. 2001). In *E. coli* the Rcs system has been demonstrated to regulate capsular polysaccharide synthesis, or the *cps* operon, in response to environmental stress and perturbations of the cellular membrane (Stout and Gottesman 1990). In particular, published data suggested that RcsC was activated when bacteria were exposed to solid surface, which may be related to its requirement for normal biofilm production. The activation of RcsC on solid surface results in the increased expression of *cps* genes (Ferrieres and Clarke 2003). Because the Rcs two component system is involved in *ysa* gene expression, we hypothesized that growth on solid surface might also affect the expression of *ysa* genes.

18.2 Materials and Methods

Strain Construction. *Y. enterocolitica* strain YVM925 (JB580v, *ysaE-lacZ*) was used in this work. YVM925 is a derivative of *Y. enterocolitica* JB580v carrying a *ysaE-lacZ* fusion on the chromosome such that the *ysa* locus stays intact (Walker and Miller 2004).

Growth Media. Two liquid culture conditions were used for these experiments. L-broth containing no NaCl (1% tryptone, 0.5% yeast extract; referred to hereafter as LB-0) was used to grow bacteria overnight at 26°C. Saturated cultures were subcultured into L-broth or into L-broth containing 290 mM NaCl (LB-290). For solid phase experiments the same media were used with the addition of 15% Bacto-Agar. Agar plates with no NaCl (15% Bacto Agar, 1% tryptone, 0.5% yeast extract; referred to hereafter as S-LB-0) and agar plates with 290mM NaCl (S-LB-290) were used.

β-galactosidase Assays. YVM925 was grown in LB-0 overnight at 26°C with nalidixic acid (20 μg/ml) and chloramphenicol (12.5 μg/ml). Cultures were diluted in LB-0 to an optical density at 600 nm (OD_{600}) of 0.2 and grown for 2 hours at 26°C on a roller drum. At this time, 2.0 ml aliquots were removed. Cells from two aliquots were collected by centrifugation and resuspended in either LB-0 or LB-290 and placed back on the roller drum for 60 minutes at 26°C. Cells from two additional aliquots were collected on a 0.45μm filter and placed on an agar plate with or without 290mM NaCl, and incubated for 60 minutes at 26°C (Ferrieres and Clarke 2003). The filters were then placed in 2.0 ml of LB-0 or LB-290 (salt concentration corresponding to concentration on plate) and vortexed to release the bacteria. β-galactosidase assays were performed as described (Miller 1992).

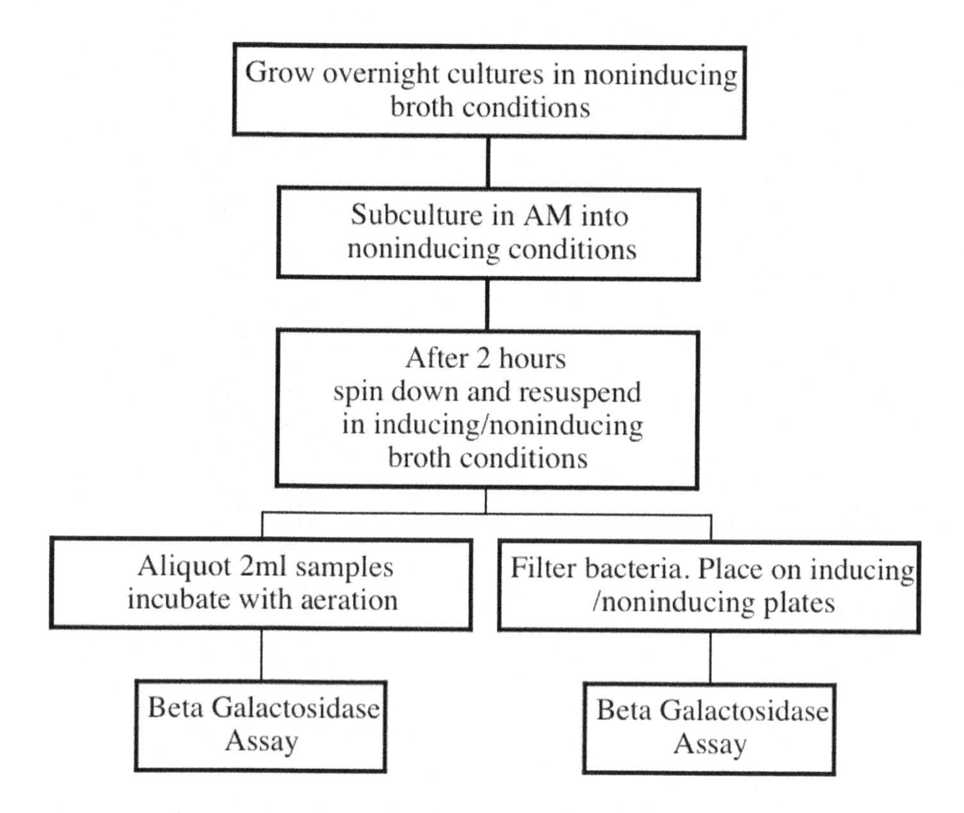

Fig. 1. Solid surface vs. liquid broth β-galactosidase protocol.

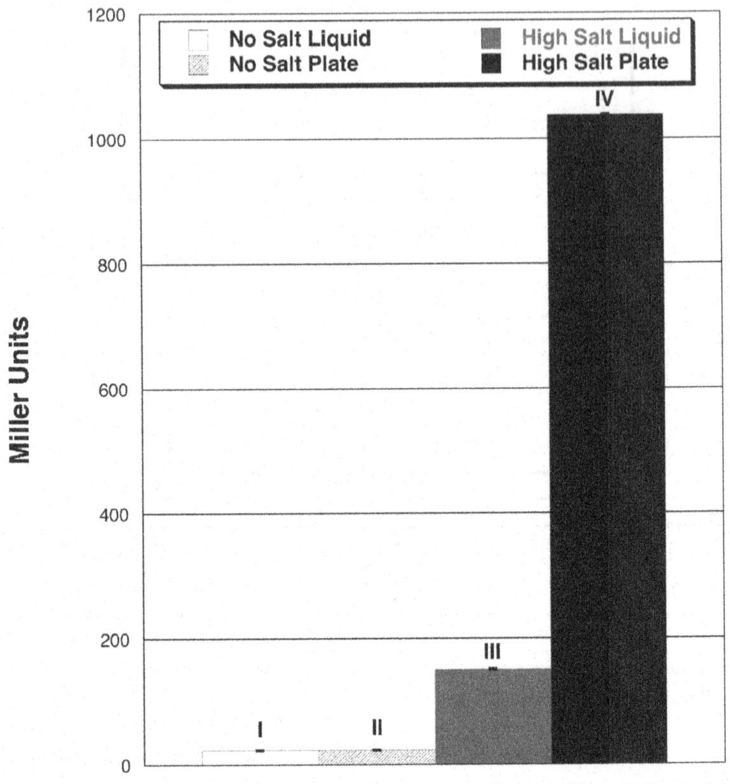

Fig. 2. *ysaE-lacZ* activation in response to varying conditions. (I) Liquid broth with 0 mM NaCl. (II) Agar plate with 0 mM NaCl. (III) Liquid broth with 290 mM NaCl. (IV) Agar plate with 290mM NaCl.

18.3 Results

To determine if the *ysa* genes were regulated by growth on a solid surface, we utilized a wild type strain of *Y. enterocolitica* harboring a chromosomal *ysaE-lacZ* transcriptional reporter fusion, YVM925 (Walker and Miller 2004). Saturated cultures of YVM925 grown in LB-0 (non-inducing conditions) were subcultured into fresh LB-0 and allowed to grow with aeration for 2 hours. At this point, samples were taken and either allowed to continue in broth culture, or transferred to nitrocellulose membranes for solid surface growth in the presence or absence of 290 mM NaCl (Fig. 1). After 60 minutes, samples were taken for *β*-galactosidase (β-gal) assays (Fig. 2). In a broth culture lacking NaCl (LB-0), *ysaE-lacZ* activity yielded 23 Miller units (MU). In a broth culture with 290 mM NaCl (LB-290), activity was measured at 151 MU, a 6.6-fold increase over that in LB-0; this data is in agreement with previous work (Walker and Miller 2004). When grown on a solid surface without salt (S-LB-0), 23 MU of activity were measured, indicating that there is no difference between broth and solid surface growth in the absence of salt. However, in the presence of salt, growth on a solid surface increased promoter activity to 1038 MU, resulting in a 6.9-fold increase over LB-290, and a nearly 45-fold increase over both LB-0 and S-LB-0. These data indicate that solid surface growth is an important signal for inducing *ysa* gene expression, and that while salt is required, solid surface growth is a potent signal for stimulation of the *ysaE* promoter.

18.4 Conclusion

There is very little known about the *ysa* TTSS. While we have tools to help us study the *ysa* genes *in vitro*, we still do not understand how this system contributes to the bacterial survival in the host. Although *in vitro* secretion of Ysps can be induced in the presence of high salt, we have identified an additional stimulus that is able to increase transcription of *ysa* genes. Exposure to solid surface in the presence of salt results in a robust activation of *ysa* genes and we hypothesize that bacterial contact with the agar plate simulates conditions bacteria might encounter in the gut of infected hosts. Thus, contact with the intestinal mucosa or biofilm formation in this environment may serve as a trigger to activate expression of the *ysa* TTSS. Using conditions *in vitro* that are more biologically relevant is an important tool for future experiments designed to look at Ysa TTSS function.

18.5 Acknowledgments

This work was supported by the National Institutes of Health research grant AI063299 (V.L.M).

18.6 References

Cornelis, G.R. (2002) *Yersinia* type III secretion: send in the effectors. J. Cell. Biol. 158, 401-408.

Ferrieres, L. and Clarke, D.J. (2003) The RcsC sensor kinase is required for normal biofilm formation in *Escherichia coli* K-12 and controls the expression of a regulon in response to growth on a solid surface. Mol. Microbiol. 50, 1665-1682.

Haller, J.C., Carlson, S., Pederson, K.J. and Pierson, D.E. (2000) A chromosomally encoded type III secretion pathway in *Yersinia enterocolitica* is important in virulence. Mol. Microbiol. 36, 1436-1446.

Matsumoto, H. and Young, G.M. (2006) Proteomic and functional analysis of the suite of Ysp proteins exported by the Ysa type III secretion system of *Yersinia enterocolitica* Biovar 1B. Mol. Microbiol. 59, 689-706.

Michiels, T., Wattiau, P., Brasseur, R., Ruysschaert, J.M. and Cornelis, G. (1990) Secretion of Yop proteins by *Yersiniae*. Infect. Immun. 58, 2840-2849.

Miller, J., (1992) *A Short Course in Bacterial Genetics*. Cold Spring Harbor Laboratory Press, Cold Spring Harbor, N.Y.

Orth, K., Palmer, L.E., Bao, Z.Q., Stewart, S., Rudolph, A.E., Bliska, J.B. and Dixon, J.E. (1999) Inhibition of the mitogen-activated protein kinase kinase superfamily by a *Yersinia* effector. Science 285, 1920-1923.

Stout, V. and Gottesman, S. (1990) RcsB and RcsC: a two-component regulator of capsule synthesis in *Escherichia coli*. J. Bacteriol. 172, 659-669.

Takeda, S., Fujisawa, Y., Matsubara, M., Aiba, H. and Mizuno, T. (2001) A novel feature of the multistep phosphorelay in *Escherichia coli*: a revised model of the RcsC --> YojN --> RcsB signalling pathway implicated in capsular synthesis and swarming behaviour. Mol. Microbiol. 40, 440-450.

Venecia, K. and Young, G.M. (2005) Environmental regulation and virulence attributes of the Ysa type III secretion system of *Yersinia enterocolitica* biovar 1B. Infect. Immun. 73, 5961-5977.

Walker, K.A. and Miller, V.L. (2004) Regulation of the Ysa type III secretion system of *Yersinia enterocolitica* by YsaE/SycB and YsrS/YsrR. J. Bacteriol. 186, 4056-4066.

Picture 19. William Goldman gives his oral presentation.

19
Polynucleotide Phosphorylase and the T3SS

Jason A. Rosenzweig[1] and Kurt Schesser

Department of Microbiology & Immunology, Miller School of Medicine, University of Miami, kschesser@med.miami.edu
[1] Current address: Division of Math Science and Technology, Nova Southeastern University

Abstract. Low temperatures as well as encounters with host phagocytes are two stresses that have been relatively well studied in many species of bacteria. The exoribonuclease polynucleotide phosphorylase (PNPase) has previously been shown to be required by several species of bacteria, including *Yersinia*, for low-temperature growth. We have shown that PNPase also enhances the ability of *Yersinia* to withstand the killing activities of murine macrophages. We have gone on to show that PNPase is required for the optimal functioning of *Yersinia*'s type three secretion system (T3SS), an organelle that injects effector proteins directly into host cells. Surprisingly, the PNPase-mediated effect on T3SS activity is independent of PNPase's ribonuclease activity and instead requires only its S1 RNA-binding domain. In stark contrast, the catalytic activity of PNPase is strictly required for enhanced growth at low temperature. Preliminary experiments suggest that the RNA-binding interface of the S1 domain is critical for its T3SS-enhancing activity. Our findings indicate that PNPase plays versatile roles in promoting *Yersinia*'s survival in response to stressful conditions.

19.1 How *Yersinia* Faces Down the Macrophage

It is well-recognized that the type 3 secretion system (T3SS) plays a central role in the pathogenesis of the yersiniae. The discovery of this protein delivery system traces back to the 1950's where it was shown that there is a unique relationship between the growth characteristics of *Yersinia pestis* (nee *Pasteurella pestis*) and virulence. Several workers, most notably Surgalla, observed an attenuation of virulence upon prolonged cultivation of *Y. pestis* at 37°C (Fukui et al. 1957). Higuchi and colleagues found that the addition of skim milk preparations to the basic growth medium forestalled this temperature-induced loss of virulence; the active compound was quickly identified as Ca^{2+} (Higuchi and Carlin 1958; Higuchi et al. 1959). It became readily apparent that millimolar levels of Ca^{2+} were required for virulent *Y. pestis* to grow at 37°C and if levels of Ca^{2+} fell below a certain threshold avirulent 'mutants' would soon predominant the cultures. What was odd though was the seemingly high frequency of this virulence to avirulence transition, calculated to be approximately 10^{-4} per bacterial generation, a value much too high to be explained by simple genetics (Higuchi and Smith 1961). This phenomenon remained enigmatic until the early 1980's when several groups reported that 'growth restriction' (i.e. the inability of virulent *Y. pestis*, as well as the enteropathogenic *Yersinia pseudotuberculosis* and *Yersinia enterocolitica*, to grow at 37°C/low Ca^{2+}) was associated with the presence of a ~70-kb extrachromosomal plasmid that in turn was shown to direct the massive secretion of *Yersinia* outer proteins, or Yops (Ben-Gurion and Schafferman 1981; Ferber and Brubaker 1981; Gemski et al. 1980; Portnoy et al. 1981 and 1984). Subsequently it was found that this appropriately-named 'virulence plasmid' encoded the components of a secretion system that directly delivers at least 6 Yops, which are also encoded on the virulence plasmid, directly into the host cell cytoplasm (Rosqvist et al. 1994). About this same time it was found that this protein delivery system, which was designated as 'type three', was in fact widely distributed among several plant- and animal-interacting bacteria, each of which delivers a unique set of proteins.

In the yersiniae, many of the components that regulate the temperature- and Ca^{2+}-sensitive functioning of the T3SS have been identified, and the process can, for practical purposes, be divided up into two stages. 'Step 1' regulation involves the temperature-dependent expression of the transcriptional activator LcrF which directs the expression of the T3SS regulon up to a certain level in the presence of millimolar Ca^{2+} (Lambert de Rouvroit et al. 1993; Hoe and Goguen 1993). Under these conditions further expression of the T3SS regulon is repressed by cytosolic LcrQ/SycH and YopD/SycD complexes (Williams and Straley 1998; Francis et al. 2001; Anderson et al. 2002). The secretion conduit opens up as extracellular Ca^{2+} levels fall to the low millimolar range leading to the export of the negative regulators LcrQ and YopD that in turn results in the full derepression of T3SS expression ('Step 2') (Pettersson et al. 1996).

A Step 2-like derepression of T3SS expression also occurs when the yersiniae interact with cultured vertebrate cells (Pettersson et al. 1996; Bartra et al. 2001). Although there is strong circumstantial evidence, it is not known whether the Ca^{2+}-sensitive sensor that is responsible for full derepression *in vitro* is the same entity that accounts for derepression *in vivo*. '*In vitro*' in this article will refer to bacteria cultivated in pure culture and '*in vivo*' will refer to bacteria co-cultivated with vertebrate cells as in infection assays. In fact there are a number of unresolved issues concerning the functioning of the T3SS *in vivo*. For example, it is almost completely unknown how T3SS activity is linked with the phenomenon of growth restriction despite the fact that these two processes are intimately coupled, at least *in vitro*. Perhaps even a more basic issue is whether growth restriction even occurs *in vivo*; there are sound arguments that the *in vitro* assay does not accurately model what occurs during an infection.

Several additional factors that are not part of the classically-defined T3SS regulon have been identified as being necessary for T3SS regulation in the yersiniae. One of the first such factors identified was the small histone-like protein YmoA, which under non-inducing conditions for T3SS gene expression is present at relatively high levels in the bacterial cell. Upon inductive conditions, YmoA is rapidly degraded and consequently its repressive affect on T3SS gene expression is lifted (Jackson et al. 2004). Recently, we identified another such factor, the exoribonuclease polynucleotide phosphorylase (PNPase) as being required for normal T3SS functioning in the pathogenic yersiniae (Rosenzweig et al. 2005).

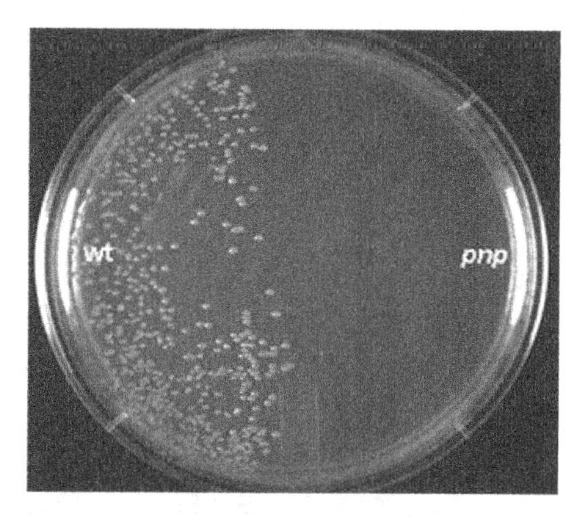

Fig. 1. Cold sensitivity of the *Y. pseudotuberculosis Δpnp* strain. The wild-type (left) and *Δpnp* (right) strains of YPIII/pIB102 were plated on Luria-Bertani medium and incubated at 5°C for 10 days (reprinted with permission from Rosenzweig et al. 2005).

19.2 The *Yersinia* Δ*pnp* Strains

The genes encoding PNPase were deleted in *Y. pestis* and *Y. pseudotuberculosis* using a recombination-based method. As expected, the resulting *Δpnp* strains were clearly deficient in the ability to grow at reduced temperature (Fig. 1). Wild type-like low-temperature growth could be restored to the *Y. pseudotuberculosis Δpnp* strain by expressing plasmid-encoded PNPase from *Escherichia coli* (Rosenzweig et al. 2005). PNPase-mediated enhancement of low-temperature growth was entirely dependent on PNPase being enzymatically active since a catalytically inactive PNPase (containing a R100D replacement) could not complement the low-temperature growth defect of the *Y. pseudotuberculosis Δpnp* strain. At least in this respect, the *Yersinia Δpnp* strains behaved as expected.

These strains were then assayed in a cell culture infection assay in which bacteria are added to cultured macrophage-like cells and the number of viable cell-associated bacteria are determined at various times after the removal of unattached bacteria. There was a notable reduction in the number of cell-associated *Δpnp* bacteria compared to wild-type cells following a 6-hour infection period (Fig. 2). Since this proliferation-based assay is sensitive to perturbations in the T3SS (e.g. *yopB* mutant in Fig. 2), the *Δpnp* strains were analyzed in a cytotoxicity assay which more directly measures the functioning of the T3SS. 'Cytotoxicity' in this context refers to the disruption of the host cell cytoskeleton that follows the T3SS-mediated delivery of the Yop virulence proteins. Cultured HeLa cells infected with wild-type *Yersinia* strains proceed from a flattened to a 'rounded-up' morphology after a 2-3-hr infection period. In contrast, the morphology of HeLa cells infected with T3SS-deficient *Yersinia* strains changes very little and resembles uninfected cells. Cells infected with either the *Y. pseudotuberculosis* or *Y. pestis Δpnp* strains displayed an intermediate level of cytotoxicity suggesting that the T3SS was functioning in *Δpnp* cells,

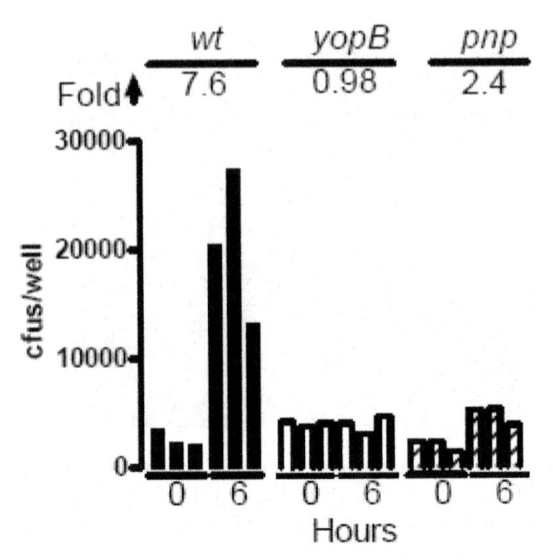

Fig. 2. Viability assay of *Y. pseudotuberculosis* strains. YPIII/pIB102 and its *ΔyopB* and *Δpnp* derivatives were added to tissue culture wells containing RAW 267 mouse macrophage-like cells at a MOI of ~0.5. Excess bacteria were removed after a 30-min attachment period, and the number of viable cell-associated bacteria was determined by plating either immediately (0 h) or 6 h later. Three independent wells were assayed per condition. The average-fold increase during the 6-h infection period is shown above.

just not as well as in wild-type cells (Rosenzweig et al. 2005). A sub-optimally functioning T3SS was clearly evident when the *Yersinia Δpnp* strains were examined in a secretion assay. In this latter assay, the *Yersinia Δpnp* strains did secrete detectable levels of Yop effector proteins, but at much reduced levels compared to the wild-type strains (Rosenzweig et al. 2005). Thus, these series of experiments show that PNPase is important for low-temperature growth as well as T3SS activity.

19.3 Critical PNPase Determinants for T3SS Activity

By testing a number of *Yersinia Δpnp* strains each possessing a different plasmid-encoded PNPase variant, it was determined that the catalytic activity of PNPase was not necessary for regulating T3SS activity (Fig. 3). This was in contrast to the cold-growth assay discussed above in which a catalytically-inactive PNPase was completely unable to restore wild-type-like growth to the *Yersinia Δpnp* strains. As shown in Fig. 3, the only PNPase variant that appeared unable to restore wild-type-like cytotoxicity to the *Δpnp* strain was one that lacked its S1 RNA-binding domain. S1 domains are found in a number of RNA-interacting proteins including a number of exo- and endoribonucleases. To test whether an S1 domain is sufficient to restore normal T3SS activity to the *Y. pestis Δ pnp* strains, several different S1 domain-

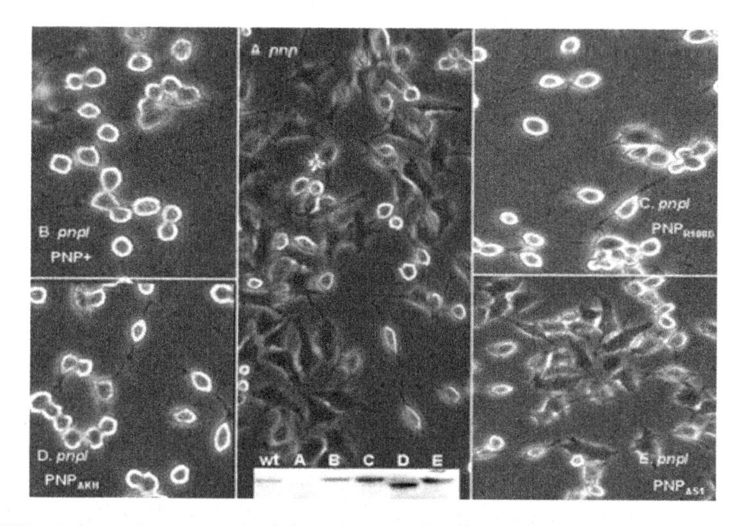

Fig. 3. PNPase determinants required for T3SS-mediated cytotoxicity. The *Y. pestis Δpnp* strain was transformed with either an empty vector (A), or vectors expressing wild-type PNPase (B), a catalytically inactive PNPase (C), or PNPase deleted in either the KH or S1 RNA binding domains (D and E, respectively). The insert shows PNPase protein levels of the various strains tested in the cytotoxicity assay. (reprinted with permission from Rosenzweig et al. 2005).

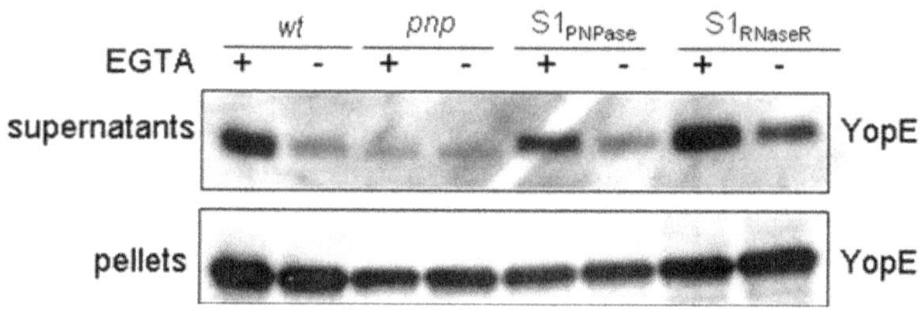

Fig. 4. S1-mediated effects on secretion of the T3SS effector YopE. The wild-type, *Δpnp*, and *Δpnp Y. pseudotuberculosis* strains transformed with the indicated S1 expression vector were propagated at 37°C in the absence of Ca^{2+}. Full T3SS activation was induced by adding the Ca^{2+} chelator EGTA which was added to half the sample (+). Twenty seconds after the addition of EGTA, YopE protein levels from both the whole cell and supernatant fractions were determined by immunoblotting (reprinted with permission from Rosenzweig et al. 2005).

encoding sequences were cloned into an expression vector and the resulting transformants were assayed in secretion and cytotoxicity assays. Quite unexpectedly, every S1 domain tested fully restored T3SS activity in the *Yersinia Δpnp* strains (Fig. 4; Rosenzweig et al. 2005). These studies have been extended by using the S1 domain from the endoribonuclease RNase E since the structure of this domain has been solved in complex with RNA (Schubert et al. 2004; Callaghan et al. 2005). Structural and biochemical analysis of RNase E's S1 domain has revealed it possesses distinct oligonucleotide and dimerization interfaces. In preliminary experiments in the authors laboratory, it appears that an intact oligonucleotide interface is essential for S1-mediated enhancement of T3SS activity in *Yersinia.*

19.4 Conclusions

As shown above, PNPase, through its S1 domain, plays a clear role in controlling T3SS activity in the *Yersinia.* One possible explanation for reduced T3SS activity in *Yersinia Δpnp* strains is simply that the components of the T3SS (e.g. secretion complexes, effectors, etc.) are expressed at reduced levels. By measuring T3SS-encoding transcripts and proteins, it does not appear that the steady-state levels of T3SS components are at lower levels in the *Yersinia Δpnp* strains compared to the wild-type strains (unpublished observations). In fact, there were no detectable differences in *yopE* promoter activity between the wild-type and *Δpnp Y. pseudotuberculosis* strains (Fig. 5). PNPase has also been linked to T3SS activity in *Salmonella*, but in this case it appears that the Salmonella *enterica Δpnp* strain possesses a hyperactive T3SS as compared to the wild-type strain (Clements et al. 2002). The accumulative data indicates that, depending on the species and/or stress (e.g. macrophages, low temperature, etc.), PNPase can play a variety of roles in enhancing bacterial survival.

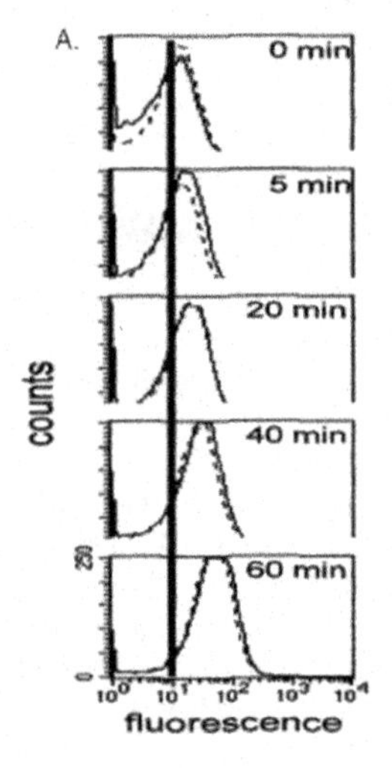

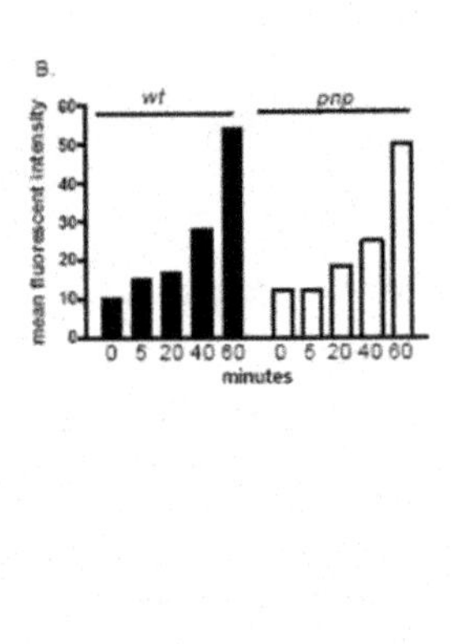

Fig. 5. *yopE* promoter activity in wild-type and *Δpnp Y. pseudotuberculosis*. Wild-type (solid trace and filled bars) and a *Δpnp* derivative (dotted trace and unfilled bars) of a YPIII/IB100 strain containing a *gfp* gene within the *yopE* transcription unit was grown and induced as described in Fig. 4. GFP levels were determined by flow cytometry at the indicated times following the addition of EGTA (reprinted with permission from Rosenzweig et al. 2005).

19.5 Acknowledgments

The authors wish to thank David Wiley, Greg Plano, and Murray Deutscher for helpful advice. Support for these studies was provided by the NIAID/NIH and the Miller School of Medicine at the University of Miami.

19.6 References

Anderson, D.M., Ramamurthi, K.S., Tam, C. and Schneewind, O. (2002) YopD and LcrH regulate expression of *Yersinia enterocolitica* YopQ by a posttranscriptional mechanism and bind to *yopQ* RNA. J. Bacteriol. 184, 1287-1295.

Bartra, S., Cherepanov, P., Forsberg, A. and Schesser, K. (2001) The *Yersinia* YopE and YopH type III effector proteins enhance bacterial proliferation following contact with eukaryotic cells. BMC Microbiol. 1, 22-33.

Ben-Gurion, R. and Shafferman, A. (1981) Essential virulence determinants of different *Yersinia* species are carried on a common plasmid. Plasmid 5, 183-187.

Callaghan, A.J., Marcaida, M.J., Stead, J.A., McDowall, K.J., Scott, W.G. and Luisi, B.F. (2005) Structure of *Escherichia coli* RNase E catalytic domain and implications for RNA turnover. Nature 437, 1187-1191.

Clements, M., Eriksson, S., Thompson, A., Lucchini, S., Hinton, J., Normark, S. and Rhen, M. (2002) Polynucleotide phosphorylase is a global regulator of virulence and persistency in *Salmonella enterica*. PNAS 99, 8784-8789.

Ferber, D.M. and Brubaker, R.R. (1981) Plasmids in *Yersinia pestis*. Infect. Immun. 2, 839-841.

Francis, M.S., Lloyd, S.A. and Wolf-Watz, H. (2001) The type III secretion chaperone LcrH co-operates with YopD to establish a negative, regulatory loop for control of Yop synthesis in *Yersinia pseudotuberculosis*. Mol. Microbiol. 42, 1075-1093.

Fukui, G.M., Ogg, J.E., Wessman, G.E. and Surgalla, M.J. (1957) Studies on the relation of cultural conditions and virulence of Pasteurella pestis. J. Bacteriol. 74, 714-717.

Gemski, P., Lazere, J.R., Casey, T. and Wohlhieter, J.A. (1980) Presence of a virulence-associated plasmid in *Yersinia pseudotuberculosis*. Infect. Immun. 28, 1044-1047.

Higuchi, K. and Carlin, C.E. (1958) Studies on the nutrition and physiology of *Pasteurella pestis*. II. A defined medium for the growth of *Pasteurella pestis*. J. Bacteriol. 75, 409-413.

Higuchi, K., Kupferberg, L.L. and Smith, J.L. (1959) Studies on the nutrition and physiology of *Pasteurella pestis*. III. Effects of calcium ions on the growth of virulent and avirulent strains of *Pasteurella pestis*. J. Bacteriol. 77, 317-321.

Higuchi, K. and Smith, J.L. (1961) Studies on the nutrition and physiology of *Pasteurella pestis*. VI. A differential plating medium for the estimation of the mutation rate to avirulence. J. Bacteriol. 81, 605-608.

Hoe, N.P. and Goguen, J.D. (1993) Temperature sensing in *Yersinia pestis*: translation of the LcrF activator protein is thermally regulated. J. Bacteriol. 175, 7901-7909.

Jackson, M.W., Silva-Herzog, E. and Plano, G.V. (2004) The ATP-dependent ClpXP and Lon proteases regulate expression of the *Yersinia pestis* type III secretion system via regulated proteolysis of YmoA, a small histone-like protein. Mol. Microbiol. 54, 1364-1378.

Lambert de Rouvroit, C., Sluiters, C. and Cornelis, G.R. (1992) Role of the transcriptional activator, VirF, and temperature in the expression of the pYV plasmid genes of *Yersinia enterocolitica*. Mol. Microbiol 6, 395-409.

Pettersson, J., Nordfelth, R., Dubinina, E., Bergman, T., Gustafsson, M., Magnusson, K.E. and Wolf-Watz, H. (1996) Modulation of virulence factor expression by pathogen target cell contact. Science 273, 1231-1233.

Portnoy, D.A., Moseley, S.L. and Falkow, S. (1981) Characterization of plasmids and plasmid-associated determinants of *Yersinia enterocolitica* pathogenesis. Infect. Immun. 31, 775-782.

Portnoy, D.A., Wolf-Watz, H., Bolin, I., Beeder, A.B. and Falkow, S. (1984) Characterization of common virulence plasmids in *Yersinia* species and their role in the expression of outer membrane proteins. Infect. Immun. 43, 108-114.

Rosenzweig, J.A., Weltman, G., Plano, G.V. and Schesser, K. (2005) Modulation of *Yersinia* type three secretion system by the S1 domain of polynucleotide phosphorylase. J. Biol. Chem. 280, 156-163.

Rosqvist, R., Magnusson, K.E. and Wolf-Watz, H. (1994) Target cell contact triggers expression and polarized transfer of *Yersinia* YopE cytotoxin into mammalian cells. EMBO J. 13, 964-972.

Schubert, M., Edge, R.E., Lario, P., Cook, M.A., Strynadka, N.C., Mackie, G.A. and McIntosh, L.P. (2004) Structural characterization of the RNase E S1 domain and identification of its oligonucleotide-binding and dimerization interfaces. J. Mol. Biol. 341, 37-54.

Williams, A.W. and Straley, S.C. (1998) YopD of *Yersinia pestis* plays a role in negative regulation of the low-calcium response in addition to its role in translocation of Yops. J. Bacteriol. 180, 350-358.

20
Roles of YopN, LcrG and LcrV in Controlling Yops Secretion by *Yersinia pestis*

Mohamad A. Hamad[1] and Matthew L. Nilles

Department of Microbiology and Immunology, School of Medicine and Health Sciences, University of North Dakota, mnilles@medicine.nodak.edu

[1] Current Address: Department of Microbiology, University of Colorado Health Science Center

Abstract. Control of Yops secretion in pathogenic Yersinia is achieved at several levels. These levels likely include transcriptional, post-transcriptional, translational and secretional controls. Secretion control appears to be mediated by two pathways. One pathway involves YopN and proteins that interact with YopN. The second pathway consists of LcrG and its interaction with LcrV. LcrV is a postive regulator of Yops secretion that exerts control over Yops secretion by negating the secretion blocking role of LcrG. However, the intersection of these two control pathways is not understood. Recent work has allowed the development of a speculative model that brings YopN-mediated and LcrG-LcrV-mediated control together in the context of the ability of the needle complex to respond to Ca^{2+}.

20.1 Type III Secretion in *Y. pestis*

Yersinia pestis, the etiologic agent of plague, harbors a 70-kb plasmid termed pCD1 that encodes a type III secretion system (T3SS) (Perry and Fetherson 1997). The T3SS is a group of structural proteins that comprise the T3S apparatus (YSC) and a set of secreted effector proteins called Yersinia outer proteins (Yops) (Cornelis 1999). *Y. pestis* utilizes this T3SS to inject Yops directly into the cytoplasm of host cells were they inhibit bacterial phagocytsis and block the activation of an early immune response (Cornelis 1999). *Y. pestis* grows at 26-28°C with normal growth kinetics and a generation time of approximately 90 minutes (Brubaker 1972). When grown at 37°C, i.e. mammalian body temperature, the growth of *Y. pestis* becomes dependent on the presence of Ca^{2+} ions (Straley et al. 1993). This Ca^{2+}-dependent growth phenotype is linked to the expression and secretion of Yops through the T3SS, which is expressed at 37°C and not at 26°C (Straley et al. 1993). At 37°C, the expression of the T3S genes is activated and *Y. pestis* assembles a functional T3S apparatus. When Yops secretion is blocked, the expression of the T3S genes is kept to a minimum and the bacteria grow normally without entering growth restriction (Straley and Bowmer 1986; Michiels et al. 1990).

20.1.1 Regulation of T3S in *Y. pestis*

The secretion of Yops by *Y. pestis* is not constitutive, instead Yops secretion is tightly regulated and is activated in response to environmental cues (Straley and Bowmer 1986; Straley 1988; Straley et al. 1993). In vivo, Yops secretion is thought to be blocked prior to eukaryotic cell contact and subsequently activated by eukaryotic cell contact (Rosqvist et al. 1994). In vitro, Yops secretion is regulated by extracellular Ca^{2+} ions and nucleotides, e.g. ATP and GTP (Michiels et al. 1990; Straley et al. 1993). Maximal Yops expression and secretion are attained in vitro by growing *Y. pestis* at 37°C in media lacking Ca^{2+} (Straley and Perry 1995). Addition of millimolar concentration of Ca^{2+} or nucleotides into the growth media blocks the secretion of Yops, represses the expression of the T3S genes, and rescues the bacteria from growth restriction (Straley et al. 1993).

When *Y. pestis* is grown at 37°C, the expression of T3S genes is activated by a transcriptional activator, LcrF. LcrQ, LcrH, and YopD control the upregulation of T3S gene expression that occurs when T3S-expression is induced. Deletion of *lcrH*, *yopD* or *lcrQ* results in maximal expression of T3S genes regardless of the Ca^{2+} concentration in the growth media, however *lcrH*, *yopD* and *lcrQ* strains retain the Ca^{2+} regulation of Yops secretion (Williams and Straley 1998; Wulff-Strobel et al. 2002; Pallen et al. 2003). The mechanism used by LcrH, LcrQ, and YopD to regulate T3S expression is unknown. Post-transcriptional regulation may occur as LcrQ, YopD, and LcrH may bind to the 5' UTR of Yops mRNAs (Anderson et al. 2002; Cambronne and Schneewind 2002).

20.1.2 Regulation of Yops Secretion by YopN

Under secretion nonpermissive conditions the secretion of Yops by *Y. pestis* is blocked by the negative regulators YopN, TyeA, SycN, YscB, and LcrG; the activation signal is unknown. Recent evidence suggests the needle complex, formed by YscF, may be involved in sensing the signal for Yops secretion (Torruellas et al. 2005). Negative regulation imposed by YopN is dependant on YopN's interaction with YscB, SycN, and TyeA (Ferracci et al. 2004; Ferracci et al. 2005), while the secretion blockage by LcrG is counteracted by LcrG's interaction with LcrV (Nilles et al. 1997; Matson and Nilles 2001).

YopN is a secreted protein that is also translocated into eukaryotic cells (Forsberg et al. 1991; Day et al. 2003). A YopN null mutant constitutively secretes Yops and hypersecretes LcrV (Forsberg et al. 1991; Skrzypek and Straley 1995). YopN interacts with the cytosolic proteins SycN, YscB, and TyeA (Day and Plano 1998; Iriarte et al. 1998; Jackson et al. 1998). The inactivation of TyeA, YscB, or SycN abolishes YopN function and results in the constitutive secretion of Yops. TyeA, a cytosolic protein, binds the C-terminus of YopN and the YopN-TyeA interaction is required to block Yops secretion (Ferracci et al. 2004). The YopN/TyeA complex is thought to block the secretion of Yops from the cytosolic face of the secretion apparatus (Ferracci et al. 2004; Ferracci et al. 2005). YscB and SycN are cytosolic chaperones of YopN (Jackson et al. 1998). YscB and SycN form a complex that binds to the N-terminus of YopN and facilitates the efficient secretion of YopN (Jackson et al. 1998; Ferracci et al. 2005). The mechanism of secretion regulation by YopN and YopN's interaction partners is not completely clear. Initially studies had suggested that YopN is surface-localized and that YopN could directly sense Ca^{2+} (Forsberg et al. 1991; Iriarte et al. 1998). However recent evidence contradicts this hypothesis. New evidence suggests that the T3S needle (Kenjale et al. 2005; Torruellas et al. 2005) probably senses the extracellular signal for secretion activation. Further strengthening a cytosolic location for YopN function, a mutant YopN that constitutively blocks Yops secretion from a cytoplasmic location in the bacteria has been described (Ferracci et al. 2005). A model for how the YopN/SycN/YscB/TyeA complex regulates Yops secretion is presented in Fig. 1 (Ferracci et al. 2005; Torruellas et al. 2005). According to this model, when conditions do not favor secretion, the YscB/SycN chaperone binds and targets the YopN/TyeA complex to the cytoplasmic face of the T3S apparatus where YopN/TyeA blocks the secretion of Yops. The mechanism of secretion blockage is not clear but partial secretion of YopN in the YopN/TyeA complex in the presence of calcium may block the T3SS. When conditions favor secretion, YopN is secreted through the T3S apparatus, which relieves the blocking activity of the YopN/TyeA complex and allows secretion of the effector Yops to occur.

20.1.3 Secretion Regulation by LcrG and LcrV

LcrV is a multifunctional protein whose activity is central for the virulence of *Y. pestis* (Perry and Fetherson 1997; Mota 2006). During a *Y. pestis* infection, LcrV mediates translocation, secretion regulation, and immunomodulation. LcrV localizes

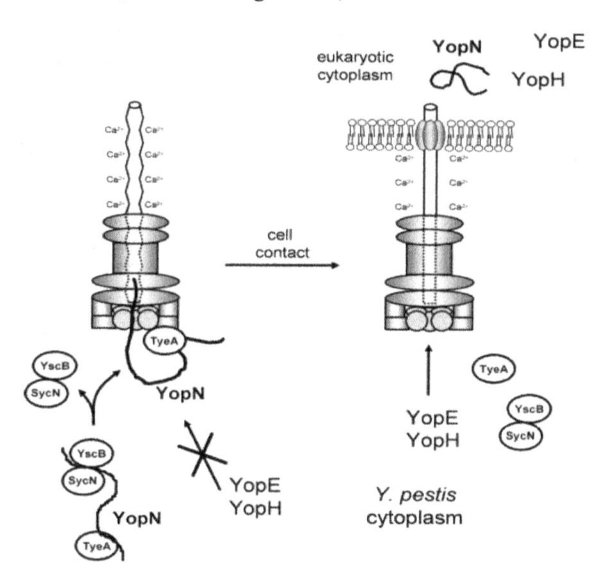

Fig. 1. A hypothetical model for the secretion regulation of Yops by YopN. In the presence of calcium, the YopN/TyeA complex is targeted to the YSC by the SycN/YscB chaperone. The YopN-TyeA complex initiates the secretion of YopN; however, secretion of YopN cannot be completed in the presence of calcium. The partially secreted YopN-TyeA complex is thought to block the T3S apparatus thus preventing the secretion of Yops. Upon secretion activation, the needle may encounter a low-calcium environment (eukaryotic cytosol or Ca^{2+} removal). YscF may sense this low Ca^{2+} environment. The YscF needle might propagate a secretion activation signal allowing YopN to be secreted. The secretion of YopN then opens the secretion apparatus. From Torruellas et al. (2005). Reprinted with permission of Blackwell Publishing Ltd.

at the bacterial surface prior to eukaryotic cell contact and forms a distinct structure at the tip of the YscF needle (Fields et al. 1999; Pettersson et al. 1999; Mueller et al. 2005). LcrV is also required along with Yops B and D for the polarized translocation of the effector Yops into the cytoplasm of targeted host cells (Nilles, Fields and Straley 1998; Fields et al. 1999). LcrV also causes immunosuppression by triggering IL-10 release by macrophages (Brubaker 2003; Overheim et al. 2005; Sing et al. 2005). Binding to the negative regulator LcrG attains regulation of Yops secretion by LcrV (Matson and Nilles 2001; Lawton et al. 2002). In addition, LcrV has been known since the mid 50's as a protective antigen and is currently being investigated in human trials as a component of a vaccine against the plague (Titball and Williamson 2004; Williamson et al. 2005).

LcrG is a small cytoplasmic protein that is required to block the secretion of Yops prior to eukaryotic cell contact or in the presence of Ca^{2+} (Skrzypek and Straley 1993; Nilles et al. 1997). An LcrG null mutant constitutively secretes Yops and hyposecretes LcrV (Skrzypek and Straley 1993; Nilles et al. 1997; Fields et al. 1999). LcrG physically interacts with LcrV in the cytosol of *Y. pestis* (Nilles et al. 1997). Disruption of the LcrG-LcrV interaction or overexpression of LcrG in a low-LcrV background both result in constitutive blockage of Yops secretion (Nilles et al. 1998; Matson and Nilles 2001). This evidence led to the LcrG-titration model (Fig. 2) that partially explains how LcrG-LcrV interaction may control Yops secretion (Matson and Nilles 2001). According to the LcrG-titration model, when culture

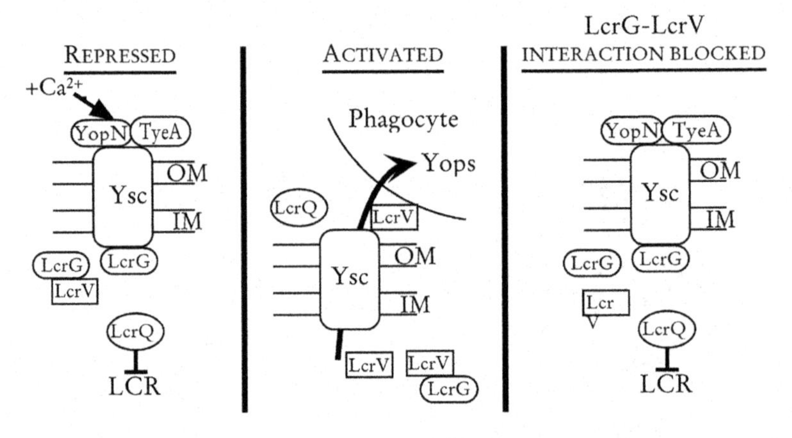

Fig. 2. LcrG titration model. In the presence of Ca^{2+}, LcrG, YopN and TyeA block the Ysc. LcrG is hypothesized to exert its blocking activity from the cytoplasm. The secretion block retains LcrQ in the cell, resulting in repression of LCR-regulated genes (Repressed). In the absence of Ca^{2+} or in the presence of eukaryotic cell contact, a block (possibly YopN) is released, allowing secretion of LcrQ (Activation). Secretion of LcrQ allows induction of LCR-regulated genes, including *lcrV*. Increased LcrV levels in the cytoplasm titrate LcrG away from the Ysc by forming a stable LcrG-LcrV complex. The removal of LcrG results in Yops and LcrV secretion and full induction of the LCR (Activated). If LcrG and LcrV cannot interact, LcrG is not titrated away from the Ysc, resulting in blockage of Yops secretion (LcrG-LcrV interaction blocked). From Matson and Nilles (2001). reprinted with permission of ASM.

conditions do not favor secretion, LcrG binds to an unidentified protein at the T3S apparatus and blocks the secretion of Yops. Upon induction of secretion LcrQ is secreted relieving the negative effects of LcrQ on Yops synthesis and allows an increase in the expression of Yops and, importantly, increases LcrV levels relative to LcrG levels. The increased LcrV binds LcrG and removes LcrG from its secretion-blocking role, resulting in the activation of Yops secretion.

20.2 Putting it all Together

A failing of the models presented for Yops secretion control by YopN in Fig. 1 and by LcrG in Fig. 2 is the lack of a link between the activities of LcrG and YopN. The development of an understanding of how LcrG, LcrV, and YopN cooperate to regulate the secretion of Yops is essential to fully understanding Yops secretion control. Recently, Hamad and Nilles demonstrated that LcrG, LcrV, and YopN are not directly involved in sensing Ca^{2+} but are required for responding to a Ca^{2+}-induced signal (Hamad 2006). These results indirectly support the proposed role of the YscF needle as the extracellular molecule involved in sensing Ca^{2+} (Torruellas et al. 2005). Interestingly, *lcrG* and *lcrGV* strains retain the ability to respond to higher Ca^{2+} concentrations by secreting reduced amounts of Yops (Hamad 2006). This ability to respond to increased Ca^{2+} is shared with a class of YscF mutants isolated by Torruellas et al. (2005). The constitutively secreting YscF mutants that respond to Ca^{2+} by secreting reduced amounts of Yops could represent a class of needle mutants that are defective in LcrG function suggesting a speculative linkage between LcrG and/or LcrV and YscF in Ca^{2+} responses. Furthermore, the ability of the *lcrG* and *lcrGV* strains to respond to high Ca^{2+} concentrations demonstrates that in the absence of LcrG, *Y. pestis* becomes less sensitive to Ca^{2+}. Therefore, if the YscF needle is involved in Ca^{2+} sensing, then LcrG may function to enhance the sensitivity of the YscF needle to extracellular Ca^{2+}. To date, no evidence suggests an LcrG association with the YscF needle complex and any linkage of LcrG to YscF function remains speculative. The analysis of YscF needle preparations by electron microscopy reveals that LcrV is found at the tip of the needles (Mueller et al. 2005) and suggests that the tip of the needle could be a site of LcrG-LcrV interaction. While LcrG has not been detected in needle preparations, that does not rule out the possibly that LcrG might regulate Yops secretion at the needle complex. Further direct experiments including immunoelectron and confocal microscopy are required to test the hypothesis that secretion control by LcrG and LcrV occurs at the tip of the YscF needle complex outside the bacterial cell. Additionally, isolation of LcrG or LcrV suppressors of YscF mutants would provide genetic evidence of a linkage between the LcrG-LcrV complex and YscF in Yops secretion control.

To dissect the roles of LcrG and LcrV in secretion control, a *ΔlcrGV3 ΔyopN* strain of *Y. pestis* was created and used as a background strain to introduce different LcrG and LcrV constructs to study how these proteins might affect the secretion of Yops in the absence of YopN (Hamad 2006; Hamad unpublished). Results obtained from this study shed light on the role of LcrG in secretion blockage relative to YopN's function. Strains lacking *lcrG* or *lcrGV* were able to respond to increased

Ca^{2+} concentrations by secreting reduced amounts of Yops only when YopN function was intact (Hamad 2006). This result suggests that when conditions do not favor secretion, LcrG contributes to secretion blockage by enhancing the blocking ability of YopN. Thus, when conditions do not favor secretion, LcrG appears to mediate secretion blockage by modulating the function of YopN. Data supporting this role of LcrG is provided by the isolation of a YopN point mutant that constitutively blocks Yops secretion (Ferracci et al. 2005). This YopN mutant's ability to block Yops secretion is dependent on TyeA but not LcrG, SycN, or YscB (Ferracci et al. 2005). Taken together the current evidence suggests that when conditions do not favor secretion, the YopN/TyeA complex serves as the molecule that physically blocks Yops secretion, while LcrG may play a role in modulating the function of the YopN/TyeA complex or LcrG may control a distinct step of the secretion process.

The mechanism by which LcrG facilitates a Yops blocking activity remains unclear. LcrG could influence YopN's function indirectly by amplifying the blocking signal induced by Ca^{2+} thus making YopN more sensitive to this signal. Another possibility is that LcrG could, directly or indirectly, modulate the interaction between YopN with interaction partners and/or the interaction between the YopN/TyeA complex with its Ysc target. When LcrG was expressed in the *ΔlcrGV3 ΔyopN* mutant, the secretion of Yops was partially blocked, suggesting that LcrG can bind a secretion-blocking target and impose a negative effect on Yops secretion in the absence of YopN. This suggests an independent role of LcrG in blocking Yops secretion. Clearly further work will be necessary to determine the influences of YopN and LcrG on each other's function as well as determining the targets of LcrG and YopN in blocking Yops secretion.

Not only are the mechanisms of secretion blockage obscure so are the mechanisms of secretion activation. When conditions favors Yops secretion, YopN's blocking role is relieved through the secretion of YopN and LcrG's role in secretion blockage is negated through the interaction of LcrG with LcrV (Nilles et al. 1998; Matson and Nilles 2001; Lawton et al. 2002; Ferracci et al. 2005). Unfortunately the mechanism that sets these events in action is unknown. According to the titration model, the removal of Ca^{2+} triggers the release of LcrQ, which results in an increase in LcrV levels (Nilles et al. 1998; Matson and Nilles 2001). This increase in LcrV levels is thought to drive the titration of LcrG by LcrV to allow for secretion activation (Nilles et al. 1998; Matson and Nilles 2001). However, the loss of LcrQ cannot simply account for secretion activation, since the secretion of Yops by an LcrQ null mutant reamins Ca^{2+} regulated (Rimpiläinen et al. 1992). In addition, inhibition of protein synthesis does not interfere with the ability of Yersinia to translocate Yops into eukaryotic cells, indicating that Yersinia is able to activate the secretion of Yops without de novo synthesis of LcrV (Lloyd et al. 2001). Thus although the interaction between LcrG and LcrV is required for secretion control, an increase in absolute LcrV levels is unlikely to be the mechanism behind LcrV's ability to bind and negate LcrG function when secretion is induced.

YopN appears to affect the function of LcrG and LcrV by influencing the interaction between LcrG and LcrV. An LcrG A16R mutant has reduced affinity for LcrV and when LcrG A16R is expressed in a *ΔlcrG3* strain the secretion of Yops is blocked (Matson and Nilles 2001; Matson and Nilles 2002), indicating that in the presence of YopN, LcrV does not bind LcrG A16R. However, when LcrG A16R was expressed with LcrV in a *ΔlcrGV3 ΔyopN* strain, LcrV negated the blocking activity of LcrG A16R. This indicates that in the absence of YopN, LcrV was able to bind and remove the blocking activity of LcrG A16R (Hamad 2006). The interaction between LcrG and LcrV occurs in the cytosol of *Y. pestis* in the presence of YopN irrespective of Ca^{2+} (Nilles et al. 1997). Potentially the interaction between LcrG and LcrV that is required for secretion control occurs at the T3S apparatus and not in the cytosol. Since LcrG and LcrV interact in the cytosol of *Y. pestis* regardless of the presence of Ca^{2+} yet the secretion of Yops occurs only in the absence of Ca^{2+}. Thus although LcrG and LcrV interact independently of YopN's presence, YopN could influence the interaction between LcrG and LcrV at the T3S apparatus where the relevant interaction between LcrG and LcrV may occur. Unfortunately, how YopN influences the function of LcrG and LcrV remains a mystery. The only known effect that YopN has on LcrG and LcrV is at the level of LcrV secretion. *Y. pestis* hyper-secretes LcrV in the absence of YopN (Skrzypek and Straley 1995; Hamad 2006). Therefore, YopN may have dual roles in blocking Yops secretion. The first role is at the level of Yops secretion, whereby YopN is required for blocking the secretion of T3S substrates when secretion is blocked. The second role of YopN is at the level of LcrV secretion, where YopN appears to negatively control the secretion of LcrV even when conditions favor Yops secretion. The significance of YopN's ability to negatively control the secretion of LcrV is unclear. Further experiments are required to determine how YopN controls the secretion LcrV and whether this event has any implications on LcrV's function in secretion control.

Based on the evidence provided above, a speculative model on how YscF, LcrG, LcrV, and YopN may control the secretion of Yops in response to Ca^{2+} is proposed (Fig. 3). In this revised model LcrG and LcrV control Yops secretion from the T3S apparatus, possibly at the tip of the YscF needle, while YopN controls secretion from the cytosolic face of the Ysc. YscF senses the presence of Ca^{2+} and transmits a blocking-signal to YopN, helping YopN to prevent the LcrG-LcrV interaction. Under these conditions LcrG's blocking role is dominant over LcrV's activating role and the YopN/TyeA complex blocks the secretion of Yops. LcrG's role in secretion blockage might possibly occur by amplifying the Ca^{2+} signal sensed by YscF, while the YopN/TyeA complex blocks the secretion of Yops from the cytosolic face of the Ysc (Ferracci et al. 2005). The removal of Ca^{2+} is sensed by the YscF needle that transmits a signal to YopN, thereby neutralizing YopN's role in preventing the LcrG-LcrV interaction. The negation of LcrG's blocking role by LcrV interaction allows the secretion of YopN, which relieves the blockage of the YopN/TyeA complex triggering Yops secretion. Current evidence largely supports this model; however further experiments are required to test the validity of this model. Future work including the identification of LcrG's and YopN's targets at the Ysc, understanding how YopN influences the function of LcrV and LcrG, and comprehending the mechanism by which the YscF needle senses and responds to the extracellular environment are required to understand the complex regulation of Yops secretion.

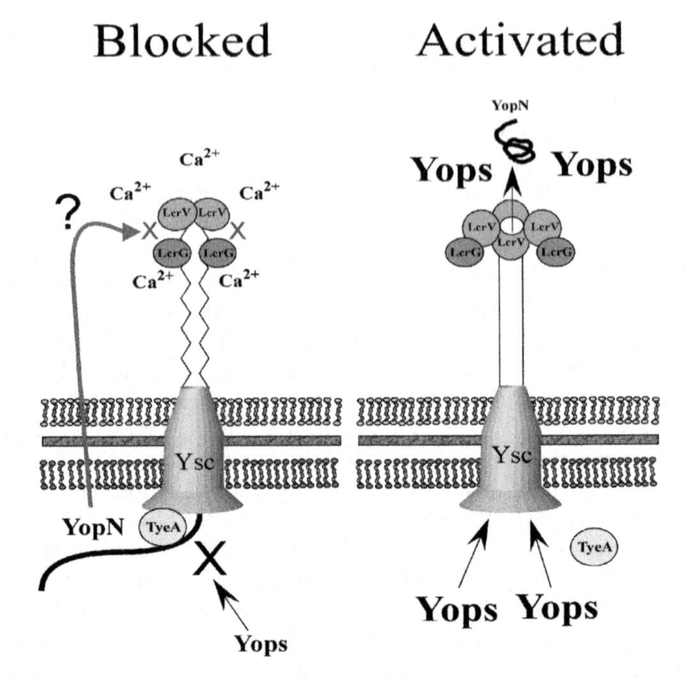

Fig. 3. A hypothetical model for the regulation of Yops secretion in *Y. pestis*. The presence of Ca^{2+} is sensed by the YscF needle that transmits a signal to YopN, allowing YopN to prevent the interaction between LcrG and LcrV. The removal of Ca^{2+} is sensed by the YscF needle that transmits a signal to YopN to neutralize YopN's role in preventing the interaction between LcrG and LcrV. Under these conditions, LcrV binds LcrG and negates LcrG's blocking role allowing the secretion of YopN. The secretion of YopN relieves the blockage of the Yopn/TyeA complex and allows for Yops secretion.

20.3 References

Anderson, D.M., Ramamurthi, K.S., Tam, C. and Schneewind, O. (2002) YopD and LcrH regulate expression of *Yersinia enterocolitica* YopQ by a posttranscriptional mechanism and bind to *yopQ* RNA. J. Bacteriol. 184, 1287-1295.

Brubaker, R.R. (1972) The genus *Yersinia*: biochemistry and genetics of virulence. Curr. Top. Microbiol. Immunol. 57, 111-158.

Brubaker, R.R. (2003) Interleukin-10 and inhibition of innate immunity to Yersiniae: roles of Yops and LcrV (V antigen). Infect. Immun. 71, 3673-3681.

Cambronne, E.D. and Schneewind, O. (2002) *Yersinia enterocolitica* type III secretion: *yscM1* and *yscM2* regulate *yop* gene expression by a posttranscriptional mechanism that targets the 5' untranslated region of *yop* mRNA. J. Bacteriol. 184, 5880-93.

Cornelis, G.R. (1999) The *Yersinia* deadly kiss. J. Bacteriol. 180, 5495-5504.

Day, J.B., Ferracci, F. and Plano, G.V. (2003). Translocation of YopE and YopN into eukaryotic cells by *Yersinia pestis yopN, tyeA, sycN, yscB* and *lcrG* deletion mutants measured using a phosphorylatable peptide tag and phosphospecific antibodies. Mol. Microbiol. 47, 807-823.

Day, J.B. and Plano, G.V. (1998) A complex composed of SycN and YscB functions as a specific chaperone for YopN in *Yersinia pestis*. Mol. Microbiol. 30, 777-788.

Ferracci, F., Day, J.B. Ezelle, H.J. and Plano, G.V. (2004) Expression of a functional secreted YopN-TyeA hybrid protein in *Yersinia pestis* is the result of a +1 translational frameshift event. J. Bacteriol. 186, 5160-5166.

Ferracci, F., Schubot, F.D., Waugh, D.S. and Plano, G.V. (2005) Selection and characterization of *Yersinia pestis* YopN mutants that constitutively block Yop secretion. Mol. Microbiol. 57, 970-987.

Fields, K. A., Nilles, M.L., Cowan, C. and Straley, S.C. (1999) Virulence role of V antigen of *Yersinia pestis* at the bacterial surface. Infect. Immun. 67, 5395-5408.

Forsberg, Å., Viitanen, A.M., Skurnik, M. and Wolf-Watz, H. (1991) "The surface-located YopN protein is involved in calcium signal transduction in *Yersinia pseudotuberculosis*." Mol. Microbiol. 5, 977-986.

Hamad, M.A. (2006) The roles of LcrG and LcrV in secretion control of Yops in *Yersinia pestis*. Ph.D. Dissertation, Department of Microbiology and Immunology, University of North Dakota, Grand Forks, ND, USA.

Iriarte, M., Sory, M.-P., Boland, A., Boyd, A.P., Mills, S.D., Lambermont, I. and Cornelis, G.R. (1998) "TyeA, a protein involved in control of Yop release and in translocation of *Yersinia* Yop effectors." EMBO J. 17, 1907-1918.

Jackson, M.W., Day, J.B. and Plano, G.V. (1998) YscB of *Yersinia pestis* functions as a specific chaperone for YopN. J. Bacteriol. 180, 4912-4921.

Kenjale, R., Wilson, J., Zenk, S.F., Saurya, S., Picking, W.L. and Blocker, A. (2005) The needle component of the type III secreton of *Shigella* regulates the activity of the secretion apparatus. J. Biol. Chem. 280, 42929-42937.

Lawton, D.G., Longstaff, C., Wallace, B.A., Hill, J., Leary, S.E., Titball, R.W. and Brown, K.A. (2002) Interactions of the type III secretion pathway proteins LcrV and LcrG from *Yersinia pestis* are mediated by coiled-coil domains. J. Biol. Chem. 277, 38714-38722.

Lloyd, S.A., Norman, M., Rosqvist, R. and Wolf-Watz, H. (2001) *Yersinia* YopE is targeted for type III secretion by N-terminal, not mRNA, signals. Mol. Microbiol. 39, 520-532.

Matson, J.S. and Nilles, M.L. (2001). "LcrG-LcrV Interaction Is Required for Control of Yops Secretion in *Yersinia pestis*." Journal of Bacteriology 183(17): 5082-91.

Matson, J.S. and Nilles, M.L. (2002) Interaction of the *Yersinia pestis* type III regulatory proteins LcrG and LcrV occurs at a hydrophobic interface. BMC Microbiol. 2, 16.

Michiels, T., Wattiau, P., Brasseur, R., Ruysschaert, J.M. and Cornelis, G. (1990) Secreation of Yop proteins by Yersiniae. Infect. Immun. 58, 2840-2849.

Mota, L.J. (2006. Type III secretion gets an LcrV tip. Trends Microbiol. 14, 197-200.

Mueller, C.A., Broz, P., Müller, S.A., Ringler, P., Erne-Brand, F., Sorg, I., Kuhn, M., Engel, A. and Cornelis, G.R. (2005). The V-antigen of *Yersinia* forms a distinct structure at the tip of injectisome needles. Science 310, 674-676.

Nilles, M.L., Fields, K.A. and Straley, S.C. (1998) The V antigen of *Yersinia pestis* regulates Yop vectorial targeting as well as Yop secretion through effects on YopB and LcrG. J. Bacteriol. 180, 3410-3420.

Nilles, M.L., Williams, A.W., Skrzypek, E. and Straley, S.C. (1997) *Yersinia pestis* LcrV forms a stable complex with LcrG and may have a secretion-related regulatory role in the low-Ca^{2+} response. J. Bacteriol. 179, 1307-1316.

Overheim, K.A., Depaolo, R.W., DeBord, K.L., MOrrin, E.M., Anderson, D.M., Green, N.M., Brubaker, R.R., Jabri, B. and Schneewind, O. (2005) LcrV plague vaccine with altered immunomodulatory properties. Infect. Immun. 73, 5152-5159.

Pallen, M.J., Francis, M.S. and Fütterer, K. (2003) Tetratricopeptide-like repeats in type-III-secretion chaperones and regulators. FEMS Microbiol. Lett. 223, 53-60.

Perry, R.D. and Fetherson, J.D. (1997) *Yersinia pestis* — etiologic agent of plague. Clin. Microbiol. Rev. 10, 35-66.

Pettersson, J., Holmström, A. , Hill, J., Leary, S., Frithz-Lindsten, E. von Euler-Matell, A., Carlsson, E. Titball, R., Forsberg, Å. and Wolf-Watz, H. (1999) The V-antigen of *Yersinia* is surface-exposed before target cell contact and involved in virulence protein translocation. Mol. Microbiol. 32, 961-976.

Rimpiläinen, M., Forsberg, Å. and Wolf-Watz, H. (1992) A novel protein, LcrQ, involved in the low-calcium response of *Yersinia pseudotuberculosis*, shows extensive homology to YopH. J. Bacteriol. 174, 3355-3363.

Rosqvist, R., Magnusson, K-E. and Wolf-Watz, H. (1994) Target cell contact triggers expression and polarized transfer of *Yersinia* YopE cytotoxin into mammalian cells. EMBO J. 13, 964-972.

Sing, A., Reithmeier-Rost, D., Granfors, K., Hill, J., Roggenkamp, A. and Heesemann, J. (2005) A hypervariable N-terminal region of *Yersinia* LcrV determines Toll-like receptor 2-mediated IL-10 induction and mouse virulence. PNAS 102, 16049-16054.

Skrzypek, E. and Straley, S.C. (1993) LcrG, a secreted protein involved in negative regulation of the low-calcium response in *Yersinia pestis*. J. Bacteriol.175, 3520-3528.

Skrzypek, E. and Straley, S.C. (1995) Differential effects of deletions in *lcrV* on secretion of V antigen, regulation of the low-Ca^{2+} response, and virulence of *Yersinia pestis*. J. Bacteriol. 177, 2530-2542.

Straley, S. (1988) The plasmid-encoded outer-membrane proteins of *Yersinia pestis*. Rev. Infect. Dis. 10, S323-S326.

Straley, S.C. and Bowmer, W.S. (1986) Virulence genes regulated at the transcriptional level by Ca^{2+} in *Yersinia pestis* include structural genes for outer membrane proteins. Infect. Immun. 51, 445-454.

Straley, S.C. and Perry, R.D. (1995) Environmental modulation of gene expression and pathogenesis in *Yersinia*. Trends Microbiol. 3, 310-317.

Straley, S.C., Plano, G.V., Skrzypek, E., Haddix, P.L. and Fields, K.A. (1993) Regulation by Ca^{2+} in the *Yersinia* low-Ca^{2+} response. Mol. Microbiol. 8, 1005-1010.

Titball, R.W. and Williamson, E.D. (2004) *Yersinia pestis* (plague) vaccines. Expert Opin. Biol. Ther. 4, 965-73.

Torruellas, J., Jackson, M.W., Pennock, J.W. and Plano, G.V. (2005) The *Yersinia pestis* type III secretion needle plays a role in the regulation of Yop secretion. Mol. Microbiol. 57, 1719-1733.

Williams, A.W. and Straley, S.C. (1998) YopD of *Yersinia pestis* plays a role in the negative regulation of the low-calcium response in addition to its role in the translocation of Yops. J. Bacteriol. 180, 350-358.

Williamson, E.D., Flick-Smith, H.C., LeButt, C. Rowland, A. Jones, S.M., Waters, E.L., Gwyther, R.J., Miller, J. Packer, P.J. and Irving, M. (2005) Human immune response to a plague vaccine comprising recombinant F1 and V antigens. Infect. Immun. 73, 3598-3608.

Wulff-Strobel, C.R., Williams, A.W. and Straley, S.C. (2002) LcrQ and SycH function together at the Ysc type III secretion system in *Yersinia pestis* to impose a hierarchy of secretion. Mol. Microbiol. 43, 411-423.

21
Identification of TyeA Residues Required to Interact with YopN and to Regulate Yop Secretion

Sabrina S. Joseph and Gregory V. Plano[1]

[1] Department of Microbiology and Immunology, University of Miami Miller School of Medicine, gplano@med.miami.edu

Abstract. The secretion of Yops via the *Yersinia* type III secretion system (T3SS) is controlled, in part, by a cytoplasmic YopN/TyeA complex. This complex is required to prevent Yop secretion in the presence of extracellular calcium and prior to contact between the bacterium and a eukaryotic cell. In this study we utilized site-directed mutagenesis to analyze the role of specific TyeA regions and residues in the regulation of Yop secretion. We identified two spatially distinct, surface-exposed regions of the TyeA molecule that were required to regulate Yop secretion. One region, identified by residues M51, F55 and P56, was required for TyeA to interact with YopN. A second region, identified by residues R19, W20 and D25 was not involved in the interaction of TyeA with YopN, but may be required for the YopN/TyeA complex to interact with the T3S apparatus in a manner that blocks Yop secretion.

21.1 Introduction

Yersinia pestis is the etiologic agent of plague, one of the most devastating diseases known (Perry and Fetherston 1997). The ability of *Y. pestis* to produce disease primarily results from its capacity to avoid or disrupt the innate defenses of its host (Cornelis 2000). This capability enables the bacterium to grow and multiply essentially unhindered. *Y. pestis* actively blocks bacterial phagocytosis and prevents the early production of proinflammatory cytokines (Fallman and Gustavsson 2005; Navarro et al. 2005). These abilities are strictly dependent upon the presence of a functional plasmid pCD1-encoded type III secretion system (T3SS). The *Yersinia* T3SS functions to inject effector proteins, termed *Yersinia* outer proteins (Yops), directly into host cells where they function to undermine cellular processes that normally prevent bacterial growth and survival.

The *Yersinia* T3SS is a complex protein secretion system that functions to transport Yops from the cytoplasm of the bacterial cell to the cytoplasm of a eukaryotic cell (Mota and Cornelis 2005). The T3S apparatus consists of a base structure that spans the bacterial inner and outer membranes and of an external needle-like structure composed of the secreted YscF protein that extends 40 to 60 nm from the bacterial surface. The injection process occurs in two distinct steps: (i) secretion of Yops across the bacterial membranes and (ii) translocation of Yops across a eukaryotic membrane. The second step in this process is dependent upon three secreted pore-forming proteins (LcrV, YopB and YopD) that form a translocation complex, or translocon, at the eukaryotic membrane (Marenne et al. 2003; Neyt and Cornelis 1999; Nilles et al. 1998; Pettersson et al. 1999). The LcrV protein forms a needle-tip complex that facilitates assembly of the needle-translocon structure at the eukaryotic membrane (Mueller et al. 2005).

The secretion of Yops is normally triggered by contact between a bacterium and a eukaryotic cell. *In vitro*, Yop secretion is blocked in the presence of millimolar levels of extracellular calcium and is triggered by the removal of extracellular calcium (Michiels et al. 1990). The regulation of Yop secretion is dependent upon at least six *Yersinia* pCD1-encoded proteins: LcrG (Matson and Nilles 2001; Skryzpek and Straley 1993), YopN (Forsberg et al. 1991; Yother and Goguen 1985), SycN (Iriarte and Cornelis 1999), YscB (Jackson et al. 1998), TyeA (Cheng and Schneewind 2000; Iriarte et al. 1998) and YscF (Torruellas et al. 2005). These proteins function to block Yop secretion in the presence of calcium *in vitro* and prior to contact with a eukaryotic cell *in vivo*. A deletion in any one of these genes results in constitutive secretion in the presence and absence of calcium and prior to contact with a eukaryotic cell (constitutive secretion [CS] phenotype). The LcrG protein forms a 1:1 complex with LcrV within the cell and assists in blocking Yop secretion when present in an excess amount over LcrV (Matson and Nilles 2001). YopN is a secreted protein that directly interacts with the cytosolic SycN/YscB chaperone and TyeA prior to its export from the bacterial cell (Cheng et al. 2001; Day et al. 2003). The SycN/YscB chaperone binds to an N-terminal region of YopN and is required for efficient YopN secretion and translocation. The TyeA protein binds to a C-terminal region of YopN and functions with YopN to block Yop secretion. Although YopN and TyeA are normally expressed as two separate proteins, an engineered YopN-TyeA fusion protein is secreted, translocated and regulates Yop secretion (Ferracci et al. 2004). Interestingly, most bacterial pathogens that employ T3SSs express and secrete a YopN/TyeA-like fusion protein, not separate YopN-like and TyeA-like proteins (Pallen et al. 2005). YscF is a secreted protein that assembles to form the extracellular needle-like structure (Hoiczyk and Blobel 2001). YscF point mutants that secrete Yops constitutively have been identified, indicating that the YscF needle, which is required for Yop secretion, also plays a role in the regulation of Yop secretion (Torruellas et al. 2005). The mechanism by which LcrG, the YopN/SycN/YscB/TyeA complex and YscF regulate Yop secretion is not understood.

Recently, several YopN mutants were identified that block Yop secretion constitutively (no secretion [NS] phenotype) (Ferracci et al. 2005). These mutants required TyeA, but did not require LcrG, the SycN/YscB chaperone, a functional YopN N-terminal secretion signal or chaperone-binding domain to block Yop secretion. These results suggest that the YopN/TyeA complex blocks Yop secretion from a cytosolic location. We hypothesize that the YscF needle/LcrV tip complex functions as a signal sensor/transmission device that controls the activity of the YopN/TyeA molecular plug.

The mechanism by which the cytosolic YopN/TyeA complex blocks Yop secretion is unknown. We hypothesize that the YopN/TyeA complex interacts with an unidentified cytosolic T3S component. If correct, this would indicate that the YopN/TyeA complex contains an essential surface-exposed binding site that could potentially be identified through genetic analyses. In this study we investigated the role of specific TyeA regions and residues in the regulation of Yop secretion using alanine-scanning mutagenesis. These studies identified a surface-exposed region of TyeA that is required to block Yop secretion, but is not required for TyeA to interact with YopN.

21.2 Materials and Methods

21.2.1 Bacterial Strains and Growth Conditions

Escherichia coli DH5α was used for routine cloning experiments (Cambau et al. 1993). *Y. pestis* and *E. coli* strains were grown in heart infusion broth (HIB) or on tryptose blood agar (TBA) plates (Difco) at 27°C and 37°C, respectively. For growth/secretion assays *Y. pestis* strains were grown with or without 2.5 mM $CaCl_2$ in TMH medium (Goguen et al. 1984) inoculated from cultures grown overnight at 27°C. Strains were grown for 1 h at 27°C and then shifted to 37°C for 5 h of growth. Bacteria carrying resistance markers were grown in the presence of the appropriate antibiotic at a final concentration of 25 μg/ml (kanamycin) or 50 μg/ml (ampicillin or streptomycin).

21.2.2 Generation of TyeA Point Mutants

The DNA fragment encoding TyeA was generated by PCR amplification of *tyeA* encoded on pCD1 using primers (TyeA-*Eco*RI; 5'- TTTGAATTCGGCAATTTTTT TCAGAGGGTAAAAC-3') and (TyeA-*Hin*dIII; 5'-TTTAAGCTTTCAATCCAAC TCACTCAATTCTTC-3'). The resulting 310 base pair PCR fragment was digested with *Eco*RI and *Hin*dIII and inserted into *Eco*RI- and *Hin*dIII-digested pBAD18, generating plasmid pTyeA. Alanine-scanning, site-directed mutagenesis of pTyeA was performed by the PCR-ligation-PCR procedure as previously described (Ali and Steinkasserer 1995; Torruellas et al. 2005). The DNA sequence of the entire *tyeA* gene present in each pTyeA derivative was confirmed by DNA sequence analysis.

21.2.3 SDS-PAGE and Immunoblotting

Bacterial cell pellets and culture supernatants were separated by centrifugation at 12,000 x g for 10 min at 4°C. Culture supernatant proteins were precipitated on ice overnight with 10% (v/v) trichloroacetic acid (TCA) and collected by centrifugation at 12,000 x g for 10 min at 4°C. Volumes of cellular fractions corresponding to equal numbers of bacteria were analyzed by SDS-PAGE and immunoblotting as described (Ferracci et al. 2005). The YopM and YopN proteins were detected with polyclonal antisera raised against purified 6X-histidine-tagged YopM or YopN proteins.

21.2.4 GST-Pulldown Experiments

Plasmid pGST-YopN$^{78\text{-}293}$ encodes a GST-YopN$^{78\text{-}293}$ fusion protein. A 648-base pair DNA fragment encoding *yopN* codons 78 to 293 was PCR amplified with primers pGST-YopN-78 (5'-ATGGCTCGAGTTAGCGACGTTGAGGAG-3') and pGST-YopN-293 (5'-TTTGAATTCTCAGAAAGGTCGTACGCCATTAGTTTT-3'), digested with *Eco*RI and inserted into *Eco*RI- and *Psh*AI-digested pET42b, generating pGST-YopN$^{78\text{-}293}$. Plasmid pGST-YopN$^{78\text{-}293}$ and pTyeA, pTyeA (R19A W20A), pTyeA (W20A D25A), pTyeA (M51A F55A) or pTyeA (S6A G10A V13A) were transformed into *E. coli* BL21. The resultant strains were grown for 3 h at 37°C after which time expression of GST-YopN$^{78\text{-}293}$ and the wild-type or mutant TyeA protein were induced by addition of IPTG (1 mM) and L-arabinose (0.2%), respectively. The bacterial cells were harvested, lysed using a French pressure cell and centrifuged at 8,000 x g for 10 min to remove unlysed cells and large debris. The resulting lysates were further centrifuged at 35,000 x g for 30 min and the supernatant fractions containing the soluble proteins were applied to glutathione sepharose columns. The soluble fractions were incubated with the glutathione sepharose beads overnight at 4°C with shaking to allow efficient binding. The columns were subsequently washed three times with Tris-HCl (20 mM), NaCl (150 mM), pH 8.0. The bound GST-YopN$^{78\text{-}293}$ and TyeA proteins were eluted with glutathione (10 mM). The soluble samples, wash fractions and elutions were collected and analyzed by SDS-PAGE and immunoblotting with antibodies specific for the GST moiety and TyeA.

21.3 Results

21.3.1 Alanine-Scanning Mutagenesis of *tyeA*

TyeA is a small 92 amino-acid residue protein that directly interacts with the C-terminus of YopN (Iriarte et al. 1998). TyeA is composed of two sets of parallel alpha helices (Schubot et al. 2005). In the YopN/TyeA complex, the C-terminal helix of YopN is found intercalated between the first and third helices of TyeA. The N-terminal helix of TyeA also makes several contacts with YopN helix nine, composed of YopN residues 212 to 222. To investigate the role of specific TyeA residues in the interaction of TyeA with YopN and in the regulation of Yop secretion, we replaced selected TyeA residues with alanine and characterized the resulting mutant TyeA proteins.

Oligonucleotide site-directed mutagenesis of plasmid pTyeA was performed to replace select TyeA amino-acid residues with alanine (Table 1). The resultant pTyeA derivatives were electroporated into a *tyeA* deletion strain and the Yop expression and secretion phenotypes of the resultant strains were determined. Thirty-one mutants with single alanine substitutions in TyeA were analyzed for Yop expression and secretion following growth in the presence or absence of 2.5 mM calcium in TMH medium for 5 h at 37°C. All of the TyeA mutants expressed a stable TyeA protein (data not shown). The parent strain *Y. pestis* KIM5-3001.P39 (Day et al. 2003), the *tyeA* deletion strain complemented with pTyeA and 25 of the 31 TyeA mutants exhibited normal calcium-regulated secretion (RS) of Yops (Table 1; data not shown). The *tyeA* deletion strain and four TyeA alanine missense mutants (W20A, D25A, F55A and P56A) showed constitutive secretion (CS) of Yops in both the presence and absence of calcium (Fig. 1A). In addition, two TyeA mutants (R19A and M51A)

Table 1. Yop Secretion Phenotypes of *Y. pestis* TyeA Point Mutants

Strain + plasmid	Yop secretion	Strain + plasmid	Yop secretion
KIM5-3001.P39 (parent)	RS	Δ*tyeA* + pTyeA (L48A)	RS
KIM-3001.P63 (Δ*tyeA*)	CS	Δ*tyeA* + pTyeA (M51A)	RS/CS
Δ*tyeA* + pTyeA	RS	Δ*tyeA* + pTyeA (R53A)	RS
Δ*tyeA* + pTyeA (Y3A)	RS	Δ*tyeA* + pTyeA (F55A)	CS
Δ*tyeA* + pTyeA (L5A)	RS	Δ*tyeA* + pTyeA (P56A)	CS
Δ*tyeA* + pTyeA (S6A)	RS	Δ*tyeA* + pTyeA (G58A)	RS
Δ*tyeA* + pTyeA (M9A)	RS	Δ*tyeA* + pTyeA (D62A)	RS
Δ*tyeA* + pTyeA (G10A)	RS	Δ*tyeA* + pTyeA (E63A)	RS
Δ*tyeA* + pTyeA (D11A)	RS	Δ*tyeA* + pTyeA (E64A)	RS
Δ*tyeA* + pTyeA (V16A)	RS	Δ*tyeA* + pTyeA (Q65A)	RS
Δ*tyeA* + pTyeA (D17A)	RS	Δ*tyeA* + pTyeA (C73A)	RS
Δ*tyeA* + pTyeA (K18A)	RS	Δ*tyeA* + pTyeA (Q74A)	RS
Δ*tyeA* + pTyeA (R19A)	RS/CS	Δ*tyeA* + pTyeA (I81A)	RS
Δ*tyeA* + pTyeA (W20A)	CS	Δ*tyeA* + pTyeA (E84A)	RS
Δ*tyeA* + pTyeA (I23A)	RS	Δ*tyeA* + pTyeA (R19A W20A)	CS
Δ*tyeA* + pTyeA (D25A)	CS	Δ*tyeA* + pTyeA (W20A D25A)	CS
Δ*tyeA* + pTyeA (E27A)	RS	Δ*tyeA* + pTyeA (M51A F55A)	CS
Δ*tyeA* + pTyeA (F33A)	RS	Δ*tyeA* + pTyeA (F55A P56A)	CS
Δ*tyeA* + pTyeA (L35A)	RS	Δ*tyeA* + pTyeA (S6A G10A V13A)	RS/CS
Δ*tyeA* + pTyeA (F44A)	RS		

RS, calcium regulated secretion; CS, constitutive secretion irrespective of calcium
NS, no secretion; RS/CS, partial secretion in the presence of calcium

had intermediate phenotypes and secreted low amounts of Yops in the presence of calcium (RS/CS phenotype). These studies identified six TyeA amino-acid residues that were required to block Yop secretion in the presence of millimolar levels of extracellular calcium. Interestingly, the identified residues mapped to two distinct regions of the TyeA molecule (Fig. 1B). TyeA residues M51, F55 and P56 are predicted to be located at the interface between TyeA and YopN (Schubot et al. 2005). In fact, the crystal structure of the YopN-TyeA complex predicts that each of the identified residues forms a direct contact point between TyeA and the C-terminal helix of YopN. These results suggest that the role of TyeA residues M51, F55 and P56 in the regulation of Yop secretion is to mediate the interaction of TyeA with YopN. On the other hand, TyeA residues R19, W20 and D25 mapped to a surface-exposed region of the TyeA molecule that was not predicted to be involved in the interaction of TyeA with YopN. Thus, this region of TyeA may be involved in interactions between the YopN-TyeA complex and the T3S apparatus that are required for the YopN/TyeA-dependent block in Yop secretion.

21.3.2 Construction and Analysis of TyeA Double and Triple Alanine-Substitution Mutants

To further analyze the role of the two identified regions of TyeA required to regulate Yop secretion (Fig. 1B), we generated a series of double alanine substitution mutants (Table 1). Oligonucleotide site-directed mutagenesis of pTyeA (W20A) and pTyeA (F55A) was performed to generate plasmids pTyeA (R19A W20A), pTyeA (W20A

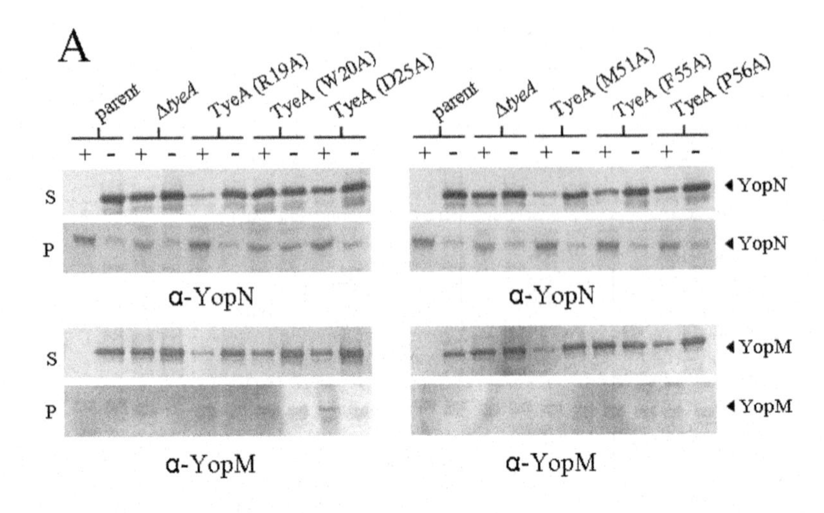

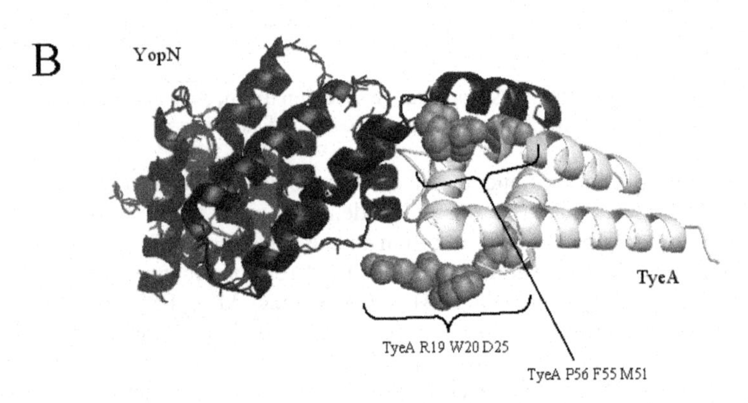

Fig. 1. Analysis of TyeA single alanine missense mutants. (A) Expression and secretion of YopN and YopM by TyeA mutants. *Y. pestis* KIM5-3001.P39 (parent), KIM-3001.P63 (Δ*tyeA*) and six TyeA alanine missense mutants were analyzed by SDS-PAGE and immunoblot analysis of cell pellet (P) and culture supernatant (S) fractions derived from bacteria grown for 5 h at 37°C in the presence and absence of 2.5 mM $CaCl_2$. Blots were probed with α-YopN and α-YopM antisera. The location of YopM and YopN are shown by arrowheads. (B) Ribbon model of the structure of the $YopN^{76-293}$–TyeA complex highlighting the two identified regions of TyeA required for regulating Yop secretion. YopN and TyeA are shown in black and light grey, respectively. The location of TyeA residues M51, F55 and P56, which form contacts with residues on helix 12 of YopN, as well as TyeA residues R19, W20 and D25 are shown (dark grey space-filled representation). This model was generated using PyMOL (DeLano 2001).

D25A), pTyeA (M51A F55A) and pTyeA (F55A P56A). Previous analyses of the YopN-TyeA crystal structure indicated that, in addition to the TyeA residues that directly interact with the C-terminal helix of YopN, TyeA amino-acid residues S6,

G10 and V13 form contacts with YopN helix number nine that may play a direct role in mediating the interaction of TyeA with YopN (Schubot et al. 2005). To investigate the role of these contact residues in the interaction of TyeA with YopN and in the regulation of Yop secretion, we used oligonucleotide site-directed mutagenesis of pTyeA to generate plasmid pTyeA (S6A G10A V13A).

The resultant pTyeA derivatives encoding TyeA double and triple alanine-substitution mutants were electroporated into the *tyeA* deletion mutant and analyzed for Yop expression and secretion, following growth in the presence or absence of 2.5 mM calcium in TMH medium for 5 h at 37°C. The parent strain showed typical calcium regulated Yop secretion; however, the *tyeA* deletion strain expressing TyeA (R19A W20A), TyeA (W20A D25A), TyeA (M51A F55A) or TyeA (F55A P56A) secreted Yops constitutively in both the presence and absence of calcium (Fig. 2A). In addition, the *tyeA* deletion strain expressing TyeA (S6A G10A V13A) exhibited a partial phenotype, secreting low amounts of Yops in the presence of calcium. These results confirm a role for the two identified TyeA regions in the regulation of Yop secretion and suggest that TyeA residues S6, G10 and V13 may also play a role in TyeA function.

21.3.3 Interaction of TyeA Mutants with YopN

The ability of TyeA (R19A W20A), TyeA (W20A D25A), TyeA (M51A F55A) and TyeA (S6A G10A V13A) to interact with YopN was evaluated using a GST-pulldown procedure and the GST-YopN$^{78\text{-}293}$ protein. The GST-YopN$^{78\text{-}293}$ protein, encoded by plasmid pGST-YopN$^{78\text{-}293}$, carries an intact TyeA-binding domain, but lacks the YopN N-terminal secretion signal and SycN/YscB chaperone-binding domain (Schubot et al. 2005). *E. coli* BL21 was transformed with plasmid pGST-YopN$^{78\text{-}293}$ and plasmid pTyeA, pTyeA (R19A W20A), pTyeA (W20A D25A), pTyeA (M51A F55A) or TyeA (S6A G10A V13A). The resultant strains were initially grown for 3 h at 37°C, at which point IPTG (1 mM) and L-arabinose (0.2%) were added to induce expression of the GST-YopN$^{78\text{-}293}$ and TyeA proteins and the cultures were incubated an additional 2 h at 37°C. Bacteria expressing the GST-YopN$^{78\text{-}293}$ protein and one of the TyeA derivatives were harvested, lysed, and the lysate centrifuged. The cleared lysates were added to glutathione sepharose columns, washed, and the bound GST-YopN$^{78\text{-}293}$ proteins (and YopN-bound TyeA proteins) eluted. The amount of GST-YopN$^{78\text{-}293}$ and TyeA in the initial lysates, the wash fractions and the elutions were determined by SDS-PAGE and immunoblot analysis (Fig. 2B). Approximately equal amounts of the GST-YopN$^{78\text{-}293}$ protein bound to, and were eluted from, each of the columns. The TyeA, TyeA (R19A W20A) and TyeA (W20A D25A) proteins efficiently bound to, and co-eluted with, the GST-YopN$^{78\text{-}293}$ protein. In contrast, the TyeA (M51A F55A) and TyeA (S6A G10A V13A) proteins did not interact or bound poorly to the GST-YopN$^{78\text{-}293}$ protein. These results indicate that the regions of TyeA defined by amino-acid residues M51, F55 and P56 and by amino-acid residues S6, G10 and V13 are required for TyeA to efficiently interact with YopN. On the contrary, the region of TyeA defined by amino-acid residues R19, W20 and D25 is required for TyeA to regulate Yop secretion, but is not required for TyeA to interact with YopN.

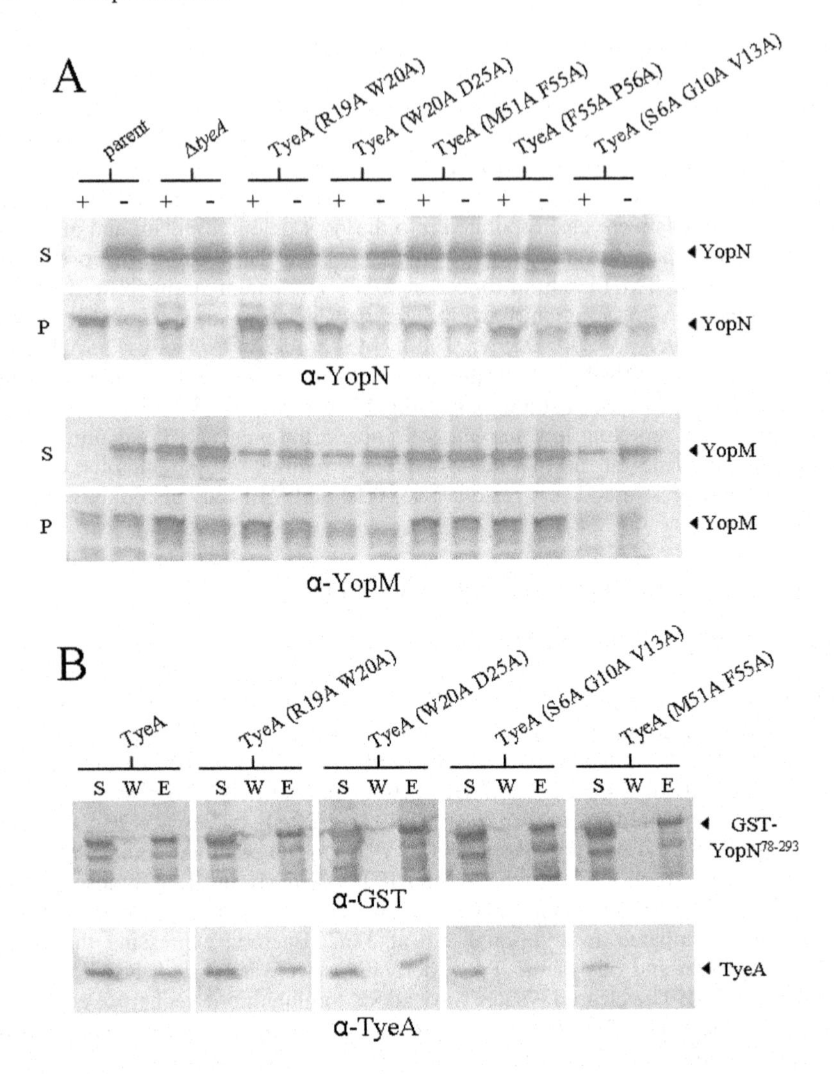

Fig. 2. Analysis of TyeA double and triple alanine missense mutants. (A) Expression and secretion of YopN and YopM by TyeA mutants. *Y. pestis* KIM5-3001.P39 (parent), KIM-3001.P63 (*ΔtyeA*), four TyeA double alanine missense mutants and one TyeA triple alanine missense mutant were analyzed by SDS-PAGE and immunoblot analysis of cell pellet (P) and culture supernatant (S) fractions using antisera specific for YopM and YopN (arrowheads). (B) Analysis of the interaction of YopN with TyeA proteins. The GST-YopN$^{78\text{-}293}$ protein and interacting (co-purifying) TyeA proteins expressed in *E. coli* BL21 were purified using glutathione sepharose. Soluble (S), wash (W) and elution (E) fractions were analyzed by immunoblotting with antisera specific for GST and TyeA.

21.4 Discussion

A cytoplasmic YopN/SycN/YscB/TyeA complex is required to block Yop secretion in the presence of millimolar amounts of extracellular calcium and prior to contact with a eukaryotic cell. Previous studies have identified YopN missense mutants that constitutevely block Yop secretion (YopN NS mutants) (Ferracci et al. 2005). Interestingly, these mutants required TyeA, but did not require LcrG or the SycN/YscB chaperone to block secretion, suggesting that the YopN/TyeA complex is the minimal complex required to prevent Yop secretion. In addition, several non-secretable, truncated YopN NS mutants that lack an N-terminal secretion signal and/or chaperone-binding domain, the $YopN^{85\text{-}293}$ (F234S) protein for example, still blocked Yop secretion constitutively (Ferracci et al. 2005). Together, these findings suggest that the YopN/TyeA complex has two independent means of interacting with the *Yersinia* T3S apparatus: (i) via its N-terminal secretion signal and chaperone-binding domain; and (ii) via the C-terminal domain of YopN complexed with the TyeA protein. This later interaction is hypothesized to be essential for the YopN/TyeA complex to block Yop secretion. The mechanism by which the cytosolic YopN/TyeA complex blocks Yop secretion is not understood; however, we hypothesize that the C-terminus of YopN complexed with TyeA must interact with a component of the T3S apparatus to block Yop secretion. Mutations that disrupt the interaction of the YopN/TyeA complex with the T3S apparatus would be expected to secrete Yops constitutively. The identification of a region of TyeA, defined by amino-acid residues R19, W20 and D25, that is required to block Yop secretion but is not required to interact with YopN, suggests that TyeA plays a direct role in blocking secretion independent of its binding to YopN. In fact, the R19, W20 and D25 amino-acid residues may be required for interactions that directly disrupt the T3S process. Conformation of this hypothesis awaits the identification of the target of the YopN/TyeA complex.

Analysis of the YopN-TyeA crystal structure revealed that the primary contacts between these proteins are mediated by hydrophobic interactions between amino-acid residues in the YopN C-terminal helix and helices one and three of TyeA (Schubot et al. 2005) (Fig. 1B). The TyeA (M51A/F55A) protein possesses alanine substitutions that disrupt hydrophobic contacts between TyeA M51 and YopN F278, as well as TyeA F55 and YopN V271. This protein did not co-purify with the GST-$YopN^{78\text{-}293}$ protein, confirming that TyeA M51 and TyeA F55 are required to form critical contacts between these two proteins. Similarly, the TyeA (S6A G10A V13A) protein carries alanine substitutions that disrupt hydrophobic contacts between TyeA S6 and YopN W216, TyeA G10 and YopN Y213 and TyeA V13 and YopN I212 (Schubot et al. 2005). The TyeA (S6A G10A V13A) protein, like the Tye (M51A F55A) protein, failed to efficiently co-purify with the GST-$YopN^{78\text{-}293}$ protein, indicating that hydrophobic contacts between the first helix of TyeA and helix nine of YopN also play an important role in the interaction of TyeA with YopN. The fact that both of these mutant TyeA proteins cannot properly regulate Yop secretion, indicates that the interaction of TyeA with YopN is required to regulate Yop secretion.

21.5 Acknowledgements

This work was supported by Public Health Service Grant AI39575 from the National Institutes of Health to G.V.P.

21.6 References

Ali, S.A. and Steinkasserer, A. (1995) PCR-ligation-PCR mutagenesis: a protocol for creating gene fusions and mutations. Biotechniques 18, 746-750.

Cambau, E. Bordon, F. Collatz, E. and Gutmann, L. (1993) Novel *gyrA* point mutation in a strain of *Escherichia coli* resistant to fluoroquinolones but not to nalidixic acid. Antimicrob. Agents. Chemother. 37, 1247-1252.

Cheng, L.W. and Schneewind, O. (2000) *Yersinia enterocolitica* TyeA, an intracellular regulator of the type III machinery, is required for specific targeting of YopE, YopH, YopM, and YopN into the cytosol of eukaryotic cells. J. Bacteriol. 182, 3183-3190.

Cheng, L.W. Kay, O. and Schneewind, O. (2001) Regulated secretion of YopN by the type III machinery of *Yersinia enterocolitica.* J. Bacteriol. 183, 5293-5301.

Cornelis, G.R. (2000) Molecular and cell biology aspects of plague. Proc. Natl. Acad. Sci. USA 97, 8778-8783.

Day, J.B. Ferracci, F. and Plano, G.V. (2003) Translocation of YopE and YopN into eukaryotic cells by *Yersinia pestis yopN*, *tyeA*, *sycN*, *yscB* and *lcrG* deletion mutants measured using a phosphorylatable peptide tag and phosphospecific antibodies. Mol. Microbiol. 47, 807-823.

DeLano, W. L. (2001) The PyMOL Molecular Graphics System. DeLano Scientific LLC, San Carlos, CA, USA.

Fallman, M. and Gustavsson, A. (2005) Cellular mechanisms of bacterial internalization counteracted by *Yersinia.* Int. Rev. Cytol. 246, 135-188.

Ferracci, F. Day, J.B. Ezelle, H.J. and Plano, G.V. (2004) Expression of a functional secreted YopN-TyeA hybrid protein in *Yersinia pestis* is the result of a +1 translational frameshift event. J. Bacteriol. 186, 5160-5166.

Ferracci, F. Schubot, F.D. Waugh, D.S. and Plano, G.V. (2005) Selection and characterization of *Yersinia pestis* YopN mutants that constitutively block Yop secretion. Mol. Microbiol. 57, 970-987.

Forsberg, A. Viitanen, A.M. Skurnik, M. and Wolf-Watz, H. (1991) The surface-located YopN protein is involved in calcium signal transduction in *Yersinia pseudotuberculosis.* Mol. Microbiol. 5, 977-986.

Goguen, J.D. Yother, J. and Straley, S.C. (1984) Genetic analysis of the low calcium response in *Yersinia pestis* mu d1(Ap *lac*) insertion mutants. J. Bacteriol. 160, 842-848.

Hoiczyk, E. and Blobel, G. (2001) Polymerization of a single protein of the pathogen *Yersinia enterocolitica* into needles punctures eukaryotic cells. Proc. Natl. Acad. Sci. USA 98, 4669-4674.

Iriarte, M. Sory, M.P. Boland, A. Boyd, A.P. Mills, S.D. Lambermont, I. and Cornelis, G.R. (1998) TyeA, a protein involved in control of Yop release and in translocation of *Yersinia* Yop effectors. EMBO. J. 17, 1907-1918.

Iriarte, M. and Cornelis, G.R. (1999) Identification of SycN, YscX, and YscY, three new elements of the *Yersinia yop* virulon. J. Bacteriol. 181, 675-680.

Jackson, M.W. Day, J.B. and Plano, G.V. (1998) YscB of *Yersinia pestis* functions as a specific chaperone for YopN. J. Bacteriol. 180, 4912-4921.

Marenne, M.N. Journet, L. Mota, L.J. and Cornelis, G.R. (2003) Genetic analysis of the formation of the Ysc-Yop translocation pore in macrophages by *Yersinia enterocolitica*: role of LcrV, YscF and YopN. Microb.Pathog. 35, 243-258.

Matson, J.S. and Nilles, M.L. (2001) LcrG-LcrV interaction is required for control of Yops secretion in *Yersinia pestis*. J. Bacteriol. 183, 5082-5091.

Michiels, T. Wattiau, P. Brasseur, R. Ruysschaert, J.M. and Cornelis, G. (1990) Secretion of Yop proteins by yersiniae. Infect. Immun. 58, 2840-2849.

Mota, L.J. and Cornelis, G.R. (2005) The bacterial injection kit: type III secretion systems. Ann. Med. 37, 234-249.

Mueller, C.A. Broz, P. Muller, S.A. Ringler, P. Erne-Brand, F. Sorg, I. Kuhn, M. Engel, A. and Cornelis, G.R. (2005) The V-antigen of *Yersinia* forms a distinct structure at the tip of injectisome needles. Science 310, 674-676.

Navarro, L. Alto, N.M. and Dixon, J.E. (2005) Functions of the *Yersinia* effector proteins in inhibiting host immune responses. Curr. Opin. Microbiol. 8, 21-27.

Neyt, C. and Cornelis, G.R. (1999) Insertion of a Yop translocation pore into the macrophage plasma membrane by *Yersinia enterocolitica*: requirement for translocators YopB and YopD, but not LcrG. Mol. Microbiol. 33, 971-981.

Nilles, M.L. Fields, K.A. and Straley, S.C. (1998) The V antigen of *Yersinia pestis* regulates Yop vectorial targeting as well as Yop secretion through effects on YopB and LcrG. J. Bacteriol. 180, 3410-3420.

Pallen, M.J. Beatson, S.A. and Bailey, C.M. (2005) Bioinformatics, genomics and evolution of non-flagellar type-III secretion systems: a Darwinian perspective. FEMS Microbiol. Rev. 29, 201-229.

Perry, R.D. and Fetherston, J.D. (1997) *Yersinia pestis*--etiologic agent of plague. Clin. Microbiol. Rev. 10, 35-66.

Pettersson, J. Holmstrom, A. Hill, J. Leary, S. Frithz-Lindsten, E. von Euler-Matell, A. Carlsson, E. Titball, R. Forsberg, A. and Wolf-Watz, H. (1999) The V-antigen of *Yersinia* is surface exposed before target cell contact and involved in virulence protein translocation. Mol. Microbiol. 32, 961-976.

Schubot, F.D. Jackson, M.W. Penrose, K.J. Cherry, S. Tropea, J.E. Plano, G.V. and Waugh, D.S. (2005) Three-dimensional structure of a macromolecular assembly that regulates type III secretion in *Yersinia pestis*. J. Mol. Biol. 346, 1147-1161.

Skryzpek, E. and Straley, S.C. (1993) LcrG, a secreted protein involved in negative regulation of the low-calcium response in *Yersinia pestis*. J. Bacteriol. 175, 3520-3528.

Torruellas, J. Jackson, M.W. Pennock, J.W. and Plano, G.V. (2005) The *Yersinia pestis* type III secretion needle plays a role in the regulation of Yop secretion. Mol. Microbiol. 57, 1719-1733.

Yother, J. and Goguen, J.D. (1985) Isolation and characterization of Ca^{2+}-blind mutants of *Yersinia pestis*. J. Bacteriol. 164, 704-711.

Part IV – Pathogenesis and Host Interactions

Picture 20. Elisabeth Carniel presents research on mouse genetic suscepticility to plague. Photograph by A. Anisimov.

Picture 21. Jurgen Heesemann gives the closing lecture. Photograph by R. Perry.

22

The Insect Toxin Complex of *Yersinia*

Nick Waterfield[1], Michelle Hares[2], Stewart Hinchliffe[3], Brendan Wren[3], and Richard ffrench-Constant[2]

[1] Department of Biology and Biochemistry, University of Bath, bssnw@bath.ac.uk
[2] School of Biological Sciences, University of Exeter
[3] London School of Hygiene and Tropical Medicine

Abstract. Many members of the *Yersinia* genus encode homologues of insect toxins first observed in bacteria that are insect pathogens such as *Photorhabdus*, *Xenorhabdus* and *Serratia entomophila*. These bacteria secrete high molecular weight insecticidal toxins comprised of multiple protein subunits, termed the Toxin Complexes or Tc's. In *Photorhabdus* three distinct Tc subunits are required for full oral toxicity in insects, that include the [A], [B] and [C] types, although the exact stochiometry remains unclear. The genomes of *Photorhabdus* strains encode multiple *tc* loci, although only two have been shown to exhibit oral and injectable activity against the Hawk Moth, *Manduca sexta*. The exact role of the remaining homologues is unclear. The availability of bacterial genome sequences has revealed the presence of *tc* gene homologues in many different species. In this chapter we review the *tc* gene homologues in *Yersinia* genus. We discuss what is known about the activity of the *Yersinia* Tc protein homologues and attempt to relate this to the evolution of the genus and of the *tca* gene family.

22.1 Introduction: What is the "Insect Toxin Complex"?

The "toxin complex" insecticidal genes were first identified in *Photorhabdus luminescens* W14, a Gram-negative insect-pathogenic member of the Enterobacteriacea, which is symbiotically associated with insect-pathogenic nematodes from the family *Heterorhabditidae*. The Tc's from *P. luminescens* W14 were shown to have both oral and injectable toxicity against caterpillar pests. These were originally separated into four native complexes each approximately 1MDa in size termed Tca, Tcb, Tcc and Tcd, which are respectively encoded by the four loci *tca, tcb, tcc* and *tcd*, each of which has multiple *tc* genes (Bowen et al. 1998). The *tcd* locus, which was shown to contain three genes, *tcdA1*, *tcdB1* and *tccC2* is now known to be part of a much larger pathogenicity island containing many tandem repeats of *tc* gene homologues (Waterfield et al. 2001). It should be noted that of the many different *tc* homologues in *P. luminescens*, only the *tca* and *tcd* loci encode proteins that have been demonstrated to be orally toxic to the moth *Manduca sexta* and it is assumed that the other homologues are specific either to different hosts or tissues. Tc operons usually have genes belonging to three homology types, '*tcdA/tcaAB/tccAB*', '*tcdB/tcaC*' and '*tccC*' types, or [A], [B] and [C] for simplicity. Tc gene operons can be classified into one of three types based on their gene complement, these being; (i) *tca* , containing [A]+[B], (ii) *tcd*, containing [A]+[B]+[C] and (iii) *tcc*, containing [A]+[C]. Figure 1 illustrates the four *tc* loci initially identified in *P. luminescens* strain W14 (ffrench-Constant and Waterfield 2006).

Initial heterologous expression experiments of genes from the *tca* and *tcd* operons in *Escherichia coli* suggested that the presence of all three distinct protein subunits is required for the formation of a complex with full oral toxicity to *M. sexta*.

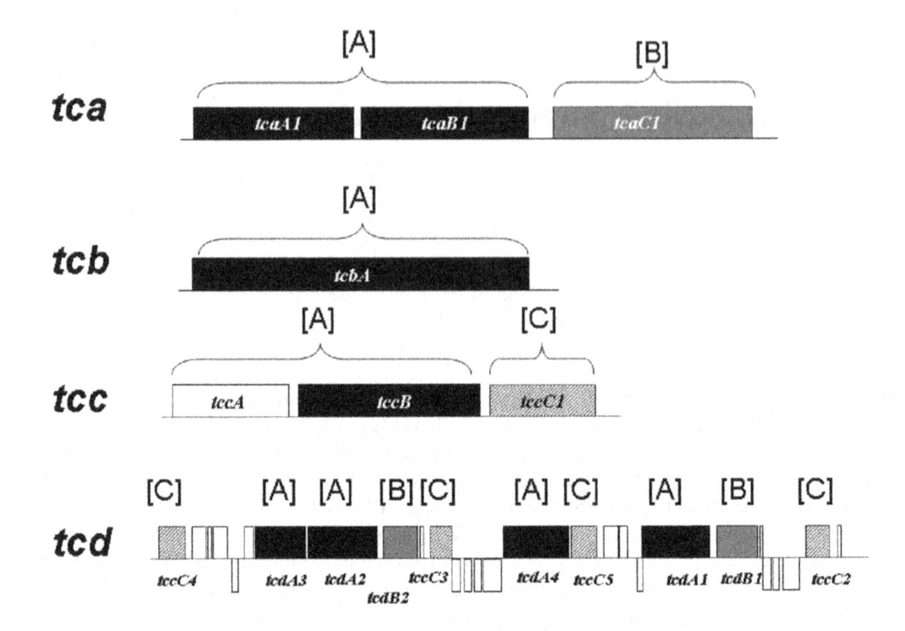

Fig. 1. The genetic organisation of four *tc* gene loci in *P. luminescens* W14. Boxes represent the open reading frames. For simplicity, the gene names are below the boxes and the homology-type of the *tc* gene is given above.

We now know that [A] and [B+C] subunits can both show toxicity independently, however *maximum* toxicity is only afforded by a complex of [A] and [B+C]. This makes a mature Tc structure a true "complex of toxins", although the exact stochiometry remains unclear (Waterfield et al. 2005).

22.2 Distribution of Toxin Complex Genes in Other Bacteria

Tc's have been demonstrated in other insect pathogens including the closely related *Xenorhabdus* (Morgan et al. 2001) (also a nematode symbiont) and the free living *Serratia entomophila.* (Dodd et al. 2006). However they have also been identified in a range of other bacteria. Not only in bacteria known to parasitise insects, such as *Pseudomonas entomophila* (Vodovar et al. 2006) and the Gram-positive *Paenibacillus* but also in cases where no such associations are known to exist. These include plant pathogenic members of the *Pseudomonas* (*syringae* and *fluorescens*), *Fibrobacter succinogens* and diverse *Burkholderia* species including the human pathogenic species. The presence of *tc* genes in diverse "soil associated" bacteria is highly suggestive that invertebrates may in fact play an as yet undefined role in the life histories of many of these bacteria. Interestingly a particular class of *tc*-homologue represented by a genetic fusion between a [B] and [C] like genes [B::C] are now turning up in a much wider range of organisms including *Shewanella*, *Rodococcus*, *Desulfotomaculum*, *Ralstonia* and even the plant pathogenic fungus *Giberella*. Of particular interest here is the discovery of *tc* gene homologues in members of the *Yersinia* (Waterfield et al. 2001).

22.3 *Yersinia tc* Gene Homologues

Genome sequencing of different *Yersinia* isolates first identified homologues of the *tca* and *tcc* genes in both *Yersinia pestis* and *Yersinia pseudotuberculosis*. Subsequent studies have confirmed the presence of *tc*-like genes in other members of the genus including some *Yersinia enterocolitica* isolates, *Yersinia mollaretii* and *Yersinia frederiksenii*. Interestingly, as is also the case with *Photorhabdus* the types of *tc* genes and their genomic context are variable across the genus. Figure 2 illustrates a comparison of the common genomic "backbone" locus which contains *tc* gene homologues in several different *Yersinia* species. Note that all the chromosomal *tc* operons are inserted into an equivalent location on the common *Yersinia* genomic

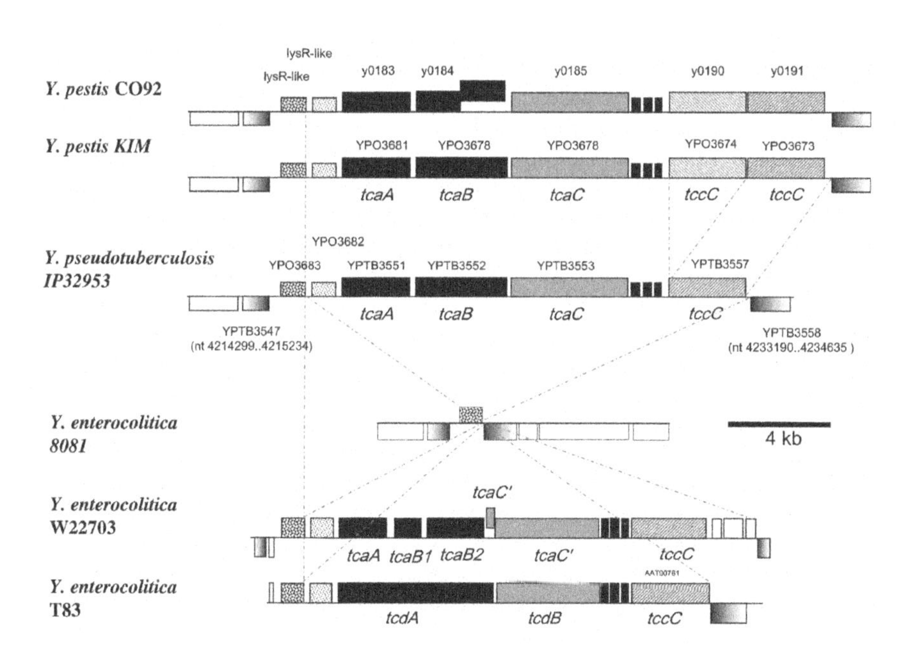

Fig. 2. The *tc* loci of the *Yersinia*. A single chromosomal locus contains the *tc* operons in diverse members of the *Yersinia*. *Y. pseudotuberculosis* and *Y. pestis* contain similar *tca* operons as expected. However *Y. enterocolitica* 8081 does not appear to have picked up this operon. Interestingly, this chromosomal locus must represent a 'hot spot' for *tc* gene horizontal acquisition as two other *Y. enterocolitica* strains have picked up different tc operon types.

backbone. This location is exemplified as between the genes YPTB3547 (nt 4214299..4215234) and YPTB3558 (nt 4233190..4234635) of the *Y. pseudotuberculosis* IP32953 genome.

22.3.1 The Chromosomal *tca* Loci of *Y. pseudotuberculosis*/ *Y. pestis* Clade

Y. pestis and *Y. pseudotuberculosis* isolates share a recent common ancestor (Achtman et al. 1999) which is reflected in conservation of the *tcaABC-tccC* gene arrangement ([A]+[B]+[C]). *Y. pestis* isolates on the other hand have a second *tccC* gene homologue and the suggestion of gene degradation in the *tcaAB* ([A]) genes, at least in the C092 isolate. Interestingly, *Y. mollaretii* also has a *tca* operon in this common genomic location, which is very similar to that of *Y. pseudotuberculosis*, suggesting that it is likely to have diverged from the common ancestor after the acquisition of the *tca* operon.

22.3.2 The *tc* Loci of *Y. enterocolitica*-like Strains

It is unlikely that the *tca*-locus is ancestral to the whole *Yersinia* genus as the equivalent *tc* genomic loci in *Yersinia bercovieri*, *Yersinia intermedia*, *Y. enterocolitica* 8081 and *Y. frederiksenii* are absent. We suggest that these strains maintain the ancestral genetic arrangement at this locus. Interestingly, *Y. enterocolitica* W22703 and *Y. enterocolitica* T83 do have *tc* operons at this location, but importantly they are dissimilar both to each other and to the *tca*-like operons of the *Y. pseudotuberculosis*-like isolates. What factors have “targeted” these different *tc* operons to this particular locus in the *Yersinia* backbone is unknown and difficult to explain. Furthermore, while *Y. frederiksenii* does not have this *tca*-locus, at least one isolate has acquired plasmid encoded close homologues of the *S. entomophila sepABC* genes (Dodd et al. 2006). This is more similar to a *tcd*-like locus, with the [A] subunit being encoded by a single continuous ORF, unlike the *tca* [A] elements which are split into two genes; *tcaA* and *tcaB*. The chromosomal *Y. enterocolitica* T83 *tc* operon also has a *tcd* organisation and in this case there is evidence to suggest they are involved in the colonisation of the mouse gut (Tennant, Skinner, Joe and Robins-Browne 2005). Conversely, the *tc* genes of *Y. enterocolitica* W22703 are only expressed at low temperatures (10°C) and have been shown to exhibit oral toxicity to insects. This operon, which has a *tca*-like organisation, also shows some evidence of undergoing degradation, with an apparent frameshift in the *tcaC* gene ([B] subunit) and the splitting of the *tcaB* gene into two separate ORFs. Experiments in this species suggest that the *tcaA* gene alone is responsible for mediating insect toxicity in this operon (Bresolin et al. 2006).

22.4 Biological Activities of the *tca* Genes From the *Y. pseudotuber*-culosis/Y. pestis Clade

The evidence from *Y. enterocolitica* suggests that *tc* genes can serve a role in either mammalian or invertebrate infections. We have therefore examined the role of the more conserved *tca*-like loci of the *Y. pseudotuberculosis/Y. pestis* clade.

22.4.1 Sequence of *tc* Genes From Different *Y. pseudotuberculosis* Strains

Microarray analysis of 10 diverse strains of *Y. pseudotuberculosis* revealed a high level of variability in *tcaB* and *tcaC* (Hinchliffe et al. 2003). We used Hi-fidelity PCR to clone *tc* ORFs from a range of different *Y. pseudotuberculosis* strains. Amino acid alignment of these sequences revealed that TcaC was reasonably well conserved between strains, but TcaB showed significant variation, especially in the region between amino acids 201-426. TcaA also showed some variability with several different C-terminal regions. These observations suggest that the *tc* genes may be under diversification selection (Hares et al. 2006).

22.4.2 Insect Oral Toxicity of *Y. pseudotuberculosis* Strains

In order to determine if any of the *Y. pseudotuberculosis* strains were capable of oral toxicity against model insects we tested them with our *M. sexta* (hawk moth) neonate oral bioassay. In order to replicate the experiments done with *Y. enterocolitica* W22703 (Bresolin et al. 2006), we tested aliquots of whole cultures grown at 10, 30 and 37°C in liquid culture (Fig. 3). We used whole live cultures as it is unknown how (or when) the *tca* genes are expressed in these strains. For example; the Tca could be secreted free into the growth medium, as in the case for both Tca and Tcd in *P. luminescens* W14, or it may remain cell associated, as in the case for Tcd in *P. luminescens* TT01 (unpublished data). Furthermore it is also possible that more specific factors, such as host cell contact, is required to trigger *tca* expression.

Interestingly we saw weak oral toxicity of several strains but no obvious correlation with growth temperature between strains (Hares et al. 2006). Obviously many different strain specific factors could contribute to these minor deleterious effects on the insects, but in no case did we observe strong oral toxicity, such as is mediated by *P. luminescens* W14 Tca. It remains possible that the *tca* genes were not being expressed in these experiments so we decided to sub-clone them for controlled heterologous expression in *E. coli* as has been done for *Photorhabdus tc* toxin genes.

22.4.3 Heterologous Expression and Insect Toxicity of *Y. pseudo*-tuberculosis and Y. pestis tca Genes

The *tca* genes from the sequenced *Y. pseudotuberculosis* IP32953 strain were PCR amplified and cloned into the arabinose inducible expression plasmid pBAD30. The *tcaAB* genes were cloned as a single fragment (the [A] subunit), but with the native SD sequence of *tcaA* replaced by an artificial optimal consensus SD sequence. In addition, the *tcaC-tccC* fragment was PCR cloned, again replacing the native *tcaC* SD with an artificial sequence (the [BC] subunit). The *Y. pestis* KIM *tc* genes were also cloned for heterologous expression in *E. coli* using the same strategy.

Manduca neonate oral bioassays suggested only very weak oral toxicity of the heterologously produced IP32953 Tc proteins when all components were present, *i.e.*; [A]+[B+C] (Fig. 4). This level was similar to that seen when whole cultures were tested. As a control, heterologously expressed Tc toxin from *P. luminescens* can be seen to be highly orally toxic to *M. sexta*. Note the *tc* genes from *Y. pestis* also show little or no effect on the *Manduca*, although their expression level was lower (Hares et al. 2006). Interestingly, while heterologously expressed *Photorhabdus* Tc toxins are also highly toxic to the rat flea vector of *Y. pestis*, *Xenopsylla cheopis*, those of either *Y. pseudotuberculosis* or *Y. pestis* KIM strains have no deleterious effect when included in blood meals. Conversely, live *Y. pseudotuberculosis* cells can cause a lethal infection when fed to the flea, although this is not affected by deletion of the *tca* loci. Similarly, a *tca* mutant in *Y. pestis* does not appear to affect

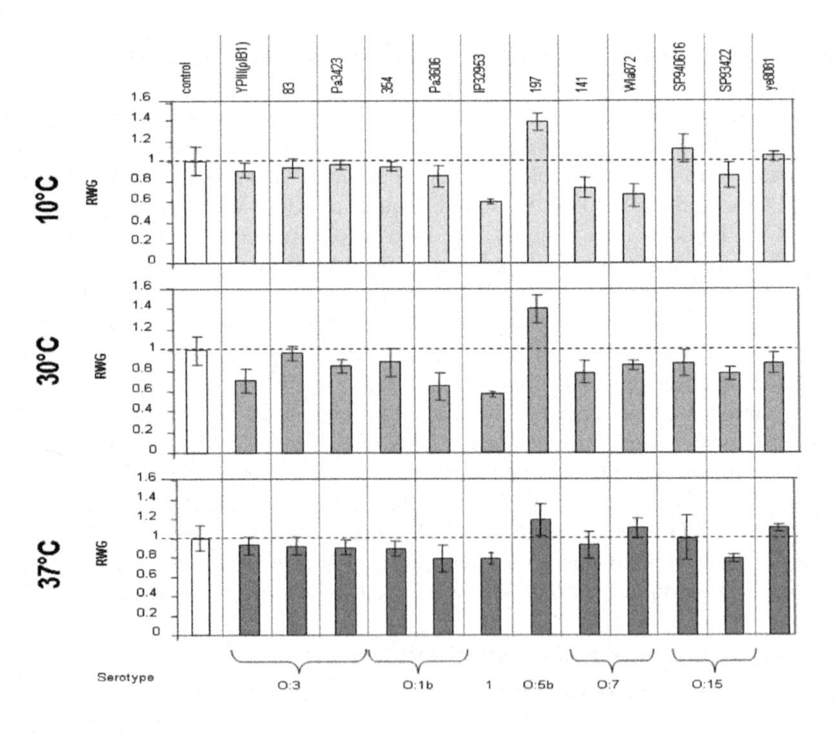

Fig. 3. Oral toxicity of overnight cultures of *Y. pseudotuberculosis* (and *Y. enterocolitica* 8081) to *M. sexta*. Toxicity relative to a control *E. coli* strain is shown (RWG).

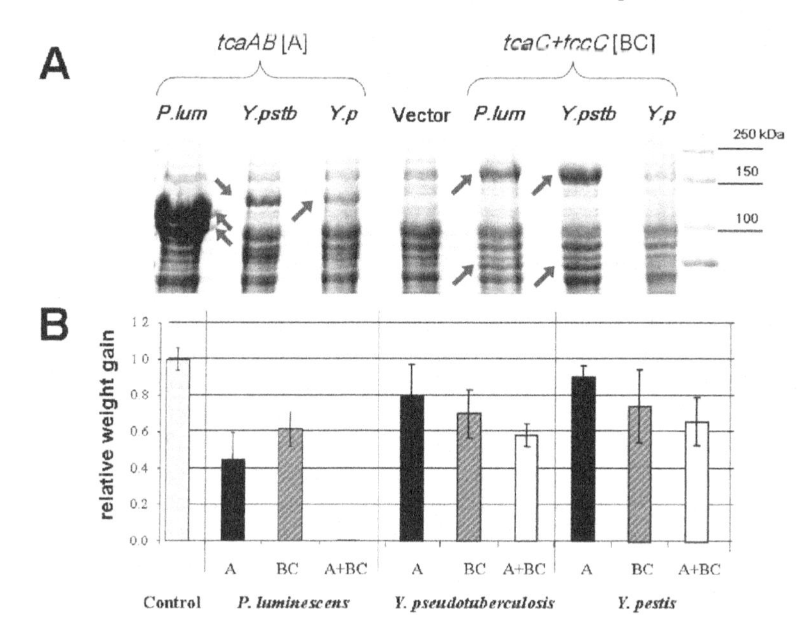

Fig. 4. (A) SDS-PAGE of cell lysates of heterologously expressed Tc proteins in *E. coli* BL21 at 15°C from *P. luminescens* (*P.lum*), *Y. pseudotuberculosis* strain IP32953 (*Y. pstb.*) and *Y. pestis* strain KIM10+ (*Y. p.*). Tc proteins are indicated by arrows. (B) *M. sexta* oral bioassay data of heterologously expressed Tc protein. The Y-axis represents mean larval weight gain relative to a vector control treatment (RWG).

the formation of biofilm on the insect proventriculus, although they are expressed after the formation of the biofilm. It has been suggested that they may play a role upon the initial entry into the mammalian host (DL Erickson and BJ Hinnebusch, personal communication). Together these observations suggest no obvious role for the *Y. pseudotuberculosis*/*Y. pestis* Tca in bacterial/insect interactions.

22.4.4 Temperature Dependence of *tca* Gene Transcription in *Y. pseudotuberculosis* IP32953

Reverse transcriptase (RT) PCR was used to determine the temperature dependence of *tc* gene transcription in *Y. pseudotuberculosis* IP32953. Interestingly, unlike the *tc* genes of *Y. enterocolitica* W22703, we saw no expression of the *tcaABC* genes at 15°C, but we did see transcription of the *tccC* gene. When the temperature was increased to either 30 or 37°C the *tcaABC* genes were also transcribed (Hares et al. 2006). This pattern of expression argues that the Tca complex is not expressed at "soil", temperatures but may be expressed in a mammalian host. The presence of the *tccC* transcript at low temperature suggests that either it has a regulatory role or it is also used for an independent function. In order to test the hypothesis that the

Y. pseudotuberculosis IP32953 Tc toxins are active in the mammalian gut we tested their effect on cultured caucasian colon adenocarcinoma (CACO-2) cells.

22.4.5 Activity of the *Y. pseudotuberculosis* Toxin Complex on Human Gut Cells

Heterologously expressed Tc toxins from *P. luminescens* W14, *Y. pseudotuberculosis* IP32953 and *Y. pestis* KIM+ were topically applied to cultured human gut cells to determine if they had any effect. Interestingly, the *Y. pseudotuberculosis* Tca caused initial membrane ruffling leading to the formation of many small vacuoles, (reminiscent of hyper pinocytosis) in over 41% of the cells examined (Fig. 5). This effect was seen only when the full [A]+[BC] complex was used. The Tca from *P. luminescens* W14 (13.5%) and *Y. pestis* (6%) had no strong effect, giving results comparable to an *E. coli* cytosolic negative control preparation (5%) (Hares et al. 2006).

Taking these observations together, we hypothesize that the *tca* operon in the *Y. pseudotuberculosis* lineage has been adapted to function in the mammalian host. In the case of *Y. pseudotuberculosis* (and the Tc's of *Y. enterocolitica* T83) they likely serve to "modify" the gut epithelium and in the case of *Y. pestis*, they may be involved in the initial invasion of the mammalian host after transmission by fleas.

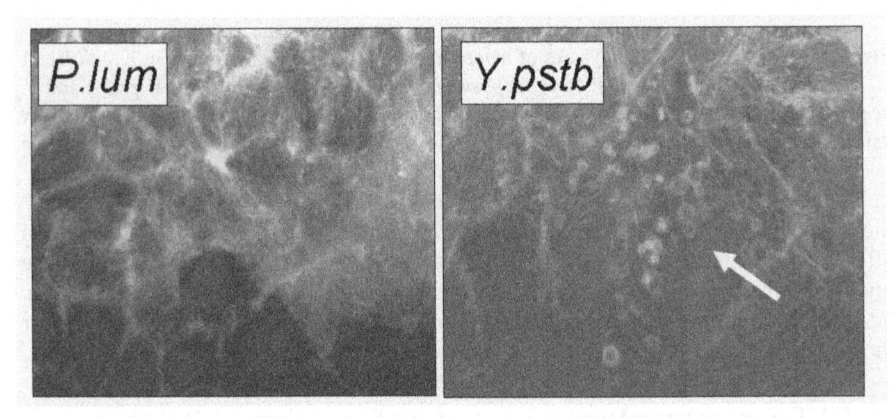

Fig. 5. Cytosolic preparations were made from heterologously expressed *P. luminescens* W14 and *Y. pseudotuberculosis* IP32953 *tcaAB* ([A]) and *tcaC+tccC* ([BC]) and topically applied to confluent mammalian CACO-2 gut cells. Cells were fixed and nuclei stained with Hoechst and the actin stained with phalloidin-rhodamine (B) Vacuoles and actin depolarisation were observed in ~40% of the cells for *Y. pseudotuberculosis* (arrow).

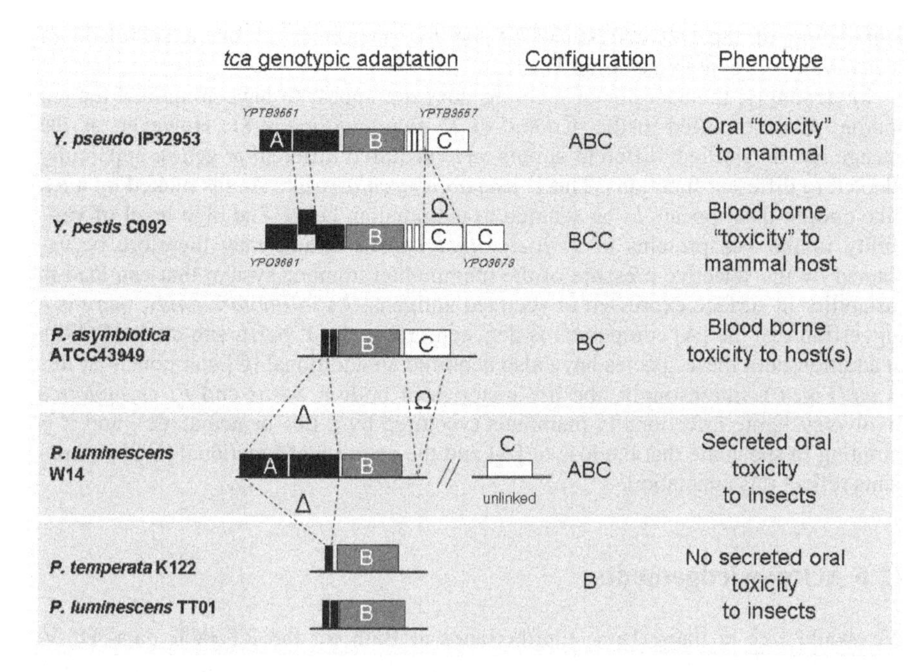

Fig. 6. A model showing the re-modeling of the tca operon in *Photorhabdus* and the *Y. pseudotuberculosis* clade of the *Yersinia*. We suggest that the loss of [A] components and the acquisition of [C] genes is related to changing selection pressures from different phenotypic requirements. Δ=deletion, Ω=insertion.

22.5 Conclusion: The Evolution of *tca* Loci

The presence of phage-associated genes in most *tc* operons and the location of the *sepABC* genes of *S. entomophila* on the conjugative plasmid pADAP (Hurst et al. 2000), suggests possible mechanisms of horizontal transfer. The demonstration that the *tc* genes of *Y. pseudotuberculosis* and *Y. enterocolitica* T83 are active against mammalian gut cells also indicates that these toxins can be adapted to more than just insect toxicity.

A comparison of the *tca* operons of the *Yersinia* with those of the *Photorhabdus* together with an understanding of the life histories of these bacteria may provide clues to the selection pressures that are driving the evolution of these complex toxins. Limited micro-array data (Marokhazi et al. 2003), genome sequencing and the analysis of *tc* containing cosmids from examples of the three different species of *Photorhabdus*; *P. luminescens* (W14 and TT01), *Photorhabdus temperata* (which are both insect pathogens) and *Photorhabdus asymbiotica* (which infects insects and man - Gerrard et al. 2006) has allowed us to build up a picture of the distribution and evolution of *tca* genes in this genus. Data suggests that the ancestral *Photorhabdus* possessed an intact *tca*-like operon. As the three species diversified there has been degradation of the [A] component of the *tca* operon in all but a sub-clade of *P. luminescens* (containing strain W14).

Interestingly, the ancestor of *Yersinia* does not appear to have contained the *tca* operon; as exemplified in the genome of *Y. enterocolitica* 8081. However, as the lineage has diversified, different strains have acquired different *tc* genes, apparently adapted to different functions. The *Y. pseudotuberculosis*-like strains picked up a *tca* -like operon that appears to be adapted to mammalian hosts. The high level of variability in the Tca proteins of *Y. pseudotuberculosis* strains may therefore be explained by the selective pressure of the mammalian immune system that can lead to variability in surface expressed or secreted antigens. As in *Photorhabdus*, there is a suggestion that the [A] component is degenerating in the *Y. pestis* sub-clade (CO92). In addition both these species have also acquired an additional [C] component at this locus (Fig. 6). Interestingly, the life histories of both *Y. pestis* and *P. asymbiotica* involve systemic infections in mammals (vectored by a flea or nematode), and it is tempting to speculate that the loss of [A] and the creation of functional [BC] components reflect this adaptation.

22.6 Acknowledgements

We would like to thank Dr I. Eleftherianos at Bath for the RT PCR data, Dr V. Pinheiro at Cambridge University for the *Y. pseudotuberculosis tc* expression constructs and Dr Andrea Dowling at Exeter for assistance with the tissue culture experiments.

22.7 References

Achtman, M., Zurth, K., Morelli, G., Torrea, G., Guiyoule, A. and Carniel, E. (1999) *Yersinia pestis*, the cause of plague, is a recently emerged clone of *Yersinia pseudotuberculosis*. Proc. Natl. Acad. Sci. USA 96, 14043-14048.

Bowen, D., Rocheleau, T.A., Blackburn, M., Andreev, O., Golubeva, E., Bhartia R. and ffrench-Constant, R. H. (1998) Insecticidal toxins from the bacterium *Photorhabdus luminescens*. Science 280, 2129-2132.

Bresolin, G., Morgan, J.A., Ilgen, D., Scherer, S. and Fuchs, T.M. (2006) Low temperature-induced insecticidal activity of *Yersinia enterocolitica*. Mol. Microbiol. 59, 503-512.

Dodd, S. J., Hurst, M.R., Glare, T.R., O'Callaghan, M. and Ronson, C.W. (2006) Occurrence of sep insecticidal toxin complex genes in *Serratia* spp. and *Yersinia frederiksenii*. Appl. Environ. Microbiol. 72, 6584-6592.

ffrench-Constant R. and Waterfield, N. (2006) An ABC guide to the bacterial toxin complexes. Adv. Appl. Microbiol. 58, 169-183.

Gerrard, J G., Clarke D.J., ffrench-Constant, R.H., Nimmo, G.R., Looke, D.F.M., Feil, E., Pearce, L. and Waterfield, N.R. (2006) A human pathogen from the invertebrate world: isolation of a nematode symbiont for *Photorhabdus asymbiotica*. Emerg. Infect. Diseases In press.

Hares, M., Pinheiro, V., Hinchliffe, S., Wren, B., Ellar, D., ffrench-Constant, R.H. and Waterfield, N.R. (2006) The toxin complex proteins of Yersinia pseudotuberculosis are active against mammalian gut cells. In preparation.

Hinchliffe, S.J., Isherwood, K.E., Stabler, R.A, Prentice, M.B. Rakin, A., Nichols, R.A., Oyston, P.C., Hinds, J., Titball, R.W. and Wren, B.W. (2003) Application of DNA microarrays to study the evolutionary genomics of *Yersinia pestis* and *Yersinia pseudotuberculosis*. Genome Res. 13, 2018-2029.

Marokhazi, J., Waterfield, N., LeGoff, G., Feil, E., Stabler, R., Hinds, J., Fodor A. and ffrench-Constant, R. H. (2003) Using a DNA microarray to investigate the distribution of insect virulence factors in strains of *Photorhabdus* bacteria. J. Bacteriol. 185, 4648-4656.

Morgan, J.A., Sergeant, M., Ellis, D., Ousley, M. and Jarrett, P (2001) Sequence analysis of insecticidal genes from *Xenorhabdus nematophilus* PMFI296. Appl. Environ. Microbiol. 67, 2062-2069.

Tennant, S.M., Skinner, N.A., Joe, A. and Robins-Browne, R.M. (2005) Homologues of insecticidal toxin complex genes in *Yersinia enterocolitica* biotype 1A and their contribution to virulence. Infect. Immun. 73, 6860-6867.

Vodovar, N., Vallenet, D., Cruveiller, S., Rouy, Z., Barbe, V., Acosta, C., Cattolico, L., Jubin, C., Lajus, A., Segurens, B., Vacherie, B., Wincker, P., Weissenbach, J., Lemaitre, B., Medigue, C. and Boccard, F. (2006) Complete genome sequence of the entomopathogenic and metabolically versatile soil bacterium *Pseudomonas entomophila*. Nat Biotechnol. 24, 673-679.

Waterfield, N., Hares, M., Yang, G., Dowling, A. and ffrench-Constant, R. (2005) Potentiation and cellular phenotypes of the insecticidal toxin complexes of *Photorhabdus* bacteria. Cell. Microbiol. 7, 373-382.

Waterfield, N.R., Bowen, D.J., Fetherston, J.D., Perry, R.D. and ffrench-Constant, R.H. (2001) The *tc* genes of *Photorhabdus*: a growing family. Trends Microbiol. 9, 185-191.

23
Twin Arginine Translocation in *Yersinia*

Moa Lavander[1,2], Solveig K. Ericsson[2], Jeanette E Bröms[1,2], and Åke Forsberg[1,2]

[1] Department of Molecular Biology, Umeå University
[2] Department of Medical Countermeasures, Division of NBC Defence, Swedish Defence Research Agency, moa.lavander@foi.se

Abstract. Bacteria utilise Twin arginine translocation (Tat) to deliver folded proteins across the cytoplasmic membrane. Disruption of Tat typically results in pleiotropic effects on e.g. growth, stress resistance, bacterial membrane biogenesis, motility and cell morphology. Further, Tat is coupled to virulence in a range of pathogenic bacteria, including species of *Pseudomonas*, *Legionella*, *Agrobacterium* and *Mycobacterium*. We have investigated this, for *Yersinia*, previously unexplored system, and have shown that the Tat pathway is functional and absolutely required for virulence of *Yersinia pseudotuberculosis*. A range of putative *Yersinia* Tat substrates have been predicted *in silico*, which together with the Tat system itself may be interesting targets for future development of antimicrobial treatments. Here we present a brief review of bacterial Tat and discuss our results concerning this system in *Yersinia.*

23.1 Bacterial Twin Arginine Translocation (Tat)

The twin arginine translocation (Tat) system has attracted a vast amount of research interest during the past decade. A lot of the curiosity surrounding the Tat system stems from its unique ability to export proteins in an already folded state across the bacterial inner membrane. The Tat pathway was originally discovered in plant chloroplasts, where is supports protein transport from stroma into thylakoids. Energized by the transmembrane H+ gradient, this system was denoted the 'ΔpH pathway' (Dalbey and Robinson 1999; Settles et al. 1997; Settles and Martienssen 1998). In bacteria, Tat was originally referred to as 'Membrane targeting and translocation' Mtt (Weiner et al. 1998) and shortly thereafter as 'Twin arginine translocation', a name that stuck (Sargent et al. 1998).

23.1.1 The Tat Translocase

In Gram-negative bacteria the typical Tat translocase, as characterised in the model organism *Escherichia coli*, is mainly built up from the components TatA, TatB and TatC encoded by the *tatABCD* operon. The remaining gene product, TatD, has been shown to be a cytoplasmic DNase redundant for Tat function, and without any known connection to the system apart from its co-localisation with the operon (Wexler et al. 2000). While TatABC are required for functional Tat, a fourth component, TatE, encoded by a monocistronic gene is redundant (Ize et al. 2002; Sargent et al. 1998; Yahr et al. 1995).

TatA and TatB are both inserted into the bacterial inner membrane and share largely similar structures: an N-terminal membrane spanning hydrophobic α-helix and a C-terminal amphipatic α-helix protruding into the bacterial cytoplasm (Palmer and Berks 2003; Porcelli et al. 2002; Sargent et al. 2001). TatC, the most conserved Tat component, has six membrane spanning domains, the N- and C-termini of the protein being located in the cytoplasm (Behrendt et al. 2004; Ki et al. 2004).

The translocation of Tat substrates is hypothesized to occur in the following manner: TatC forms a 1:1 complex with TatB that may include low levels of TatA, although the bulk of TatA forms a homooligomeric channel, all in the inner membrane (Bolhuis et al. 2000; de Leeuw et al. 2002; Oates et al. 2003; Porcelli et al. 2002; Sargent et al. 2001). TatC is responsible for recognition of the Tat signal sequence (see below) and when a substrate is present the TatA complex is tran siently recruited to TatBC, providing a conduit for substrate export to the periplasmic space (Alami et al. 2003; Jongbloed et al. 2000). The translocation event occurs independently of ATP and is energised by the transmembrane proton gradient (Yahr and Wickner 2001). Recent findings have shown that TatA forms complexes of different sizes, probably assembled from modules of three or four TatA subunits. This could allow the Tat translocation channel to adapt to the size of the substrate at hand (Gohlke et al. 2005; Oates et al. 2005).

23.1.2 The Tat Signal Peptide

Substrates exported via the Tat system are recognised by an N-terminal signal sequence, between 26 and 58 amino acids long, compared to 18-26 amino acids for the classical Sec signal peptides (Berks 1996; von Heijne 1985). Further, while the Sec signal peptide is devoid of consensus sequences, Tat substrates typically carry a (S/T)-RR-x-FLK motif (Berks 1996). This conserved sequence has a very strong amino acid bias, the twin arginines (RR) giving the system its name being the most invariable, but not completely irreplaceable for Tat function (Ize et al. 2002). Substrates with atypical signal sequences, including variations of the signature twin arginines have been found: *Salmonella enterica* TtrB has a motif, S-KR-Q-FLQ, where one of the two arginines is missing (Hinsley et al. 2001). Also, RNR instead of the typical RR is seen for the Rieske Fe/S protein, an essential component of the photosynthetic electron transport chain of chloroplasts in *Spinacia oleracea* (Molik et al. 2001). Similar to Sec, the Tat signal peptide is commonly cleaved off after successful export to the periplasmic space (Berks 1996).

23.1.3 Faith and Function of the Tat Substrates

In contrast to Sec, Tat exports substrates that acquire their folding within the bacterial cytoplasm (Bogsch et al. 1998; Pugsley 1993). The Tat pathway is believed to have evolved for delivery of cofactor-containing proteins across the plasma-membrane, and the vast majority of Tat substrates in *E. coli* undergo cytosolic cofactor incorporation (Berks et al. 2003; Halbig et al. 1999; Palmer et al. 2005). Cofactorless substrates also exist, as exemplified by halophilic archaea. These organisms have extremely high salt levels in their cytosol, and rapid folding is required to prevent protein damage under these conditions. As a result, the majority of exported proteins in these archaea take the Tat route (Hutcheon and Bolhuis 2003; Rose et al. 2002). A third variant of Tat substrates are proteins that lack the typical signal peptide, but still are routed to Tat by a "piggy-back" mechanism. In other words, they form heteromeric complexes in the cytoplasm with Tat-signal containing partners (Palmer et al. 2005).

Proteins translocated by Tat to the periplasm either remain there, are integrated into the outer or inner membrane or secreted to the extrabacterial milieu (Hatzixanthis et al. 2003; Ochsner et al. 2002; Sargent et al. 1998; Sargent et al. 2002).

Tat has been shown to be connected to the ability to cause disease in a number of bacteria, for example phospholipase toxins exported across the outer membrane via type II secretion in *Legionella pneumophila* and *Pseudomonas aeruginosa* have been shown to be Tat substrates (Rossier and Cianciotto 2005; Voulhoux et al. 2001). Disruption of Tat by mutagenesis commonly results in strains displaying pleiotropic defects. Apart from impaired phospholipase delivery, a *P. aeruginosa* Tat mutant was also deficient for motility, biofilm formation, ability to endure osmotic stress, respiration under anaerobic conditions, iron-metabolism and was also shown to be attenuated for virulence in a rat lung model (Ochsner et al. 2002).

23.2 *Yersinia* Twin Arginine Translocation

We have shown that both *Yersinia pestis* and *Yersinia pseudotuberculosis* express a functional Tat system, which in *Y. pseudotuberculosis* is required for motility, acid resistance and virulence. Several putative Tat substrates can further be identified within the genomes of both *Yersinia* species. Future investigations will identify which are true Tat substrates and in what way they contribute to the observed phenotypes (Lavander et al. 2006).

23.2.1 *Yersinia* Tat is Functional

The genome sequences of *Y. pseudotuberculosis* strain IP32953 (Chain et al 2004), and *Y. pestis* KIM5 (Deng et al. 2002) reveal a chromosomal region encoding four genes with high homology to the *E. coli* genes *tatA, tatB, tatC* and *tatD* (Fig. 1), with similar genetic organisation to that of *E. coli* (Berks et al. 2000), and probably sharing a common promoter. Further, the *Yersinia* genomes also contain the monocistronic gene encoding TatE, which has overlapping functions with TatA and is believed to be the result of gene duplication (Sargent et al. 1998).

Functionality of the *Yersinia* Tat systems could be confirmed utilising a GFP reporter fused to the *E. coli* Tat signal peptide from TorA (Ding and Christie 2003). This reporter was introduced in *Y. pseudotuberculosis* IP32953 and *Y. pestis* KIM5 as well as in *tatC* mutants derived from these two strains. Thus, specific targeting of GFP to the periplasmic space could be observed for both *Y. pseudotuberculosis* and *Y. pestis*. In contrast, with the GFP reporter expressed in the *tatC* mutant strains, the

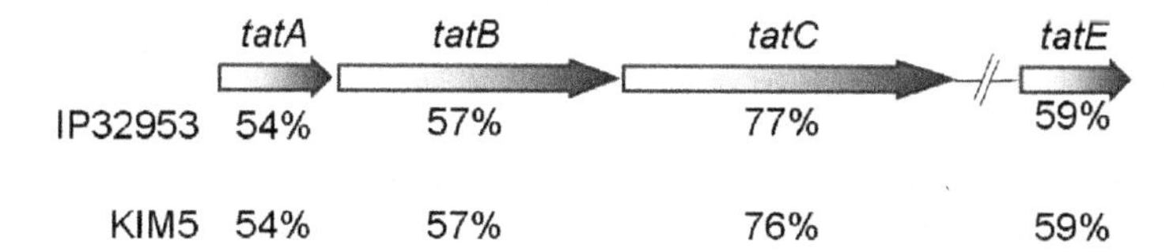

Fig. 1. Schematic representation of the *tatABC* operon and the monocistronic *tatE* gene of *Y. pseudotuberculosis* IP32953 and *Y. pestis* KIM5. The amino acid identity shared with the corresponding gene products in *E. coli* K12 is indicated. Genebank accession numbers were IP32953/KIM: YP_068804/NP_667790 (*tatA*), YP_068805/NP_667791 (*tatB*), YP_068806/NP_667792 (*tatC*), and YP_069628/NP_668496 (*tatE*). The amino acid identities between the *Yersinia* proteins and their *E. coli* K12 homologues were determined with pBlast.

fluorescence was restricted to the bacterial cytoplasm, revealing that Tat is non-functional for these strains. This shows that *Yersinia* indeed encodes a functional Tat system and that this system can be disrupted by deletion of *tatC* (Lavander, et al. 2006). Not surprisingly, the closely related *Yersinia. enterocolitica* also encodes a complete Tat system (Sanger institute; www.sanger.ac.uk/Projects/Y_enterocolitica), although its functionality remains to be confirmed.

23.2.2 A Connection Between Tat and Type III Secretion?

Our first incentive to study the *Yersinia* Tat system was due to an intriguing observation of a connection between Tat and the type III secretion (T3S) system as reported by Michael Vasil and coworkers (unpublished results). Their studies on the opportunistic pathogen *P. aeruginosa* revealed that T3S requires a functional Tat system. Despite being pathogens of dissimilar life styles, with very different outcomes of infection, the virulence plasmid encoded T3S system (T3SS) of *Yersinia* is remarkably similar to the chromosomally encoded system of *P. aeruginosa*, with a number of components being functionally interchangeable between the two systems (Bröms et al. 2003a, b; Frithz-Lindsten et al. 1998). With Tat in the inner membrane and the T3S apparatus spanning the bacterial envelope, crosstalk between the two systems also in *Yersinia* was a distinct possibility, in particular since PscO, the YscO homologue, of *P. aeruginosa* was considered a putative Tat substrate (M. Vasil, unpublished results). As seen in Table 1 both these proteins possess a pair of twin arginines (bold) in the N-terminus.

Table 1. N-terminal amino acid sequences of PscO and YscO

Protein	Residue 1-36
PscO	1-MSLALLLRV**RR**LRLDRAERAQGRQLLRVRAAAQEHT---
YscO	1-MI**RR**LHRVKVLRVERAEKAIKTQQACLQAAHRRHQE---

Genebank accession numbers: PscO: NP_250387; YscO: CAF25411.

However, with increased understanding of the Tat signal peptide, more precise predictions of putative substrates can now be made. Analysis of these two putative signal sequences with the TatP software suggests that they are not true Tat substrates (Bendtsen et al. 2005; www.cbs.dtu.dk/services/TatP). This implies that the Tat system of *P. aeruginosa* may affect the function of the T3SS in some other way than by impairing secretion of PscO. Yet, it is important to consider that the TatP software is based on the current knowledge of Tat which is mainly derived from studies of *E. coli*. It is conceivable that the Tat signal sequence can vary more widely, than this program allows. Nonetheless, our search for a Tat-T3SS connection in *Yersinia* turned out negative. The *tatC* mutant was indistinguishable from the isogenic wild type strain with respect to Yop expression and secretion, ability to induce a YopE mediated cytotoxic effect on HeLa cells as well as to block phagocytosis by J774.1 macrophage-like cells (Lavander et al. 2006).

23.2.3 *Yersinia* Tat is Required for Virulence, Motility and Acid Resistance

The *Y. pseudotuberculosis tatC* strain was investigated for a range of phenotypes, but the pleiotropic effects seen for *tat* mutants in other bacterial backgrounds were largely absent. A decreased resistance to acidic conditions (pH3) could be observed for the *Y. pseudotuberculosis tatC* mutant as well as impaired motility and loss of flagellation. The most stunning feature of the *tatC* strain was however the severe attenuation it displayed in the mouse infection model (Lavander et al. 2006).

Oral infection of mice revealed the *Y. pseudotuberculosis tatC* mutant to be highly attenuated for virulence and impaired for colonisation of lymphoid tissue, i.e. the Peyer's patches and spleen. Since the mutant also exhibited increased *in vitro* sensitivity to acid, a possible explanation to the severe attenuation could be that the bacteria were eradicated by the low pH within the mouse gastrointestinal tract. Further, the loss of motility could also have a negative effect on the ability to efficiently invade the intestinal epithelium. To resolve this issue, the virulence of the *tatC* strain was studied by infection via the intraperitoneal route. The results showed that, also for this route, the Tat system is essential for virulence. Fascinatingly, the *tatC* mutant was equally attenuated as the virulence plasmid cured strain, i.e. with a 50% infection dose of: 3.7×10^7 bacteria/ml for the *tatC* strain and 2.2×10^7 for the plasmid cured strain compared to 2.2×10^3 for wild type infections. The acid sensitivity is an interesting phenotype also for this mode of infection, since it may deprive *Y. pseudotuberculosis* of its ability to survive and multiply within macrophages. Still, the *tatC* mutant survived equally well as the wild type strain inside J774.1 macrophage-like cells. This suggests that *Yersinia* Tat is redundant for intracellular survival (Lavander et al. 2006).

23.2.4 *Yersinia* Tat Substrates in Relation to Virulence

In silico methods (Bendtsen et al. 2005) were used to identify putative Tat substrates in the genomes of *Y. pestis* and *Y. pseudotuberculosis*. Altogether there are 30 predicted substrates for *Y. pestis* KIM5 and *Y. pseudotuberculosis* IP32953; 18 of which share no homology to *E. coli* Tat substrates. Further there are 4 putative substrates that are unique to *Y. pseudotuberculosis* and 7 unique to *Y. pestis*. This is slightly surprising, given the fact that *Y. pestis* has a condensed genome compared to *Y. pseudotuberculosis* (Chain et al. 2004). The putative *Yersinia* Tat substrates are suggested to be involved in respiration, bacterial growth and division, iron acquisition and transport (e.g. of carbohydrates). Further there is a carbonic anhydrase, a proline-specific amino peptidase, a CueO homologue (involved in copper homeostasis) and a sulfatase, and lastly a group of predicted substrates of unknown function (Lavander, et al. 2006).

Some of the predicted Tat substrates are promising candidates that could explain of the phenotypes of the *tatC* mutant. The resistance to acidic conditions could possibly be mediated by the carbonic anhydrase (CA) identified in the genomes of both *Y. pestis* and *Y. pseudotuberculosis*. CAs are zinc metalloenzymes, found in both pro- and eukaryotes, that catalyze the reversible hydration of CO_2 [$CO_2+H_2O \leftrightarrow HCO_3^-+H^+$]. They are known to be involved in many physiological processes and have been proposed to contribute to intracellular survival of bacteria. Interestingly, a recent publication on *Helicobacter pylori* demonstrated a periplasmic CA that was required for *in vitro* acid resistance (Marcus et al. 2005; Smith and Ferry 2000).

Other interesting putative Tat substrates are those involved in iron acquisition, an ability absolutely crucial for colonisation of a host. These include YbtP and YbtQ which both have been shown to be essential for iron uptake by Yersiniabactin, and a *Y. pestis ybtP* mutant strain is avirulent via the subcutaneous route (Fetherston et al. 1999). Further, the proteins of unknown functions are of course also interesting for future investigations of their potential roles in *Yersinia* virulence.

23.3 Future Perspectives

Specialised systems for protein trafficking comprise essential virulence mechanisms in many bacterial species. We have focussed on the Tat system, and its potential role in *Yersinia* virulence. In other bacterial pathogens where Tat has been studied, the system has been found to be involved in important processes like cell envelope biogenesis, biofilm formation, assembly of the electron-transport chain, motility and resistance to various types of environmental stress. The pleiotropic defects and the fact that Tat deficient strains often are impaired for *in vitro* growth, makes it difficult to determine the direct contribution of Tat substrates to virulence in these bacteria. Intriguingly, such pleiotropic effects are principally absent in the *Y. pseudotuberculosis tatC* mutant. This strain displays normal *in vitro* growth under a wide range of culture conditions, wild type cell morphology and is largely unaffected for stress resistance. Moreover, Tat was revealed to be absolutely required for the ability of *Y. pseudotuberculosis* to cause a systemic infection in mice via both the oral and intraperitoneal routes, indicating that this bacterium encodes Tat delivered proteins that are directly involved in virulence. The *in silico* screen for putative Tat substrates turned up a number of interesting proteins that may contribute to the ability of *Yersinia* to cause disease. Proteomic comparison of fractions of the extracytoplasmic bacterial compartments derived from the *tatC* mutant and the isogenic wild type strain will contribute to the search for Tat substrates. Verification and complementation of these screens will determine which proteins are i) true substrates of *Yersinia* Tat, and ii) important for virulence.

Our findings that the Tat pathway is required for the ability of *Y. pseudotuberculosis* to cause a systemic infection in mice has implications also for *Y. pestis* as these species are closely related (Achtman et al. 1999), and cause similar symptoms in the mouse infection model. We have verified that *Y. pestis* KIM5 also expresses a functional Tat system and most of the predicted Tat substrates are shared between both genomes, making it quite possible that Tat substrates are involved in virulence also in *Y. pestis*. Work to identify the actual Tat substrates of *Yersinia* and their involvement in the infection process is currently underway. Additionally, as Tat is required for virulence, not only in *Yersinia* but in a number of pathogenic bacteria, this system constitutes an interesting target for development of novel generic antimicrobial treatments.

23.4 Acknowledgments

Professor Michael Vasil at the Department of Microbiology, University of Colorado Health Sciences Centre, Denver, contributed with valuable discussions and is also thanked for sharing data prior to publication. This work was funded by the JC Kempe Memorial Fund, the Wenner-Gren foundations and the Swedish Research Council.

23.5 References

Achtman, M., Zurth, K., Morelli, G.,. Torrea, G., Guiyoule, A. and Carniel, E. (1999) *Yersinia pestis*, the cause of plague, is a recently emerged clone of *Yersinia pseudotuberculosis*. Proc. Natl. Acad. Sci. U S A 96,14043-8.

Alami, M., Luke, I., Deitermann, S., Eisner, G., Koch, H.G., Brunner, J. and Muller, M. (2003) Differential interactions between a twin-arginine signal peptide and its translocase in *Escherichia coli*. Mol. Cell 12, 937-946.

Behrendt, J., Standar, K., Lindenstrauss, U. and Bruser, T. (2004) Topological studies on the twin-arginine translocase component TatC. FEMS Microbiol Lett 234, 303-308.

Bendtsen, J. D., Nielsen, H., Widdick, D., Palmer, T. and Brunak, S. (2005). Prediction of twin-arginine signal peptides. BMC Bioinformatics 6,167

Berks, B.C. (1996) A common export pathway for proteins binding complex redox cofactors? Mol. Microbiol. 22, 393-404.

Berks, B. C., Sargent, F. and Palmer, T. (2000). The Tat protein export pathway. Mol. Microbiol. 35,260-74.

Berks, B.C., Palmer, T. and Sargent, F. (2003) The Tat protein translocation pathway and its role in microbial physiology. Adv. Microb. Physiol. 47, 187-254.

Bogsch, E.G., Sargent, F., Stanley, N.R., Berks, B.C., Robinson, C. and Palmer, T. (1998) An essential component of a novel bacterial protein export system with homologues in plastids and mitochondria. J. Biol. Chem. 273, 18003-18006.

Bolhuis, A., Bogsch, E.G. and Robinson, C. (2000) Subunit interactions in the twin-arginine translocase complex of *Escherichia coli*. FEBS Lett. 472, 88-92.

Bröms, J.E., Forslund, A.L., Forsberg, Å. and Francis, M.S. (2003a) Dissection of homologous translocon operons reveals a distinct role for YopD in type III secretion by *Yersinia pseudotuberculosis*. Microbiology 149, 2615-2626.
Bröms, J.E., Sundin, C., Francis, M.S. and Forsberg, Å. (2003b) Comparative analysis of type III effector translocation by *Yersinia pseudotuberculosis* expressing native LcrV or PcrV from *Pseudomonas aeruginosa*. J. Infect. Dis. 188, 239-249.
Chain, P.S., Carniel, E., Larimer, F.W., Lamerdin, J., Stoutland, P.O., Regala, W.M., Georgescu, A.M., Vergez, L.M., Land, M.L., Motin, V.L., Brubaker, R.R., Fowler, J., Hinnebusch, J., Marceau, M., Medigue, C., Simonet, M., Chenal-Francisque, V., Souza, B., Dacheux, D., Elliott, J.M., Derbise, A., Hauser, L.J. and Garcia, E. (2004) Insights into the evolution of *Yersinia pestis* through whole-genome comparison with *Yersinia pseudotuberculosis*. Proc. Natl. Acad. Sci. U S A 101, 13826-13831.
Dalbey, R.E. and Robinson, C. (1999) Protein translocation into and across the bacterial plasma membrane and the plant thylakoid membrane. Trends Biochem. Sci. 24, 17-22.
de Leeuw, E., Granjon, T., Porcelli, I., Alami, M., Carr, S.B., Muller, M., Sargent, F., Palmer, T. and Berks, B.C. (2002) Oligomeric properties and signal peptide binding by *Escherichia coli* Tat protein transport complexes. J. Mol. Biol. 322, 1135-1146.
Deng, W., Burland, V., Plunkett, G., 3rd, Boutin, A., Mayhew, G.F., Liss, P., Perna, N.T., Rose, D.J., Mau, B., Zhou, S., Schwartz, D.C., Fetherston, J.D., Lindler, L.E., Brubaker, R.R., Plano, G.V., Straley, S.C., McDonough, K.A., Nilles, M.L., Matson, J.S., Blattner, F.R. and Perry, R.D. (2002) Genome sequence of *Yersinia pestis* KIM. J. Bacteriol. 184, 4601-4611.
Ding, Z. and Christie, P.J. (2003). *Agrobacterium tumefaciens* twin-arginine-dependent translocation is important for virulence, flagellation, and chemotaxis but not type IV secretion. J. Bacteriol. 185,760-771.
Fetherston, J.D., Bertolino, V.J. and Perry, R.D. (1999) YbtP and YbtQ: two ABC transporters required for iron uptake in *Yersinia pestis*. Mol. Microbiol. 32, 289-299.
Frithz-Lindsten, E., Holmstrom, A., Jacobsson, L., Soltani, M., Olsson, J., Rosqvist, R. and Forsberg, Å. (1998) Functional conservation of the effector protein translocators PopB/YopB and PopD/YopD of *Pseudomonas aeruginosa* and *Yersinia pseudotuberculosis*. Mol. Microbiol. 29, 1155-1165.
Gohlke, U., Pullan, L., McDevitt, C.A., Porcelli, I., de Leeuw, E., Palmer, T., Saibil, H.R. and Berks, B.C. (2005) The TatA component of the twin-arginine protein transport system forms channel complexes of variable diameter. Proc. Natl. Acad. Sci. U S A 102, 10482-10486.
Halbig, D., Wiegert, T., Blaudeck, N., Freudl, R. and Sprenger, G.A. (1999) The efficient export of NADP-containing glucose-fructose oxidoreductase to the periplasm of *Zymomonas mobilis* depends both on an intact twin-arginine motif in the signal peptide and on the generation of a structural export signal induced by cofactor binding. Eur. J. Biochem. 263, 543-551.
Hatzixanthis, K., Palmer, T. and Sargent, F. (2003) A subset of bacterial inner membrane proteins integrated by the twin-arginine translocase. Mol. Microbiol. 49, 1377-1390.
Hinsley, A.P., Stanley, N.R., Palmer, T. and Berks, B.C. (2001) A naturally occurring bacterial Tat signal peptide lacking one of the 'invariant' arginine residues of the consensus targeting motif. FEBS Lett. 497, 45-49.
Hutcheon, G.W. and Bolhuis, A. (2003) The archaeal twin-arginine translocation pathway. Biochem. Soc. Trans. 31, 686-689.
Ize, B., Gerard, F., Zhang, M., Chanal, A., Voulhoux, R., Palmer, T., Filloux, A. and Wu, L.F. (2002) In vivo dissection of the Tat translocation pathway in *Escherichia coli*. J. Mol. Biol. 317, 327-335.

Jongbloed, J.D., Martin, U., Antelmann, H., Hecker, M., Tjalsma, H., Venema, G., Bron, S., van Dijl, J.M. and Muller, J. (2000) TatC is a specificity determinant for protein secretion via the twin-arginine translocation pathway. J. Biol. Chem. 275, 41350-41357.

Ki, J.J., Kawarasaki, Y., Gam, J., Harvey, B.R., Iverson, B.L. and Georgiou, G. (2004) A periplasmic fluorescent reporter protein and its application in high-throughput membrane protein topology analysis. J. Mol. Biol. 341, 901-909.

Lavander, M., Ericsson, S.K., Bröms, J.E. and Forsberg, Å. (2006). The twin arginine translocation system is essential for virulence of *Yersinia pseudotuberculosis*. Infect. Immun. 74,1768-76.

Marcus, E.A., Moshfegh, A.P., Sachs, G. and Scott, D.R. (2005) The periplasmic alpha-carbonic anhydrase activity of *Helicobacter pylori* is essential for acid acclimation. J. Bacteriol. 187, 729-738.

Molik, S., Karnauchov, I., Weidlich, C., Herrmann, R.G. and Klosgen, R.B. (2001) The Rieske Fe/S protein of the cytochrome b6/f complex in chloroplasts: missing link in the evolution of protein transport pathways in chloroplasts? J. Biol. Chem. 276, 42761-42766.

Oates, J., Mathers, J., Mangels, D., Kuhlbrandt, W., Robinson, C. and Model, K. (2003) Consensus structural features of purified bacterial TatABC complexes. J. Mol. Biol. 330, 277-286.

Oates, J., Barrett, C.M., Barnett, J.P., Byrne, K.G., Bolhuis, A. and Robinson, C. (2005) The *Escherichia coli* twin-arginine translocation apparatus incorporates a distinct form of TatABC complex, spectrum of modular TatA complexes and minor TatAB complex. J. Mol. Biol. 346, 295-305. Epub 2004 Dec 2013.

Ochsner, U.A., Snyder, A., Vasil, A.I. and Vasil, M.L. (2002) Effects of the twin-arginine translocase on secretion of virulence factors, stress response, and pathogenesis. Proc. Natl. Acad. Sci. U S A 99, 8312-8317.

Palmer, T. and Berks, B.C. (2003) Moving folded proteins across the bacterial cell membrane. Microbiology 149, 547-556.

Palmer, T., Sargent, F. and Berks, B.C. (2005) Export of complex cofactor-containing proteins by the bacterial Tat pathway. Trends Microbiol. 13, 175-180.

Porcelli, I., de Leeuw, E., Wallis, R., van den Brink-van der Laan, E., de Kruijff, B., Wallace, B.A., Palmer, T. and Berks, B.C. (2002) Characterization and membrane assembly of the TatA component of the *Escherichia coli* twin-arginine protein transport system. Biochemistry 41, 13690-13697.

Pugsley, A.P. (1993) The complete general secretory pathway in gram-negative bacteria. Microbiol. Rev. 57, 50-108.

Rose, R.W., Bruser, T., Kissinger, J.C. and Pohlschroder, M. (2002) Adaptation of protein secretion to extremely high-salt conditions by extensive use of the twin-arginine translocation pathway. Mol. Microbiol. 45, 943-950.

Rossier, O. and Cianciotto, N.P. (2005) The *Legionella pneumophila* tatB gene facilitates secretion of phospholipase C, growth under iron-limiting conditions, and intracellular infection. Infect. Immun. 73, 2020-2032.

Sargent, F., Bogsch, E.G., Stanley, N.R., Wexler, M., Robinson, C., Berks, B.C. and Palmer, T. (1998) Overlapping functions of components of a bacterial Sec-independent protein export pathway. Embo J. 17, 3640-3650.

Sargent, F., Gohlke, U., De Leeuw, E., Stanley, N.R., Palmer, T., Saibil, H.R. and Berks, B.C. (2001) Purified components of the *Escherichia coli* Tat protein transport system form a double-layered ring structure. Eur. J. Biochem. 268, 3361-3367.

Sargent, F., Berks, B.C. and Palmer, T. (2002) Assembly of membrane-bound respiratory complexes by the Tat protein-transport system. Arch. Microbiol. 178, 77-84.

Settles, A.M., Yonetani, A., Baron, A., Bush, D.R., Cline, K. and Martienssen, R. (1997) Sec-independent protein translocation by the maize Hcf106 protein. Science 278, 1467-1470.

Settles, A.M. and Martienssen, R. (1998) Old and new pathways of protein export in chloroplasts and bacteria. Trends Cell Biol. 8, 494-501.
Smith, K.S. and Ferry, J.G. (2000) Prokaryotic carbonic anhydrases. FEMS Microbiol. Rev. 24, 335-366.
Weiner, J.H., Bilous, P.T., Shaw, G.M., Lubitz, S.P., Frost, L., Thomas, G.H., Cole, J.A. and Turner, R.J. (1998) A novel and ubiquitous system for membrane targeting and secretion of cofactor-containing proteins. Cell 93, 93-101.
Wexler, M., Sargent, F., Jack, R.L., Stanley, N.R., Bogsch, E.G., Robinson, C., Berks, B.C. and Palmer, T. (2000) TatD is a cytoplasmic protein with DNase activity. No requirement for TatD family proteins in sec-independent protein export. J. Biol.Chem. 275, 16717-16722.
von Heijne, G. (1985) Signal sequences. The limits of variation. J. Mol. Biol. 184, 99-105.
Voulhoux, R., Ball, G., Ize, B., Vasil, M.L., Lazdunski, A., Wu, L.F. and Filloux, A. (2001) Involvement of the twin-arginine translocation system in protein secretion via the type II pathway. Embo J. 20, 6735-6741.
Yahr, T.L., Hovey, A.K., Kulich, S.M. and Frank, D.W. (1995) Transcriptional analysis of the *Pseudomonas aeruginosa* exoenzyme S structural gene. J. Bacteriol. 177, 1169-1178.
Yahr, T.L. and Wickner, W.T. (2001) Functional reconstitution of bacterial Tat translocation in vitro. Embo J. 20, 2472-2479.

Picture 22. Dancing at the banquet. Photograph by R. Perry.

24
Using Every Trick in the Book: The Pla Surface Protease of *Yersinia pestis*

Marjo Suomalainen, Johanna Haiko, Päivi Ramu, Leandro Lobo, Maini Kukkonen, Benita Westerlund-Wikström, Ritva Virkola, Kaarina Lähteenmäki, and Timo K. Korhonen

Faculty of Biosciences, General Microbiology, University of Helsinki,
timo.korhonen@helsinki.fi

Abstract. The Pla surface protease of *Yersinia pestis*, encoded by the *Y. pestis*-specific plasmid pPCP1, is a versatile virulence factor. In vivo studies have shown that Pla is essential in the establishment of bubonic plague, and in vitro studies have demonstrated various putative virulence functions for the Pla molecule. Pla is a surface protease of the omptin family, and its proteolytic targets include the abundant, circulating human zymogen plasminogen, which is activated by Pla to the serine protease plasmin. Plasmin is important in cell migration, and Pla also proteolytically inactivates the main circulating inhibitor of plasmin, α_2-antiplasmin. Pla also is an adhesin with affinity for laminin, a major glycoprotein of mammalian basement membranes, which is degraded by plasmin but not by Pla. Together, these functions create uncontrolled plasmin proteolysis targeted at tissue barriers. Other proteolytic targets for Pla include complement proteins. Pla also mediates bacterial invasion into human endothelial cell lines; the adhesive and invasive charateristics of Pla can be genetically dissected from its proteolytic activity. Pla is a 10-stranded antiparallel β-barrel with five surface-exposed short loops, where the catalytic residues are oriented inwards at the top of the β-barrel. The sequence of Pla contains a three-dimensional motif for protein binding to lipid A of the lipopolysaccharide. Indeed, the proteolytic activity of Pla requires rough lipopolysaccharide but is sterically inhibited by the O antigen in smooth LPS, which may be the selective advantage of the loss of O antigen in *Y. pestis.* Members of the omptin family are highly similar in structure but differ in functions and virulence association. The catalytic residues of omptins are conserved, but the variable substrate specificities in proteolysis by Pla and other omptins are dictated by the amino acid sequences near or at the surface loops, and hence reflect differences in substrate binding. The closest orthologs of Pla are PgtE of *Salmonella* and Epo of *Erwinia*, which functionally differ from Pla. Pla gives a model of how a horizontally transferred protein fold can diverge into a powerful virulence factor through adaptive mutations.

24.1 Pla as a Member of the Omptin Family of Aspartic Proteases

Surface-associated proteolysis has for long been recognized as an important feature in migration of eukaryotic cells across tissue barriers that are formed by extracellular matrices (ECM) and basement membranes (BM). This holds for migration of phagocytes to infected tissue sites as well as migration of metastatic tumor cells into the circulation and secondary tissue sites (Plow et al. 1999; Myöhänen and Vaheri

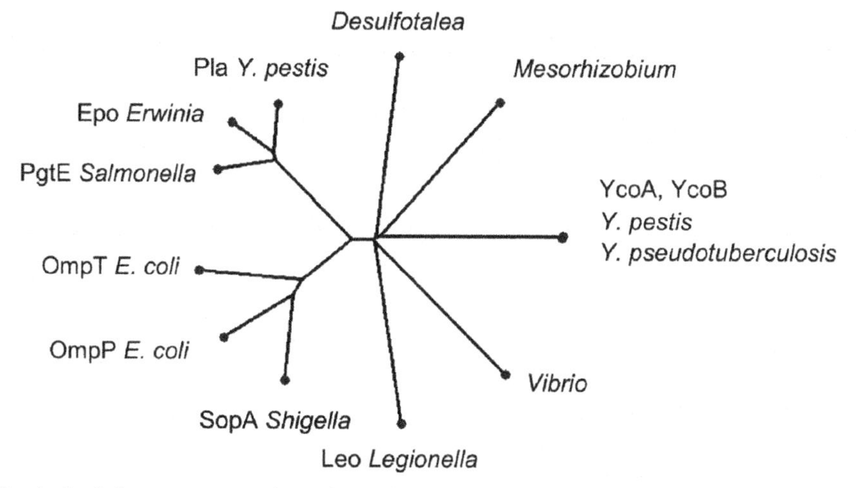

Fig. 1. A cladogram presentation of omptin protein sequences. The omptin sequences were from the MEROPS database (http//:merops.sanger.ac.uk), except for PgtE which was from Guina et al. 2000. The protein sequences were aligned by the Clustal W programme (http://www.ebi.ac.uk/clustalw/index.html) and the tree was drawn with the Phylodraw software (http://pearl.cs.pusan.ac.kr/phylodraw).

2004). Much less attention has been focused on the mechanisms of how invasive bacteria spread across tissue barriers of the host. Secreted hydrolytic enzymes, e.g. proteases and lipases, that damage tissue structures and barriers are obvious virulence factors of a number of bacterial pathogens. However, several pathogens, such as most species in the *Enterobacteriaceae* do not express active surface-bound or secreted proteolytic enzymes. Recent evidence has indicated that these bacteria turn themselves into proteolytic organisms by using host derived proteolytic systems. For this aspect, *Yersinia pestis* and its Pla surface protease make a prime example of how bacteria engage the host-derived plasminogen (Plg) system to cause tissue damage and to enhance bacterial spread.

The Pla surface protease is a member of the omptin family of transmembrane β-barrels in the outer membrane of Gram-negative bacteria. These proteins share a high identity in their predicted amino acid sequence, a similar size of 290-301 amino acids, and the same predicted protein fold (discussed in more detail in section 24.3 below). At present, the omptin family contains 13 members from 9 Gram-negative bacterial genera (Fig. 1). The individual omptins share ca. 40-50% sequence identity to any member of the family, and three subfamilies are obvious. Pla forms a subfamily with PgtE of *Salmonella* and Epo (also called PlaA) of *Erwinia*; these proteins share 74% sequence identity. The *pla* gene is located in the *Y. pestis*-specific plasmid pPCP1 (Sodeinde and Goguen 1989), and *epo* (*plaA*) is also encoded on a plasmid (McGhee et al. 2002), whereas *pgtE* has a chromosomal location (Yu and Hong 1986). The *ompT* gene – encoding the prototype omptin OmpT - of *Escherichia coli* is located in a cryptic prophage (Grodberg et al. 1988), *ompP* in F-plasmid (Matsuo et al. 1999), and *sopA* in the large virulence plasmid of *Shigella* (Egile et al. 1997).

The *ycoA* and *ycoB* genes are chromosomal, and YcoA and YcoB (*Yersinia* chromosal omptin) form a subfamily with 99% sequence identity (Fig. 1). Thus the *Y. pestis* genome contains two omptin genes, *ycoA* shared with *Yersinia pseudotuberculosis* and the *Y. pestis*-specific *pla*. The individual omptins differ in their in vitro functions (target specificity in proteolysis, adhesiveness, invasiveness) as well as in regulation of their expression. The omptin fold appears to have spread horizontally amongst Gram-negative bacteria infecting humans, animals, or plants, and the functional differences probably result from mutations and adaptation to the life style of the host bacterium.

24.2 Virulence Functions of Pla

Pla is an essential virulence factor in bubonic plague, which is an extracellular zoonotic infection where the bacteria invade from the subcutaneous infection site to lymphatic tissue and multiply in the lymph nodes (Perry and Fetherston, 1997). Of the individual omptins, Pla stands out as a dramatic virulence factor. Early studies indicated that the presence of the pPCP1 plasmid and the *pla* gene are associated with the invasive character of plague (Beesley et al. 1967; Ferber and Brubaker 1981). Deletion of *pla* in pPCP1 attenuates *Y. pestis* by a millionfold in a subcutaneous infection model (Sodeinde et al. 1992), and recent in vivo evidence shows that the *pla* gene is needed for the establishment of bubonic plague by *Y. pestis* (Sebbane et al. 2006a) and is one the most highly expressed genes in the bubo (Sebbane et al. 2006b).

We have expressed in *E. coli* most of the omptins shown in Fig. 1 and compared their in vitro functions in an attempt to determine what makes Pla such a potent virulence factor. We have not identified a single function (substrate specificity in proteolysis, adhesion, or invasion) that would be shared by all omptins, it rather seems that the omptins have diverged functionally and adapted to support the life style of their host bacterium. In short, the results show that the virulence role of Pla is dependent on: 1) its high capacity to engage the human proteolytic Plg system to damage tissue barriers, 2) its efficient adhesive and invasive characteristics, 3) surface architecture of *Y. pestis* that allows high Pla activity in extracellular conditions, i.e. outside host cells, and 4) high constitutive transcription of the *pla* gene in extracellular conditions.

A main virulence function of Pla is the activation of Plg, a circulating, abundant mammalian precursor of plasmin, which is a potent serine protease with numerous physiological and pathological functions (Plow et al. 1999; Myöhänen and Vaheri, 2004). A major function of plasmin is to dissolve fibrin clots, it also is important in cell migration through its capacity to digest laminin of basement membranes and to activate precursors of mammalian collagenases and gelatinases (pro-MMPs; matrix metalloproteinases) to their active forms that further damage the tissue barriers. A central role of the Plg/plasmin system in plague is supported by the finding that Plg-deficient mice show increased resistance to the disease (Goguen et al. 2000). Activation of Plg results from a single cleavage of the molecule between residues Arg561 and Val562 to give the two-chain plasmin, it is noteworthy that Pla and PgtE cleave

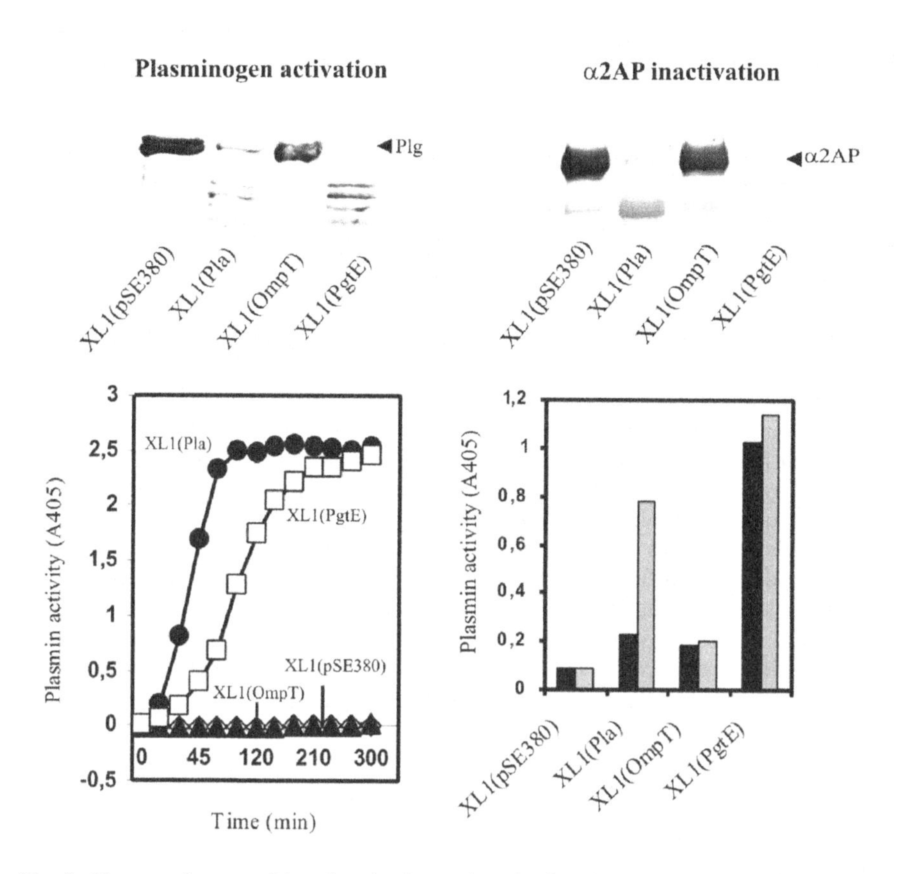

Fig. 2. Cleavage (top panels) and activation or inactivation (lower panels) of the human Plg (left) and α2AP (right) by recombinant *E. coli* XL1 expressing *pla*, *ompT*, or *pgtE* cloned into the vector plasmid pSE380. In the cleavage assays, the bacteria were incubated with the target protein, which was then analyzed by Western blotting. In assays for Plg activation (bottom left), bacteria and Plg were incubated with a chromogenic plasmin substrate, whose degradation was measured by spectrophotometry. In assays for α2AP inactivation (bottom right), bacteria and α2AP were first incubated for 2 h (black bars) or 5 h (grey bars), plasmin was then added, and its activity was measured after a 90-min incubation. PgtE has inactivated nearly all α2AP in 2 h, whereas inactivation by Pla remains partial after 5 h, the inactivation of the antiprotease by OmpT remains close to the control assay with pSE380. The original data has been published in Kukkonen et al. 2001 and in Lähteenmäki et al. 2005a.

the plasmin chains further to smaller, angiostatin-like fragments and that OmpT of *E. coli* is very poor in the cleavage as well as activation of Plg (Fig. 2). Cleavage of Plg by Pla leads to formation of plasmin activity, which is detectable with Pla more rapidly than with PgtE; however, PgtE cleaves the Plg chain slightly more rapidly (Fig 2). Our hypothesis is that PgtE has more than one initial cleavage site within the Plg molecule, the other one(s) leading to enzymatically inactive Plg fragments. This is supported by our ongoing research showing that many of the other omptins cleave the Plg molecule without forming active plasmin. It follows that one of the important virulence properties of Pla is the rapid cleavage of Plg, which precisely mimics that of mammalian Plg activators.

Plasmin is a potent protease, and its formation as well as activity are tightly controlled in the mammalian body at various levels. Circulating plasmin is rapidly inactivated by circulating antiproteases, of which α2-antiplasmin (α2AP) is the main inhibitor of soluble plasmin (Lijnen and Collen 1995). α2AP binds to the plasmin molecule and reduces the half-life of soluble plasmin to 0.1s. Pla overcomes this control mechanism simply by cleaving and inactivating α2AP (Kukkonen et al. 2001), and thus creates uncontrolled plasmin proteolysis. Figure 2 shows the cleavage and inactivation of α2AP by recombinant *E. coli* expressing Pla, PgtE or OmpT. Both Pla and PgtE inactivate the antiprotease, with PgtE being slightly more efficient. Our conclusion is that destroying the control system is more important for the intracellular pathogen *Salmonella* as it can rely – at least partially - on Plg activation performed by activated and migrating phagocytes (Lähteenmäki et al. 2005b). Proteolytic targets for Pla also include complement proteins (Sodeinde et al. 1992). Whether this leads to inactivation of the complement cascade has not been shown, and the biological significance of this function remains open as cells of *Y. pestis* are resistant to serum killing, due to e.g. binding of the inhibitory C4bp protein onto the bacterial cell surface (Ngampasutadol et al. 2005). PgtE and OmpT degrade α-helical antimicrobial peptides (Stumpe et al. 1998; Guina et al. 2000), which could increase bacterial infectivity in the host; however, this function has not been observed with *Y. pestis*.

Pla also has nonproteolytic functions in adhesion and invasion of *Y. pestis*. It is an adhesin with affinity to BMs, where the main target is the major glycoprotein of BM, laminin (Lähteenmäki et al. 1998; Lobo 2006). Conflicting results have been reported for Pla binding to collagens, which could indicate that Pla actually binds to gelatins, i.e. denaturated collagens, which are present in variable amounts in the commercial collagen preparations that were used (Kienle et al. 1992; Lähteenmäki et al. 1998; Lobo 2006). Laminin is not a target for Pla proteolysis but is efficiently degraded by plasmin. In concert, Plg activation, cleavage of α2AP, and bacterial adherence to laminin creates uncontrolled proteolysis which is directed onto a susceptible target in BMs (Lähteenmäki et al. 2005a). Pla is a more efficient adhesin to BM than the other omptins (Kukkonen et al. 2004). Cowan et al. (2000) reported that pPCP1-positive *Y. pestis* invades HeLa cells from a cervical cancer, and we observed that recombinant *E. coli* with Pla invaded human umbical vein endothelial cells (HUVECs) as well as the endothelial-like cell line ECV304 (Lähteenmäki et al. 2001; Kukkonen et al. 2004). The invasion is inhibited by small amounts of fetal calf serum and by normal human serum (unpublished), which indicates that it is not functional from the luminal side of blood vessels. At present, the biological mechanisms and significance of the Pla-mediated invasion remain open.

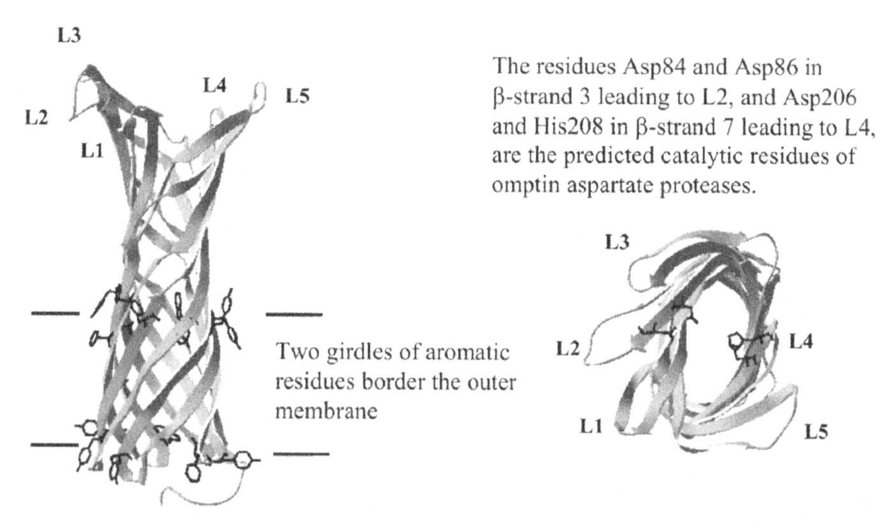

Fig. 3. The β-barrel structure of Pla modeled using the coordinates of the OmpT crystal structure (Vandeputte-Rutten et al. 2001) as a template, the Swiss-Model homology modeling server (http://www.expasy.org/swissmod/SWISS-MODEL.html), and the Swiss-Pdb viewer 3.7 (SP4) programme. On the left is a side view with the outer membrane borders and the five surface loops L1-L5 indicated. On the right, a top view with the four catalytic residues indicated.

24.3 Structure-Function Relationships in Pla

Resolution of the crystal structure of OmpT led to reclassification of omptins as apartate proteases (Vandeputte-Rutten et al. 2001). The structure of OmpT serves as a template to model other omptins, including Pla (Kukkonen et al. 2004). Pla is a β-barrel with 10 antiparallel β-strands connected by four short periplasmic turns and five extracellular loops L1-L5 (Fig. 3). The barrel is long, ca. (70 Å) and vase-shaped and protrudes ca. 40 Å from the lipid bilayer, with the outermost loops located just above the lipopolysaccharide (LPS) core region. The catalytic residues Asp84, Asp86, Asp206, and His208 are located in an acidic groove at the top of the barrel (Fig. 3). They are oriented inwards in the barrel and bordered by the mobile, short loops L1-L5. The catalytic residues are conserved in the omptin sequences, excepting YcoA and YcoB which have Asn84 instead of Asp84 and are classified in the MEROPS database as nonprotease members of the family, due to this sequence difference. However, there is no experimental evidence showing that YcoA/B really are nonproteolytic, neither have their adhesive or invasive properties been assessed.

Omptins cleave polypeptide targets after basic residues but differ substantially in recognition of protein substrates. Correct substrate binding is dictated by the loop structures. In general, the β-barrel is a stable membrane-embedded structure, which

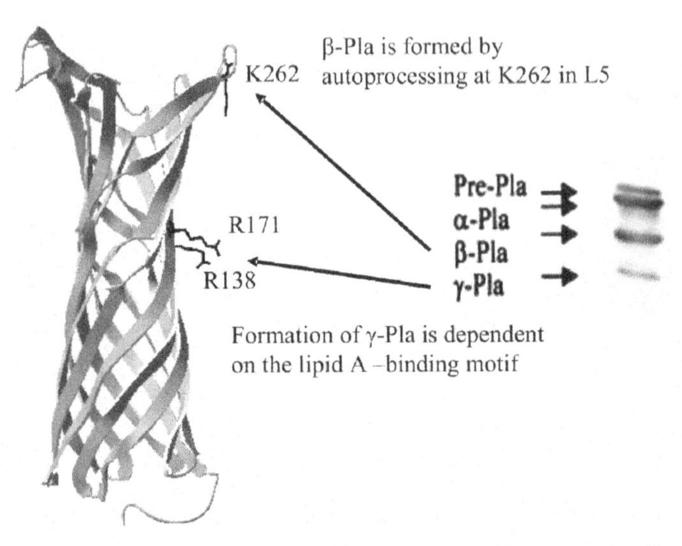

Fig. 4. Isoforms of the Pla molecule. Western blotting of recombinant *E. coli* cell wall preparations obtained by sonication reveals four forms of the Pla molecule. The pre-Pla is the unprocessed form and α-Pla is the mature form of the protein. β-Pla is formed by autoprocessing at Lys262 in L5, and formation of γ-Pla is dependent on the presence of the lipid A–binding motif formed by Arg138 and R171. The original data was published in Kukkonen et al. 2001, 2004.

tolerates large deletions or insertions in surface loops (Schulz 2000), and this seems to provide the basis for the variable proteolysis specificities of the omptins. We have been able to change the substrate specificity of OmpT and to turn it into an efficient Plg activator and α2AP inactivator by cumulative substitutions of OmpT surface residues with the corresponding residues of Pla (Kukkonen et al. 2001). Reverse substitutions rendered Pla incapable in both functions.

Another important feature in Pla functions is the dependency on LPS. Omptin sequences contain a consensus motif for protein binding to lipid A (Vandeputte-Rutten 2001; Kukkonen et al. 2004). This motif occurs in several prokaryotic and eukaryotic proteins that bind LPS and is formed by basic amino acids – Arg138 and Arg171 in Pla (Fig. 4) - that bind to phosphates in lipid A (Ferguson et al. 2000). Purified, detergent-solubilized Pla requires addition of LPS to be active (Kukkonen et al. 2004). On the other hand, Pla and PgtE are sterically inhibited by smooth LPS with a long O chain; this was seen both with recombinant bacteria expressing Pla or PgtE as well as by successful reactivation of purified His_6-Pla with rough but not by smooth LPS. *Y. pestis* is genetically rough and lacks smooth LPS (Skurnik et al. 2000), and our conclusion is that full activity of Pla is the selective advantage for loss of the O antigen in *Y. pestis* (Kukkonen et al. 2004). Isolates of *Salmonella enterica* invariably express smooth LPS, and *S. enterica* overcomes this problem by shortening the LPS to a rough type inside macrophages, where PgtE is fully active (Lähteenmäki et al. 2005b). Thus these two bacterial species obviously are able to utilize their omptins in different environments, *Y. pestis* in extracellular conditions but *S. enterica* inside or immediately after being released from macrophages. The substitution of Arg138 and Arg171 renders PgtE and Pla enzymatically inactive and abolishes formation of the β- and the γ-forms, but at present it is not known exactly how this is associated with enzymatic activity of Pla/PgtE and whether these protein forms are formed within the outer membrane.

Formation of β-Pla is prevented by catalytic-residue substitutions as well as by substitution of the residue Lys262 at L5 (Fig. 4), indicating that it is formed by auto-processing (Kukkonen et al. 2001). Change of other lysines at L5 does not prevent the autoprocessing, which indicates that Lys262 is the cleavage site. Autoprocessing is also prevented by certain substitutions near the catalytic site, these residues probably function in self recognition by Pla. Different substrate specificities of the omptins are also reflected in the fact that the autoprocessing site in Pla is different from those in PgtE and OmpT (Kramer et al. 2000; Kukkonen et al. 2004). It was earlier assumed that formation of β-Pla and γ-Pla represent activation processes similar to those in mammalian Plg activators; however, prevention of β-Pla formation does not affect Plg activation by Pla (Kukkonen et al. 2001), and the biological significance of the autoprocessing remains open.

24.4 Conclusions

The Pla surface protease/adhesin is an essential and multifunctional virulence factor in the plague disease process. Its high virulence potential results in part from its versatile character in generating uncontrolled and targeted proteolysis as well as being an invasin. Important also is the inactivation of the O antigen genes in *Y. pestis*, which exemplifies how a loss of one virulence function (the synthesis of the O antigen) is compensated by the subsequent high activity of another virulence function (Pla). *Y. pestis* and *S. enterica* also demonstrate how a coordinated control and modification of the cell wall and surface proteolysis is obtained by two pathogens having very different life styles and pathogenic mechanisms.

The omptin family has most likely resulted from horizontal gene transfer and subsequent genetic adaptation to gain new or modified functions. Our on-going work has shown that the gain of novel functions in omptins can result from a few amino acid substitutions at critical protein regions. The omptin barrel with its five mobile loops seems an adjustable template to gain novel functions. The nearest ortholog of Pla is Epo of the plant pathogen *Erwinia.* The gene encoding *epo* (*plaA*) is located in a 36-kb mosaic plasmid, which contains homologs to transposases and integrases as well as to genes encoding surface proteins of several bacterial species (McGhee et al. 2002). In functions and regulation, Pla, PgtE and Epo show similarities as well as differences, and our hypothesis is that they have diverged from a common omptin ancestor to serve the different lifestyles of the host bacteria.

Pla has clearly been demonstrated to be an essential virulence factor in bubonic plague. So far these studies have used deletion mutants only, and a challenging task will be to create Pla derivatives impaired in one function only so that the relative virulence roles of specific Pla functions can be addressed.

24.5 Acknowledgements

We thank the Academy of Finland (grant numbers 105824, 211300, 80666, 201967, the Microbes and Man Research Programme, Network of Excellence in Europathogenomics, project number 110716), and the University of Helsinki for financial support.

24.6 References

Beesley, E.D., Brubaker, R.R., Janssen, W.A. and Surgalla, M.J. (1967) Pesticins. III. Expression of coagulase and mechanisms of fibrinolysis. J. Bacteriol. 94, 19-26.

Cowan, C., Jones, H.A., Kaya, Y.H., Perry, R.D. and Straley, S.C. (2000) Invasion of epithelial cells by *Yersinia pestis*: evidence for a *Y. pestis*-specific invasion. Infect. Immun. 68, 4523-4530.

Egile, C., d'Hauteville, H., Parsot, C., and Sansonetti, P.J. (1997) SopA, the outer membrane protease responsible for polar localization of IcsA in *Shigella flexneri*. Mol. Microbiol. 23, 1063-1073.

Ferber, D.M. and Brubaker, R.R. (1981) Plasmids in *Yersinia pestis*. Infect. Immun. 31, 839-841.

Ferguson, A.D., Welte, W., Hofmann, E., Lindner, B., Holst, O., Coulton, J.W. and Diedrerichs, K. (2000) A conserved structural motif for lipopolysaccharide recognition by prokaryotic and eukaryotic proteins. Struct. Fold. Des. 8, 585-592.

Goguen, J.D., Bugge, T. and Degen, J.L. (2000) Role of the pleiotropic effects of plasminogen deficiency in infection experiments with plasminogen-deficient mice. Methods 21, 179-183.

Grodberg, J., Lundrigan, M.D., Toledo, D.L., Mangel, W.F. and Dunn, J.J. (1988) Complete nucleotide sequence and deduced amino acid sequence of the *ompT* gene of *E. coli* K-12. Nucleic Acids Res. 16, 1209.

Guina, T., Yi, E.C., Wang, H., Hackett, M. and Miller, S.I. (2000) A PhoP-regulated outer membrane protease of *Salmonella enterica* serovar Typhimurium promotes resistance to alpha-helical antimicrobial peptides. J. Bacteriol. 182, 4077-4086.

Kienle, Z., Emödy, L., Svanborg, C. and O'Toole, P.W. (1992) Adhesive properties conferred by the plasminogen activator of *Yersinia pestis*. J. Gen. Microbiol. 138, 1679-1687.

Kramer, R.A., Zandwijken, D., Egmond, M.R. and Dekker, N. (2001) In vitro folding, purification and characterization of *E. coli* outer membrane protease OmpT. Eur. J. Biochem. 267, 885-893.

Kukkonen, M., Lähteenmäki, K., Suomalainen, M., Kalkkinen, N., Emödy, L., Lång, H. and Korhonen T.K. (2001) Protein regions important for plasminogen activation and inactivation of α_2-antiplasmin in the surface protease Pla of *Yersinia pestis*. Mol. Microbiol. 40, 1097-1111.

Kukkonen, M., Suomalainen, M., Kyllönen, P., Lähteenmäki, K., Lång, H., Virkola, R., Helander, I.M., Holst, O. and Korhonen, T.K. (2004) Lack of O-antigen is essential for plasminogen activation by *Yersinia pestis* and *Salmonella enterica*. Mol. Microbiol. 51, 215-225.

Lähteenmäki, K., Kukkonen, M. and Korhonen, T.K. (2001) The Pla surface protease/adhesin of *Yersinia pestis* mediates bacterial invasion into human endothelial cells. FEBS Lett. 504, 69-72.

Lähteenmäki, K., Virkola, R., Sarén, A., Emödy, L. and Korhonen, T.K. (1998) Expression of plasminogen activator Pla of *Yersinia pestis* enhances bacterial attachment to the mammalian extracellular matrix. Infect. Immun. 66, 5755-5762.

Lähteenmäki, K., Edelman, S. and Korhonen, T.K. (2005a) Bacterial metastasis: the host plasminogen system in bacterial invasion. Trends Microbiol. 13, 79-85.

Lähteenmäki, K., Kyllönen, P., Partanen, L. and Korhonen, T.K. (2005b) Antiprotease inactivation by *Salmonella enterica* released from infected macrophages. Cell. Microbiol. 7, 529-538.

Lobo, L.A. (2006) Adhesive properties of the purified plasminogen activator Pla of *Yersinia pestis*. FEMS Microbiol. Lett. 262, 158-162.

Lijnen, H.R. and Collen, D. (1995) Mechanisms of physiological fibrinolysis. Bailliere´s Clinical Haematology 8, 277-290.

Matsuo, E., Sampei, G., Mizobuchi, K. and Ito, K. (1999) The plasmid F OmpP protease, a homolue of OmpT, as a potential obstacle to *E. coli*-based protein production. FEBS Lett. 461, 6-8.

McGhee, G.C., Schnabel, E.L., Maxson-Stein, K., Jones, B., Stromberg, V.K., Lacy, G.H. and Jones A.L. (2002) Relatedness of chromosomal and plasmid DNAs of *Erwinia pyrifoliae* and *Erwinia amylovora*. Appl. Environ. Microbiol. 68, 6182-6192.

Myöhänen, H. and Vaheri, A. (2004) Regulation and interactions in the activation of cell-associated plasminogen. Cell. Mol. Life Sci. 61, 2840-2858.

Ngampasutadol, J., Ram, S., Blom, A.M., Jarva, H., Jerse, A.E., Lien, E., Goguen, J., Gulati, S. and Rice, P.A. (2005) Human C4b-binding protein selectively interacts with *Neisseria gonorrhoeae* and results in species-specific infection. Proc. Natl. Acad. Sci. USA 102, 17142-17147.

Perry, R.D. and Fetherston, J.D. (1997) *Yersinia pestis* – etiological agent of plague. Clin. Microbiol. Rev. 10, 35-66.

Plow, E.F., Ploplis, V.A. , Carmeliet, P. and Collen D. (1999) Plasminogen and cell migration in vivo. Fibrinol. Proteol. 13, 49-53.

Schulz, G.E. (2000) β-barrel membrane protens. Curr. Opin. Struct. Biol. 10, 443-462.

Sebbane, F., Jarrett, C.O., Gardner, D., Long, D. and Hinnebusch, B.J. (2006a). Role of the *Yersinia pestis* plasminogen activator in the incidence of distinct septicemic and bubonic forms of flea-borne plague. Proc. Natl. Acad. Sci. USA 103, 5526-5530.

Sebbane, F., Lemaite, N., Sturdevant, D.E., Rebeil, R., Virtaneva, K., Porcella, S.F. and Hinnebusch, B.J. (2006b) Adaptive response of *Yersinia pestis* to extracellular effectors of innate immunity during bubonic plague. Proc. Natl. Acad. Sci. USA 103, 11766-11771.

Skurnik, M., Peippo, A. and Ervelä, E. (2000) Characterization of the O-antigen gene clusters of *Yersinia pseudotuberculosis* and the cryptic O-antigen cluster of *Yersinia pestis* shows that the plague bacillus is most closely related to and has evolved from *Y. pseudotuberculosis* serotype O1b. Mol. Microbiol. 37, 316-330.

Sodeinde, O.A. and Goguen J.D. (1989) Nucleotide sequence of the plasminogen activator gene of *Yersinia pestis*: relationship to *ompT* of *E. coli* and gene E of *Salmonella typhimurium*. Infect. Immun. 57, 1517-1523.

Sodeinde, O.A., Subrahmanyam, Y.V.B.K., Stark, K., Quan, T., Bao, Y. and Goguen, J.D. (1992) A surface protease and the invasive character of plague. Science 258, 1004-1007.

Stumpe, S., Schmid, R., Stephens, D.L., Georgiu, G. and Bakker, E.P. (1998) Identification of OmpT as the protease that hydrolyzes the antimicrobial peptide protamine before it enters growing cells of *E. coli*. J. Bacteriol. 180,4002-4006.

Vandeputte-Rutten, L., Kramer, R.A., Kroon, J., Dekker, N., Egmond, M.R. and Gros, P. (2001) Crystal structure of the outer membrane protease OmpT from *E. coli* suggests a novel catalytic site. EMBO J. 20, 5033-5039.

Yu, C.Q. and Hong, J.S. (1986) Identification and nucleotide sequence of the activator gene of the externally induced phosphoglycerate transport system of *Salmonella typhimurium*. Gene 45, 51-57.

Picture 23. Attendees at the banquet. Photograph by R. Perry.

25
Invasion and Dissemination of *Yersinia enterocolitica* in the Mouse Infection Model

Konrad Trülzsch, Mark F. Oellerich, and Jürgen Heesemann

Max von Pettenkofer Institute for Hygiene and Medical Microbiology, Ludwig Maximillians University, truelzsch@mvp.uni-muenchen.de, heesemann@mvp.uni-muenchen.de

Abstract. *Yersinia enterocolitica* is one of the most common causes of food borne gastrointestinal disease. After oral uptake yersiniae replicate in the small intestine, invade Peyer's patches of the distal ileum and disseminate to spleen and liver. In these tissues and organs yersiniae replicate extracellularly and form exclusively monoclonal microabscesses. Only very few yersiniae invade Peyer's patches and establish just a very few monoclonal microabscesses. This is due to both *Yersinia* and host specific factors.

25.1 Introduction

Yersinia enterocolitica and *Yersinia pseudotuberculosis* cause food-borne gastrointestinal disease in humans. While *Y. pseudotuberculosis* is primarily an animal pathogen that only rarely causes disease in humans, *Y. enterocolitica* is one of the most common causes of gastrointestinal disease in the moderate and subtropical climates of the world. It may present as enteritis, terminal ileitis, or mesenteric lymphadenitis (pseudoappendicitis) with watery or sometimes bloody diarrhoea. In patients with iron overload states (e.g. haemochromatosis, haemolytic anemia) a systemic infection can ensue leading to focal abscess formation in liver and spleen (Bottone 1997). A similar disease results after oral infection of mice with yersiniae replicating in the small intestine, invading Peyer's patches (PPs) of the distal ileum, and disseminating to liver and spleen. Extracellular replication of yersiniae in these tissues and organs leads to the formation of microabscesses. This is made possible by the injection of *Yersinia* outer proteins (Yops) by a type three secretion system (T3SS) which paralyses phagocytes of the innate immune system (reviewed in Heesemann et al. 2006). The precise mechanisms that lead to the formation of microabscesses *in vivo* are not clear. Invasion of PPs is mediated by several non-fimbrial adhesins such as invasin and the *Yersinia* adhesion A (YadA). Both of these surface proteins interact with β1 integrins of eukaryotic cells and are believed to mediate adherence and invasion of M cells (reviewed in Isberg and Barnes 2001 and Heesemann et al. 2006). Invasin is able to directly bind β1 integrins of host cells (Isberg and Leong 1990; Leong et al. 1990), whereas YadA interacts with extracellular matrix (ECM) proteins such as collagen and fibronectin and with host cell β1 integrins by ECM bridging (Heise and Dersch 2006; Eitel and Dersch 2002). β1 integrins are expressed on the apical surface of M cells but not by enterocytes (Clark et al. 1998). PP invasion is therefore believed to be mediated by M cells of the follicle-associated epithelium. Invasin is the most important invasion factor of *Yersinia* and is known to be essential for early invasion of PPs in the mouse oral infection model (Marra and Isberg 1996; Pepe and Miller 1993). The surface exposed C-terminal region of invasin consists of five globular domains (D1-5) that protrude 18 nm from the bacterial surface in *Y. pseudotuberculosis* (Hamburger et al. 1999). *Y. enterocolitica* invasin lacks the D2 self association domain. The adhesion unit that is responsible for the high affinity interaction with β1 integrins is formed by the D4-D5 domains. Preferential invasion of M cells by *Y. pseudotuberculosis* has been demonstrated by microscopy of mouse ligated gut loops (Clark and Jepson 2003). M cells were found to carry multiple adherent/invading yersiniae suggesting translocation of multiple bacteria to submucosal tissue (Clark et al. 1998). After translocation across the mucosal barrier by M cells, yersiniae disseminate from PPs to mesenteric lymph nodes (Grutzkau et al. 1990; Hanski et al. 1989; Simonet et al. 1990). Further dissemination to spleen and liver probably does not occur via PPs and lymph nodes. Recently it was shown that organized intestinal lymphoid tissue was not required for dissemination of yersiniae to spleen and liver (Handley et al. 2005). It was furthermore demonstrated that yersiniae colonizing spleens and livers were derived from the gut lumen but not mesenteric lymph nodes (Barnes et al. 2006). The precise mechanism of dissemination from the gut lumen to the liver and spleen are not well understood at this time. Early after infection, microabscesses formed by yersiniae in liver and spleen of mice consist primarily of neutrophils (Autenrieth et al. 1993; Carter 1975). During the course of infection, lesions are populated by mononuclear cells and exhibit a granulomatous character (Autenrieth et al. 1993). We have recently studied abscess formation in the oral mouse infection model using RFP- (red fluorescent protein) and GFP- (green fluorescent protein) expressing yersiniae. We were able to show that oral *Y. enterocolitica* infection of mice leads to monoclonal microabscess formation in PPs, spleen and liver. Furthermore experiments with red and green fluorescing yersiniae revealed that only very few yersiniae were able to invade PPs from the gut lumen and that both *Yersinia* and the host contribute to this phenomenon.

25.2 Monoclonal Abscess Formation by *Yersinia*

In the mouse oral infection model yersiniae disseminate from the gut lumen to PPs of the distal ileum, lymph nodes, liver and spleen (Trülzsch et al. 2004). In these tissues and organs, yersiniae replicate predominately extracellularly and form microabscesses. Recently we asked the question whether many bacteria were required for microabscess formation or if a single bacterium was sufficient to initiate abscess formation in PPs, liver, and spleen. To answer this question we infected mice orally with an equal mixture of red and green fluorescing yersiniae (expressing RFP or GFP) (Oellerich et al. 2007). At different time points after infection, organs were removed, sectioned, and abscesses were analysed by fluorescence microscopy. These experiments revealed that *Yersinia* microabscesses in PPs, liver, and spleen were

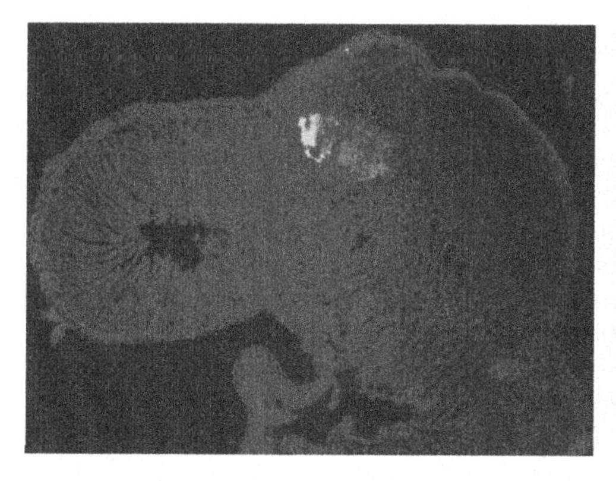
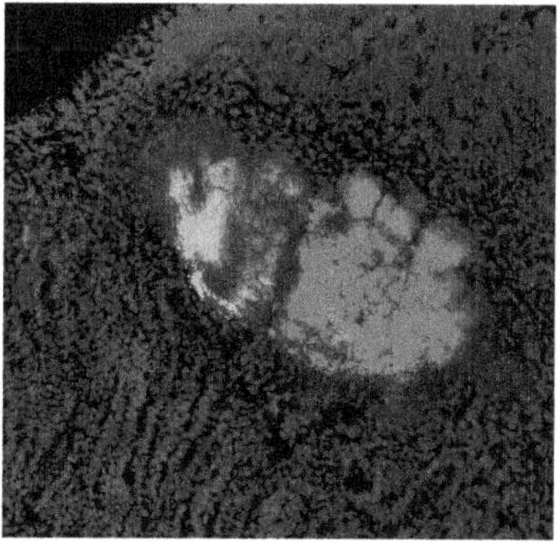

Fig. 1. Typical cryosection (10 μm) of a DAPI stained PP from a Balb/c mouse infected orally with an equal mixture of 10^9 red and green fluorescing *Y. enterocolitica*, 5 days post infection. One red and one green monoclonal microabscess can be seen within the PP. (See color plate.)

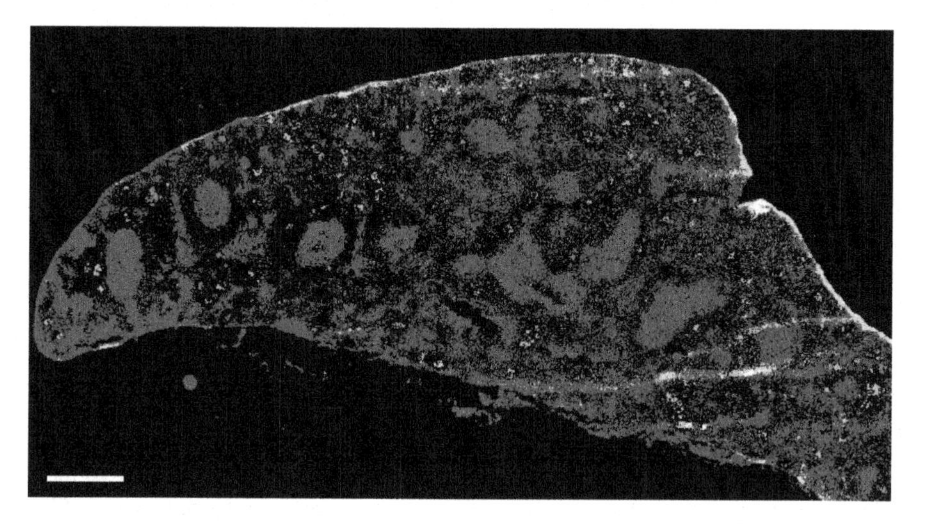

Fig. 2. Typical cryosection (10 μm) of a DAPI stained spleen from a Balb/c mouse infected orally with an equal mixture of 10^9 red and green fluorescing yersiniae, 5 days post infection. Hundreds of red and green fluorescing monoclonal microabscesses can be seen throughout the organ. (See color plate.)

monoclonal as evidenced by the fact that microabscesses were exclusively single colored after infection with a mixture of red and green fluorescing yersiniae (Fig. 1, Fig. 2) (Oellerich et al. 2007). Two scenarios are conceivable when multiple yersiniae invade PPs or disseminate to liver and spleen: either multiple bacteria need to associate prior to forming an abscess or a single bacterium could be sufficient to initiate abscess formation. Recently we demonstrated monoclonal abscess formation according to the latter scenario. Only very few monoclonal microabscesses were observed in any given PP, whereas hundreds of monoclonal microabscesses were seen in cross sections of liver and spleen tissue 5 days after infection (Fig. 2). These results showed that bacterial dissemination from the gut lumen to liver and spleen was much more efficient than dissemination of bacteria from the gut lumen to PPs. This is in line with a recent publication demonstrating that yersiniae colonizing spleen and liver were derived from a replicating pool of bacteria in the intestine rather than disseminating via PPs and lymph nodes (Barnes et al. 2006). Monoclonal abscess formation in liver and spleen is presumably due to single yersiniae disseminating from the gut lumen to these organs. Single yersiniae may be trapped in the capillary vessels of liver and spleen which could subsequently be plugged by the proliferating yersiniae leading to monoclonal abscess formation.

25.3 Clonal Invasion of Peyer´s Patches

The fact that only very few (1-4) monoclonal microabscesses form in each PP, even at a high oral infection dose, suggests clonal invasion of PPs by *Yersinia* (Oellerich et al. 2007). This observation could be the result of many yersiniae initially invading a given PP with only very few bacteria surviving the initial encounter with the host immune response. This latter possibility has been suggested by the electron microscopic observation of multiple yersiniae invading M cells (Clark et al. 1998; Clark and Jepson 2003). However these experiments were performed using the murine ligated gut loop model which cannot be compared to *in vivo* oral infections. To support the finding that it is limited invasion of PPs which is responsible for our observation, we performed infection experiments with neutropenic and oxidative burst deficient mice. Neutrophils are known to be critical in early host defense against *Yersinia*. It has been shown that the initial inactivation of yersiniae that implant in liver and spleen during the first few hours of infection is primarily the feat of neutrophils (Conlan 1997). Besides accumulating in liver and spleen early after infection, neutrophils have been demonstrated in PPs within 24 h post infection (Carter 1975). Therefore mixed infections with red and green fluorescing yersiniae were performed using neutropenic mice (Oellerich et al. 2007). If many yersiniae were to invade PPs and only a few to survive the influx of neutrophils, an increase in the number of microabscesses per PP for neutropenic vs immunocompetent mice would be expected. The same should hold for mice impaired in the oxidative burst ($p47^{phox-/-}$ mice). Infection of these mice with *Yersinia* resulted in essentially the same picture seen with immunocompetent mice (Oellerich et al. 2007). These findings support the clonal invasion hypothesis. If the observation of just a few clones of *Yersinia* establishing microabscesses in a certain PP was solely the result of many yersiniae invading PPs with just a few surviving the initial encounter with the immune response, then a 10 fold higher infection dose would be expected to lead to a 10 fold higher number of monoclonal microabscesses per PP. Recently we were able to show that this was not the case lending support to the clonal invasion hypothesis (Oellerich et al. 2007). Clonal invasion would imply that signature-tagged mutagenesis would not be a suitable tool to identify attenuated yersiniae in mouse PPs. In fact a study on the dissemination of signature-tagged *Y. pseudotuberculosis* mutants in the mouse model previously noted that barriers must exist that limit the number of bacteria that are able to reach mesenteric lymph nodes and spleen. Such barriers were responsible for the failure of signature-tagged mutagenesis to identify attenuated mutants of *Y. pseudotuberculosis* in mesenteric lymph nodes and spleen (Mecsas et al. 2001). However, dissemination of these mutants to PPs was not studied.

25.4 Sequential Invasion of Peyer´s Patches is Inhibited by the Host

The simplest explanation for clonal invasion would be limited contact between *Yersinia* and M cells in the mouse model. Clonal invasion of PPs could furthermore be a *Yersinia* specific characteristic, the result of the host response to *Yersinia* infection, or a combination of both. Invasin is the most important invasion factor for PPs (Pepe and Miller 1993) and is known to bind β-integrins of M cells (Clark et al. 1998). From *in vitro* studies it is known that invasin expression is high at ambient temperature and down-regulated at the host temperature of 37°C (Pepe et al. 1994). It was therefore of interest to determine whether yersiniae also down-regulate invasin expression in the gut lumen after oral infection. Recently we demonstrated, by Western blotting of the small intestinal content, that this is the case (Oellerich et al. 2007). The lack of invasin expression in the small intestine obviously restricts early bacterial invasion to a short time period after oral uptake during which invasin is still present on the bacterial surface. Another possible explanation for clonal invasion could be that *Yersinia* actively prevents invasion of PPs by injecting Yops into M cells thereby paralyzing these cells and preventing *Yersinia* uptake. To look into this possibility we recently performed co-infection experiments with YopH, -O, -P, -E, -M, -T, and -Q mutants expressing RFP and GFP (Trülzsch et al. 2004). These experiments however revealed a similar number of monoclonal microabscesses per PP as seen for wild type yersiniae (unpublished results). Besides several conceivable *Yersinia* specific factors, the host response to infection could be limiting *Yersinia* invasion of PPs. To look into this possibility and to determine if *Yersinia* was able to invade and form microabscesses in previously abscessed PPs, we performed sequential infection experiments, orally inoculating mice with green fluorescing yersiniae followed by red fluorescing yersiniae two days later (Oellerich et al. 2007). These experiments revealed that yersiniae orally inoculated two days after a primary *Yersinia* infection, preferentially invaded those PPs that were not initially abscessed. The freshly inoculated yersiniae of the successive infection were invasion competent since they invaded and replicated in "naïve" PPs but showed severely reduced ability to establish microabscesses in previously abscessed PPs. This indicates that the host severely limits sequential infection of PPs and is obviously one important reason why only very few monoclonal microabscesses are seen in a certain PP after infection with a high bacterial dose. Presumably PPs are only permissive for invasion of multiple yersiniae if they are invaded concomitantly. This hypothesis is supported by the fact that with logarithmically increasing infection doses, the number of microabscesses per PP increases only in a linear fashion. Possibly a signal generated locally in a certain PP shuts off antigen sampling and *Yersinia* uptake by M cells of that PP only. Alternatively it is possible that further yersiniae invade but are rapidly eliminated by the "activated" PP. Very rarely were previously abscessed PPs invaded and abscessed by a subsequent *Yersinia* infection. Possibly only the invasin/β-integrin mediated invasion process is inhibited by the host with residual invasion taking place by alternate mechanisms such as YadA, Ail or inter-epithelial dendritic cells transporting yersiniae to the subepithelium, which has been demonstrated for *Salmonella* (Vazquez-Torres et al. 1999). Finally we have shown that some yersiniae replicating in the gut lumen continue to invade PPs between days 2 and 5 of infection. However the number of microabscesses per PP remains constant during this time period supporting the finding that the host limits invasion of previously abscessed PPs.

25.5 Summary

In summary we have recently demonstrated monoclonal abscess formation in PPs, spleen, and liver by *Yersinia* in the mouse oral infection model indicating that single yersiniae are able to initiate abscess formation. Only very few *Yersinia* cells establish microabscesses in PPs of the small intestine presumably due to clonal invasion of PPs. This is probably due to both the host severely limiting sequential infection of PPs and *Yersinia* down-regulating invasin expression in the small intestinal lumen.

25.6 References

Autenrieth, I.B., Vogel, U., Preger, S., Heymer, B. and Heesemann, J. (1993) Experimental *Yersinia enterocolitica* infection in euthymic and T-cell-deficient athymic nude C57BL/6 mice: comparison of time course, histomorphology, and immune response. Infect. Immun. 61, 2585-2595 .

Barnes, P.D., Bergman, M.A., Mecsas, J. and Isberg, R.R. (2006) *Yersinia pseudotuberculosis* disseminates directly from a replicating bacterial pool in the intestine. J. Exp. Med. 203, 1591-1601.

Bottone, E.J. (1997) *Yersinia enterocolitica*: the charisma continues. Clin. Microbiol. Rev. 10, 257-276.

Carter, P.B. (1975) Pathogenecity of *Yersinia enterocolitica* for mice. Infect. Immun. 11, 164-170 .

Clark, M.A., Hirst, B.H. and Jepson, M.A. (1998) M-cell surface beta1 integrin expression and invasin-mediated targeting of *Yersinia pseudotuberculosis* to mouse Peyer's patch M cells. Infect. Immun. 66, 1237-1243.

Clark, M.A. and Jepson, M.A. (2003) Intestinal M cells and their role in bacterial infection. Int. J. Med. Microbiol. 293, 17-39 .

Conlan, J.W. (1997) Critical roles of neutrophils in host defense against experimental systemic infections of mice by *Listeria monocytogenes*, *Salmonella typhimurium*, and *Yersinia enterocolitica*. Infect. Immun. 65, 630-635.

Eitel, J. and Dersch, P. (2002) The YadA protein of *Yersinia pseudotuberculosis* mediates high-efficiency uptake into human cells under environmental conditions in which invasin is repressed. Infect. Immun. 70, 4880-4891.

Grutzkau, A., Hanski, C., Hahn, H. and Riecken, E.O. (1990) Involvement of M cells in the bacterial invasion of Peyer's patches: a common mechanism shared by *Yersinia enterocolitica* and other enteroinvasive bacteria. Gut 31, 1011-1015 .

Hamburger, Z.A., Brown, M.S., Isberg, R.R. and Bjorkman, P.J. (1999) Crystal structure of invasin: a bacterial integrin-binding protein. Science 286, 291-295.

Handley, S.A., Newberry, R.D. and Miller, V.L. (2005) *Yersinia enterocolitica* invasin-dependent and invasin-independent mechanisms of systemic dissemination. Infect. Immun. 73, 8453-8455.

Hanski, C., Kutschka, U, Schmoranzer, H.P., Naumann, M., Stallmach, A., Hahn, H., Menge, H. and Riecken, E.O. (1989) Immunohistochemical and electron microscopic study of interaction of *Yersinia enterocolitica* serotype O8 with intestinal mucosa during experimental enteritis. Infect. Immun. 57, 673-678.

Heesemann, J., Sing, A. and Trülzsch,K. (2006) *Yersinia*'s stratagem: targeting innate and adaptive immune defense. Curr. Opin. Microbiol 9, 55-61.

Heise, T. and Dersch, P. (2006) Identification of a domain in *Yersinia* virulence factor YadA that is crucial for extracellular matrix-specific cell adhesion and uptake. Proc. Natl. Acad. Sci. USA 103, 3375-3380.

Isberg, R.R. and Barnes, P. (2001) Subversion of integrins by enteropathogenic *Yersinia*. J. Cell. Sci. 114, 21-28.

Isberg, R.R. and Leong, J.M. (1990) Multiple beta 1 chain integrins are receptors for invasin, a protein that promotes bacterial penetration into mammalian cells. Cell 60, 861-871.

Leong, J.M., Fournier, R.S. and Isberg, R.R. (1990) Identification of the integrin binding domain of the *Yersinia pseudotuberculosis* invasin protein. EMBO J. 9, 1979-1989.

Marra, A. and Isberg, R.R. (1996) Analysis of the role of invasin during *Yersinia pseudotuberculosis* infection of mice. Ann. N.Y. Acad. Sci. 797, 290-292.

Mecsas, J., Bilis, I. and Falkow, S. (2001) Identification of attenuated *Yersinia pseudotuberculosis* strains and characterization of an orogastric infection in BALB/c mice on day 5 postinfection by signature-tagged mutagenesis. Infect. Immun. 69, 2779-2787.

Oellerich, M., Jacobi, C., Freund, S., Niedung, K., Bach, A., Heesemann, J. and Trülzsch, K. (2007) *Yersinia enterocolitica* infection of mice reveals clonal invasion and abscess formation. submitted for publication.

Pepe, J.C., Badger, J.L. and Miller, V.L. (1994) Growth phase and low pH affect the thermal regulation of the *Yersinia enterocolitica inv* gene. Mol. Microbiol 11, 123-135.

Pepe, J.C. and Miller, V.L. (1993) *Yersinia enterocolitica* invasin: a primary role in the initiation of infection. Proc. Natl. Acad. Sci. USA 90,6473-6477.

Simonet, M., Richard, S. and Berche, P. (1990) Electron microscopic evidence for in vivo extracellular localization of *Yersinia pseudotuberculosis* harboring the pYV plasmid. Infect. Immun. 58, 841-845.

Trülzsch, K., Sporleder, T., Igwe, E.I., Russmann, H. and Heesemann, J. (2004) Contribution of the major secreted yops of *Yersinia enterocolitica* O:8 to pathogenicity in the mouse infection model. Infect. Immun. 72, 5227-5234.

Vazquez-Torres, A., Jones-Carson, J., Baumler, A.J., Falkow, S., Valdivia, R., Brown, W., Le, M., Berggren, R., Parks, W.T. and Fang, F.C. (1999) Extraintestinal dissemination of Salmonella by CD18-expressing phagocytes. Nature 401, 804-808.

26
The Ysa Type 3 Secretion System of *Yersinia enterocolitica* Biovar 1B

Glenn M. Young

Department of Food Science and Technology University of California, Davis, gmyoung@ucdavis.edu

Abstract. *Yersinia enterocolitica* biovar 1B maintains two distinct and independently operating type 3 secretion (T3S) systems with the capacity to translocate toxic effector proteins into mammalian cells. Each of these T3S systems plays a role in the outcome of an infection by influencing different stages of infection. Recent investigations of the Ysa T3S system have revealed it is important for *Y. enterocolitica* survival during the gastrointestinal phase of infection. This sets this system apart from the Ysc T3S system which is important for systemic infections. Identification of the effector proteins has provided insight on how the Ysa T3S system modulates *Y. enterocolitica* interactions with the host. In part, the Ysa T3S system targets the innate immune response to suppress the ability of the host to rapidly clear an infection.

26.1 Type 3 Secretion Systems of *Yersinia enterocolitica* Biovar 1B

Yersinia enterocolitica is a heterogeneous species with a wide-ranging potential for causing human illness (Carniel and Mollaret 1990). Epidemiological investigations have classified this bacterium into six biovars (1A, 1B, and 2-5) defined by specific repertoires of microbiological phenotypes, biochemical activities and somatic O-antigens (Cornelis et al. 1987). The highly pathogenic biovar 1B strains are distinguished from other biovars by the severity of human illnesses they cause, lethality to experimentally infected mice and propensity to cause mammalian cell cytotoxicity (Carniel and Mollaret 1990; Denecker et al. 2002).

In infants, *Y. enterocolitica* biovar 1B commonly causes an acute Yersiniosis, a disease that initially manifests as a gastroenteritis with abdominal pain and diarrhea (Bottone 1999). Commonly, the bacterium will invade beyond gastrointestinal sites to systemically infect visceral tissues and overrun the immune response resulting in a lethal outcome. Patients that develop a systemic infection have a 50% mortality rate (Cover and Aber 1989). Both the gastrointestinal and systemic phases of Yersiniosis can be recapitulated in experimentally infected inbred mice and aspects of infections can be modeled *in vitro* by infection of cultured mammalian cells (Carter and Collins 1974a, b; Hanski et al. 1989). This has facilitated efforts to examine and decipher mechanisms that influence the progress of a *Y. enterocolitica* biovar 1B infection.

At the host pathogen interface, *Y. enterocolitica* biovar 1B depends on contact-dependent type 3 secretion (T3S) to deliver toxic effector proteins into host cells. T3S systems promote bacterial-directed alterations of cellular and host immunological responses (Galán and Wolf-Watz 2006). Biovar 1B strains maintain two distinct, independently operating, T3S systems (Haller et al. 2000; Snellings et al. 2001). The Ysa system is encoded by the YSA pathogenicity island (YSA-PI) and the Ysc system is encoded by a plasmid, pYV/pCD1, carried by nearly all virulent isolates of the genus *Yersinia* (Foultier et al. 2002; Portnoy et al. 1984). Central to each of these T3S systems is a secretion apparatus and the collection of toxic effector proteins it delivers. Our knowledge of how each of these T3S systems contributes to virulence differs widely.

Efforts over the past two decades determined that proteins targeted by the Ysc T3S system, called effector Yops, represent a group of virulence factors conserved among numerous isolates of *Y. enterocolitica*, *Y. pseudotuberculosis* and *Y. pestis* (Marenne et al. 2004). Once delivered into host cells, the Yops alter host cell activities through modification of a wide spectrum of biochemical activities. Effector Yops translocated into host cells include the Rho-family GTPase activator YopE, the protein tyrosine phosphatase YopH, the nuclear-localized protein YopM, the serine-threonine kinase YopO/YpkA, the protein acetylase YopP/YopJ, and the protease YopT (Marenne et al. 2004). Collectively, delivery of Yop effectors by the Ysc T3S system strongly influences the outcome of systemic infections by disengaging various cellular and host responses, including inhibition of phagocytosis by macrophages and polymorphonuclear leukocytes (PMNs); suppression of T and B lymphocyte activation; and alteration of cytokine production by dendritic cells, epithelial cells, endothelial cells, macrophages and PMNs (Marenne et al. 2004). It is well established that the Ysc T3S systems plays an important role in systemic infections, but only recently has the role of the Ysa T3S system been investigated (Matsumoto and Young 2006; Venecia and Young 2005). These studies, which were carried out by Young and colleagues have been especially informative by revealing the Ysa T3S system influences bacterial-host interactions during the gastrointestinal phase of infection (Matsumoto and Young 2006; Venecia and Young 2005). This has enabled our efforts to be refocused on deciphering the mechanisms through which the Ysa T3S system affects pathogenesis by investigating the activity of the effector proteins.

26.2 The YSA Pathogenicity Island

The YSA pathogenicity island (YSA-PI) was initially described by Pierson and colleagues during an effort in which they sequenced DNA fragments from a randomly generated library of genomic material from *Y. enterocolitica* biovar 1B strain 8081 (Haller et al. 2000). The initial sequence analysis identified a 167 bp chromosomal locus that was 46% identical to *lcrD*, a gene that encodes a component of the plasmid-encoded Ysc T3S system. A larger chromosomal region was then cloned; the nucleotide sequenced was determined and it was found to represent a genetic element restricted to the highly pathogenic biovar 1B group of *Y. enterocolitica* (Fig. 1). The locus was designated Ysa (*Yersinia* secretion apparatus) when computational annotation suggested it encoded for a bona fide protein T3S secretion system (Haller et al. 2000). Genes of the YSA-PI are arranged into two divergently oriented clusters (Fig. 1). Genetic and functional analysis revealed a set of Ysa-dependent secreted

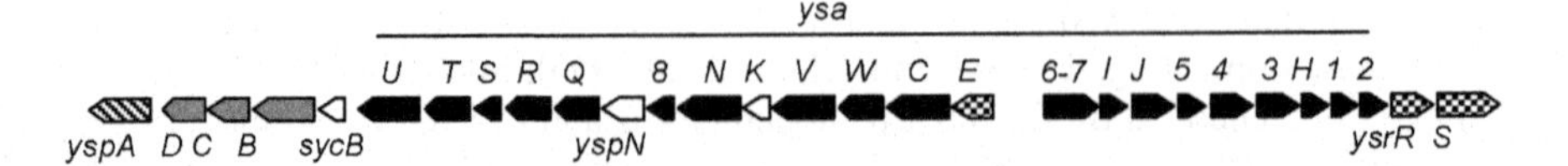

Fig. 1. The YSA pathogenicity island, found only in biovar 1B isolates of *Y. enterocolitica*, is a cluster of genes that encodes structural and regulatory components of the Ysa T3S system. The transcriptional orientation of each gene is indicated by the arrowhead. Genes are shaded to indicate the function of the encoded protein: black, structural components of the protein secretion apparatus; grey, components of the translocon; white, chaperones and assembly facilitators; hatched, transcriptional regulators; and slashed, a Ysp effector.

proteins produced by *Y. enterocolitica* biovar 1B when cultured under specialized-conditions (Haller et al. 2000; Young and Young 2002a, b). Subsequent investigations expanded the genomic analysis to reveal that the YSA-PI maps to the plasticity zone, a chromosomal region rich with virulence-associated genes specific to highly pathogenic *Y. enterocolitica* biovar 1B (Foultier et al. 2002; Iwobi et al. 2003; Thomson et al. 2006).

26.3 The Ysa T3S System Influences *Y. enterocolitica* Infection of Gastrointestinal Tissues

Analysis of human biopsies indicates that the terminal ileum is the primary site where *Y. enterocolitica* infection manifests. Infected tissues often exhibit necrotic areas infiltrated with bacteria, PMNs and mononuclear cells (Bradford et al. 1974; Gleason and Patterson 1982; Vantrappen et al. 1977). While gastrointestinal infection is frequently self-limiting, it can progress systemically in patients who lack robust immune responses. Post-mortem analysis of individuals that succumbed to a systemic infection revealed that vital organ failure was due to an overwhelming bacterial load. Oral infection of mice with *Y. enterocolitica* biovar 1B results in an infection that follows a pattern similar to that seen in human infections and is marked by a distinct gastrointestinal infection followed by a systemic, and ultimately fatal, spread of bacteria to vital organs (Carter 1975a; Carter 1975b).

We recently examined the contribution of the Ysa T3S system to *Y. enterocolitica* infection using the mouse model of Yersiniosis (Venecia and Young 2005). In preparation for our analysis, a library of mutants was generated that harbored transposon insertion mutations in genes required for expression of a functional Ysa T3S system, including loci that encode structural components of the T3S apparatus and transcription factors. From this mutant collection, a representative mutant (*ysaU*) was tested for its ability to colonize gastrointestinal tissues following oral infection of BALB/c mice (Venecia and Young 2005). It was found that the mutant displayed a 100-fold reduction in its ability to colonize the terminal ileum during the first 24 hours of infection, a timeframe where the bacterium was restricted to gastrointestinal sites. There was no difference in the frequency of mice infected with the *ysaU* mutant to develop systemic infections as determined by LD_{50} analysis and quantification

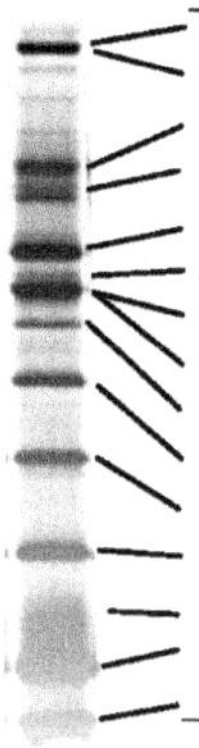

Proteins secreted by the Ysa T3S system

Designated Name	Mass (kDa)	Protein Feature/ Predicted Function	Homologue	Reference
YspA	68.5			Foultier et al. 2002
YspB	72.2	translocon	IpaB (*Shigella flexneri*)	Foultier et al. 2002
YspL	79.9	two different 15 and 17 amino acid repeats		Matsumoto and Young 2006
YspN	38.7	secretion apparatus	InvJ/SpaN (*Salmonella enterica*)	Matsumoto and Young 2006
YspC	48.1	translocon	IpaC (*Shigella flexneri*)	Foultier et al. 2002
YspY	51.2			Matsumoto and Young 2006
YspF	39.7			Matsumoto and Young 2006
YspP	44.8	protein tyrosine phosphatase	PTPase domain of YopH (*Yersinia* spp.)	Matsumoto and Young 2006
YspE	43.5			Matsumoto and Young 2006
YspD	38.9	translocon	IpaD (*Shigella flexneri*)	Foultier et al. 2002
YopN	32.6			Young and Young 2002
YopP	32.4	cysteine protease		Young and Young 2002
YspI	21.2	three tandem 12 amino acid repeats		Matsumoto and Young 2006
YopE	23.0	Rho GTPase activating protein		Young and Young 2002
YspK	22.6	serine/threonine kinase	OspG (*Shigella flexneri*)	Matsumoto and Young 2006

Fig. 2. The secreted proteome of the Ysa T3S system. The complete set of secreted Ysps collected from a culture of *Y. enterocolitica* biovar 1B, strain 8081, was separated by electrophoresis through a 12.5% SDS-polyacylamine gel. Proteins were visualized by staining with silver. The location of each Ysp is indicated with relevant biochemical features.

of bacterial load in vital organs such as the liver and spleen (Venecia and Young 2005). This suggested that the Ysa T3S system contributes primarily to the gastrointestinal phase of infection. Additional experiments evaluated a variety of Ysa T3S-defective mutants for the ability to competitively colonize gastrointestinal tissues by coinfection of mice with mixtures containing equivalent numbers of Ysa T3S mutants and wild type *Y. enterocolitica* (Venecia and Young 2005). This analysis included strains with mutations inactivating genes encoding Ysa T3S apparatus components (*ysaR, ysaU, ysaV* and *orf6-7*) and transcription factors regulating Ysa gene expression (*ysrR* and *rcsB*). Adding further support for the idea that the Ysa T3S system affects gastrointestinal infection, the results from this analysis revealed each of these Ysa T3S-defective strains was 10-100 fold less competitive for terminal ileum colonization.

26.4 The Ysa T3S System Secretome

The pathogenic properties conferred by a T3S system are dictated by activities of the toxic effector proteins it delivers to targeted host cells. Therefore, identification of the proteins exported by the Ysa T3S system was a major research priority (Fig. 2). Early investigations were successful in establishing conditions that induce Ysa T3S system expression with concomitant secretion of the Ysp proteins (Venecia and Young 2005; Young and Young 2002a, Young and Young b). This involved the cultivation of *Y. enterocolitica* in a high ionic strength, nutrient rich medium at 26°C. While these laboratory conditions clearly do not directly reflect the host environment,

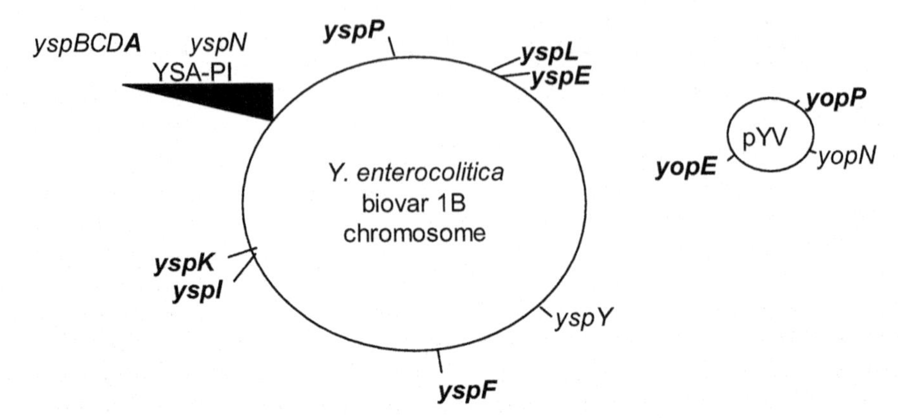

Fig. 3. The relative location of *ysp* genes on the *Y. enterocolitica* biovar 1B, strain 8081, chromosome and on plasmid pYV. Genes that encode translocated effectors are shown in bold text.

the *in vitro* induction of Ysa T3S of proteins was instrumental to gaining perspective on the complete repertoire of Ysps.

The first proteins determined to be Ysps by virtue of their being substrates exported by the Ysa T3S system unexpectedly turned out to be YopE, YopN and YopP, three effector proteins also exported by the Ysc T3S system (Young and Young 2002a). They were identified when the profile of Ysp proteins secreted by wild type *Y. enterocolitica* and a pYV-cured mutant were compared. Because *yopE*, *yopN* and *yopP* are located on pYV, loss of this plasmid consequently eliminated secretion of these three Ysps by the Ysa T3S system. Ultimately, confirmation of YopE, YopN and YopP as Ysa T3S system substrates was attained through a series of genetic and biochemical approaches (Young and Young 2002a). Subsequently, another study independently confirmed that YopE was a secreted by the Ysa T3S system (Foultier et al. 2003). Thus it became clear that the Ysa and Ysc T3S systems share a subset of effectors thereby implicating some effectors in playing a wider role in virulence of *Y. enterocolitica* biovar 1B than previously suspected. Consistent with this idea, it was demonstrated that *Y. enterocolitica* biovar 1B delivers YopP, which suppresses macrophage production of the pro-inflammatory mediator TNFα, into cultured macrophages by utilizing either the Ysa or Ysc pathways (Young and Young 2002a).

Another study focused on characterizing Ysps predicted from analysis of genes within the YSA-PI. That study successfully demonstrated that the *yspBCDA* locus encoded four secreted proteins (Foultier et al. 2003). YspB, YspC and YspD appear to be orthologs of proteins previously described in other bacteria that form the T3S translocon, a portion of the T3S apparatus that forms a pore in the mammalian cell membrane and serves as a passage for the translocation of effectors. YspA however does not share obvious identity to other proteins suggesting it may be an effector.

Recently, a comprehensive proteomics approach defined the complexity of the entire Ysa T3S system secretome (Matsumoto and Young 2006). To characterize secreted proteins, *Y. enterocolitica* was cultivated under conditions that induce the

Ysa T3S system and Ysps released into the surrounding medium were collected (Fig. 2). The proteins were separated by gel electrophoresis and analyzed by a variety of bioanalytical procedures, including mass spectrometry of trypsin-generated peptide fragments and amino acid sequence determination (Matsumoto and Young 2006). Computational analysis was then used to match the peptide mass and amino acid sequence data with predicted open reading frames of the recently completed *Y. enterocolitica* biovar 1B genomic sequence. This effort identified eight additional Ysps while also confirming the seven previously characterized proteins (Matsumoto and Young 2006). These results were further corroborated by two independent approaches. First, each *ysp* gene was inactivated to confirm it was necessary for the production of a corresponding secreted Ysp. Second, each Ysp was confirmed to contain a T3S secretion signal by demonstrating Ysa T3S-dependent export of Ysp-'CyaA chimeras (Matsumoto and Young 2006). The chimeras were created by fusing the amino-terminal half of the Ysp to the secretion signal-lacking activity domain, 'CyaA, of the *Bordetella pertusis* adenylate cyclase protein (Sory and Cornelis 1994; Sory et al. 1995). Strikingly, the genes encoding the fifteen Ysps are dispersed throughout the genome (Matsumoto and Young 2006). Five are located within the YSA-PI, three are located on pYV and the remaining eight map to different chromosomal sites (Fig. 3). The lack of co-localization of Ysp genes suggests that the evolution of *Y. enterocolitica* biovar 1B probably involved a multitude of genetic lateral transfer events. As a disparate group of proteins, the complexity of the Ysps suggests that *Y. enterocolitica* biovar 1B encounters situations that place selective pressure on maintaining the Ysa T3S system.

26.5 Translocated Effectors of the Ysa T3S System

The availability of Ysp-'CyaA chimeras provided the opportunity to examine Ysa T3S-dependent polypeptide delivery into host cells (Matsumoto and Young 2006). Cultured human epithelial HeLa cells were infected with *Y. enterocolitica* strains that each expressed one of the Ysp-'CyaA chimeric proteins under conditions that induce expression of the Ysa T3S system. Upon delivery of the Ysp-'CyaA fusion to the host cell cytosol, 'CyaA synthesis of cyclic AMP (cAMP) is induced due to its activation by calmodulin (Sory and Cornelis 1994). Measurement of a sharp increase in cAMP levels following infection of *Y. enterocolitica* expressing YspA, YspE, YspF, YspI, YspK, YspL, YspP, YopE and YopP chimeras indicated each of these proteins is a translocated effector of the Ysa T3S system. As expected, translocation of each of these chimeras was blocked when the Ysa T3S system was inactivated by a mutation of *ysaV*, encoding an essential component of the Ysa T3S apparatus (Matsumoto and Young 2006; B. Young and G. Young, unpublished results 2007).

Each of these nine effector proteins has been evaluated to assess their role in *Y. enterocolitica* virulence (Matsumoto and Young 2006). YopE and YopP are required for the normal progression of both the gastrointestinal and systemic phases of Yersiniosis in experimentally infected mice. This results are consistent with YopE and YopP serving as substrates for both the Ysa and Ysc T3S systems (Young and Young 2002a). Furthermore, in agreement with the other Ysps being exclusive substrates of the Ysa T3S system, mutants lacking the ability to produce these effectors were defective in bacterial colonization of gastrointestinal tissues.

The well-characterized phenotypes of YopE and YopP provide some mechanistic insight on how the Ysa T3S system affects pathogenesis. YopE is known to block phagocytic cells from internalizing bacteria and counteracts proinfammatory cytokine production by deactivating Rho-family GTPases, including RhoA, Rac-1 and CD42 (Aepfelbacher 2004; Viboud and Bliska 2005). YopP is a potent inhibitor of the innate immune response. It suppresses cytokine production and induces macrophage apoptotic death by blocking activation of the MAP kinase and NF-κB signaling pathways (Mukherjee et al. 2006; Viboud and Bliska 2005).

YspK and YspP are the only other effectors for which a biochemical function has been investigated. They appear to be involved with modification of host cellular processes that are controlled by protein phosphorylation. Both of these effectors have been purified and their enzymatic activities examined *in vitro*. YspK phosphorylates small peptides (Matsumoto and Young 2006) and displays autokinase activity (Matsumoto and Young, unpublished results 2007). YspP was found to cleave the artificial phosphatase substrate *p*-nitrophenyl phosphate with an activity that was most pronounced under acidic conditions (Matsumoto and Young 2006). These results, along with the similarity of YspK and YspP to other proteins, hints at the role they play in pathogenesis. YspK is 91% identical to the *Shigella flexneri* T3S effector OspG, another kinase whose cellular substrate has not been determined but was recently shown to interfere with NF-κB signaling to down regulate the innate immune response (Kim et al. 2005; Matsumoto and Young 2006). YspP shares similarity with a variety of protein tyrosine phosphatases, including the YopH phosphatase secreted by the Ysc T3S system (Bliska et al. 1991). Given the plethora of cellular responses that are affected by protein phosphorylation states, one might expect that YspP will influence signaling pathways that control the immune response.

YspA, YspE, YspF, YspI and YspL are the five remaining effectors which present an exploratory challenge because they are not homologous to other proteins for which a function has been established. YspI and YspL each have a series of amino acid repeats providing some circumstantial evidence of how they may act (Matsumoto and Young 2006). Repeated domains in proteins frequently contribute to inter- and intra-molecular associations leading to the speculative idea that YspI and YspL target specific host cell proteins. Clearly, identification of the effectors translocated by the Ysa T3S system presents opportunities to investigate aspects of *Y. enterocolitica* biovar 1B interactions with the host that otherwise would remain elusive.

26.6 Secreted Facilitators of the Ysa T3S System

The delivery of polypeptides into a eukaryotic cell by *Y. enterocolitica* requires transport across three biological membranes; the bacterial cytoplasmic and outer membranes, and the host cell cytosolic membrane. This process occurs in a spatially restricted location demarcated by the site of bacterial attachment. It is well recognized from studies of other T3S systems that the translocon, a channel through which effector proteins pass across the host cell membrane, consists of two or three different subunit-types (Buttner and Bonas 2002). These translocon subunits are delivered to the host cell membrane by the T3S apparatus. When T3S is induced in the absence of a host cell, these proteins are released into the surrounding medium. YspB, YspC and YspD share similarity to other translocon proteins (Foultier et al. 2002; Matsumoto and Young 2006) and recent studies have established they are essential for translocation of effectors by the Ysa T3S system (Young and Young, unpublished results 2007). In agreement with the hypothesis that these proteins form the Ysa T3S translocon, deletion of *yspB*, *yspC* or *yspD* was found to occlude delivery of both YopE and YspA. Reintroduction of a functional copy of the deleted gene fully restored translocation. Additional preliminary studies using fluorescence microscopy revealed that YspB, YspC and YspD localize to sites where there is intimate contact between the bacterium and host cell (Young and Young, unpublished results 2007).

Another protein that plays an important role in protein translocation is the "molecular ruler" that controls the length of the T3S apparatus needle (Tamano et al. 2002). Improper control of needle elongation has deleterious effects on effector translocation. The best characterized of this family of proteins is InvJ, a component of the *Salmonella* SPI-1 T3S system (Marlovits et al. 2006). The "molecular ruler" for the Ysa T3S system appears to be YspN, a protein that shares a low, but significant, level of similarity to other needle-length control proteins. While it remains to be formally established that the length of the Ysa needle is controlled by YspN, mutational analysis has already established this YspN is required for effector translocation (Matsumoto and Young 2006).

26.7 Transcriptional Control of the Ysa T3S System

YSA-PI genes are not expressed by *Y. enterocolitica* biovar1B cultivated in standard-types of growth media suggesting that specific host environmental cues signal the onset of infection and induction of the Ysa T3S system. Efforts to model these host signals have been elusive, but have lead to the fortuitous observation that a combination of non-physiological parameters promote expression of Ysa and Ysp genes (Venecia and Young 2005). Induction *in vitro* occurs in response to cultivation of *Y. enterocolitica* in a high ionic strength, nutrient rich growth medium at 26°C. It is not clear what specific signal is sensed by the bacterium but, presumably, these conditions simulate the stresses or environmental permutations experienced by the bacterium during an infection. This nonetheless serves a utilitarian purpose by providing an opportunity to gain insight on transcription factors that govern gene transcription within the YSA-PI. We conducted a series of studies to define elements required for Ysa regulation which revealed several transcription factors play important roles, including Crp, RscB and YsrR (Petersen and Young 2002; Venecia and Young 2005).

One study focused on defining genes that affect both the Ysa and Ysc T3S systems as well as the flagellar T3S system, a secretion pathway required for assembly of the flagellar organelle and secretion of the non-flagellar extracellular phospholipase YplA (Petersen and Young 2002). That particular effort defined the cAMP receptor protein (Crp) as a transcription factor required for the expression of each of the three different T3S pathways. This transcription factor senses intracellular levels of the signaling molecular cAMP, which fluctuates in bacteria in response to changes in metabolic potential and osmolarity (Botsford and Harman 1992). Crp may be important for coupling expression of the Ysa T3S system to nutrient stress imposed by growth in the host. Another study determined that the RcsB and YsrR transcription factors also contribute to Ysa transcriptional control (Venecia and Young 2005). In *E. coli* and *Salmonella enterica*, RcsB is the response regulator of the tripartite RcsCDB phosphorelay signal transduction pathway that responds to stresses that affect cell envelope integrity (Erickson and Detweiler 2006). RcsB also appears to act independently of RcsCD to modulate acid resistance (Castanie-Cornet et al. 2007). It remains to be proven, but it is likely that RscB has similar roles in *Y. enterocolitica* suggesting cell envelop stress or acid stress may be other predisposing factors for the induction of the Ysa T3S system. YsrR is a transcription factor that appears to be the response regulator of a two component phosphorelay with the sensor kinase YsrS (Venecia and Young 2005). These two proteins are encoded by a tandem pair of genes located at one extremity of the YSA-PI. It was first thought that YsrSR responded specifically to high concentrations of NaCl (Walker and Miller 2004), but this seems unlikely since KCl serves as an equally potent inducer (Venecia and Young 2005). Furthermore, *ysrS* was defined as a gene whose transcription is itself subject to transcriptional control in response to NaCl and KCl. Thus the mechanism of signal perception by YsrS is a topic open for further investigation. The common thread that appears to connect Crp, RscB and YsrS is environmental stress.

26.8 Concluding Remarks

Many unanswered questions about the Ysa T3S system remain open to further exploration. The current studies provide tantalizing evidence indicating the Ysa T3S system affects the outcome of an infection by influencing bacterial-host interactions during the earliest stages of gastrointestinal infection. Delineating the functional attributes of each Ysp effector will be an especially interesting line of inquiry that will establish the mechanisms through which *Y. enterocolitica* biovar 1B targets host immune responses. The recent discovery of this group of proteins clearly serves a foundation on which future efforts will be executed. Central to defining these mechanisms of host manipulation will be efforts devoted to determining the cell-types that are targeted by the Ysa T3S system during an infection. In the broader sense, there is significant interest in understanding what ecological advantage exists for *Y. enterocolitica* biovar 1B to have acquired and maintained this complex protein secretion pathway.

26.9 Acknowledgements

Members of G. M. Young's research group are gratefully acknowledged for each of their contributions to the results presented at the *Yersinia* Symposium and within this article. This work was supported, in part, by a grant from the National Institutes of Health, National Institute of Allergy and Infectious Diseases, R21 AI156042.

26.10 References

Aepfelbacher, M. (2004) Modulation of Rho GTPases by type III secretion system translocated effectors of *Yersinia*. Rev. Physiol. Biochem. Pharmacol. 152, 65-77.

Bliska, J.B., Guan. K.L., Dixon. J.E. and Falkow, S. (1991) Tyrosine phosphate hydrolysis of host proteins by an essential *Yersinia* virulence determinant. PNAS 88, 1187-1191.

Botsford, J.L. and Harman, J.G. (1992) Cyclic AMP in prokaryotes. Microbiol. Rev. 56, 100-122.

Bottone, E.J. (1999) *Yersinia enterocolitica*: overview and epidemiologic correlates. Microbes Infect. 1, 323-333.

Bradford, W.D., Noce, P.S. and Gutman, L.T. (1974) Pathologic features of enteric infection with *Yersinia enterocolitica*. Arch. Pathol. 98, 17-22.

Buttner, D. and Bonas, U. (2002) Port of entry--the type III secretion translocon. Trends Microbiol. 10, 186-192.

Carniel, E. and Mollaret, H.H. (1990) Yersiniosis. Comp. Immunol. Microbiol. Infect. Dis. 13, 51-58.

Carter, P.B. (1975a) Oral *Yersinia enterocolitica* infection of mice. Amer. J. Pathol. 81, 703-705.

Carter, P.B. (1975b) Pathogenicity of *Yersinia enterocolitica* for mice. Infect. Immun. 11, 164-170.

Carter, P.B. and Collins, F.M. (1974a) Experimental *Yersinia enterocolitica* infection in mice: kinetics of growth. Infect. Immun. 9, 851-857.

Carter, P.B. and Collins, F.M. (1974b) The route of enteric infection in normal mice. J. Exp. Med. 139, 1189-1203.

Castanie-Cornet, M.P., Treffandier, H., Francez-Charlot, A., Gutierrez, C. and Cam, K. (2007) The glutamate-dependent acid resistance system in *Escherichia coli*: essential and dual role of the His-Asp phosphorelay RcsCDB/AF. Microbiology 153, 238-246.

Cornelis, G., Laroche, Y., Balligand, G., Sory, M.-P. and Wauters, G. (1987) *Y. enterocolitica*, a primary model for bacterial invasiveness. Rev. Infect. Dis. 9, 64-87.

Cover, T.L. and Aber, R.C. (1989) *Yersinia enterocolitica*. N. Engl. J. Med. 321, 16-24.

Denecker, G., Totemeyer, S., Mota, L.J., Troisfontaines, P., Lambermont, I., Youta, C., Stainier, I., Ackermann, M. and Cornelis, G.R. (2002) Effect of low- and high-virulence *Yersinia enterocolitica* strains on the inflammatory response of human umbilical vein endothelial cells. Infect. Immun. 70, 3510-20.

Erickson, K.D. and Detweiler, C.S. (2006) The Rcs phosphorelay system is specific to enteric pathogens/commensals and activates *ydeI*, a gene important for persistent *Salmonella* infection of mice. Mol. Microbiol. 62, 883-894.

Foultier, B., Troisfontaines, P., Muller, S., Opperdoes, F.R. and Cornelis, G.R. (2002) Characterization of the *ysa* pathogenicity locus in the chromosome of *Yersinia enterocolitica* and phylogeny analysis of Type III secretion systems. J. Mol. Evol. 55, 37-51.

Foultier, B., Troisfontaines, P., Vertommen, D., Marenne, M.-N., Rider, M., Persot, C. and Cornelis, G.R. (2003) Identification of substrates and chaperone from the *Yersinia enterocolitica* 1B Ysa type III secretion system. Infect. Immun. 71, 242-253.

Galan, J.E. and Wolf-Watz, H. (2006) Protein delivery into eukaryotic cells by type III secretion machines. Nature 444, 567-573.

Gleason, T.H. and Patterson, S.D. (1982) The pathology of *Yersinia enterocolitica* ileocolitis. Amer. J. Sur. Pathol. 6, 347-355.

Haller, J.C., Carlson, S., Pederson, K.J. and Pierson, D.E. (2000) A chromosomally encoded type III secretion pathway in *Yersinia enterocolitica* is important in virulence. Mol. Microbiol. 36, 1436-1446.

Hanski, C., Kutschka, U., Schmoranzer, H.P., Naumann, M., Stallmach, A., Hahn, H., Menge, H. and Riecken, E.O. (1989) Immunohistochemical and electron microscopic study of interaction of *Yersinia enterocolitica* serotype O:8 with intestinal mucosa during experimental enteritis. Infect. Immun. 57, 673-678.

Iwobi, A., Heesemann, J., Garcia, E., Igwe, E., Noelting, C. and Rakin, A. (2003) Novel virulence-associated type II secretion system unique to high-pathogenicity Yersinia enterocolitica. Infect. Immun. 71, 1872-1879.

Kim, D.W., Lenzen, G., Page, A.-L., Legrain, P. and Sansonetti, P.J. (2005) The *Shigella flexneri* effector OspG interferes with innate immune responses by targeting ubiquitin-conjugating enzymes. PNAS 102, 14046-14051.

Marenne, M.-N., Mota, L.J. and Cornelis, G.R. (2004) The pYV plasmid and the Ysc-Yop type III secretion system. In: E. Carniel E and B.J. Hinebusch (Eds.), Yersinia: *Molecular and Cellular Biology*. Horizon Press, Norfolk, pp. 319-348.

Marlovits, T.C., Kubori, T., Lara-Tejero, M., Thomas, D., Unger, V.M. and Galan, J.E. (2006) Assembly of the inner rod determines needle length in the type III secretion injectisome. Nature 441, 637-40.

Matsumoto, H. and Young, G.M. (2006) Proteomic and functional analysis of the suite of Ysp proteins exported by the Ysa type III secretion system of *Yersinia enterocolitica* Biovar 1B. Mol. Microbiol. 59, 689-706.

Mukherjee, S., Keitany, G., Li, Y., Wang, Y., Ball, H.L., Goldsmith, E.J. and Orth, K. (2006) *Yersinia* YopJ acetylates and inhibits kinase activation by blocking phosphorylation. Science 312, 1211-1214.

Petersen, S. and Young, G.M. (2002) Essential role for cAMP and its receptor protein in *Yersinia enterocolitica* virulence. Infect. Immun. 70, 3665-3672.

Portnoy, D.A., Wolf-Watz, H., Bolin, I., Beeder, A.B. and Falkow, S. (1984) Characterization of common virulence plasmids in *Yersinia* species and their role in the expression of outer membrane proteins. Infect. Imunn. 43, 108-114.

Snellings, N.J., Popek, M. and Linder, L.E. (2001) Complete DNA sequence of *Yersinia enterocolitica* Serotype O:8 low-calcium response plasmid reveals a new virulence plasmid-associated replicon. Infect. Immun. 69, 4627-4638.

Sory, M.-P. and Cornelis, G.R. (1994) Translocation of hybrid YopE-adenylate cyclase from *Yersinia enterocolitica* into HeLa cells. Mol. Microbiol. 14, 583-594.

Sory, M.-P., Boland, A., Lambermont, I., Cornelis, G.R. (1995) Identification of the YopE and YopH domains required for secretion and internalization into the cytosol of macrophages, using the *cyaA* gene fusion approach. PNAS 92, 11928-12002.

Tamano, K., Katayama, E., Toyotome, T., Sasakawa, C. (2002) *Shigella* Spa32 is an essential secretory protein for functional type III secretion machinery and uniformity of its needle length. J. Bacteriol. 184, 1244-1252.

Thomson, N.R., Howard, S., Wren, B.W., Holden, M.T., Crossman, L., Challis, G.L., Churcher, C., Mungall, K., Brooks, K., Chillingworth, T., Feltwell, T., Abdellah, Z., Hauser, H., Jagels, K., Maddison, M., Moule, S., Sanders, M., Whitehead, S., Quail, M.A., Dougan, G., Parkhill, J. and Prentice, M.B. (2006) The complete genome sequence and comparative genome analysis of the high pathogenicity *Yersinia enterocolitica* strain 8081. PLoS Genet 2, e206.

Vantrappen, G., Ponette, E., Geboes, K. and Bertrand, P. (1977) *Yersinia* enteritis and enterocolitis: gastroenterological aspects. Gastroenterol. 72, 220-227.

Venecia, K. and Young, G.M. (2005) Environmental regulation and virulence attributes of the Yas type III secretion system of *Yersinia entercolitica* Biovar 1B. Infect. Immun. 73, 5961-5977.

Viboud, G.I. and Bliska, J.B. (2005) *Yersinia* outer proteins: role in modulation of host cell signaling responses and pathogenesis. Annu. Rev. Microbiol. 59, 69-89.

Walker, K.A. and Miller, V.L. (2004) Regulation of the Ysa type III secretion system of *Yersinia enterocolitica* by YsaE/SycB and YsrS/YsrR. J. Bacteriol. 186, 4056-4066.

Young, B.M. and Young, G.M. (2002a) Evidence for targeting of Yop effectors by the chromosomally encoded Ysa type III secretion system of *Yersinia enterocolitica*. J. Bacteriol. 184, 5563-5571.

Young, B.M. and Young, G.M. (2002b) YplA is exported by the Ysc, Ysa and flagellar type III secretion systems of *Yersinia enterocolitica*. J. Bacteriol. 184, 1324-1334.

Picture 24. Dancing to the music of Impact. Photograph by R. Perry.

27

A Rationale for Repression and/or Loss of Motility by Pathogenic *Yersinia* in the Mammalian Host

Scott A. Minnich and Harold N. Rohde

Department of Microbiology, Molecular Biology, and Biochemistry, University of Idaho, sminnich@uidaho.edu

Abstract. Pathogenic yersiniae either repress flagella expression under host conditions (*Yersinia enterocolitica* and *Yersinia pseudotuberculosis*) or have permanently lost this capability by mutation (*Yersinia pestis*). The block in flagella synthesis for the enteropathogenic *Yersinia* centers on *fliA* (σ^F) repression. This repression ensures the downstream repression of flagellin structural genes which can be cross-recognized and secreted by virulence type III secretion systems. *Y. pestis* carries several flagellar mutations including a frame shift mutation in *flhD*, part of the flagellar master control operon. Repression of flagellins in the host environment may be critical because they are potent inducers of innate immunity. Artificial expression of flagellin in *Y. enterocolitica* completely attenuates virulence, supporting the hypothesis that motility is a liability in the mammalian host.

27.1 Introduction

Motility for *Yersinia enterocolitica* and *Yersinia pseudotuberculosis* is temperature-regulated between the narrow range of 30°C (Mot$^+$) and 37°C (Mot$^-$). In contrast, *Yersinia pestis* has been classed as a nonmotile organism since it was first isolated at the turn of the last century. Temperature is a key environmental cue used by these facultative pathogens during the host adaptational response. Exposure to 37°C (host temperature) results in major phenotypic changes involving numerous suites of genes that undergo repression, such as motility, and activation, evidenced by induction of a number of chromosomal and virulence plasmid (pYV) loci. In this manuscript we would like to summarize our efforts in dissecting the mechanism of temperature regulation, particularly as it applies to motility, and discuss how understanding the temperature-regulation of motility has shed additional light on understanding virulence of *Yersinia* and related Gram-negative pathogens.

27.1.1 Flagellum Biosynthesis

Flagellum biosynthesis in Gram negative enteric pathogens involves the coordinated expression of ca. 50 genes. Much of our knowledge of flagellar biosynthesis and regulation has been derived from the *Salmonella typhimurium* and *Escherichia coli* models (reviewed in Mcnab 2004; Aldridge and Hughes 2002) and the pathogenic *Yersinia* fit this same general pattern of hierarchal gene expression. Motility is regulated by a number of environmental and physiological conditions, but initiation of flagella synthesis proper begins with expression of the *flhDC* operon or class I genes. This operon encodes two transcriptional activators that form a heterotetramer required for induction of class II flagellar genes. Class II genes encode basal body and hook structural genes as well as two regulators of class III genes. These class III gene regulators are denoted as FliA or σ^F, a specific sigma factor that directs RNA polymerase to the conserved promoter sequence of flagellar class III genes (TAA-N_{15}-GCCGATA), and FlgM, the anti- σ^F factor. FliA and FlgM are synthesized with other class II genes, but FliA is prevented from activating class III genes by FlgM until the basal body hook structure is completed whereupon FlgM is secreted through this structure to the extracellular environment. This depletion of FlgM relieves FliA repression and promotes FliA-dependent induction of Class III flagellar genes which include flagellin(s), chemotaxis, and motor genes. Of these, flagellin is highly expressed and accounts for ca. 95% of the mass of the flagellum. Flagellin monomers are exported from the cytosol through the hollow core of the basal body hook complex spanning the inner to outer membrane, and are polymerized at the growing distal tip of the flagellum filament. *Y. enterocolitica* expresses three highly similar flagellin genes, designated *fleA*, *fleB*, and *fleC* that are positioned tandemly on the chromosome, each of which is expressed from its own FliA-dependent promoter (Kapatral and Minnich 1995.)

27.2 *Yersinia* Temperature Regulation: The Nature of the Cellular Thermostat

In the early 1990s we initiated experiments to address two basic questions involving temperature regulation of gene expression in pathogenic yersiniae. First we wished to understand the mechanism of temperature-induced signal transduction. In other words, this question focused on understanding the nature of the bacterium's 'thermostat' and how a change in temperature is recognized and processed at the genetic level. Second, we desired to understand if there was an underlying deeper rationale for repressing motility under host conditions beyond a simple explanation of energy conservation and/or repressing flagellin due to its high antigenic properties.

Our original strategy to address the first question above was to use transposon and chemical mutagenesis to determine if a class of motility mutants could be isolated that demonstrated motility at the nonpermissive temperature of 37°C. Additionally, we also set out to generate *phoA* reporter fusions to high-temperature induced genes such as members of *yop* (*Yersinia* outer protein) and *ysc* (*Yersinia* secretion) gene families. Such reporter fusions would allow us to screen for mutants expressing these genes at the nonpermissive low temperature of <30°C. We reasoned that mutants over-riding normal temperature regulation in either system should lead to the genetic components of the thermostat, particularly if such mutations displayed pleiotropy for other temperature-regulated loci.

Serendipitously we were lucky in identifying such a pleiotropic mutant at the initial stage of constructing strains to carry out these experiments. To generate mini-Tn*phoA* mutants by conjugation, we arbitrarily decided to use novobiocin resistance as a selectable marker in our *Y. entercolitica* recipient target strain. A number of spontaneous novobiocin resistant isolates of *Y. enterocolitica* 8081v were selected at 25°C, and we noted a class of these drug resistant mutants displayed spontaneous autoaggregation when grown in LB broth at 25°C. Because autoaggregation is a phenotype associated with growth at 37°C, these mutants were further characterized for motility and Yop secretion. We found this class of mutants was nonmotile by motility agar assay, did not make flagellin protein as assayed by western blot using the 15D8 anti-flagellin monoclonal antibody (Feng et al. 1990) and that this mutant class was constitutive for Yop secretion. Therefore, it appeared that this mutant displayed a 'locked-on' high temperature phenotype.

Because novobiocin targets DNA gyrase activity, we examined the effect of temperature on DNA supercoiling using pACYC as a reporter of supercoiling (topoisomer distribution) during temperature shifts. There was a direct correlation with the timing of temperature-regulated gene expression and temperature-induced changes in supercoiling (Rohde et al. 1994). However, the observed net change in temperature-induced DNA supercoiling was greater than the theoretical predicted change over the temperature range examined suggesting another component was contributing to temperature regulation. Several reports in the literature noted that DNAs with intrinsic bends underwent conformational changes at the mean temperature of 37°C (Chan et al. 1990). Using a novel gel system to identify intrinsically bent DNA allowed us to determine that the pYV plasmid was enriched for regions of intrinsic DNA bending and that several promoters on pYV that were induced by temperature were bracketed by these regions of DNA bending. This work and a report by the Cornelis laboratory that mutations in the histone-like gene, *ymoA*, produced a similar phenotype to our novobiocin resistant mutants (Cornelis et al. 1991), led to the overall model that DNA was acting as the cellular thermostat (Rohde et al. 1999).

27.2.1 Other Factors in Temperature Regulation

Given this model of temperature-regulation, there is still another component to this process yet to be identified. Virulence genes, such as the *yops* and *ysc* are transcriptionally activated immediately when cells are exposed to 37°C, and flagellin genes are immediately repressed (Kapatral et al. 1996). As such, we envision virulence gene promoters as 'spring-loaded', responding immediately to a shift to mammalian temperature by localized DNA rearrangement. In contrast, re-setting the 'spring' in a temperature downshift takes considerable time. We found that restoration of flagellum biosynthesis, monitored by appearance of flagellins, is cell-cycle dependent, requiring ca. 2.5 cell generations (Rohde et al. 1994). The finding that *dam* methylase affects both *yop* and flagellar gene expression is intriguing and it may well be that selective methylation of specific promoters in each system contributes to temperature regulation (Julio et al. 2002). One could envision that temperature activated DNA methylation at 37°C may set the switch for a given promoter, and that 2 generations would be required to titrate out methylated DNA (fully methylated>hemi-methylated>unmethylated) upon a temperature-downshift. Further support for this type of scenario is evidenced by differential *Bam*H1 restriction patterns of pYV isolated from cells grown at 37°C vs. 25°C (our unpublished observation). *Bam*H1 restriction sites have an internal *dam* methylation site, GATC.

These early studies provided insight into the mechanism of temperature regulation, but most importantly, this initial work showed that motility and Yop expression were coordinately regulated by temperature in a reciprocal manner, i.e. there was a direct genetic link between these two seemingly disparate systems. This observation provided the impetus to address our second major question as to why flagellum biosynthesis is temperature regulated.

27.3 Is There a Direct Link Between Flagellar and Virulence Secretion Systems?

Several key observations were reported during the early 1990s suggesting additional ties between flagellar and Yop systems. First, David Mullin (Tulane University) and Austin Newton (Princeton University) published that the *Caulobacter crescentus flbF* gene (now designated *flhA*) showed sequence similarity to *Y. pseudotuberculosis lcrD* encoded on pYV. (Sanders et al. 1992; Ramakrishnan et al. 1991) LcrD is required for Yop secretion. Lucy Shapiro's laboratory, also working on *C. crescentus*, soon thereafter identified additional *C. crescentus* flagellar genes showing sequence similarity to *ysc* genes on the pYV of pathogenic *Yersinia* (Zhuang and Shapiro 1995). The interesting fact regarding these similarities and their homologs in the *S. typhimurium* flagellar system was that the proteins encoded were not part of the flagellum engine structure, but appeared to be involved in protein secretion. The Cornelis lab also reported that the secreted Yop proteins were exported independent of the *secA*-dependent general secretory pathway (GSP), in part evidenced by lack of N-terminal signal sequence processing (Michiels and Cornelis 1991). This secretion pattern is also reflected in bacterial flagellin export which is GSP-independent. Finally, Roqvist et al. (1994) showed that Yop proteins were exported from discrete foci on the surface of *Y. pseudotuberculosis* in contrast to the heretofore accepted model that Yops were distributed uniformly to the outer membrane and released by limiting calcium ion concentrations (in vitro) or contact with target host cells (in vivo). Thus the following pieces of a developing puzzle linking Yop and flagellar systems were falling into place: (i) the Yop and flagellar systems were regulated in an opposing or reciprocal manner by temperature; (ii) a class of mutants showed this regulation was coordinate; (iii) both systems had a set of genes displaying sequence similarity; (iv) proteins exported by each system were GSP-independent; (v) secretion of Yops was directed from discrete foci on the cell surface similar to flagellin export from the basal body.

27.3.1 A Testable Model

These similarities between Yop and flagellin regulation and secretion led us to hypothesize in 1994 that the simplest explanation to account for these similarities was that the flagellum was being used for Yop secretion at 37°C. This idea was in part based on considering the flagellum not only an organelle for cell motility, but viewing it as a highly efficient and dedicated protein secretory device. When considering the amount of protein (flagellin) exported by the basal body it seemed reasonable to suggest that this organelle might be adaptable to secrete nonflagellar proteins under certain conditions. This hypothesis made several testable predictions. First, it predicted that mutations in basal body structural genes of *Y. enterocolitica* would disrupt motility as well as Yop export. Second, the model predicted that a subset of flagellar genes would have to be expressed at 37°C even though the cells are nonmotile at this temperature. Finally, it predicted that nonmotile *Y. pestis* which secretes Yops similar to *Y. enterocolitica* and *Y. pseudotuberculosis*, would have to possess flagellar genes, a subset of which would have to be expressed at 37°C. To test these possibilities, Michael Smith in my laboratory set about making a library of mini-Tn*lacZ* transcriptional fusions to *Y. enterocolitica* flagellar genes. Using the Tn*lacZ* reporter system would enable us to address the first two predictions simultaneously; nonmotile transposon insertion mutants could be tested for their inability to secrete Yops at 37°C, and identified flagellar promoter-*lacZ* fusions generated could be tested for flagellar gene expression at 37°C by β-galactosidase assays.

Using this approach, over one hundred mini-Tn*lacZ* nonmotile mutants were isolated from a screen of several thousand transposon insertions. These nonmotile mutants were first screened to verify that they had not lost pYV and therefore had the *yop* regulon intact. These nonmotile transposon mutants were then assayed for their ability to secrete Yops by shifting cultures to 37°C in media chelated for calcium ion and examining cell free culture supernates for Yop proteins. Eight of these nonmotile mutants were also negative for Yop secretion at 37°C (Fig. 1). Two of the mini-Tn*lacZ* insertion sites were identified; one was localized to a rod protein flagellar gene and the other was mapped as being intergenic between *fliB* and *fliA*. This latter mutant therefore seemed to be a regulatory mutation and because it was not within a structural coding region of a flagellar gene, we reasoned that a spontaneous suppressor might be obtainable. This nonmotile mutant was stabbed into motility agar and screened for motile flares. Several flares were observed after 3-4 days incubation and these suppressors were shown to be flagellum positive and restored for Yop secretion (Fig. 1, lanes 3 and 4). This class of nonmotile mutant that failed to secrete Yops was consistent with the hypothesis that the flagellum was directly involved in Yop export. More importantly, the monmotile mutant between *fliA* and *fliB* and its cognate suppressor showed a single mutation knocked out both phenotypes, and a spontaneous suppressor restored both phenotypes.

27.3.2 Temperature Regulation of *Y. enterocolitica* Flagella Genes

The set of nonmotile mini-Tn*lacZ* fusions was also screened for *lacZ* expression on LB X-gal plates to identify blue colonies signifying the *lacZ* was regulated by a flagellar promoter. Positive clones were analyzed to identify the flagellar gene insertionally inactivated using DNA sequence analysis. β-galactosidase assays were performed on representative genes identified in each class of the regulatory tier of the flagellar regulon, (classes 1-3). These results are presented in Fig. 2. We found that the class I operon, *flhDC*, showed a 3-4 fold decrease in expression at 37°C compared to 30°C. A translational *lacZ* fusion was also constructed for this operon;

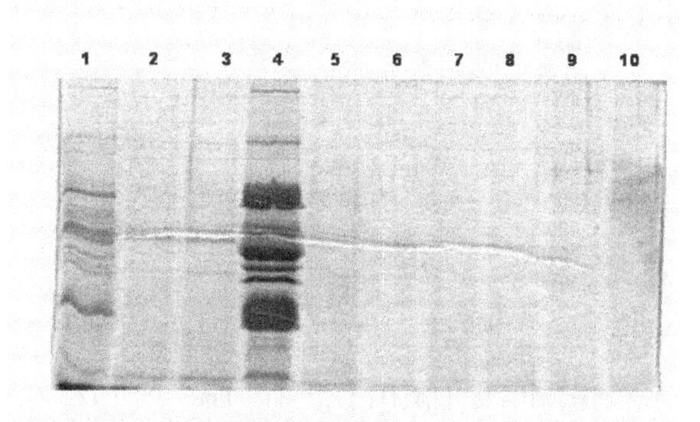

Fig. 1. A subset of Tn*lacZ* insertion mutants screened for loss of motility also demonstrate loss of Yop secretion. Lane 1 is a cell supernate from the positive control, wild-type *Y. enterocolitica* 8081v induced for Yop secretion by growth at 37°C in medium chelated for calcium ion. Lanes 2-3 and 5-10 are independent nonmotile isolates that show loss of Yop secretion under inducing conditions. Lane 3 is the supernate derived from a Tn*lacZ* insertion mapped to the *fliB/fliA* intergenic region. Lane 4 is a spontaneous revertant of this mutant which showed restored motility and Yop secretion.

we found no significant difference in the expression levels of β-galactosidase between high and low temperature (Smith 2000). These experiments showed that, at the transcriptional level, there is a modest temperature regulation of flagellar class I genes, but levels are still very high at 37°C. No temperature regulation was observed at the translational level.

Several flagellar class 2 genes were likewise analyzed for temperature regulation. These included *flhA* ,*flhB, fliA, flgM, flgK* and *flgE*. Several of these are shown in Fig. 2. The temperature regulation of class II flagellar genes is more pronounced showing anywhere from 4 to 10 fold regulation, with the exception of *fliA* which showed showed between 14 and 20 fold regulation. Interestingly, these class II genes show similar levels of repression in *E. coli* and *S. typhimurium fliA* mutants (Kutsukake 1997; Clarke and Sperandio, 2005). This is noteworthy for two reasons. First, there may be in *Yersinia*, like *E. coli* and *S. typhimurium*, some type of positive feedback on class II gene expression by FliA. We have not examined this possibility. Secondly, the higher level of *fliA* repression by temperature seems to point to this gene as the 'lynch-pin' for flagellar temperature regulation. The more recent microarray data by Horne and Pruss (2006) is consistent with this finding that implies *Y. enterocolitica* temperature-regulation of motility is conferred by repression of *fliA*, the positive regulator of class III genes including flagellin. *lacZ* fusions to one of the flagellin genes, *fleB*, shows a 25 fold repression at 37°C. Likewise, no flagellin protein is detectable by western blot from cells grown at 37°C. We conclude that temperature regulation of motility in *Y. enterocolitica* is primarily centered on repression of class III genes which includes flagellin, the major secreted type III protein of the flagellar system. These findings are consistent for the first two predictions of the hypothesis that the flagellum could be used to secrete virulence

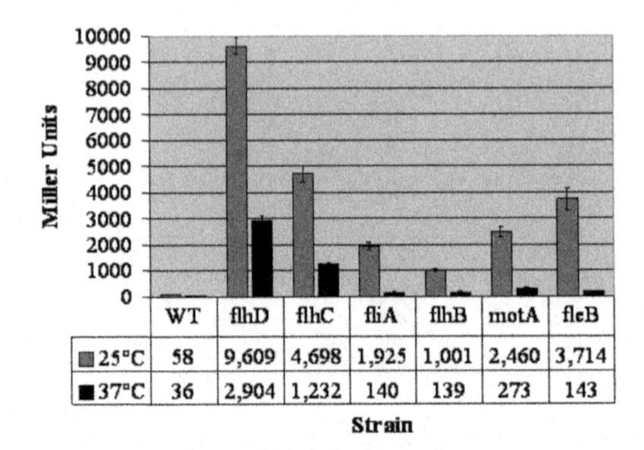

	WT	flhD	flhC	fliA	flhB	motA	fleB
25°C	58	9,609	4,698	1,925	1,001	2,460	3,714
37°C	36	2,904	1,232	140	139	273	143

Fig. 2. β-galactosidase assays on Tn*lacZ* transcriptional fusions to a representative set of *Y. enterocolitica* flagellar genes. Cells for each mutant were grown to the same stage of growth at either 25°C or 37°C and the amount of β-galactosidase was assayed and is displayed in Miller units. WT denotes background levels of β-galactosidase in the wild-type control strain. The *flhD* and *flhC*, class 1 flagellar genes, show a 3.3 and 3.8 fold decrease in expression at 37°C respectively. The *fliA* gene, encoding σ^F, shows a ca. 14 fold repression at 37°C. The class II *flhB* gene shows a 10-fold reduction in expression at 37°C, and *fleB*, flagellin structural gene, shows a 26-fold reduction in expression at 37°C. Other class II genes (not shown) exhibited anywhere from 5 to 10 fold repression at 37°C. Transcriptional repression of motility thus appears centered on blocking flagellin transcription, the major exported type III protein.

factors since mutations in the flagellar system were found to affect Yop protein secretion, and that flagellar genes remained expressed at 37°C.

27.3.3 Examination of *Y. pestis* Flagellar Genes

The question regarding whether or not *Y. pestis* harbored flagellar genes was also examined. Before the DNA genomic sequences were completed for *Y. pestis* strains CO92 and KIM, we used *Y. enterocolitica* flagellar genes as probes for Southern blotting against *Y. pestis* DNA under high stringency hybridization conditions. All probes tested were positive. Using reverse transcriptase PCR, Vinayak Kapatral in my laboratory, showed that transcripts for *Y. pestis flhD* (class 1 flagellar gene), *fliA* and *flgM* (class 2 flagellar genes) were detectable from RNA isolated from *Y. pestis* KIM6 grown at 25°C and 37°C (Fig 3). However, no transcript was detectable for flagellin genes at either temperature, consistent with a nonmotile phenotype. Likewise, immunoblots with the monoclonal antibody 15D8, which recognizes flagellin from both *Y. enterocolitica* and *Y. pseudotuberculosis*, did not react with total protein isolated from *Y. pestis*. Thus, these preliminary data were also consistent with our hypothesis since *Y. pestis* harbors flagellar genes, and by RT-PCR a subset of flagellar genes showed evidence of transcription at 37°C. It is now evident from genomic sequencing that the remnants of two flagellar systems are present on the *Y. pestis* chromosome.

To further verify the potential role of flagellar and virulence protein export, functional expression of *Y. pestis flhDC* operon was examined. DNA sequence analysis of the *Y. pestis flhDC* clone showed high similarity for other enteric *flhD*C genes; however, a single T insertion generating a frameshift mutation was identified in *flhD*. To determine if this mutation negated FlhDC function we used a complementation assay.

Mike Smith generated a nonmotile *Y. enterocolitica flhDC* deletion mutant (Smith 2000). A plasmid bearing the *Y. pestis flhDC* operon was then introduced into the *Y. enterocolitica* Δ*flhDC* strain to determine if the *Y. pestis* fragment could complement the motility defect. Complementation was not observed using a soft agar motility plate assay. However, by prolonged incubation of these plates suppressors, evidenced as motility flares originating from the nonmotile colony stabs, were obtained. DNA sequencing of the plasmid-borne *Y. pestis flhDC* insert from one of these suppressors showed a 5 base pair duplication by replication 'slippage' just downstream of the T insertion which restored the proper reading frame and FlhD function. It is interesting to note that the same *flhD* mutation identified in *Y. pestis* KIM6 is also present in the *Y. pestis* CO92 strain. Genomic analysis of other *Y. pestis* KIM6 flagellar genes identified two additional deletion mutations in basal body structural genes.

In summary, *Y. pestis* was motile at some point in its history but acquired a mutation in *flhD*. It appears that some flagellar genes are expressed based on our RT-PCR analysis but overall the pathway is blocked by additional acquired mutations. As we argue below, this loss of motility confers a selective advantage to *Y. pestis* and may contribute significantly to virulence. Regarding the original hypothesis of flagellar and Yop functional overlap, these observations were enigmatic; flagellar genes were present as predicted but a mutation in *flhD* (class I gene) was inconsistent. We have not pursued this point further, so the question of *Y. pestis* class II gene expression remains unresolved.

At this point, all the data appeared consistent with an overlap between flagellar gene and Yop export. To further confirm this point we employed the *Y. enterocolitica* Δ*flhDC* mutation constructed to look at *Y. pestis flhDC* function. A mutation in this class I operon should knock out both systems if the hypothesis was correct. This Δ*flhDC* strain is nonmotile, but when examined for Yop secretion we found this mutation had no effect on Yop export; Yops were secreted just as efficiently as the wild-type motile strain. These experiments were conducted at the same time as reports of isolation of the *S. typhimurium* 'needle-like' type III secretion apparatus, which are also made in nonmotile mutants of *S. typhimurium* (Kubori et al. 1998). Hence, our original hypothesis was negated, Yop and flagellin export operate as separate parallel type III secretion systems (TTSS).

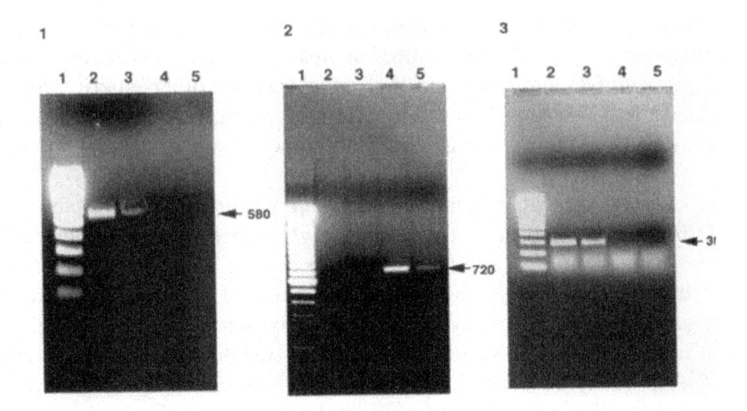

Fig. 3. Reverse transcriptase assays on RNA isolated from *Y. pestis* grown at 25°C and 37°C shows expression of *flhC, fliA*, and *flgM*. In each panel (1-3) the first lane is an RNA molecular weight marker. RT-PCR was performed using a set of primers designed to amplify *flhC* as a 500 bp fragment based on the DNA sequence. Product for *flhC* of the predicted size is shown in panel 1 with lane 2 representing RNA from 25°C cultures; lane 3, RNA isolated from 37°C cultures; lanes 4 and 5 are control lanes with no RNA (lane 4) or RNA treated with DNase free RNase. Panel 2 is an RT-PCR experiment using primers based on the *Y. pestis fliA* gene sequence. Lanes 2 and 3 are controls similar to those in panel 1. Lane 4 is RT-PCR product generated from RNA isolated from 25°C cultures and lane 5 is product generated from RNA isolated at 37°C. For both lanes 4 and 5, the PCR product is of the predicted size. Panel 3 is an RT-PCR experiment using primers based on the *Y. pestis flgM* gene sequence. Lanes 2 and 3 represent predicted size product from using RNA template isolated from 25°C and 37°C cultures respectively. Panel 3 lanes 4 and 5 are control lanes as described above.

27.3.4 Substrate Reciprocity Between Type III Secretion Systems

Part of this model linking virulence protein and flagellar secretion was resurrected when Glenn Young identified a secreted *Y. enterocolitica* phospholipase, YplA, that was under *flhDC* regulation. We made our set of *Y. enterocolitica* flagellar gene mutants available to test whether YplA utilized the flagellar apparatus for secretion. In an elegant study, it was shown that not only YplA was secreted by the flagellum, but a number of additional nonflagellar proteins, designated Fops, can use the flagella secretory system for export (Young et al. 1999). This was the first report that the flagellum could be used to secret virulence factors. Concurrently, Dorthy Pierson identified an additional *Y. enterocolitica* TTSS on the chromosome, termed the Ysa TTSS, that was induced in vitro by high salt (0.3 M NaCl) at temperatures $\leq$30°C (Haller et al. 2000). At this point the function of the Ysa TTSS is not known. Glenn Young has shown the YplA can be exported from not only the flagellum but also the Ysa and Ysc systems (Young and Young 2002). This work, and the earlier work by Anderson et al. (1999) showed that the type-III protein secretion signal of a given protein can be cross-recognized by multiple TTSS in a given organism, or be recognized if expressed in a heterologous host.

These observations caused us to consider the possibility that the reciprocal coordinate temperature regulation of flagellar and Yop systems was necessary due to the potential of substrate reciprocity between the systems. Although it is clear that the *Y. enterocolitica* flagellar, Ysa, and Ysc TTSS are separate parallel systems, the observed regulatory segregation of these systems by specific environmental factors might ensure prevention of substrate reciprocity of their respective secreted protein products. If flagellin is cross-recognized by other TTSS it could compete for export. The more recent recognition that bacterial flagellins fall into a class of pathogen associated molecular pattern (PAMP) molecules interacting with specific surface bound and intracellular toll-like receptors (TLRs) lends itself to a further potential reason why pathogenic *Yersinia* repress motility in the host (reviewed by Steiner 2006).

To test whether or not flagellin is recognized by the Ysa and Ysc TTSS, we cloned one of the *Y. entercolitica* flagellin genes, *fleB*, and placed it under control of the inducible p*tac* promoter. Two *Y. enterocolitica* nonmotile strains, *flhD*::Ω and *flhB*::*lacZ*, were used to examine FleB secretion under various environmental conditions. Both strains tested are nonmotile and do not synthesize flagellins as verified by immunoblots with flagellin specific monoclonal antibody prior to transformation with pACYC-p*tac*-*fleB*. Using these strains we found that FleB can be exported from either the Ysa or Ysc TTSS as shown in Fig. 4. At 37°C, FleB is exported but only when calcium is limiting, the same conditions required for Yop export. Expressing FleB at 37°C in a pYV-cured strain under limiting calcium conditions did not result in FleB export, consistent with dependency on the Ysc TTSS. At 25°C, FleB is exported, but only when the salt concentration is raised to 0.3 M, conditions conducive to expression of the Ysa TTSS and repression of flagellum biosynthesis. We conclude that FleB, if expressed, is recognized by any operative TTSS in the cell. As such, segregation of expression of these TTSS by mutually exclusive environmental signals prevents TTSS substrate cross-recognition.

27.3.5 Flagellin Expression in the Host Attenuates *Y. enterocolitica*

Does artificial expression of flagellin under host conditions affect virulence? To test this possibility, we compared the morbidity and mortality of wild-type *Y. entercolitica* and *Y. enterocolitica* expressing *fleB* from p*tac* in the mouse model using oral inoculation (n=10 for each group of animals). The results from this preliminary experiment showed 100% mortality in the control mice receiving wild-type *Y. enterocolitica* strain 8081v, and complete attenuation of the isogenic strain expressing FleB. Presently, we are determining if attenuation is due to cytokine induction via TLR-5 or Ipaf stimulation, competitive inhibition of Yop secretion, or a combination of both.

27.4 Summary

In summary, these experiments outline the rationale for *Y. enterocolitica* and *Y. pseudotuberculosis* reciprocal temperature regulation of Yops and flagellar type III

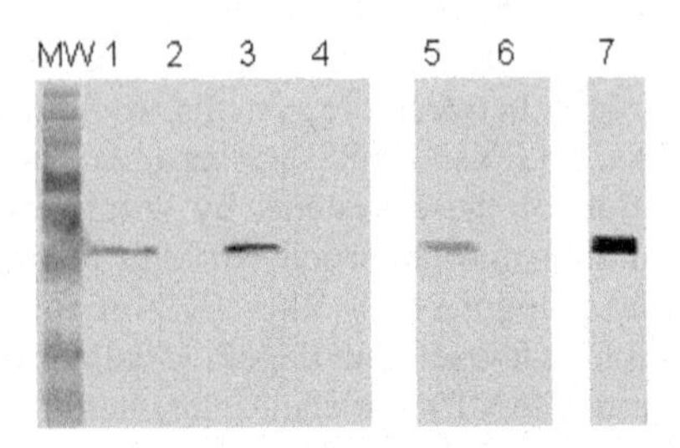

Fig. 4. *Y.enterocolitica* flagellin protein (FleB) is exported from both the Ysa and Ysc TTSS. These immunoblots were prepared from concentrated cell-free supernates probed with flagellin-specific monoclonal antibody showing secretion of FleB (flagellin) from cells grown under Ysa or Ysc TTSS permissive and nonpermissive conditions. MW denotes molecular weight markers. Lanes 1 and 2 are cell-free supernates derived from a *Y. enterocolitica flhD*⁻ nonmotile mutant with *fleB* artificially expressed from the p*tac* promoter; grown at 37°C under Yop permissive conditions, lane 1 (chelated for calcium ion) and nonpermissive conditions, lane 2 (calcium ion present). Lanes 3-4; *Y. enterocolitica flhD*⁻ mutant (nonmotile) secreting FleB at 25°C from medium containing 0.3 M NaCl (Ysa permissive conditions) or no NaCl (Ysa repressive conditions). Lane 5-6 are supernates from an *flhB::lacZ* nonmotile isolate under Yop export permissive conditions and nonpermissive conditions respectively. Lane 7 is the wild-type *Y. enterocolitica* grown at 37°C with p*tac-fleB* expression under calcium limiting conditions.

systems. We and others have shown specifically that temperature acts primarily on the flagellum regulon by shutting down class III (flagellin) expression, most likely by inhibition of transcription of σ^F (*fliA*). This repression of flagellin is necessary because as shown, flagellin, if expressed, can be recognized and transported from the Ysc TTSS along with Yops. This has a dramatic attenuating affect in vivo probably via induction of innate immunity. Without repression of flagellin in the host environment, flagellin would be injected into phagocytes, the very cells primed to recognize this PAMP. We find this explanation consistent with similar findings regarding the *Yersinia* temperature modification of lipid A, a potent TLR-4 inducer. *Y. pestis* modifies lipid A at 37°C by converting it from the hexa-acylated to the tetra-acylated form. Mounting et al. (2006) recently reported that artificial expression of hexa-acylated lipid A by *Y. pestis* also has a dramatic attenuating effect in vivo. Thus, both modification of lipid A and either permanent loss of flagellin synthesis (*Y. pestis*), or temperature regulation of flagellin synthesis (*Y. enterocolitica* and *Y. pseudotuberculosis*) reflect strategies these organisms may have developed to evade the innate immune system of the mammalian host.

Temperature regulation of motility or permanent loss by mutation is also reflected in other Gram-negative pathogens such as *Bordetella, Shigella*, and *Legionella*. It further signifies an additional consideration in understanding pathogenesis. One of the surprises in the comparative genomics of the closely related organisms, *Y. pseudotuberculosis* ('mild' pathogen) and *Y. pestis* (serious pathogen), is the number of mutations that have accumulated in *Y. pestis* relative to *Y. pseudotuberculosis* (Chain et al. 2004). As these authors suggest pathogenesis may not only reflect gain of new genetic capability, but loss of information. This certainly appears to be the case with flagellum biosynthetic capability. These observations further suggest that a complete understanding of pathogenesis may entail what we consider as 'Reverse' Molecular Koch's postulates. In other words a full understanding of pathogenesis will not only require the traditional approach of identifying virulence genes by mutational analysis, but will entail complementing those genes naturally lost by mutation, or those transcriptionally repressed in the host, to determine if their expression has an attenuating effect as seen with flagellin expression and LPS modification.

27.5 Acknowledgements

This work was supported by NIH grant 5U54AI57141-03 and the Idaho Agricultural Experiment Station.

27.6 References

Aldridge, P. and Hughes, K.T. (2002) Regulation of flagellar assembly. Curr. Opin. Microbiol. 5, 160-165.

Anderson, D.M., Fouts, D.E., Collmer, A. and Schneewind, O. (1999) Reciprocal secretion of proteins by the bacterial type III machines of plant and animal pathogens suggests universal recognition of mRNA targeting signals. Proc. Natl. Acad. Sci. USA. 96, 12839-12843.

Chain, P.S.G., Carniel, E., Larimer, F.W., Lamerdin, J., Stoutland, P.O., Regala, W.M., Georgescu, A.M., Vergez, L.M., Land, M.L., Motin, V.L., Brubaker, R.R., Fowler, J., Hinnebusch, J., Marceau, M., Medigue, C., Simonet, M., Chenal-Francisque, V., Souza, B., Dacheux, D., Elliott, J.M., Derbise, A., Hauser, L.J. and Garcia, E. (2004) Insights into the evolution of *Yersinia pestis* through whole-genome comparison with *Yersinia pseudotuberculosis*. Proc. Natl. Acad. Sci. USA 101, 13826-13831.

Chan, S.S., Breslauer, K.J., Hogan, M.E., Kessler, D.J., Austin, R.H., Ojemann, J., Passner, J.M. and Wiles, N.C. (1990) Physical studies of DNA premelting equilibria in duplexes with and without homo dA.dT tracts: correlations with DNA bending. Biochemistry. 29, 6161-6171.

Clarke, M. B. and Sperandio, V. (2005) Transcriptional regulation of *flhDC* by QseBC and sigma (FliA) in enterohaemorrhagic *Escherichia coli*. Mol. Microbiol. 57, 1734-1749.

Cornelis, G. R., Sluiters, C., Delor I., Geib, D., Kaniga, K., Lambert de Rouvroit, C., Sory, M.P., Vanooteghem, J.C. and Michiels, T. (1991) *ymoA*, a *Yersinia enterocolitica* chromosomal gene modulating the expression of virulence functions. Mol. Microbiol. 5, 1023-1034.

Feng P., Sugasawara R.J. and Schantz, A. (1990) Identification of a common enterobacterial flagellin epitope with a monoclonal antibody. J. Gen. Microbiol. 136, 337-342.

Haller, J.C., Carlson, S., Pederson, K.J., and Pierson, D.E. (2000) A chromosomally encoded type III secretion pathway in *Yersinia enterocolitica* is important in virulence. Mol. Microbiol. 36, 1436-1446. Erratum in: Mol. Microbiol (2002) 44, 599.

Horne, S.M. and Pruss, B.M. (2006) Global gene regulation in *Yersinia enterocolitica*: effect of FliA on the expression levels of flagellar and plasmid-encoded virulence genes. Arch. Microbiol. 185, 115-126.

Julio, S.M., Heithoff, D.M., Sinsheimer, R.L., Low, D.A. and Mahan, M.J. (2002) DNA adenine methylase overproduction in *Yersinia pseudotuberculosis* alters *YopE* expression and secretion and host immune responses to infection. Infect. Immun. 70, 1006-1009

Kapatral, V. and Minnich, S.A. (1995) Co-ordinate, temperature-sensitive regulation of the three *Yersinia enterocolitica* flagellin genes. Mol Microbiol. 17(1):49-56.

Kapatral, V., Olson, J.W., Pepe, J.C., Miller, V.L. and Minnich, S.A. (1996) Temperature-dependent regulation of *Yersinia enterocolitica* Class III flagellar genes. Mol. Microbiol. 19, 1061-1071

Kubori, T., Matsushima, Y., Nakamura, D., Uralil, J., Lara-Tejero, M., Sukhan, A., Galan, J.E. and Aizawa, S.I. (1998) Supramolecular structure of the *Salmonella typhimurium* type III protein secretion system. Science. 280, 602-605.

Kutsukake, K. (1997) Autogenous and global control of the flagellar master operon, *flhD*, in *Salmonella typhimurium*. Mol. Gen. Genet. 254, 440-448.

Mcnab, R.M. (2004) Type III flagellar protein export and flagellar assembly. Biochim. Biophys. Acta. 1694, 207-217.

Michiels T. and Cornelis, G.R. (1991) Secretion of hybrid proteins by the *Yersinia* Yop export system. J. Bacteriol. 173, 1677-1685.

Montminy S.W., Khan, N., McGrath, S., Walkowicz, M.J., Sharp, F., Conlon, J.E., Fukase, K., Kusumoto, S., Sweet, C., Miyake, K., Akira, S., Cotter, R.J., Goguen, J.D. and Lien, E. (2006) Virulence factors of *Yersinia pestis* are overcome by a strong lipopolysaccharide response. Nat. Immunol. 7, 1066-1073

Ramakrishnan G., Zhao J.L. and Newton A. (1991)The cell cycle-regulated flagellar gene *flbF* of *Caulobacter crescentus* is homologous to a virulence locus (*lcrD*) of *Yersinia pestis*. J. Bacteriol. 173, 7283-7292

Rohde, J.R., Fox, J.M. and Minnich, S.A. (1994) Thermoregulation in *Yersinia enterocolitica* is coincident with changes in DNA supercoiling. Mol. Microbiol. 12, 187-199.

Rohde J.R., Luan, X.S., Rohde, H., Fox J.M. and Minnich S.A. (1999) The *Yersinia enterocolitica* pYV virulence plasmid contains multiple intrinsic DNA bends which melt at 37 degrees C.J. Bacteriol. 181, 4198-4204

Rosqvist, R., Magnusson, K.E. and Wolf-Watz, H. (1994) Target cell contact triggers expression and polarized transfer of *Yersinia* YopE cytotoxin into mammalian cells.EMBO J. 13, 964-972.

Sanders, L.A, Van Way, S. and Mullin, D.A. (1992) Characterization of the *Caulobacter crescentus flbF* promoter and identification of the inferred FlbF product as a homolog of the LcrD protein from a *Yersinia enterocolitica* virulence plasmid. J. Bacteriol. 174, 857-866.

Smith, J.M. (2000) Genetic Regulation of type III secretion systems in Yersinia enterocolitica. Ph.D. Dissertation, University of Idaho

Steiner, T.S. (2006) How flagellin and toll-like receptor 5 contribute to enteric enfection. Infect. Immun. 2006 Nov 21; [Epub ahead of print]

Young G.M., Schmiel D.H. and Miller V.L. (1999) A new pathway for the secretion of virulence factors by bacteria: the flagellar export apparatus functions as a protein-secretion system. Proc. Natl. Acad. Sci. USA 96, 6456-6461.

Young, B.M. and Young, G.M. (2002) Evidence for targeting of Yop effectors by the chromosomally encoded Ysa type III secretion system of *Yersinia enterocolitica*. J. Bacteriol. 184, 5563-5571.

Picture 25. James Bliska introduces Scott Minnich. Photograph by A. Anisimov.

28
Disparity Between *Yersinia pestis* and *Yersinia enterocolitica* O:8 in YopJ/YopP-Dependent Functions

Ayelet Zauberman, Baruch Velan, Emanuelle Mamroud, Yehuda Flashner, Avigdor Shafferman, and Sara Cohen

Department of Biochemistry and Molecular Genetics, Israel Institute for Biological Research, cohens@iibr.gov.il

Abstract. YopP in *Y. enterocolitica* and YopJ in *Y. pseudotuberculosis*, have been shown to exert a variety of adverse effects on cell signaling leading to suppression of cytokine expression and induction of programmed cell death. A comparative *in vitro* study with *Y. pestis* and *Y. enterocolitica* O:8 virulent strains shows some critical disparity in YopJ/YopP-related effects on immune cells. Involvement of yopJ in virulence was evaluated in mouse model of bubonic plague.

28.1 Introduction

The genus *Yersinia* includes three pathogenic species to humans, *Yersinia pestis*, *Yersinia enterocolitica* and *Yersinia pseudotuberculosis*. These species are closely related genetically, yet differ significantly in their pathogenesis. The molecular basis for the difference in virulence has been the subject of a long debate. The many differences revealed in the chromosomal coding sequences of these species together with further annotation of plasmids unique to *Y. pestis*, may lead to a better understanding of the differences in pathogencity of these strains. Thorough examination of potential functional differences between homologue genes may be another avenue for deciphering the evolution of *Y. pestis* from enteropathogens into a highly virulent strain.

All three *Yersinia* species share a common ~70kb plasmid which is essential for virulence. The plasmids encode a type III secretion system (TTSS), comprised of a secretion apparatus, chaperones and several secreted effectors (Yops) exhibiting a high degree of sequence similarity (Cornelis and Wolf-Watz 1997). The Yop effectors are translocated into target cells and affect various signaling pathways to subvert the host innate immunity (Cornelis 2002; Mota and Cornelis 2005; Ruckdeschel 2002; Viboud and Bliska 2005; Zhang and Bliska 2005). Such effects have been mainly demonstrated for the enteropathogenic *Yersinia* and the limited information available on *Y. pestis* is based essentially on studies with attenuated strains (Cornelis, 2000; Weeks et al. 2002). Three *Y. pestis* TTSS genes *lcrF*, *yopJ* and *yopH* are being studied in our laboratory using the EV76 vaccine strain and a virulent *Y. pestis* strain Kimberley 53 strain (Flashner et al. 2004; Mamroud, 2006; Velan et al. 2006; Zauberman et al. 2006). Here we present a study evaluating *Y. pestis* YopJ-related effects on macrophage cells (Zauberman et al. 2006) and compare them to those of YopP of *Y. enterocolitica* WA O:8 virulent strain.

When comparing *Y. pestis* and *Y. enterocolitica* effects on cells, several strategies have been employed to overcome the potential differences in the interaction of the two pathogens with the host cell surface. These include: a) the use of *Y. pestis* cells which are not enveloped by the F1 capsule (based on the differential expression kinetics of TTSS and F1 upon temperature shift of bacteria from 28°C to 37°C), b) cell infection at conditions which favor intimate contact of both strains with macrophages and finally c) evaluation of YopJ or YopP functions in the context of a *Y. pestis* bacterial cell.

The *Y. enterocolitica* effector YopP and the *Y. pseudotuberculosis* homologue, YopJ, were shown to affect host cells by suppression of cytokine production and induction of apoptosis, through inhibition of MAP kinase and NF-κB pathways (Boland and Cornelis 1998; Orth et al. 1999, Zhang et al. 2005; Palmer et al. 1998; Ruckdeschel et al. 1998; Schesser et al. 1998) and activation of upstream caspases (Denecker et al. 2001). In our studies we have evaluated these activities in *Y. pestis* and compared them to those of *Y. enterocolitica* under similar experimental conditions and have observed similarities as well as disparities (Zauberman et al. 2006; Fig. 1).

28.2 MAPK Signaling

Activation and suppression of MAPK signaling was followed by monitoring phosporylation of MAPK family members. Phosphorylation profiles revealed similarities in the pattern and time course of the effects exerted by the two *Yersinia* species on J774A.1 cells. The *Y. pestis* YopJ-dependent suppression of MAPK signaling activation, shown in the three family members: ERK1/2, JNK and p38, was rapid upon infection with bacteria pre-grown at 37°C. Moreover, infection with *Y. pestis* pre-grown at 28°C led also to effective down-regulation of MAPK signaling, yet the kinetics were different and the full extent of the effect was reached after longer periods of interaction with macrophages. Thus, *Y. pestis* YopJ acts in a similar manner to its two enteropathogenic homologues in down regulating MAPK pathways in macrophages (Palmer et al. 1999; Palmer et al. 1998, Ruckdeschel et al. 1997, Boland et al. 1998).

28.3 NF-κB Signaling

NF-κB is another key regulator of the inflammatory response. Infection of cells with the plasmid cured strains of *Y. pestis* and of *Y. enterocolitica* led to NF-κB activation, suggesting that *Y. pestis*, like *Y. enterocolitica*, can activate NF-κB nuclear translocation. Nevertheless, in contrast to *Y. enterocolitica* O:8, suppression of this activation was not observed with the plasmid-harboring EV76 strain (Zauberman et al. 2006) nor with the fully virulent Kimberley 53 strain (Fig. 1B). It thus appears that the Yop translocon of *Y. pestis*, as opposed to that of *Y. enterocolitica* O:8 (Ruckdeschel et al. 1998; Ruckdeschel et al. 2001b), is not capable of suppressing the induction of NF-κB activation under the same infection conditions. Interestingly, in this aspect, the virulent *Y. pestis* strain resembles serotype O:9 of *Y. enterocolitica* strain (Ruckdeschel et al. 2001a).

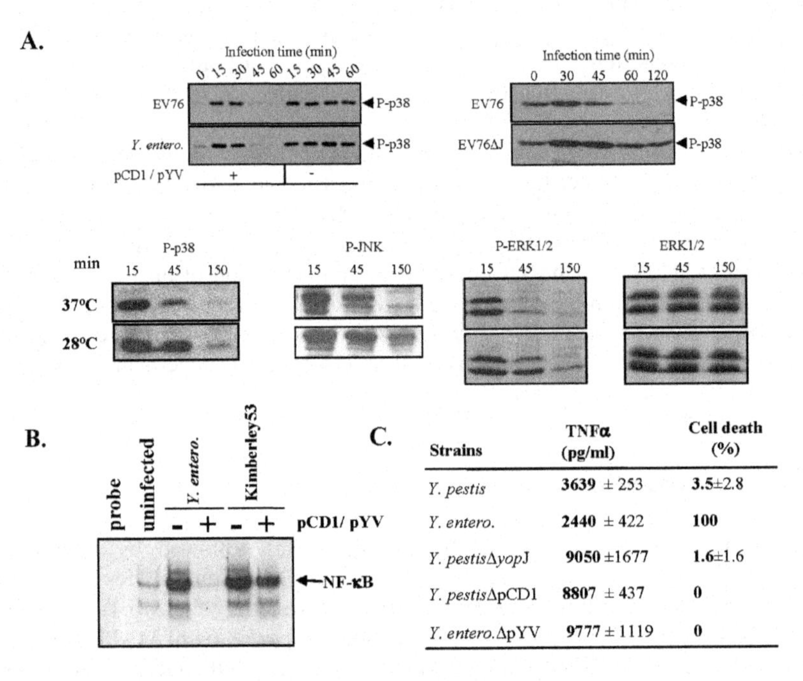

Strains	TNFα (pg/ml)	Cell death (%)
Y. pestis	**3639** ± 253	**3.5**±2.8
Y. entero.	**2440** ± 422	**100**
*Y. pestis*Δ*yopJ*	**9050** ±1677	**1.6**±1.6
*Y. pestis*ΔpCD1	**8807** ± 437	**0**
*Y. entero.*ΔpYV	**9777** ± 1119	**0**

Fig. 1. Similarity in the effects of *Y. pestis* and *Y. enterocolitica* O:8 on phosphorylation of MAPK proteins and TNF-α secretion and disparity in the effects on NF-κB DNA-binding activity and on induction of cell death in infected macrophages. *Yersinia* strains, pre-grown at 37°C for 3 hrs, were used to infect J774A.1 cells at an MOI of 50-100 by the overlay method and the following effects were monitored (Zauberman et al. 2006): (A) MAPK phosphorylation: Infected macrophages were incubated for the indicated period of time and cell lysates were analyzed by immunoblotting with antibodies against phospho-specific forms of p38 (P-p38), JNK (P-JNK), or ERK1/2 (P-ERK1/2). As a loading control, filters were re-probed with standard anti-ERK1/2 antibodies. Note that both JNK and ERK1/2 have two isoforms. This portion of Panel A was reproduced from (Zauberman et al. 2006) with the permission of ASM. For comparison a similar experiment was done with bacteria pre-grown at 28°C. (B) NF-κB Electro Mobility Shift Assays: Cells were exposed for 90 min to the indicated *Yersinia* strains and nuclear lysates were tested for interaction with NF-κB specific DNA probe. (C) TNF-α secretion and cell death: Secretion of TNF-α by infected cells was monitored by ELISA 2hrs after initiation of infection. Cell death was determined by LDH release 6 hours after an infection period of 1 h.

28.4 TNF-α secretion

Suppression of TNF-α secretion, another adverse function attributed to YopJ/YopP in *Yersinia* species (Ruckdeschel 2002), was evaluated by monitoring secretion of TNF-α from infected macrophage cells. *Y. pestis* was found to suppress secretion in a similar manner to *Y. enterocolitica* WA O:8 (Fig. 1C). These results are in accordance with the observed suppression of MAPK pathways which are known as important regulators of TNF-α production in response to bacterial infection.

28.5 Induction of Apoptosis

The interrelationship between suppression of NF-κB, MAPK activation and induction of macrophage apoptosis by enteropathogenic *Yersinia*, is well documented (Zhang et al. 2005, Ruckdeschel et al. 2002). To further evaluate this relationship we compared the apoptotic effect conferred by *Y. pestis*, to that of *Y. enterocolitica* O:8 – two strains which are highly different in their ability to suppress NF-κB activation (Fig. 1B). Cell infection conditions were similar to those used to study effects on NF-κB and MAPK signaling. As expected, the cytotoxicity conferred by *Y. enterocolitica* WA O:8 was rapid and very efficient (100% cell death), exhibiting characteristic features of an apoptotic process. In contrast, no cytotoxicity was observed in *Y. pestis* infected macrophages and the markers of apoptosis were practically absent. Condensed chromatin formations (DAPI staining) were not detected in the *Y. pestis* infected macrophages and the number of annexin-V stained cells was lower than 2.5% as compared to 63% in *Y. enterocolitica* O:8 infected macrophages. Nevertheless, *Y. pestis*-induced cytotoxcicity can be induced by specific experimental conditions. Infection with centrifugation, which is believed to synchronize cell association, was shown previously to promote cell death by *Y. pestis* (Weeks et al. 2002). Side by side comparison of such infection conditions with conditions where contact of the bacteria with cells is less extensive (overlay of a bacterial suspension on cells which results in about a three fold lower *Y. pestis*/cell association) reveals again the disparity between the two *Yersinia* species. An extensive contact with the macrophage (infection by impaction) and high MOI was required for *Y. pestis* in order to exert measurable, YopJ dependent, cell death (Fig. 2A, compare upper to lower panel). A similar pattern was obtained by monitoring activation of the caspase pathway (Zauberman et al. 2006). Overproduction of YopJ (100-fold higher compared to the endogenous level) also led to higher cytotoxicity.

The inferiority of *Y. pestis*, relative to *Y. enterocolitica*, in triggering host cell death could result from differences in surface composition (i.e. LPS composition and presence of specific adhesins) which may influence the nature and effectiveness of the interactions between bacteria and macrophages, and/or from intrinsic differences between the YopJ and YopP effectors. YopJ and YopP of the bacterial strains used here differ in 17 amino acids, yet none of them appear to reside in the putative active sites - the amino acids previously implicated in suppression of NF-κB or promotion of apoptosis, like the protease related catalytic triad (Orth et al. 2000) and the arginine at position 143 which is common to the high-virulence *Y. enterocolitica* O:8 serotypes (Ruckdeschel et al. 2001a).

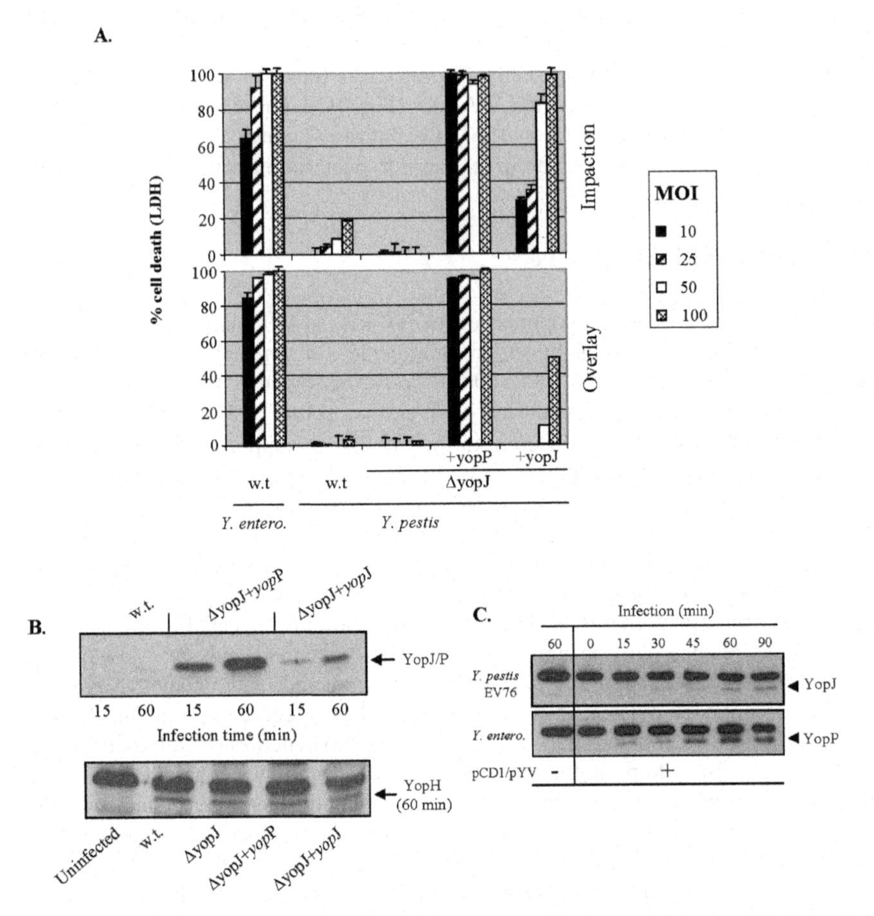

Fig. 2. YopP is more efficient than YopJ in translocation into macrophages and in induction of cell death upon infection with *Y. pestis* recombinant strains expressing either YopP or YopJ. Cell death and translocation of YopP/J into RAW264.7 cells were evaluated by synchronized infection (impaction of bacteria onto cells by centrifugation) with *Y. enterocolitica*, *Y. pestis* EV76 strains and Δ*yopJ* derivative mutants. (A) Cytotoxicity was evaluated by monitoring LDH release 6 hours following infection with different MOI (10-100) (gentamicin was added one hour post infection - upper panel). For comparison cells were infected by the overlay method (bacteria were washed out one hour post infection - lower panel). (B and C) Kinetics of YopJ or YopP translocation into target cells by the various strains was evaluated by monitoring their presence in cell lysates using immunoblotting with a common anti-peptide antibody. As a control translocation of YopH was examined. Panels B and C are reproduced from (Zauberman et al. 2006) with the permission of ASM.

To evaluate the contribution of the effectors nature to *Yersinia*-induced cytotoxicity, YopP and YopJ were expressed in the same genetic background of a *Y. pestis* Δ*yopJ* strain under an heterologous promoter, yielding strains which express one of the two effectors at comparable high levels – about two orders of magnitude higher than the endogenous YopJ levels (Zauberman et al. 2006). Over-expression of YopJ in the *Y. pestis* augmented significantly *Y. pestis* mediated cytotoxicity. A more pronounced increase in cytotoxicity was observed, however, when the heterologous effector YopP was over-expressed in *Y. pestis* background (Fig. 2A). This difference between the two bacterial strains is striking when infection was conducted using the overlay procedure, which could be more relevant for simulation of *in vivo* conditions. The *Y. pestis* strain expressing YopP is actually as effective as the parent *Y. enterocolitica*, while under these conditions native *Y. pestis* fails to trigger cell death.

28.6 Conclusions

These results indicate that the difference in cell death induction by *Y. pestis* and *Y. enterocolitica* O:8 is mainly due to the intrinsic differences between the YopJ and YopP effectors. Indeed examining the efficiency of YopP/YopJ translocation from *Y. pestis* cells into macrophage cells, clearly indicates that the amount of YopP in the cells was higher than that of YopJ (Fig. 2B). YopP was also translocated by *Y. entericolitica* more rapidly than YopJ by *Y. pestis* (Fig. 2C). Even though differences in stability of the effectors within the cells cannot be ruled out, the translocation kinetic profile and the lack of detectable degradation products suggest that this is not a major contributor to the observed accumulation difference in the intracellular effectors. It is worth noting that differences in the efficiency of YopP translocation between two serotypes of *Y. enterocolitica* were previously shown (Denecker et al. 2002). Yet, these differences were not as striking as those shown here between YopP and YopJ.

In conclusion, YopP carried either in a *Y. enterocolitica* O:8 or in a *Y. pestis* background is more amenable to translocation from the bacterium into the host cell. Consequently, YopP more rapidly reachs cellular concentrations that can trigger cell death. It appears that different thresholds of cytosolic effector concentrations are required for initiating the various YopJ related cellular effects. While the slow accumulation of YopJ in the target cells results in limited apoptosis and inefficient signaling of the apoptosis-related pathways, it is sufficient to confer a high level of suppression of MAPK pathways and TNF-α secretion.

While the nature of YopJ/YopP effectors appears to be a major contributor to translocation efficiency, other mechanisms could also be involved in determining concentrations and effects of these effectors within the target cells. The higher sequence similarity between the YopJ effectors of *Y. pestis* and *Y. pseudotuberculosis*, (the latter differ from the *Y. pestis* YopJ of Kimberley53 as well as CO92 strain, only at amino acid at position 206 - E instead of K) and the reported effectiveness of the latter in apoptosis induction (Monack et al. 1997; Zhang and Bliska 2003), although less than that of *Y. enterocolitica* O:8 (Ruckdeschel et al. 2001a), could indeed argue in favor of this. Yet, it is worth mentioning that tight contact is required for

Y. pseudotuberculosis to exert its cytotoxicity (Monack et al. 1997). To summarize, the compilation of the studies on *Yersinia*-induced apoptosis (Reviewed by Zhang and Bliska 2005), with data presented here on the role of YopP/YopJ nature in determining translocation efficiency, suggest an intricate balance of various factors in determining the extent to which this effect is manifested.

The limited ability of *Y. pestis* to induce apoptosis may have a role in enhancing the intracellular life cycle of *Y. pestis* using macrophages/ dentritic cells as shielded vehicles en route to the lymphatic target (Cavanaugh and Randall 1959). Yet, as shown here, limited apoptosis is observed when the level of YopJ in the target cell is increased (either by using a high bacterial load in a synchronized infection mode, or by over-expression of exogenous YopJ in *Y. pestis*). This may be of relevance *in vivo* when a high bacterial level is reached, such as during late stages of the disease. Indeed, Lemaitre et al. (2006) have observed recently some level of YopJ-dependent apoptosis in cells from bubo of rats taken at the terminal stage of the disease.

Evaluation of YopJ and YopP contribution to *Y. pestis* and *Y. enterocolitica* O:8 pathogenesis in animal models is much more complicated due to the differences in the mode of infection and nature of the diseases. Few reports indicate the contribution of YopP/J to enteropathogenic *Yersinia* virulence using the mouse oral infection model. Trulzsch et al. (2004) reported that a *yopP* mutant of *Y. enterocolitica* O:8 was somewhat attenuated. In a similar mouse model, Monack et al. (1998) reported a 64-fold increase in the LD_{50} of a *Y. pseudotuberculosis* strain by inactivating *yopJ*. Nevertheless, Galyov et al. (1994) reported that a *yopJ* mutation had no effect on the virulence of *Y. pseudotuberculosis*. Evaluating the role of *Y. pestis* YopJ *in vivo* revealed that inactivation of the gene did not affect virulence in mice (Straley and Bowmer 1986). These experiments were conducted with the conditional virulent Δ*pgm* KIM strain in a mouse model of systemic infection. To evaluate the requirement of YopJ in a model mimicking natural infection, we have used the highly virulent Kimberley53 strain (a Biovar O strain; Flashner et al. 2004) in a mouse model of bubonic plague. The subcutaneous LD_{50} of the Kimberley53Δ*yop*J strain was 3 CFU, similar to that of the wild type strain (1-3 CFU). A similar result was observed using a rat model of bubonic plague infecting with the 195/P strain (Lemaitre et al. 2006). The involvement of YopJ in *Y. pestis* virulence using other modes of infection or animal models is still to be determined.

28.7 References

Boland, A. and Cornelis, G. (1998) Role of YopP in suppression of tumor necrosis factor alpha release by macrophages during *Yersinia* infection. Infect. Immun. 66, 1878-1884.

Cavanaugh, D.C. and Randall, R. (1959) The role of multiplication of *Pasteurella pestis* in mononuclear phagocytes in the pathogenesis of flea-borne plague. J. Immunol. 83, 348-363.

Cornelis, G. and Wolf-Watz, H. (1997) The *Yersinia* Yop virulon: a bacterial system for subverting eukaryotic cells. Mol. Microbiol. 23, 861-867.

Cornelis, G.R. (2000) Molecular and cell biology aspects of plague. PNAS 97, 8778-8783.

Cornelis, G.R. (2002) *Yersinia* type III secretion: send in the effectors. J. Cell Biol. 158, 401-408.

Denecker, G., Declercq, W., Geuijen, C.A., Boland, A., Benabdillah, R., van Gurp, M., Sory, M.P., Vandenabeele, P. and Cornelis, G.R. (2001) *Yersinia enterocolitica* YopP-induced apoptosis of macrophages involves the apoptotic signaling cascade upstream of Bid. J. Biol. Chem. 276, 19706-19714.

Denecker, G., Totemeyer, S., Mota, L.J., Troisfontaines, P., Lambermont, I., Youta, C., Stainier, I., Ackermann, M. and Cornelis, G.R. (2002) Effect of low- and high-virulence *Yersinia enterocolitica* strains on the inflammatory response of human umbilical vein endothelial cells. Infect. Immun. 70, 3510-3520.

Flashner, Y., Mamroud, E., Tidhar, A., Ber, R., Aftalion, M., Gur, D., Lazar, S., Zvi, A., Bino, T., Ariel, N., Velan, B., Shafferman, A. and Cohen, S. (2004) Generation of *Yersinia pestis* attenuated strains by signature-tagged mutagenesis in search of novel vaccine candidates. Infect. Immun. 72, 908-915.

Galyov, E.E., Hakansson, S. and Wolf-Watz, H. (1994) Characterization of the operon encoding the YpkA Ser/Thr protein kinase and the YopJ protein of *Yersinia pseudotuberculosis.* J. Bacteriol. 176, 4543-4548.

Lemaitre, N., Sebbane, F., Long, D. and Joseph Hinnebusch, B. (2006) *Yersinia pestis* YopJ suppresses tumor necrosis factor alpha induction and contributes to apoptosis of immune cells in the lymph node but is not required for virulence in a rat model of bubonic plague. Infect. Immun. 74, 5126-5131.

Mamroud, E., Flashner, Y., Tidhar, A., Zauberman, A., Aftalion, M., Shafferman, A., Velan, B. and Cohen, S. (2006) Evidence for in vivo trans complementation of tissue colonization deficiency of *Y. pestis ΔyopH* by co-infection with the virulent wild-type strain. 9th International Symposium on *Yersinia.* Lexington, Kentucky, USA, abstract C63.

Monack, D., Mecsas, J., Ghori, N. and Falkow, S. (1997) *Yersinia* signals macrophages to undergo apoptosis and YopJ is necessary for this cell death. PNAS 94, 10385-10390.

Monack, D.M., Mecsas, J., Bouley, D. and Falkow, S. (1998) *Yersinia*-induced apoptosis *in vivo* aids in the establishment of a systemic infection of mice. J. Exp. Med. 188, 2127-2137.

Mota, L.J. and Cornelis, G.R. (2005) The bacterial injection kit: type III secretion systems. Ann, Med. 37, 234-249.

Orth, K., Palmer, L., Bao, Z., Stewart, S., Rudolph, A., Bliska, J. and Dixon, J. (1999) Inhibition of the mitogen-activated protein kinase kinase superfamily by a *Yersinia* effector. Science 285, 1920-1923.

Orth, K., Xu, Z., Mudgett, M.B., Bao, Z.Q., Palmer, L.E., Bliska, J.B., Mangel, W.F., Staskawicz, B. and Dixon, J.E. (2000) Disruption of signaling by *Yersinia* effector YopJ, a ubiquitin-like protein protease. Science 290, 1594-1597.

Palmer, L., Pancetti, A., Greenberg, S., and Bliska, J. (1999) YopJ of *Yersinia* spp. is sufficient to cause downregulation of multiple mitogen-activated protein kinases in eukaryotic cells. *Infect Immun.* 67:708-716.

Palmer, L.E., Hobbie, S., Galan, J.E., and Bliska, J.B. (1998) YopJ of *Yersinia pseudotuberculosis* is required for the inhibition of macrophage TNF-α production and downregulation of the MAP kinases p38 and JNK. Mol. Microbiol. 27, 953-965.

Ruckdeschel, K., Harb, S., Roggenkamp, A., Hornef, M., Zumbihl, R., Kohler, S., Heesemann, J. and Rouot, B. (1998) *Yersinia enterocolitica* impairs activation of transcription factor NF-κB: involvement in the induction of programmed cell death and in the suppression of the macrophage tumor necrosis factor alpha production. J. Exp. Med. 187, 1069-1079.

Ruckdeschel, K., Richter, K., Mannel, O. and Heesemann, J. (2001a) Arginine-143 of *Yersinia enterocolitica* YopP crucially determines isotype-related NF-κB suppression and apoptosis induction in macrophages. Infect. Immun. 69, 7652-7662.

Ruckdeschel, K., Mannel, O., Richter, K., Jacobi, C.A., Trulzsch, K., Rouot, B. and Heesemann, J. (2001b) *Yersinia* outer protein P of *Yersinia enterocolitica* simultaneously blocks the nuclear factor-κ B pathway and exploits lipopolysaccharide signaling to trigger apoptosis in macrophages. J. Immunol. 166, 1823-1831.

Ruckdeschel, K. (2002) Immunomodulation of macrophages by pathogenic *Yersinia* species. Arch. Immunol. Ther. Exp. 50, 131-137.

Schesser, K., Spiik, A., Dukuzumuremyi, J., Neurath, M., Pettersson, S. and Wolf-Watz, H. (1998) The *yopJ* locus is required for *Yersinia*-mediated inhibition of NF-κB activation and cytokine expression: YopJ contains a eukaryotic SH2-like domain that is essential for its repressive activity. Mol. Microbiol. 28, 1067-1079.

Straley, S.C. and Bowmer, W.S. (1986) Virulence genes regulated at the transcriptional level by Ca^{+2} in *Yersinia pestis* include structural genes for outer membrane proteins. Infect. Immun. 51, 445-454.

Trülzsch, K., Sporleder, T., Igwe, E.I., Russmann, H. and Heesemann, J. (2004) Contribution of the major secreted Yops of *Yersinia enterocolitica* O:8 to pathogenicity in the mouse infection model. Infect. Immun. 72, 5227-5234.

Velan, B., Bar-Haim, E., Zauberman, A., Mamroud, E., Shafferman, A. and Cohen, S. (2006) Discordance in the effects of *Yersinia pestis* on dendritic cell functions: induction of maturation and paralysis of migration. Infect. Immun. 74, 6365-6376.

Viboud, G.I. and Bliska, J.B. (2005) *Yersinia* outer proteins: role in modulation of host cell signaling responses and pathogenesis. Annu. Rev. Microbiol. 59, 69-89.

Weeks, S., Hill, J., Friedlander, A. and Welkos, S. (2002) Anti-V antigen antibody protects macrophages from *Yersinia pestis*-induced cell death and promotes phagocytosis. Microb. Pathog. 32, 227-237.

Zauberman, A., Cohen, S., Mamroud, E., Flashner, Y., Tidhar, A., Ber, R., Elhanany, E., Shafferman, A. and Velan, B. (2006) Interaction of *Yersinia pestis* with macrophages: limitations in YopJ-dependent apoptosis. Infect. Immun. 74, 3239-3250.

Zhang, Y. and Bliska, J.B. (2003) Role of Toll-like receptor signaling in the apoptotic response of macrophages to *Yersinia* infection. Infect. Immun. 71, 1513-1519.

Zhang, Y. and Bliska, J.B. (2005) Role of macrophage apoptosis in the pathogenesis of *Yersinia*. Curr. Top. Microbiol. Immunol. 289, 151-173.

Zhang, Y., Ting, A.T., Marcu, K.B. and Bliska, J.B. (2005) Inhibition of MAPK and NF-κB pathways is necessary for rapid apoptosis in macrophages infected with *Yersinia*. J. Immunol. 174, 7939-7949.

Picture 26. The band gets a round of applause at the banquet. Photograph by R. Perry.

Part V – Molecular Epidemiology and Detection

Picture 27. Alexandre Leclercq gives a talk on three IS-RFLP analysis. Photograph by A. Anisimov.

29
3 IS-RFLP: A Powerful Tool for Geographical Clustering of Global Isolates of *Yersinia pestis*

A.J.L. Leclercq, G. Torrea, V. Chenal-Francisque, and E. Carniel

Yersinia Research Unit, National Reference Laboratory and WHO Collaborative Center for *Yersinia*, Institut Pasteur, alexlec@pasteur.fr

Abstract. Multiple copies of several classes of insertion sequences (IS) are found in the genome of *Yersinia* pestis, the causative agent of bubonic and pneumonic plague. We used the genetic instability generated by these IS to develop a method (designated 3IS-RFLP) based on the restriction fragment length polymorphism of the IS*100*, IS*285* and IS*1541* elements for studying *Y. pestis* strains of worldwide origin. We show that 3IS-RFLP is a powerful tool to group *Y. pestis* isolates according to their geographical origin, and therefore that this method may be valuable for investigating the origin of new or re-emerging plague foci or for addressing forensic issues.

29.1 Introduction

Yersinia pestis, the causative agent of plague, is primarily a rodent pathogen. Bubonic plague is transmitted subcutaneously from rodents to humans by the bite of infected fleas. Human-to-human transmission of pneumonic plague occurs via infected respiratory droplets.

The rise in human plague cases since the beginning of the 1990s and the reappearance of the disease in countries where no cases were reported for several decades have led to categorizing plague as a reemerging disease (World Health Organization 2004).

Having the capability to trace *Y. pestis* isolates might be critical (i) to identify the source of plague cases appearing in a previously plague-free region, (ii) in a quiescent focus where the disease re-emerged after several decades of silence, to determine if the bacillus has been re-imported or if this focus has been reactivated, and (iii) to identify strains used as a biological terrorism weapon.

The species *Y. pestis* is divided into three classical biovars, and more recently into four phylogenetic sub-branches (Achtman et al. 2004). Since biovar Antiqua and Medievalis are restricted to a few geographical areas in Africa (1.ANT) and Asia (2.MED and 2. ANT), the identification of their biovar is sufficient to identify the origin of the strains. In contrast, Orientalis strains have spread globally during the third pandemic and represent the vast majority of the *Y. pestis* isolates responsible for human or animal outbreaks today.

The main aim of this study was thus to find a method which could group the Orientalis strains according to their geographical focus. To this end, a RFLP method based on the analysis of the profiles generated after hybridization of the DNA of *Y.*

pestis strains of worldwide origin with three insertion sequences [IS*100*, IS*285*, and IS*1541*] present in multiple copies on the *Y. pestis* genome, was developed and evaluated (Torrea et al. 2006).

29.2 Materials and Methods

Sixty one *Y. pestis* strains of biovar Orientalis (43 isolates), Medievalis (6 isolates) and Antiqua (12 isolates), from the strain collection of the *Yersinia* Research Unit (Institut Pasteur) and isolated from 17 countries on four continents between 1908 and 1994 were analyzed.

Total DNA extraction of each isolate was performed as described previously (Carniel et al. 1989). Five micrograms of each sample were digested overnight at 37°C with *Eco*RI (IS*100*-RFLP) and *Hin*dIII (IS*285*- and IS*1541*-RFLP) before being loaded onto 0.8% agarose gels and subjected to electrophoresis for 24h (IS*285*- and IS*1541*-RFLP) or 26h (IS*100*-RFLP). Alkaline denaturation, neutralization, and transfer of total DNA onto nylon filters (Hybond N+; Amersham, England) with a VacuGene apparatus (Pharmacia LKB Biotechnology, Uppsala, Sweden) were performed as previously described (Guiyoule et al. 1994). A portion of IS*100*, IS*1541* and IS*285* coding sequences was amplified by PCR as previously described (Guiyoule et al. 1994). Probes were labelled with horse radish peroxidase using the ECL gene detection system (Amersham).

The hybridization patterns obtained with each IS were scanned and the computerized data were analyzed using the BioNumerics software version 4.0 (Applied Maths, Kortrijk, Belgium) and a position tolerance of 1.8. Hybridizing bands corresponding to IS located on unstable genetic elements (the plasmids pFra (101 kb), pYV (70.5 kb), and pPla (9.6 kb) or the 102 kb chromosomal *pgm* locus) were removed from the profiles. Cluster analyses were done by the unweighted pair group method with average linkages (UPGMA), using the Dice coefficient to analyze the similarities of the banding patterns.

Strain clustering, resulting from the data obtained with these three IS-RFLPs used individually or in combination, were evaluated and compared. The discriminatory power of each analysis was determined, using the Simpson's index (D) of diversity (Hunter and Gaston 1988).

29.3 Results

The studies were first set-up by: (i) choosing the most appropriate restriction enzymes and hybridization conditions for each IS, (ii) checking that the results were reproducible when done by different investigators at different times, (iii) defining a window in which all bands were clearly distinguishable for all profiles, and (iv) within the windows, identifying and removing the hybridization fragments corresponding to IS located on unstable genetic elements.

As expected, *Y. pestis* strains of biovar Antiqua and Medievalis fell into specific clusters with most of the IS-RFLP methods tested (data not shown).

Within the Orientalis branch, none of the three IS-RFLPs used individually allowed an efficient strain clustering based on their geographical origin (data not shown). The combination two-by-two of the IS-RFLP data moderately improved their grouping but many strains isolated from the same country remained scattered in various branches of the UPGMA tree (data not shown).

In contrast, the combination of the three IS-RFLP data (3IS-RFLP) resulted in a robust clustering of most Orientalis isolates according to their geographical origin. Strains from Namibia, Germany, South-East Asia, USA, Brazil, and North East Africa (Morocco and Senegal) were efficiently grouped into distinct clusters (Fig. 1). Within the South-East Asia branch, two sub-branches corresponding to isolates from Burma and Vietnam were delineated. The main exception was the strains from Madagascar which were found in different branches of the tree, but nonetheless, all recent isolates (>1989) clustered in one sub-branch (Fig. 1).

IS-RFLP also exhibited an extremely high discriminatory power (Table 1), with 3IS-RFLP allowing to distinguish 59 individual patterns for the 61 strains analyzed. In addition to be used for *Y. pestis* strain grouping, 3IS-RFLP may also be a valuable tool for strain typing.

Table 1. Discriminatory power of each IS-RFLP technique applied to 61 *Y. pestis* strains of worldwide origin.

IS-RFLP	Number of profiles	Discrimination index (D)[a]
IS*100*	47	0.987
IS*1541*	35	0.968
IS*285*	38	0.963
IS*100*+IS*1541*	58	0.998
IS*100*+IS*285*	55	0.995
IS*1541*+IS*285*	52	0.993
IS*100*+IS*1541*+IS*285* (3IS)	59	0.999

[a] D based on Simpson's index of diversity (Hunter and Gaston 1988).

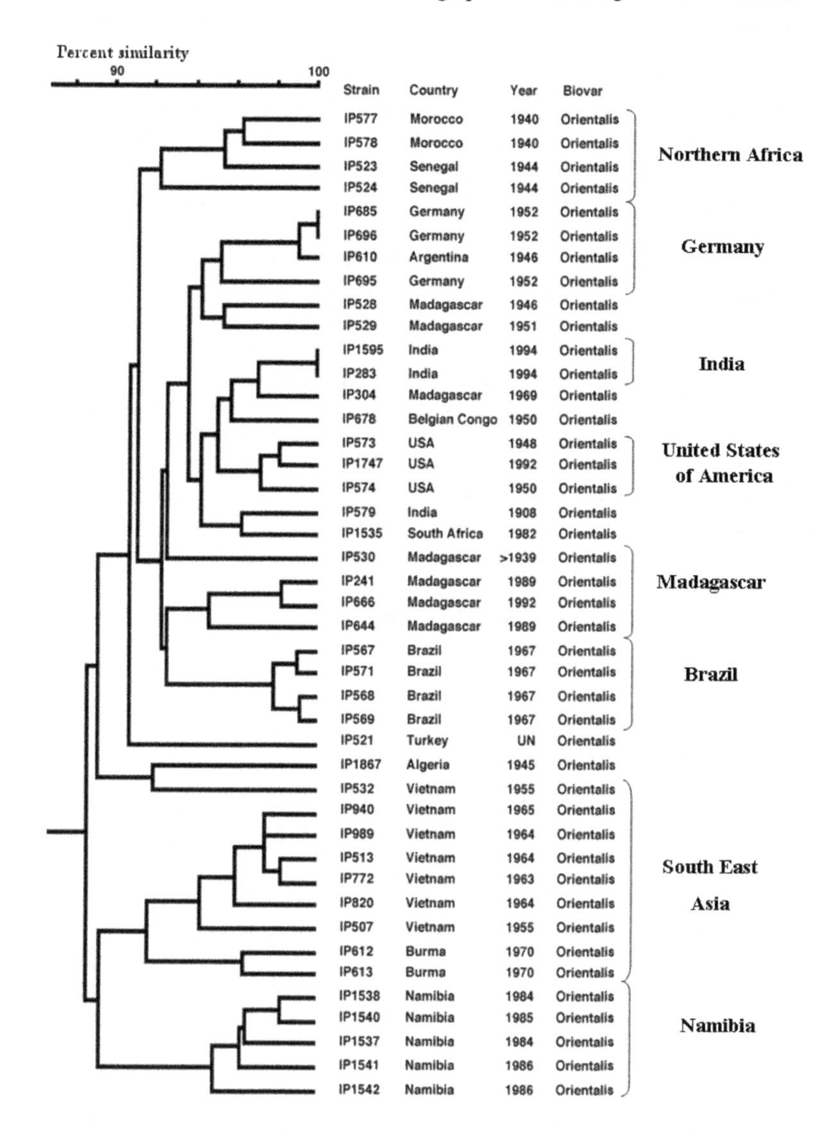

Fig. 1. 3IS-RFLP UPGMA dendrogram of *Y. pestis* strains of biovar Orientalis of worldwide origins (adapted from Torrea et al. 2006).

29.4 Discussion

Because *Y. pestis* is a recently emerged pathogen, its genetic polymorphism is extremely limited. However, the genomic instability generated by the numerous insertion sequences present in its genome creates a polymorphism of the hybridizing restriction fragments which can be used to subtype this otherwise relatively clonal species.

We show in this study that, even within the Orientalis group, which is genetically highly homogeneous since it corresponds to the recent (1894) spread of one clone over the world, 3IS-RFLP could differentiate individual isolates within a given focus.

But most interestingly, this technique could efficiently group these strains according to their geographical origin. This grouping was sensitive enough to allow, within a geographical zone (such as South-East Asia or North Africa), clustering of the isolates based on the country from where they were isolated. In some instances, such as Madagascar, it seems that 3IS-RFLP could even group *Y. pestis* strains at a regional or local scale.

3IS-RFLP may thus be a valuable technique for investigating plague outbreaks and addressing forensic issues. Any new *Y. pestis* isolate may be included into our database and compared with strains isolated worldwide to determine its geographical origin.

29.5 References

Achtman, M., Morelli, G., Zhu, P., Wirth, T., Diehl, I., Kusecek, B., Vogler, A.J., Wagner D.M., Allender, C.J., Easterday, W.R., Chenal-Francisque, V., Worsham, P., Thomson, N.R., Parkhill, J., Lindler, L.E., Carniel, E. and Keim, P. (2004) Microevolution and history of the plague bacillus, *Yersinia pestis*. Proc. Natl. Acad. Sci. U.S.A., 101, 17837-17842.

Carniel, E., Mercerau-Puijalon, O. and Bonnefoy, S. (1989) The gene coding for the 190,000-dalton iron-regulated protein of *Yersinia* species is present only in the highly pathogenic strains. Infect. Immun. 57, 1211-1217.

Guiyoule, A., Grimont, F., Iteman, I., Grimont, P.A.D., Lefèvre, M. and Carniel, E. (1994) Plague pandemics investigated by ribotyping of *Yersinia pestis* strains. J. Clin. Microbiol. 32, 634-641.

Hunter, P.R. and Gaston, M.A. (1988) Numerical index of the discriminatory ability of typing systems: an application of Simpson's index of diversity. J. Clin. Microbiol. 26, 2465-2466.

Torrea, G., Chenal-Francisque, V., Leclercq, A. and Carniel, E. (2006) Efficient tracing of global isolates of *Yersinia pestis* by restriction fragment length polymorphism analysis using three insertion sequences as probes. J. Clin. Microbiol. 44, 2084-2092.

World Health Organization (2004) Human plague in 2002 and 2003. Wkly. Epidemiol. Rec. 79, 301-306.

30

Analysis of the Three *Yersinia pestis* CRISPR Loci Provides New Tools for Phylogenetic Studies and Possibly for the Investigation of Ancient DNA

Gilles Vergnaud[1,2], Yanjun Li[3], Olivier Gorgé[2], Yujun Cui[3], Yajun Song[3], Dongsheng Zhou[3], Ibtissem Grissa[1], Svetlana V. Dentovskaya[4], Mikhail E. Platonov[4], Alexander Rakin[5], Sergey V. Balakhonov[6], Heinrich Neubauer[7], Christine Pourcel[1], Andrey P. Anisimov[4], and Ruifu Yang[3]

[1] Univ Paris-Sud, Institut de Génétique et Microbiologie, gilles.vergnaud@igmors.u-psud.fr
[2] Division of Analytical Microbiology, Centre d'Etudes du Bouchet
[3] Institute of Microbiology and Epidemiology, Academy of Military Medical Sciences
[4] State Research Center for Applied Microbiology and Biotechnology
[5] Max von Pettenkofer-Institute of Hygiene and Medical Microbiology
[6] Antiplague Research Institute of Siberia and Far East
[7] Friedrich-Loeffler Institute

Abstract. The precise nature of the pathogen having caused early plague pandemics is uncertain. Although *Yersinia pestis* is a likely candidate for all three plague pandemics, the very rare direct evidence that can be deduced from ancient DNA (aDNA) analysis is controversial. Moreover, which of the three biovars, Antiqua, Medievalis or Orientalis, was associated with these pandemics is still debated. There is a need for phylogenetic analysis performed on *Y. pestis* strains isolated from countries from which plague probably arose and is still endemic. In addition there exist technical difficulties inherent to aDNA investigations and a lack of appropriate genetic targets. The recently described CRISPRs (clustered regularly interspaced short palindromic repeats) may represent such a target. CRISPR loci consist of a succession of highly conserved regions separated by specific "spacers" usually of viral origin. To be of use, data describing the mechanisms of evolution and diversity of CRISPRs in *Y. pestis*, its closest neighbors, and other species which might contaminate ancient DNA, are necessary.

The investigation of closely related *Y. pestis* isolates has revealed recent mutation events in which elements constituting CRISPRs were acquired or lost, providing essential insight on their evolution. Rules deduced represent the basis for subsequent interpretation. In the present study, the CRISPR loci from representative *Y. pestis* and *Yersinia pseudotuberculosis* strains were investigated by PCR amplification and sequence analysis. The investigation of this wider panel of strains, including other subspecies or ecotypes within *Y. pestis* and also *Y. pseudotuberculosis* strains provides a database of the existing CRISPR spacers and helps predict the expected CRISPR structure of the *Y. pestis* ancestor. This knowledge will open the way to the development of a spoligotyping assay, in which spacers can be amplified even from highly degraded DNA samples.

The data obtained show that CRISPR analysis can provide a very powerful typing tool, adapted to the systematic, large-scale genotyping of *Y. pestis* isolates, and the creation of international typing databases. In addition, CRISPRs do constitute a very promising new tool and genetic target to investigate ancient DNA. The corresponding genetic targets are small (<70bp), present in multiple copies (usually more than 10), highly conserved and specific. In addition, the assay can be run in any laboratory. Interpretation of the data is not dependent on accurate sequence data.

30.1 Introduction

In the past few years, and owing in part to the availability of whole genome sequence data from many bacterial species including different strains from the same species, a high number of polymorphisms sources, and consequently of typing methods, has emerged. Regarding *Yersinia pestis*, these include MLST (Multiple Loci Sequence Typing) (Achtman et al. 1999), SNPs (Single Nucleotide polymorphism) (Achtman et al. 2004), MLVA (Multiple Loci VNTR Analysis) (Achtman et al. 2004; Pourcel et al. 2004), and CRISPR (Clustered Regularly Interspaced Short Palindromic Repeats) analysis (Pourcel et al. 2005).

These new methods are likely to replace the previous pattern-comparison methods (such as IS typing by southern blotting, or pulsed-field gel electrophoresis) which are more expensive and not fully appropriate to the creation of international databases. One among the emerging methods takes advantage of the polymorphism of particular structures, the CRISPR loci. CRISPRs are well-defined structures (Fig. 1). They are present in many bacteria and in most archaea, sometimes in multiple copies. The CRISPR structure itself is usually surrounded by CRISPRs-associated genes (*cas* genes) (Jansen et al. 2002). CRISPRs have been shown to be transcribed, and the transcription product is processed into micro-RNAs (Tang et al. 2002). New spacers are not synthesized *de novo*, but are copied from existing DNA sequences (Pourcel et al. 2005). The vast majority of known spacers lack any similarity with currently available sequences. However, when similarities exist, they most often correspond to short portions of mobile elements such as phages. These observations have led to the suggestion that CRISPRs were a defense-mechanism against genetic aggressions (Mojica et al. 2005; Pourcel et al. 2005; Lillestol et al. 2006; Makarova et al. 2006).

Simple evolution rules have been proposed for CRISPR which open the way to phylogenetic investigations (Pourcel et al. 2005; Lillestol et al. 2006): (1) new spacers are acquired in a polarized way from one extremity adjacent to the leader sequence which acts as a transcription promoter; (2) losses may occur randomly along the array; (3) the probability of acquisition of the same spacers independently is extremely low.

The analysis of CRISPR has already played an important role in investigating the epidemiology of the major human pathogen, *Mycobacterium tuberculosis* (the corresponding typing method is called "spoligotyping"). A database containing the typing information from thirty thousands isolates has been built (Brudey et al. 2006). Although this represents only a very small fraction of TB isolates worldwide, the database is by far the largest existing typing database for a bacterial pathogen. One reason for this is that the method was sufficiently robust, easy to run at a reasonable cost, so that many laboratories could produce data easy to share and eventually merge. Another reason was that the resulting data did make sense and enabled the definition of large families of strains. A third reason was the relatively simple situation of the CRISPR locus in *M. tuberculosis*. The locus is apparently inactive, it does not acquire new spacers, so that a fixed and limited set of relevant spacers could be defined to produce "spoligotyping membranes".

The availability of a similar approach for *Y. pestis* would be of use for at least two reasons. The first one is that it would allow for the large-scale screening of *Yersinia pseudotuberculosis* strains in search of *Y. pestis* closest neighbors, as well as for the systematic routine typing of current *Y. pestis* collections and new isolates. The second one is that CRISPRs represent potentially very interesting tools for the investigation of ancient DNA (aDNA). The nature of the early plague pandemics is still controversial, and one reason for this situation may be the lack of appropriate genetic targets for *Y. pestis* aDNA investigation.

Y. pestis CRISPR loci are still active (Pourcel et al. 2005) and able to acquire new spacers (in contrast with the *M. tuberculosis* CRISPR). Therefore, it is necessary to list the repertoire of existing CRISPR spacers. If this repertoire eventually turns out to be very large, as in some bacteria, then the development of such an assay will necessitate the use of DNA chips which are able to deal with a larger number of spacers compared to the current spoligotyping assay format. The first repertoire of

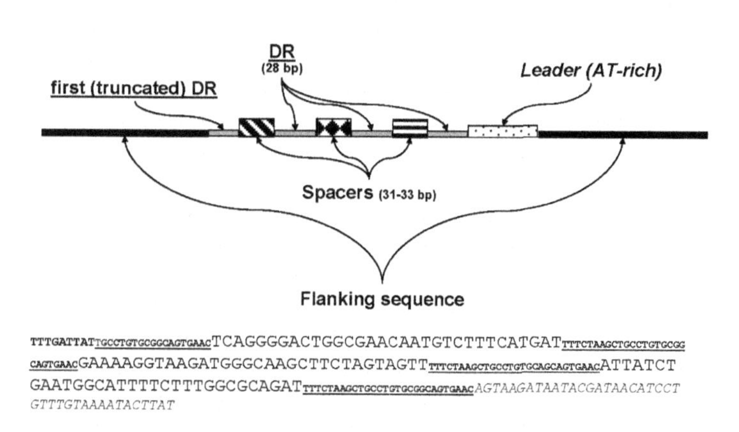

- Observed in many bacterial genomes and most archaea
- Three CRISPRs are present in *Y. pestis* and *Y. pseudotuberculosis*

Fig. 1. Organization of the CRISPR region. The CRISPR locus generally begins with a truncated direct repeat (DR) unit, followed by a succession of spacers and DRs. The leader sequence is presumed to act as the promoter region for transcription. In *Y. pestis* and *Y. pseudotuberculosis* CRISPRs, the DR is 28 bp long, and the spacers are 31 to 33 bp long. This may be different in other CRISPRs, and the observed range so far for DRs is 24 to 47 bp. Different spacers are totally unrelated. In the sample sequence, DRs are underlined, spacers are in larger typeset, the start of the leader sequence is in italics.

CRISPR spacers was deduced from the analysis of only a small part of *Y. pestis* genetic diversity (Pourcel et al. 2005). Less than 40 spacers were identified, in the three *Y. pestis* CRISPR loci, suggesting that perhaps the repertoire within a larger collection might remain tractable.

The detailed field investigations of *Y. pestis* natural foci in the former Soviet Union and China provide a relatively complete view of *Y. pestis* diversity for these countries. In addition to the *Y. pestis* main subspecies *pestis*, which is highly pathogenic for humans, five subspecies have been defined based upon biochemical analyses, geographic distribution, and favored host. These subspecies are called *caucasica, altaica, hissarica, ulegeica, talassica* (reviewed by (Anisimov et al. 2004)) and a similar biovar Microtus was proposed for isolates from *Microtus brandti* and *Microtus fuscus* in China (Song et al. 2004). The collective "pestoides" name is also used. Biochemical analysis suggests that the *caucasica* subspecies represents the oldest lineage. In addition a unique example of an African pestoides (the so-called "Angola" strain) has been described and the genome is currently being sequenced. Investigation on this strain using a number of molecular typing methods suggests that it represents a lineage which is even older than the *caucasica* lineage (Achtman et al. 2004).

In the present study, a representative collection of strains was investigated in order to produce a library of the most frequent spacers present within *Y. pestis*.

30.2 CRISPRs as a Potential Tool for Large-scale Screening of *Yersinia pseudotuberculosis* Strains Most Closely Related to *Yersinia pestis*

Whereas *Y. pestis* evolution studies attracts some attention and interest, mainly for biodefense purposes (microbial forensics), similar investigations in *Y. pseudotuberculosis* are much more limited. One reason for this is the huge diversity existing in the *Y. pseudotuberculosis* species, as compared to *Y. pestis*, and the very large number of *Y. pseudotuberculosis* isolates. A MLST assay was used by Achtman et al. to better define the relative position of *Y. pestis* and *Y. pseudotuberculosis* (Achtman et al. 1999). In this study, approximately 2 kb of sequence data were produced for a small number of diverse *Y. pseudotuberculosis* strains. The data obtained showed that all *Y. pestis* strains investigated were identical, whereas *Y. pseudotuberculosis* strains were quite diverse. Figure 2, based on published data, illustrates this finding. Although this work clearly confirms the recent emergence of *Y. pestis*, it also shows the distance between the two species. Many isolates representing intermediate evolutionary steps are missing. Isolating such intermediates will be necessary to understand the emergence of *Y. pestis*. MLST is a very powerful method, however it is not presently adapted to the large-scale and low-cost screening required to undertake

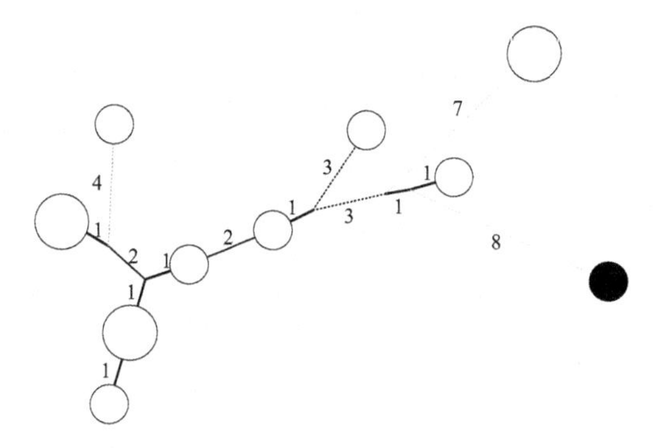

Fig. 2. Twelve *Y. pseudotuberculosis* and 36 *Y. pestis* strains compared by MLST analysis. MLST data (Achtman et al. 1999) were re-analysed and presented in a way which illustrates the close relationship between the two species and within *Y. pestis* in terms of point mutations observed within approximately 2 kb of sequence. The number of point mutations occurring along each segment is indicated (open circles: *Y. pseudotuberculosis* strains; full circle, 36 diverse *Y. pestis* strains).

such a study. CRISPRs typing may offer an alternative but additional data is needed to validate this approach.

30.3 CRISPRs as a Tool for Ancient DNA Investigations

Ancient DNA analysis is probably the only way to prove the role if any of a given pathogen in long past pandemics. However, aDNA investigations raise a number of very challenging technical issues due to DNA degradation, rarity, contamination, and chemical alterations (Prentice et al. 2004; Shapiro et al. 2006). The quality and conditions of corpse conservation are obviously key parameters and varies according to local geographic context (soil nature, average humidity, temperature, etc.). For instance, investigation of aDNA in search of *M. tuberculosis* traces in mummies (Zink et al. 2003) is more favorable than the search for *Y. pestis* traces in remains from collective graves and direct soil burial (Gilbert et al. 2004). A key element in aDNA investigations is the choice of bacterial genetic targets which take into account the characteristics of aDNA (Table 1). One major aspect is that because the sample will be usually contaminated with foreign DNA, the target should have a strong phylogenetic content (Shapiro et al. 2006). The other aspects are the size of the target and the copy number, which should be respectively as small and as high as possible. Finally, data analysis should not be extensively dependent upon the accuracy of sequence

Table 1. Ancient DNA analysis challenges

Wish list: the appropriate target for aDNA analysis	
Because aDNA is:	The appropriate genetic target should be:
Fragmented	Small
Present in low amounts	Present in multiple-copies, stable
Contaminated	With a high phylogenetic content
Subject to erroneous nucleotide incorporation	And interpretation not dependant upon sequence accuracy
Y. pestis CRISPRs satisfy all these criteria	

data. There is no consensus so far on the choice of the genetic target in the case of *Y. pestis*. Plasmid targets are sometimes used, for their relatively high copy number. However, plasmids have a low phylogenetic value as they are usually acquired by horizontal transfer from different species. In a more recent investigation (Drancourt et al. 2004), a polymorphic tandem repeat (VNTR) was targeted but some of the weaknesses of this approach were subsequently demonstrated (Vergnaud 2005). The investigators were able to amplify a particular VNTR allele from ancient remains from the first two pandemics. Among the *Y. pestis* strains investigated so far, only strains from biovar Orientalis possess this VNTR allele. Consequently the authors concluded that all three pandemics were caused by Orientalis. However the phylogenetic evidence is too weak, because the strain collection investigated is by far not representative of the diversity of *Y. pestis*. Indeed studying an enlarged collection, the same "Orientalis" allele was observed in Antiqua and Medievalis strains as well (Yang and colleagues, unpublished data).

30.4 CRISPRs Diversity Within *Y. pestis* and *Y. pseudotuberculosis*

Y. pestis and *Y. pseudotuberculosis* contain three CRISPRs called YP1, YP2 and YP3 (Pourcel et al. 2005) (Fig. 3A). Since CRISPR loci can also be considered to some extent as polymorphic tandem repeats, they have been designated, respectively, ms06, ms76 and ms77 (Le Flèche et al. 2001; Pourcel et al. 2005). The number of motifs (one DR and one spacer) in an allele is easily deduced from the size of the PCR product obtained by using flanking primers. These PCR products can be sequenced to identify the spacers (Fig. 3A-3B).

It is this approach, applied to many very closely related isolates, some of which differed at the CRISPR loci, which led to the current model of evolution for CRISPR (Pourcel et al. 2005) (Fig. 4).

RS 28bp Sp 32-33bp
VNTRyp2769ms06 CRISPR YP1
GTTACAAAATGCGCTTCCGCTCGC**AATTTTGCTCCCCAAATAGCAT**CAGCACATGGCCCA
*tttgattatTGCCTGTGCGGCAGTGAAC*TCAGGGGACTGGCGAACAATGTCTTTCATGAT a
*TTTCTAAGCTGCCTGTGCGGCAGTGAAC*GAAAAGGTAAGATGGGCAAGCTTCTAGTAGTT b
*TTTCTAAGCTGCCTGTGCGGCAGTGAAC*ATTATCTGAATGGCATTTTCTTTGGCGCAGAT c
*TTTCTAAGCTGCCTGTGCGGCAGTGAAC*TCGCCATTCCGTGAACCTGAGCGCGTTCGCGA d
*TTTCTAAGCTGCCTGTGCGGCAGTGAAC*ATATTCTCGAGCGATAGCAATAGCCATTCCAC e
*TTTCTAAGCTGCCTGTGCGGCAGTGAAC*TCGGTCAAACAAATTTAGGCGACGATTTAACA f
*TTTCTAAGCTGCCTGTGCGGCAGTGAAC*AAAAAGAATTTGGGATTAAAGTTACCCATCAG g
*TTTCTAAGCTGCCTGTGCGGCAGTGAAC*TCAATGCCTGAATCTCTGGCGTGATAGCTGCGG h
*TTTCTAAGCTGCCTGTGCGGCAGTGAAC*AGTAAGATAATACGATAACATCCTGTTTGTAA
AA**TACTTATTTCGCTAATGGGGAAA**AAACCCTTTTTTTAGACCACCGATAACCACAATGT
AAAATCAATGAGTTAGCAGTAGCTAAAAAAATAGGGTCAGAACATAACTCATAATAAAAC

yp2895 CRISPR YP2
CAGGTAGATGCCTTCCGATCTC**AATCAGCCACGCTCTGTCTA**GTGCAGTCGCTGGTCGTG
GCGTTGGCCTACCAGCAGGAGGCGCAGGCCGGGGCCGCGCTGGCGCACAGACAGTGACCC
*tctaTAAGCTGCCTGTGCGGCAGTGAAC*TCTGTACGCATACCGCCATCTTGCATCAGTCT a2
*TTTCTAAGCTGCCTGTGCGGCAGTGAAC*AGCAAAAATCTTAATTACATCTGATGATTTCGG b2
*TTTCTAAGCTGCCTGTGCGGCAGTGAAC*TTTACGGCACGGCGAAAGATTCGGTTCTTGTC c2
*TTTCTAAGCTGCCTGTGCGGCAGTGAAC*TTCTGGATAGGACAAATAGGATGATTGTATCAG d2
*TTTCTAAGCTGCCTGTGCGGCAGTGAAC*AACGAACCCACGTAGAATTGCCATCACCGCCGG e2
*TTTCTAAGCTGCCTGTGCGGCAGTGAAC*AGTAAGATAATACGGGTAACAGACTGTTTGTAA
AATAATTCTTTCGCCAAAGGGTAAAAAATGATTTTTTTTA**ACCCTCGGTAAGCAGGATAT**
AAAATCAATGAGTTAGCCATAGCTAAAAAAATAGGGTCAAAAAATGATTCCCCTGATGCG

yp1773 CRISPR YP3
AATAT**GCCAAGGGATTAGTGAGTTAA**TATTTGCAGATAAAACGCCGCCAGAGAGCTGAGA
*ttattggGCTGCCTGTGCGGCAGTGAAC*GTTATACCCCGCGCAGGGAGTGAAGCGTTGAC a3
*TTTCTAAGCTGCCTGTGCGGCAGTGAAC*TTAAGTTCTTTTTGTCAGCATCTTTAATAAATA b3
*TTTCTAAGCTGCCTGTGCGGCAGTGAAC*CTGAAATACAAATAAAATAAATCGTCGAACATA c3
*TTTCTAAGCTGCCTGTGCGGCAGTGAAC*GTAAGATAATACGGATAACCCGATGTTTATCAA
ATGAG**CAATGGCGCAAAATGCGTAAA**AACCCTTTTTTTAGTGAAATACCTGAGTAGCATA
AAAATCAATACGTTAGTCATAGTGATAAAAAGAGGGTCACAAGAATCGGGGGGGACGTAA

3A

Strain 195P

*tttgattatTGCCTGTGCGGCAGTGAAC*ATATTCTCGAGCGATAGCAATAGCCATTCCAC e
*TTTCTAAGCTGCCTGTGCGGCAGTGAAC*TCGGTCAAACAAATTTAGGCGACGATTTAACA f
*TTTCTAAGCTGCCTGTGCGGCAGTGAAC*AAAAAGAATTTGGGATTAAAGTTACCCATCAG g
*TTTCTAAGCTGCCTGTGCGGCAGTGAAC*TCAATGCCTGAATCTCTGGCGTGATAGCTGCGG h
*TTTCTAAGCTGCCTGTGCGGCAGTGAAC*AGTAAGATAATACGATAACATCCTGTTTGTAA

Strain Java9

*tttgattatTGCCTGTGCGGCAGTGAAC*TCAGGGGACTGGCGAACAATGTCTTTCATGAT a
*TTTCTAAGCTGCCTGTGCGGCAGTGAAC*GAAAAGGTAAGATGGGCAAGCTTCTAGTAGTT b
*TTTCTAAGCTGCCTGTGCGGCAGTGAAC*ATTATCTGAATGGCATTTTCTTTGGCGCAGAT c
*TTTCTAAGCTGCCTGTGCGGCAGTGAAC*TCGCCATTCCGTGAACCTGAGCGCGTTCGCGA d
*TTTCTAAGCTGCCTGTGCGGCAGTGAAC*ATATTCTCGAGCGATAGCAATAGCCATTCCAC e
*TTTCTAAGCTGCCTGTGCGGCAGTGAAC*TCGGTCAAACAAATTTAGGCGACGATTTAACA f
*TTTCTAAGCTGCCTGTGCGGCAGTGAAC*AAAAAGAATTTGGGATTAAAGTTACCCATCAG g
*TTTCTAAGCTGCCTGTGCGGCAGTGAAC*TCAATGCCTGAATCTCTGGCGTGATAGCTGCGG h
*TTTCTAAGCTGCCTGTGCGGCAGTGAAC*ACGTCATCCTGAAGGCTAGGCAGCTCGGCTTC 0
*TTTCTAAGCTGCCTGTGCGGCAGTGAAC*AGTAAGATAATACGATAACATCCTGTTTGTAA

Strain CEB02-449

*tttgattatTGCCTGTGCGGCAGTGAAC*TCAGGGGACTGGCGAACAATGTCTTTCATGAT a
*TTTCTAAGCTGCCTGTGCGGCAGTGAAC*GAAAAGGTAAGATGGGCAAGCTTCTAGTAGTT b
*TTTCTAAGCTGCCTGTGCGGCAGTGAAC*ATTATCTGAATGGCATTTTCTTTGGCGCAGAT c
*TTTCTAAGCTGCCTGTGCGGCAGTGAAC*TCGCCATTCCGTGAACCTGAGCGCGTTCGCGA d
*TTTCTAAGCTGCCTGTGCGGCAGTGAAC*ATATTCTCGAGCGATAGCAATAGCCATTCCAC e
*TTTCTAAGCTGCCTGTGCGGCAGTGAAC*TCGGTCAAACAAATTTAGGCGACGATTTAACA f
*TTTCTAAGCTGCCTGTGCGGCAGTGAAC*AAAAAGAATTTGGGATTAAAGTTACCCATCAG g
*TTTCTAAGCTGCCTGTGCGGCAGTGAAC*TCAATGCCTGAATCTCTGGCGTGATAGCTGCGG h
*TTTCTAAGCTGCCTGTGCGGCAGTGAAC*ACGTCATCCTGAAGGCTAGGCAGCTCGGCTTC 0
*TTTCTAAGCTGCCTGTGCGGCAGTGAAC*GAAATTGTGGGTGTAGATGTTGCAGACGCCTC V
*TTTCTAAGCTGCCTGTGCGGCAGTGAAC*TCTGACGTTGCCTGTGTTGCCGCTCTCGTATT W
*TTTCTAAGCTGCCTGTGCGGCAGTGAAC*AGTAAGATAATACGATAACATCCTGTTTGTAA

3B

(Continued)

Previous work by Yang and colleagues, based upon deletion analysis of a significant number of strains from China, led to the currently available view of relationships between the different biovars of *Y. pestis* (Zhou et al. 2004a; Zhou et al. 2004b). This is illustrated in Fig. 5. In this view, the Orientalis lineage branched out of the Antiqua biovar earlier than the Medievalis biovar.

Biovar Orientalis

CO92	a	b	c	d	e	f	g	h			
					e	f	g	h			
	a	b	c	d		f	g	h			
	a	b	c	d	e	f	g	h	l		
	a	b	c	d	e	f	g	h	r		
	a	b	c	d	e	f	g	h	y		
	a	b	c	d	e	f	g	h	z		
	a	b	c	d	e	f	g	h	q		
	a	b	c	d	e	f	g	h	s		
	a	b	c	d	e	f	g	h	x		
	a	b	c	d	e	f	g	h	u		
	a	b	c	d	e	f	g	h	o		
	a	b	c	d	e	f	g	h	o	p	
	a	b	c	d	e	f	g	h	o	v	w

Fig. 4. CRISPR YP1 variations observed within biovar Orientalis. The list of different CRISPR YP1 alleles observed within Orientalis isolates illustrates the pattern of variations and mode of evolution of CRISPR structures. Interstitial losses are compensated by polarized insertions.

Fig. 3. (Continued)

Fig. 3. Sequence of CRISPR alleles. A- the three CRISPR loci in the CO92 genome sequence: CO92 is a *Y. pestis* strain belonging to the Orientalis biovar. The three loci contain, respectively, eight, five and three spacers. CRISPR sequence data can be coded by giving each spacer a name. Following the nomenclature proposed by Pourcel et al. (Pourcel et al. 2005) a combination of letters and figures is used. Spacers from CRISPR locus 2 and 3 are identified by the 2 and 3 added to the spacer name. Spacers are given names as they are discovered. Spacer 'a' from locus YP1 is unrelated to spacer 'a2' from locus YP2 or 'a3' from locus YP3. In the initial report (Pourcel et al. 2005), less than 26 spacers were observed at each locus. Spacers after 'z' are numbered starting at spacer 27. B- CRISPR YP1 in three Orientalis isolates: three different alleles illustrate the main features of CRISPR evolution. Within the Orientalis biovar, the vast majority of isolates shares an identical "abcdefgh" CRISPR YP1 allele. In some rare instances, differences are observed. The independent analysis done by MLVA does not suggest that these rare alleles belong to specific lineages within Orientalis. On the contrary, isolates with an otherwise identical MLVA type may show CRISPR YP1 differences, demonstrating that these mutation events are of very recent origin (Pourcel et al. 2005). Here an interstitial deletion event is observed in strain 195P, resulting in the loss of the 4 contiguous spacers a, b, c, d. Addition of one or more spacers is observed in strains Java9 and CEB02-449.

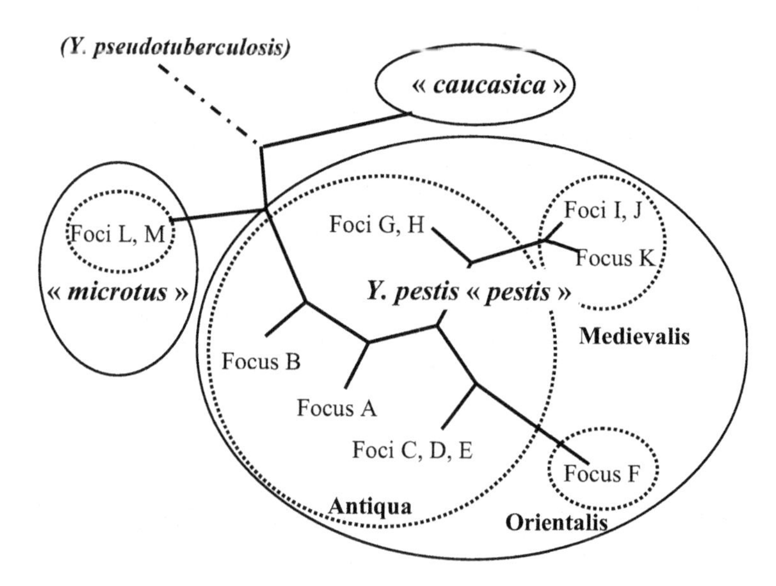

Fig. 5. Current view of relationships between some *Y. pestis* "subspecies". Deletion analysis within *Y. pestis* "*pestis*" with respect to *Y. pestis* "*microtus*" lead to this current view of *Y. pestis* evolution. The relative position of African Antiqua strains is unknown. The naming of Chinese foci is as described in Zhou et al. (Zhou et al. 2004a).

A set of diverse strains from each of the main foci identified in China (Zhou et al. 2004a) and of some foci from the former Soviet Union (Anisimov et al. 2004) including strains from the different *Y. pestis* subspecies (with the exception of *talassica*) was selected accordingly for YP1 sequence analysis. More than 80 representative YP1 alleles were investigated and more than 40 new YP1 spacers were identified. The current total number of YP1 spacers is 71. Interestingly the number of new spacers is relatively low, given the very significant increase in the diversity of *Y. pestis* strains investigated. In particular, the non-*pestis* "subspecies" including the "Angola strain" share a number of previously identified spacers (spacers labeled a, b, c, d, e, f, 37). New lineages were uncovered in particular within the A and B "Antiqua" foci (Fig. 6).

In *Y. pseudotuberculosis* the CRISPR polymorphism is very large with several hundred spacers identified to date (Gorge et al. unpublished). Spacers 'a', 'b', and 'c' were observed in a couple of strains. In agreement with previous findings, the

- CRISPR YP1 in *Y. pestis* :
  - pestoides (« Angola ») a.c.d.37.
  - *caucasica* (including pestoides F) a.b.c.d.e.m.n.
  - *altaica, hissarica, microtus* a.d.f.
  - *pestis* :

```
Antiqua :     a.b.c.d.e.f.37.38.39.40.41.50.
              a.b.c.d.e.f.37.38.39.40.41.50.51.
              a.  c.d.e.f.37.38.39.40.41.42.43.44.45.
              a.  c.d.e.f.37.38.39.40.   42.43.44.45.
              a.b.c.d.  f.37.   39.      42.46.
              a.  c.d.  f.37.   39.      42.46.47.48.
              a.  c.d.  f.37.   39.      42.46.47.48.49.
              a.b.c.d.e.f.g. and "a.b.c.d.e.f.g.+"
Orientalis :  a.b.c.d.e.f.g.h. and "a.b.c.d.e.f.g.h.+"
Medievalis :  a.b.c. and "a.b.c.+"
```

Fig. 6. CRISPR YP1 alleles observed across *Y. pestis.* A few representative alleles are indicated. Allele codes are aligned to illustrate differences resulting from interstitial deletion or progressive addition of spacers from the right end. Spacer 37 is observed in the "Angola strain" which indicates that the combination "a.b.c.d.e.f.37" was already present in the *Y. pestis* ancestor.

majority of the spacers for which an origin could be found corresponds to a prophage. This strengthens the hypothesis that the remaining spacers correspond to presently unknown viruses.

30.5 Conclusions

The present work provides a significantly enlarged view of the diversity of CRISPR spacers within *Y. pestis* intraspecies groups. Seventy CRISPR YP1 spacers have now been uncovered, and these are likely to represent the most frequently occurring spacers. Some very recently acquired and rare spacers present in only a few isolates will probably be identified in the future, but they would not significantly increase the validity of a future spoligotyping assay for *Y. pestis*. Consequently it will be possible to develop a very efficient typing assay when the other two (and less variable) CRISPR loci will have been similarly investigated.

We anticipate that such an assay will help in deciphering the phylogeny of *Y. pestis* and in identifying closely related *Y. pseudotuberculosis* strains. In addition, the investigation of CRISPRs in aDNA should greatly improve our knowledge of the agent responsible for the different pandemics.

30.6 Acknowledgments

Research by SD, MP, SB, AA, was partially supported by the International Science and Technology Center Project #2426.

30.7 References

Achtman, M., Morelli G., Zhu, P., Wirth, T., Diehl, I., Kusecek, B., Vogler, A.J., Wagner, D.M., Allender, C.J., Easterday, W.R., Chenal-Francisque, V., Worsham, P., Thomson, N.R., Parkhill, J., Lindler, L.E., Carniel, E. and Keim, P. (2004) Microevolution and history of the plague bacillus, *Yersinia pestis*. Proc. Natl. Acad. Sci. USA 101, 17837-17842.

Achtman, M., Zurth, K., Morelli, G., Torrea, G., Guiyoule, A. and Carniel, E. (1999) *Yersinia pestis*, the cause of plague, is a recently emerged clone of *Yersinia pseudotuberculosis* [published erratum appears in Proc. Natl. Acad. Sci. USA 2000 97, 8192]. Proc. Natl. Acad. Sci. USA 96, 14043-14048.

Anisimov, A.P., Lindler, L.E. and Pier, G.B. (2004) Intraspecific diversity of *Yersinia pestis*. Clin. Microbiol. Rev. 17, 434-464.

Brudey, K., Driscoll, J.R., Rigouts, L., Prodinger, W.M., Gori, A., Al-Hajoj, S.A., Allix, C., Aristimuno, L., Arora, J., Baumanis, V., Binder, L., Cafrune, P., Cataldi, A., Cheong, S., Diel, R., Ellermeier, C., Evans, J.T., Fauville-Dufaux, M., Ferdinand, S., Garcia de Viedma, D., Garzelli, C., Gazzola, L., Gomes, H.M., Guttierez, M.C., Hawkey, P.M., van Helden, P.D., Kadival, G.V., Kreiswirth, B.N., Kremer, K., Kubin, M., Kulkarni, S.P., Liens, B., Lillebaek, T., Ho, M.L., Martin, C., Martin, C., Mokrousov, I., Narvskaia, O., Ngeow, Y.F., Naumann, L., Niemann, S., Parwati, I., Rahim, Z., Rasolofo-Razanamparany, V., Rasolonavalona, T., Rossetti, M.L., Rusch-Gerdes, S., Sajduda, A., Samper, S., Shemyakin, I.G., Singh, U.B., Somoskovi, A., Skuce, R.A., van Soolingen, D., Streicher, E.M., Suffys, P.N., Tortoli, E., Tracevska, T., Vincent, V. Victor, T.C. Warren, R.M., Yap, S.F., Zaman, K., Portaels, F., Rastogi, N. and Sola, C. (2006) *Mycobacterium tuberculosis* complex genetic diversity: mining the fourth international spoligotyping database (SpolDB4) for classification, population genetics and epidemiology. BMC Microbiol. 6, 23.

Drancourt, M., Roux, V., Dang, L.V., Tran-Hung, L., Castex, D., Chenal-Francisque, V., Ogata, H., Fournier, P-E., Crubézy, E. and Raoult, D. (2004) Genotyping, Orientalis-like *Yersinia pestis*, and plague pandemics. Emerg. Infect. Dis. 10, 1585-1592.

Gilbert, M.T., Cuccui, J., White, W., Lynnerup, N., Titball, R.W., Cooper, A. and Prentice, M.B. (2004) Absence of *Yersinia pestis*-specific DNA in human teeth from five European excavations of putative plague victims. Microbiology 150, 341-354.

Jansen, R., Embden, J.D., Gaastra, W. and Schouls, L.M. (2002) Identification of genes that are associated with DNA repeats in prokaryotes. Mol. Microbiol. 43, 1565-1575.

Le Flèche, P., Hauck, Y., Onteniente, L., Prieur, A. Denoeud, F., Ramisse, V., Sylvestre, P., Benson, G. Ramisse, F. and Vergnaud, G. (2001) A tandem repeats database for bacterial genomes: application to the genotyping of *Yersinia pestis* and *Bacillus anthracis*. BMC Microbiol. 1, 2.

Lillestol, R.K., Redder, P., Garrett, R.A. and Brugger, K. (2006) A putative viral defence mechanism in archaeal cells. Archaea 2, 59-72.

Makarova, K.S., Grishin, N.V., Shabalina, S.A., Wolf, Y.I. and Koonin, E.V. (2006) A putative RNA-interference-based immune system in prokaryotes: computational analysis of the predicted enzymatic machinery, functional analogies with eukaryotic RNAi, and hypothetical mechanisms of action. Biol. Direct 1, 7.

Mojica, F.J., Diez-Villasenor, C., Garcia-Martinez, J. and Soria, E. (2005) Intervening sequences of regularly spaced prokaryotic repeats derive from foreign genetic elements. J. Mol. Evol. 60, 174-182.

Pourcel, C., Andre-Mazeaud, F., Neubauer, H., Ramisse, F. and Vergnaud, G. (2004) Tandem repeats analysis for the high resolution phylogenetic analysis of *Yersinia pestis*. BMC Microbiol. 4, 22.

Pourcel, C., Salvignol, G. and Vergnaud, G. (2005) CRISPR elements in *Yersinia pestis* acquire new repeats by preferential uptake of bacteriophage DNA, and provide additional tools for evolutionary studies. Microbiology 151, 653-663.

Prentice, M.B., Gilbert, T. and Cooper, A. (2004) Was the Black Death caused by *Yersinia pestis*? Lancet Infect. Dis. 4, 72.

Shapiro, B., Rambaut, A. and Gilbert, M.T. (2006) No proof that typhoid caused the Plague of Athens (a reply to Papagrigorakis et al.). Int. J. Infect. Dis. 10, 334-335; author reply 335-336.

Song, Y., Tong, Z., Wang, J., Wang, L., Guo, Z., Han, Y., Zhang, J., Pei, D., Zhou, D., Qin, H., Pang, X., Zhai, J., Li, M., Cui, B., Qi, Z., Jin, L., Dai, R., Chen, F., Li, S., Ye, C., Du, Z., Lin, W., Yu, J., Yang, H., Huang, P. and Yang, R. (2004) Complete genome sequence of *Yersinia pestis* strain 91001, an isolate avirulent to humans. DNA Res. 11, 179-197.

Tang, T.H., Bachellerie, J.P., Rozhdestvensky, T., Bortolin, M.L., Huber, H., Drungowski, M., Elge, T., Brosius, J. and Huttenhofer, A. (2002) Identification of 86 candidates for small non-messenger RNAs from the archaeon *Archaeoglobus fulgidus*. Proc. Natl. Acad. Sci. USA 99, 7536-7541.

Vergnaud, G. (2005) *Yersinia pestis* genotyping. Emerg. Infect. Dis. 11: 1317-1318; author reply 1318-1319.

Zhou, D., Han Y., Song, Y., Huang, P. and Yang, R. (2004a) Comparative and evolutionary genomics of *Yersinia pestis*. Microbes Infect. 6, 1226-1234.

Zhou, D., Han, Y., Song, Y., Tong, Z., Wang, J., Guo, Z., Pei, D., Pang, X., Zhai, J., Li, M., Cui, B., Qi, Z., Jin, L., Dai, R., Du, Z., Bao, J., Zhang, X., Yu, J., Wang, J., Huang, P. and Yang, R. (2004b) DNA microarray analysis of genome dynamics in *Yersinia pestis*: insights into bacterial genome microevolution and niche adaptation. J. Bacteriol. 186, 5138-5146.

Zink, A.R., Sola, C., Reischl, U., Grabner, W., Rastogi, N., Wolf, H. and Nerlich, A.G. (2003) Characterization of *Mycobacterium tuberculosis* complex DNAs from Egyptian mummies by spoligotyping. J. Clin. Microbiol. 41, 359-367.

31

Enrichment of *Yersinia pestis* from Blood Cultures Enables Rapid Antimicrobial Susceptibility Determination by Flow Cytometry

Ida Steinberger-Levy[1], Eran Zahavy[2], Sara Cohen[1], Yehuda Flashner[1], Emanuelle Mamroud[1], Moshe Aftalion[1], David Gur[1], and Raphael Ber[1]

[1] Department of Biochemistry and Molecular Genetics, Israel Institute for Biological Research, ber@iibr.gov.il
[2] Department of Infectious Diseases

Abstract. Mortality from plague is high if not treated with the proper antibiotics within 18-24 hours after onset of symptoms. The process of antibiotic susceptibility determination of *Yersinia pestis* isolated from blood samples may extend from 4 to more than 7 days, since the *in vitro* growth is very slow. To accelerate this process, we developed an enrichment protocol as well as a non-standard yet reliable method for rapid antibiotic susceptibility analysis of *Y. pestis* from blood cultures using flow cytometry technology. This rapid method is applicable to blood cultures containing low levels of *Y. pestis*.

31.1 Introduction

Following exposure to *Yersinia pestis*, illness symptoms begin after 1-8 days, depending on variables such as route of exposure, infection dose and patient's immunocompetence. Plague is characterized by rapid progress and high mortality rate (40-60% in bubonic, 30-50% in septicemic and ~100% in pneumonic plague) if not treated within 18-24 hours after symptoms onset (Pollitzer 1954; WHO 1999; Inglesby et al. 2000). Therefore early determination of antimicrobial susceptibility is essential for efficient antibiotic therapy.

Bacteremia is frequently observed in plague patients. *Y. pestis* load in patient's blood in the acute phase of disease is variable, spanning from none to higher than 10^7 cfu/ml (Pollitzer 1954; Butler et al. 1976; Butler 1983; Perry and Fetherston 1997) with significant correlation between level of bacteremia and death (Butler et al. 1976). Patients with 100 cfu/ml in the blood had a higher mortality rate than patients with less than 100 cfu/ml, although the survival of a patient that reached bacteremic levels of 10^7 cfu/ml has been reported (Butler et al. 1976). Massive bacteremia is usually a clear sign for death (Butler 1983).

Since plague, and especially the pneumonic form, is characterized by a fast progression of the disease, it is recommended that 4 blood samples be collected at 30-min. intervals before initiation of antibiotic therapy of a suspected plague patient. Because of the severity of the disease, antibiotic therapy should begin, based on clinical or epidemiological suspicions, as soon as possible even before revealing the

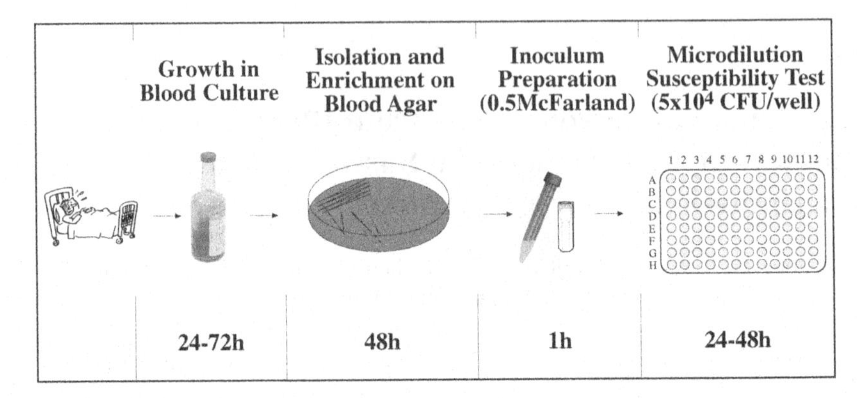

Fig. 1. The standard procedure for antibacterial susceptibility determination of blood derived *Y. pestis*. A procedure recommended by the CDC and CLSI (CDC 2000; CLSI 2006).

antimicrobial susceptibility of the bacteria (Perry and Fetherston 1997; WHO 1999). To ensure the appropriate antibiotic therapy for plague patients, antimicrobial susceptibility must be determined as quickly as possible. However, in contrast to the rapid progress *in vivo*, the *in vitro* growth of *Y. pestis* is very slow, hence the determination of antibacterial susceptibility of bacteria derived from blood culture may extend from 4 to 7 days (Fig. 1). This procedure includes 1-3 days for the blood culture to become positive, 2 days for colony (1-2 mm) formation (Perry and Fetherston 1997; Inglesby et al. 2000) and 1-2 days for the antimicrobial susceptibility analysis (Clinical and Laboratory Standards Institute/ CLSI 2006).

Efforts have been made to develop procedures that would shorten the time needed for antimicrobial susceptibility determination, especially for slow growing pathogens. Flow Cytometry (FCM) technology has been used to study the effect of different antimicrobial agents on prokaryotes and has been suggested as a rapid method for antimicrobial susceptibility determination of *Escherichia coli, Klebsiella pneumoniae, Staphylococcus aureus* and *Mycobacterium tuberculosis* (Steen 2000; Alvarez-Barrientons et al. 2000). However, direct monitoring of bacterial growth in blood culture using the FCM technology is still a challenge since blood components (cells and plasma) quench the fluorescence signal of the stained bacteria (Mansour et al. 1985).

In this report we describe a procedure that was developed for rapid determination of antimicrobial susceptibility of *Y. pestis* from blood cultures, in a non-standard yet reliable test that yields MIC values equal to standard microdilution assay. This procedure includes separation of bacteria from the blood culture components and enrichment in optimal growth conditions using Cation-Adjusted Mueller Hinton Broth (CAMHB), a medium recommended by the CLSI for standard antimicrobial susceptibility tests. Finally, direct monitoring of bacterial counts is performed using FCM technology for both inoculum adjustment and antimicrobial susceptibility determination using the microdilution test platform.

31.2 Materials and Methods

31.2.1 Bacterial Strains and Growth Conditions

Two *Y. pestis* strains were used in this study: the avirulent vaccine strain EV76 and the fully virulent Kimberley53 strain (Flashner et al. 2004). Bacteria were isolated on rich selective BIN agar (Ber et al. 2003), suspended in PBS or in CAMHB medium (BBL), and the optical density was measured in a spectrophotometer.

For growth in blood culture, fresh human blood (10ml) was spiked with 100 ul of a *Y. pestis* suspension to yield 10^1-10^6 cfu/ml final concentrations in Bactec plus$^+$ Aerobic/F vials (Becton Dickinson, USA). Blood cultures were grown at 37°C, 150 rpm. Bacterial growth was monitored by plating the appropriate dilutions on Brain Heart Infusion Agar (BHIA) and incubating the plates for 48h at 28°C.

31.2.2 Separation of *Y. pestis* From Blood Components Using Vacutainer Serum Separation Tubes (VSST) and Enrichment in CAMHB

Blood culture samples (2-10 ml, taken by syringe) were injected into VSST (Becton Dickinson, USA) and centrifuged at 1700g for 10 min. The bacterial layer on the separating gel was washed with PBS, centrifuged at 700g for 15 min. and resuspended in CAMHB (1:10 dilution of the original concentration). This bacterial culture was grown at 28°C, 200 rpm for enrichment.

31.2.3 *Y. pestis* Growth and Viability Monitoring Using FCM

Samples in the enrichment culture were monitored by FCM following staining with a live/dead BacLight viability kit (Invitrogen, USA). FCM analysis was performed on a FACSCalibur (B&D, CA) using logarithmic amplified side scatter (SSD) and forward scatter detectors (FSD). Live bacteria were identified by their light scatter analysis and their FL1 increased fluorescence after SSD/FSD and FL1/SSD gating. Bacterial counts by FCM were verified by cfu counting.

31.2.4 Antibiotic Susceptibility Tests

Susceptibility to doxycycline (Sigma) was determined by a standard microdilution test performed according to the CLSI guidelines (CLSI 2006). The microdilution test was also analyzed by monitoring the tested bacteria using FCM analysis for plotting bacterial cell counts versus time. Etest (AB Biodisk, Sweden) was performed on Mueller Hinton Agar (MHA) (DIFCO) according to the manufacturer and CDC recommendations (CDC 2000). Plates were incubated at 28°C for 24 h.

31.3 Results

31.3.1 Separation of *Y. pestis* From Blood Components Using VSST

A reliable antimicrobial susceptibility test should be conducted using a defined number of logarithmic bacteria (CDC 2000; CLSI 2006). Yet, some clinical laboratories perform the antimicrobial susceptibility test with bacterial cells taken directly from the positive blood culture, skipping the prolonged agar isolation step. This direct test is performed with no pre-determination and adjustment of bacterial concentration since blood components (blood cells and plasma chromophores) prevent conventional spectral measurements. We demonstrated that disregarding the above requirements may yield inaccurate susceptibility results, especially with slow growing bacteria (Fig. 2). We determined the MIC value of *Y. pestis* sampled during growth in blood culture, using the Etest method. Different MIC values were observed when using those samples. While the standard MIC value of doxycycline (DC) for EV76 is 0.75 μg/ml, direct tests from early stage blood cultures (at bacterial concentrations of $5x10^5$/ml) resulted in a 2-fold lower MIC value. Furthermore, even a 3 fold lower

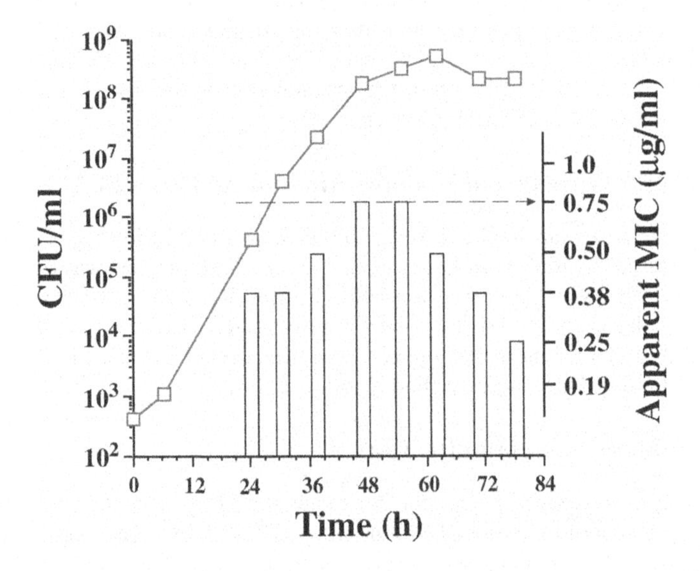

Fig. 2. MIC values of non-standardized samples taken directly from *Y. pestis* blood culture. A blood culture was inoculated with $4x10^2$ cfu/ml of *Y. pestis* EV76 and incubated at 37°C, 150 rpm. Samples were taken for the determination of MIC values of doxycycline (DC) using the Etest method. The curve represents growth of EV76 in blood culture; columns represent MIC values obtained for the samples taken at the indicated time points. Dashed arrow indicates the MIC obtained for EV76 under standard Etest conditions. At right: DC Etest strip scale representing MIC values (μg/ml).

MIC value was obtained when using stationary phase blood cultures (60-78 h after inoculation) even though the bacterial concentrations were in the appropriate range (10^8 cfu/ml). Expected MIC values were obtained only when the blood culture was in the late exponential growth phase and contained about 1-2 x 10^8 cfu/ml *Y. pestis,* as required by the CDC or CLSI guidelines. This phenomenon of inaccurate MIC determination is of more concern in cases of strains having MIC values that are close to the categorical end point value, as the change in the apparent MIC value measured by the non-standard test may alter the interpreted sensitivity category of those strains. Thus, in order to correctly measure MIC values, the bacterial concentration and appropriate physiological growth phase must be adjusted prior to the antimicrobial susceptibility tests. For this purpose, an isolation step and change of growth medium step must be introduced to eliminate the blood components that prevent direct bacterial monitoring.

We showed that rapid separation of the bacterial cells from blood components can be achieved using the VSST, routinely used in clinical laboratories for serum separation from blood cells. When a blood culture containing *Y. pestis* was centrifuged in the VSST, the bacteria accumulated on the surface of the gel barrier, separating them from the red blood cells that penetrate through the gel (Fig. 3). The efficiency of this step was 50-100% when the bacterial concentration was in the range of 10^5-10^9 cfu/ml. Bacteria were collected, washed and suspended in CAMHB, the standard medium required by the CLSI guidelines for liquid-based antimicrobial susceptibility tests. This separation process enabled both enhanced growth rate and direct monitoring of viable bacteria by FCM (sections 31.3.2-31.3.3).

31.3.2 Enrichment of Blood Culture-Derived *Y. pestis* in Optimal Growth Conditions

The growth of *Y. pestis* in the blood culture vial at 37°C is a major rate-limiting step in the standard process of antimicrobial susceptibility determination. It is well known that the optimal *in vitro* growth temperature for *Y. pestis* is 28°C. Thus, for culture enrichment, we incubated the VSST isolated bacteria at 28°C in CAMHB, the recommended medium for antimicrobial susceptibility determination tests. The growth rate of *Y. pestis* in this media was higher than the growth rate in the blood culture (Fig. 4). Furthermore, even when the initial concentration of the bacteria in the enrichment culture was very low (~10 cfu/ml), it reached the concentration needed for microdilution susceptibility test (10^6 cfu/ml) within 19 h, 10 h before the original blood culture (Fig. 4). Another advantage of the VSST separation is that FCM analysis can be performed on the CAMHB enrichment culture, to verify that the cells are in the logarithmic phase of growth required for susceptibility test inocula by the standard guidelines (CLSI 2006), as shown in section 31.3.3.

A

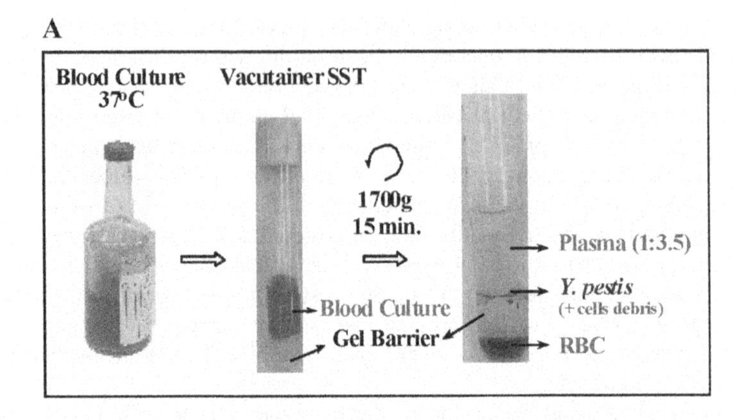

B

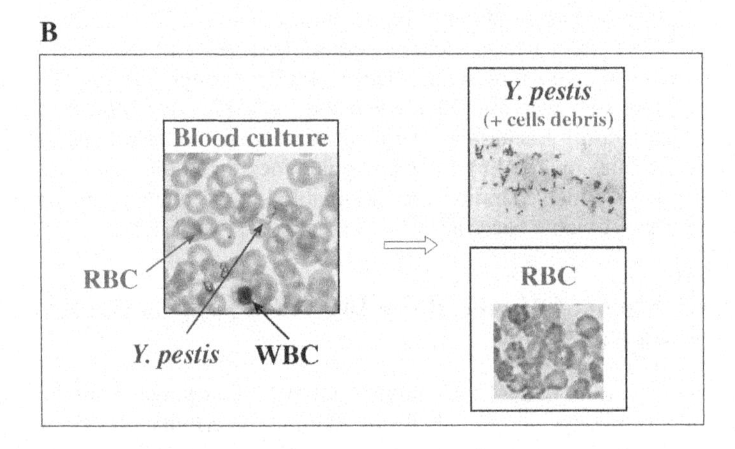

Fig. 3. *Y. pestis* separation from blood culture using the Vacutainer Serum Separation Tube. Panel A. Separation of *Y. pestis* EV76 from blood culture was done using the VSST as described in Materials and Methods. Panel B. Slides representing samples from the blood culture before and after VSST separation, were fixed by methanol for 10 min., giemsa stained (SIGMA, USA) and viewed (1000 x) with an Axiovert 200 inverted light microscope (ZEISS).

31.3.3 Flow Cytometry Analysis for Counting Viable Bacteria and Determination of Antibiotic Susceptibility

Since we were able to separate bacteria from blood components using the VSST technique, we evaluated the application of FCM technology for determination of viable bacterial counts even at a low concentration of bacteria in the blood culture. We found that following the vacutainer separation step and the dilution in CAMHB, FCM can be used for the detection and the analysis of *Y. pestis* (Fig. 5-D). The VSST step performed prior to the FCM was necessary in order to overcome the interference attributed to both the high number of red blood cells (RBCs) (RBC concentration in

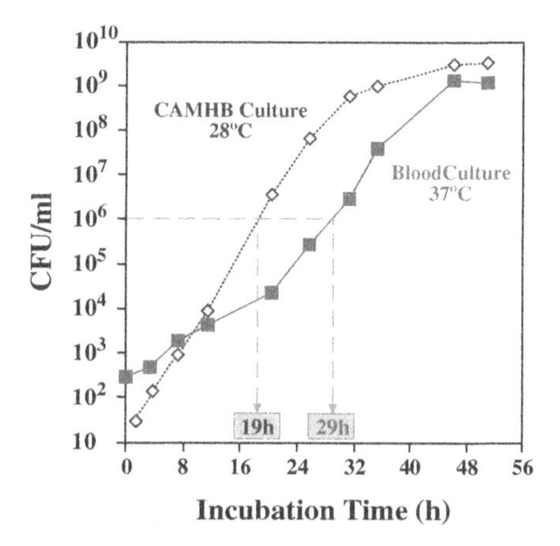

Fig. 4. Growth of *Y. pestis* in a blood culture and in enriched medium (CAMHB) following separation from blood components using the VSST. A Blood culture was spiked with *Y. pestis* Kimberley53 and grown at 37°C, 150 rpm (squares). Separation by VSST was done as described in Materials and Methods. Bacteria were diluted 1:10 in CAMHB and incubated at 28°C, 200 rpm (diamond). Growth curves were determined by measuring cfu counts.

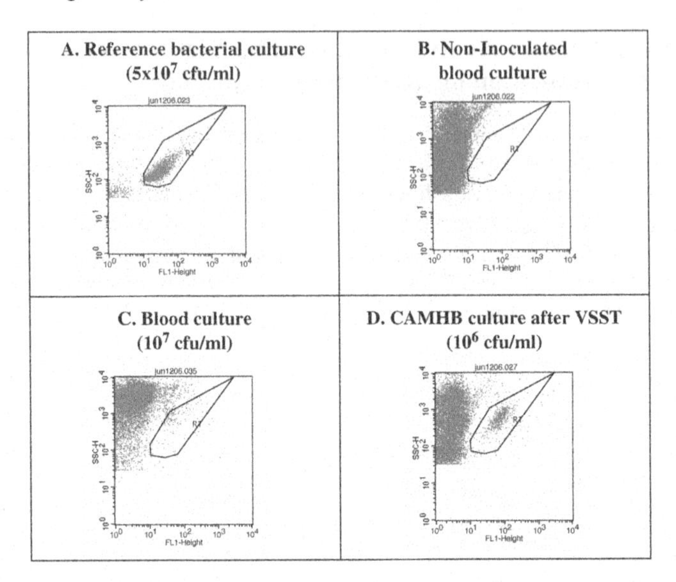

Fig. 5. Quantitation of *Y. pestis* using FCM technology. Viable bacteria were labeled fluorescently using a BacLight kit as described in Materials and Methods. FCM gating parameters for *Y. pestis* were first developed in broth conditions. For this purpose a reference suspension was prepared from *Y. pestis* plated on BHIA, incubated for 48h at 28°C and resuspended in CAMHB (A). Blood cultures were prepared by adding 10 ml of human blood with (C and D) or without (B) *Y. pestis* to Bactec plus$^+$ Aerobic/F vials. After incubation of the blood culture vials at 37°C, 150 rpm, samples were stained either directly (C) or after separation using the VSST procedure (B and D). FCM-based bacterial counts were validated by cfu counting.

human blood is 10^9/ml) and the quenching of the bacterial fluorescence signal by plasma components (chromophores etc.) (Fig. 5C). After VSST separation, FCM can be used to determine the concentration of a fluorescence-labeled bacterial suspension even at about 10^5 cfu/ml. Thus, as the inoculum for microdilution susceptibility tests requires only $5x10^4$ cfu/0.1ml wells, and since a total of 10^6 bacteria are sufficient for categorical determination of 3 different antibiotic agents in a microdilution test, this amount can be monitored at an early step of the enrichment. This concentration is 100 fold lower than the amount of bacteria needed for turbidity adjustment (0.5 McFarland) prior to use in a standard turbidity based test.

We evaluated the FCM technology also for monitoring growth directly in the microdilution assay and compared it to the standard turbidity monitoring. The high sensitivity of the FCM analysis enabled the detection of *Y. pestis* growth as soon as 2h after the start of the antimicrobial susceptibility test (Fig. 6) and revealed the antibiotic effect much sooner compared to the turbidity-based analysis of the microdilution assay. Yet, to accomplish a reliable antimicrobial susceptibility test (CLSI based criteria), an incubation period of 14 hours was chosen as the earliest time for determination of the MIC value by FCM, 10 hours shorter than the 24 hour incubation period required in the standard microdilution test. The same MIC values were obtained for both the EV76 strain and Kimberley53 (0.75 μg/ml and 0.5 μg/ml,

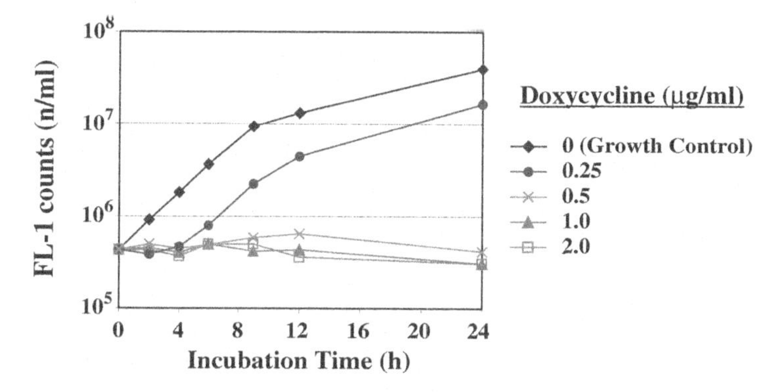

Fig. 6. Monitoring of doxycycline microdilution susceptibility test using FCM. Growth of *Y. pestis* EV76 was monitored in the CAMHB enrichment medium until a concentration of 10^6 cfu/ml was achieved, allowing a direct 1:1 dilution of the culture into microdilution wells containing 50 μl of the appropriate concentration of doxycycline or CAMHB for growth control, to yield an initial concentration of $5x10^4$ cfu/well. The plate was incubated at 28°C and the response of the cultures to the antibiotic was plotted as FCM-based positive viable counts (n/ml) versus time, for the indicated doxycycline concentrations (Materials and Methods). Standard microdilution susceptibility test was performed as a control in accordance to CLSI guidelines (CLSI 2006).

respectively) in both FCM-based and turbidity-based antimicrobial susceptibility tests. Moreover, the 14 hour time point was found to fit for MIC value determination of other antimicrobial agents (data not shown). The *in vitro* sensitivity of *Y. pestis* to doxycycline was corroborated with *in vivo* susceptibility (Ber et al. data not shown) in a mouse model of bubonic plague (Flashner et al. 2004).

The process of *Y. pestis* separation from blood culture components enables the otherwise impossible monitoring and counting of viable bacteria by FCM.

31.4 Discussion

The microdilution test is one of the standard antimicrobial susceptibility methods and it is based on monitoring turbidity of bacterial growth in the presence or absence of antimicrobial agents. This test requires a low amount of bacteria relative to other antimicrobial susceptibility methods, as an inoculum of 10^6 cfu/ml allows a direct 1:1 dilution to yield the required standard concentration of $5x10^4$ cfu/well. This amount of bacteria (10^6) is sufficient to determine the susceptibility category (Sensitive/Intermediate/Resistant) of up to 3 different antimicrobial agents. However, adjustment of the bacterial concentration from blood cultures to the required 0.5 McFarland, as well as direct monitoring of bacterial growth using optical turbidity measurement, is impossible due to the presence of RBCs and chromophores in the blood culture. Therefore, a time-consuming step (2 days) of isolation on agar plates is traditionally required prior to the preparation of a calibrated inoculum and performing the antimicrobial susceptibility test. Since plague progresses rapidly, this delay may be critical for efficient antibiotic therapy.

The aim of the present study was to develop a rapid yet reliable method for determination of antimicrobial susceptibility of *Y. pestis* growing in blood cultures, which would be applicable also for samples collected from suspected plague patients having low level bacteremia. Hence, we present an alternative rapid procedure that enables the separation of bacteria from the red blood cells and chromophores of the blood culture, using VSST. The separated bacteria are transferred to CAMHB (appropriate also for antimicrobial susceptibility tests) and incubated at 28°C instead of 37°C for enrichment. The next step is evaluating the bacterial concentration, adjustment to the required inoculum and monitoring bacterial growth in a microdilution-like antimicrobial susceptibility assay using FCM technology.

Using FCM analysis for quantitation of the enriched bacteria reduces significantly the time required for growth to the minimal amount of bacteria (10^6) required for the microdilution assay compared to the minimal amount of 10^8 cfu/ml needed for the standard turbidity-based measurement.

In addition, the FCM reduces also the time which is required for the microdilution test itself and the MIC value is obtained in an incubation period as short as 14h (Fig. 6) compared to the 24h needed for the standard CLSI turbidity–based microdilution test (CLSI 2006). Moreover, correct MIC values were obtained for other tested bacteriostatic and bactericidal antimicrobial agents by the FCM monitoring (data not shown). This new process yields reliable susceptibility results as verified by identical MIC values obtained by both the FCM method and by the standard microdilution test with both the avirulent EV76 strain and virulent Kimberley53 strain. However, further validation with other *Y. pestis* strains having different MIC values and *in vitro* growth rates should be done.

The removal of quenching chromophores and masking background using the VSST separation technique allowed, for the first time, the application of FCM for accurate, rapid and reliable antimicrobial susceptibility analysis of bacteria grown in blood culture without the need for prior isolation on agar. These results add to

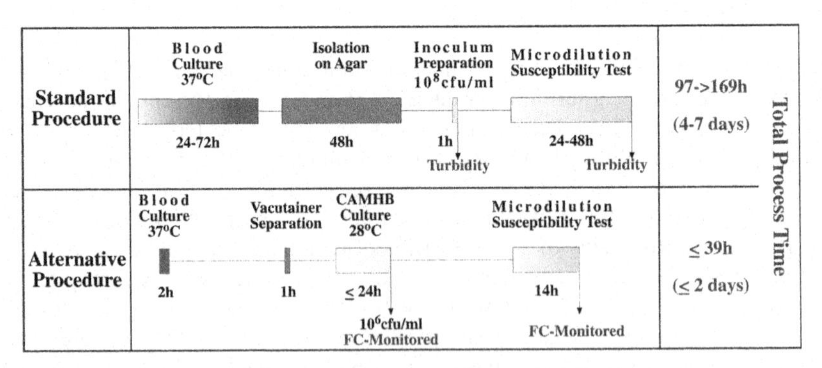

Fig. 7. Time scale comparison of the antibiotic susceptibility determination assay of *Y. pestis* isolated from blood cultures using the standard and the alternative procedures.

previously published reports describing the use of FCM technology for rapid and reliable antimicrobial susceptibility determination of a wide range of antimicrobial agents against various bacteria (Steen 2000; Alvarez-Barrientons et al. 2000). Due to high cost and the need for FCM experienced personnel, FCM technology still does not constitute a real part of clinical laboratories, however, future refinements may overcome these obstacles allowing the use of FCM in routine testing.

In summary, we show that the whole procedure we developed for the separation and enrichment of *Y. pestis* from blood cultures, together with the reliable FCM-mediated inoculum preparation and antimicrobial susceptibility determination, reduce analysis time from between 4 to 7 days to less than 2 days, even for low-level bacteremic blood samples (Fig. 7). Application of this concept in the clinical laboratory would allow improved treatment for suspected plague patients and could be applicable for the antimicrobial susceptibility determination of other blood born slow growing bacterial pathogens as well.

31.5 References

Alvarez-Barrientons, A., Arroyo, J., Canton, R., Nombela, C. and Sanchez-Perez, M. (2000) Application of flow cytometry to clinical microbiology. Clin. Microbiol. Rev. 13, 167-195.

Ber, R., Mamroud, E., Aftalion, M., Tidhar, A., Gur, D., Flashner, Y. and Cohen, S. (2003) Development of an improved selective agar medium for isolation of *Yersinia pestis*. Appl. Environ. Microbiol. 69, 5787–5792.

Butler, T., Levine, J., Linh, N., Chau, D., Adickman, M. and Arnold, K. (1976) *Yersinia pestis* infection in Vietnam. II: Quantitative blood cultures and detection of endotoxin in the cerebrospinal fluid of patients with meningitis. J. Infect. Dis. 133, 493-498.

Butler, T. (1983) Plague and other *Yersinia* infections. In: Greenough W. B. III and Merigan T. C. (Eds.), *Current topics in infectious disease*. Plenum Medical Book and Company, New York.

CDC (2000) Antimicrobial susceptibility tests for *Yersinia pestis*. In: *Laboratory manual of plague diagnostic tests*. pp. 34-38.

CLSI (2006) Performance standards for antimicrobial susceptibility testing: sixteenth informational supplement. M100-S16.

Flashner, Y., Mamroud, E., Tidhar, A., Ber, R., Aftalion, M., Gur, D., Lazar, S., Zvi, A., Bino, T., Ariel, N., Velan, B., Shafferman, A. and Cohen, S. (2004) Generation of *Yersinia pestis* attenuated strains by signature-tagged mutagenesis in search of novel vaccine candidates. Infect. Immun. 72, 908-915.

Inglesby, T., Dennis, D., Henderson, D., Bartlett, J., Ascher, M., Eitzen, E., Fine, A., Friedlander, A., Hauer, J., Koerner, J., Layton, M., McDade, J., Osterholm, M., O'Toole, T., Parker, G., Perl, T., Russell, P., Schoch-Spana, M. and Tonat, K. (2000) Plague as biological weapon- Medical and public health management. J. of American Medical Association. 283, 2281-2290.

Mansour, J., Robson, J., Arndt, C. and Schulte, T. (1985) Detection of *Eschcerichia coli* in blood using flow cytometry. Cytometry 6, 186-190.

Perry, R. and Fetherston, J. (1997) *Yersinia pestis*- Etiologic agent of plague. Clin. Microbiol. Rev. 10, 35-66.

Pollitzer, R. (1954) Plague. WHO Monograph series 22. WHO, Geneva, Swizerland.

Steen, H. (2000) Flow cytometry of bacteria: glimps from past with a view to the future. J. Microb. Methods.42, 65-74.
WHO (1999) Treatment of Plague. In: *Plague manual: Epidemiology, distribution, surveillance and control.* pp. 55-62.

Picture 28. Dancing at the banquet. Photograph by R. Perry.

32

Development and Evaluation of a Single Tube Nested PCR Based Approach (STNPCR) for the Diagnosis of Plague

Gerlane Souza, Frederico Abath, Nilma Leal, Alexandra Farias, and Alzira Almeida

Centro de Pesquisas Aggeu Magalhães, gerlane@cpqam.fiocruz.br

Abstract. The performance of a single-tube nested-PCR (STNPCR) technique was evaluated for plague diagnosis in comparison to conventional (one step) and two step nested PCR (NPCR). Assays were carried out with primers targeting the gene *caf1* that encodes the *Yersinia pestis* F1 antigen. For STNPCR inner primers were immobilized onto the inside of the microtube caps and after the first amplification they were eluted by inversion of the tube. This procedure avoids opening the tube, reducing the risks of false-positive results by cross-contamination. The immobilized primers are stable for several months at –20°C, thus, the tubes can be prepared beforehand and stored until use. STNPCR was more sensitive than conventional PCR, and less sensitive than NPCR. This drawback is compensated by a lower risk of cross-contamination. The experiments with infected animals showed that NPCR and STNPCR were able to produce positive results in all samples tested, despite contamination with other organisms. In contrast, conventional PCR yielded positive results in a smaller number of samples. Three out of 62 culture-negative rodents from plague areas, were positive by STNPCR. In conclusion, the PCR approaches evaluated, particularly NPCR and STNPCR have potential to be used as alternative tools in epidemiological surveys of plague. Furthermore, as the results can be obtained quickly (less than 24 hour), these techniques could be useful in emergency situations in which the rapidity in diagnosis is essential for adoption of immediate measures of control.

32.1 Introduction

Plague surveillance requires efficient and rapid diagnostic methods to test different types of biological samples. The gold standard for *Yersinia pestis* diagnosis is the isolation of the bacteria by culture (Chu 2000). However, this procedure has several limitations: it is time consuming; it can be hampered by contamination of samples with other organisms; and the usual procedures for shipment of samples can make the bacteria unviable for culture. Some molecular methods based on PCR and nested-PCR (NPCR) have been proposed for the diagnosis of plague (Campbell et al. 1993; Engelthaler et al. 1999; Hinnebush and Schawn 1993; Leal et al. 1996; Leal and Almeida 1999; Norkina et al. 1994; Rahalison et al. 2000; Tsukano et al. 1996). NPCR is a two-step procedure in which the products of a first PCR using outer primers are subjected to re-amplification with a second set of inner primers located within the previously amplified sequence. Although the two-step nested PCR is more sensitive than conventional PCR, an inherent drawback is the need to open tubes after the first round of amplification to transfer products to a tube for a further PCR amplification reaction that utilizes a different primer pair or to introduce new reagents and/or primers (Picken et al. 1996). This process dramatically increases the risk of cross-contamination of negative samples with amplicons derived from positive specimens during the first round of amplification, and false positive PCR results are the main concern of most clinical diagnostic laboratories (Gookin et al. 2002; Herrmann et al. 1996; Nagl et al. 2001).

Thus, we developed and evaluated a single tube nested PCR (STNPCR) based upon immobilization of the internal primers onto the inside of the microtube caps (Abath et al. 2002). This format is less prone to cross-contamination, more suitable for field use, and more cost-effective. In addition, this novel method was compared with conventional PCR and NPCR.

32.2 Materials and Methods

32.2.1 Bacterial Culture and Preparation of DNA

Y. pestis P. PB 881 (CPqAM/FIOCRUZ collection) was grown either on brain heart infusion broth (BHI, Difco) or blood agar base (BAB, Difco) at 28°C and was used throughout the experiments.

Bacterial DNA was extracted by centrifugation of 1 ml of broth culture; the pellet was washed and subsequently homogenized in 500 µl of TE (10 mM Tris-HCl, 1 mM EDTA pH 8.0), 10 µl of lysozyme (10 mg/ml) and 10 µl of proteinase K (5 mg/ml), and incubated at 60°C for 20 minutes. Then, 100 µl of STE (2.5% SDS, 10 mM Tris-HCl pH 8.0, 0.25 M EDTA) was added and the suspension was incubated for 15 minutes at 60°C, followed by 5 minutes at room temperature and 5 minutes on ice. The suspension was neutralized with 130 µl of 7.5 M ammonium acetate (CH_3COONH_4), kept on ice for 15 minutes and centrifuged. The supernatant was transferred to a new tube and DNA was extracted with phenol-chloroform, and precipitated with isopropanol. The precipitate was resuspended in 10 µl of a solution containing 10 mg/ml RNAse and stored at –20°C.

32.2.2 Primers

The primers targeted the gene *caf1* that encodes the *Y. pestis* F1 antigen (Galyov et al. 1990). The external primers (5'-cagttccgttatcgccattgc-3' and 5'-tattggttagatacggttacggt-3') have been described before (Norkina et al. 1994). The internal primers (5'-ttggaactattgcaactgcta-3' and 5'-ttagatacggttacggtta-3') have been published before by our group (Leal et al. 1996).

32.2.3 PCR Procedures

32.2.3.1 Conventional PCR

For conventional PCR a 25 μl PCR reaction mixture was prepared containing 10 mM Tris-HCl, 50 mM KCl, 1.5 mM $MgCl_2$, 0.2 mM of each dNTP, 20 pmol of each primer, 1 U of Taq DNA polymerase (Amersham Biosciences, Uppsala, Sweden). Amplifications were carried out in 0.5 ml PCR tubes by using a Biometra Trio thermocycler™. Programs were run for 30 cycles, consisting of denaturation at 92°C for 1 min, annealing at 55°C for 1 min, and extension at 72°C for 1 min, followed by 7 min at 72°C.

32.2.3.2 Analysis of the PCR Products

PCR products (10 μl) were separated by electrophoresis in 1% agarose gels, and ethidium bromide stained gels were visualized and photographed under ultraviolet light using the MP4+ Polaroid System™ (Sigma, St. Louis, MA, USA).

32.2.3.3 NPCR

NPCR was performed using the same reaction mixture and cycling conditions as reported for conventional PCR. Twenty pmol of outer primers were used in the first PCR, and 20 pmol of internal primers were used in the second PCR. Two μl of the products of the first PCR were used as templates for the second PCR. Amplification products were analyzed as described.

32.2.3.4 STNPCR

STNPCR assays were carried out in 0.5 ml PCR tubes using a Biometra Trio thermocycler™, essentially as described previously (Abath et al. 2002). Briefly, the reactions were optimized for concentrations of Taq DNA polymerase, deoxynucleoside triphosphates, Mg^{2+}, and outer and inner primers ratio. For the optimization of primer-pair ratio, STNPCR assays were set up using 20 ng of purified *Y. pestis* DNA as template, and various primer-pair ratios (data not shown). The relative amounts of long, short and intermediate amplicons were determined by agarose gel electrophoresis so that the best yield of the targeted product could be achieved. Based on these experiments an outer primer set/internal primer set ratio of 1:10 (2 pmol:20 pmol) was employed in all subsequent experiments. Ten μl containing 20 pmol of inner primers with traces of bromophenol blue were previously immobilized onto the inside of the microtube cap by incubating the tubes at 37°C until the solution had dried. The first amplification round consisted of 15 cycles (90°C for 1 min, 55°C for 1 min and 72°C for 1 min), whereas the second round of amplification consisted of 45 cycles. The first amplification round of the STNPCR was performed in a 50 μl volume containing 20 mM Tris-HCl, 100 mM KCl, 3 mM $MgCl_2$, 0.4 mM of each dNTP, 2 pmol of outer primers, 2 U of Taq DNA polymerase (Amersham

Biosciences, Uppsala, Sweden). After the first round PCR, the thermocycler was paused at 92°C, and the closed tubes were inverted several times to dissolve the inner primers inside the cap, briefly centrifuged, and returned to the machine for the second round of amplification.

32.2.4 Assessment of the Sensitivity of the PCR Assays

A $\log_{10}$ dilution series in sterile deionized water containing one ng to 100 fg of DNA was used to asses the detection limit of the PCR procedures. An overnight broth culture of the strain was serially diluted (ten-fold) in normal saline. The suspensions were boiled to expose the DNA and 10 µl of dilutions containing two to 2,000 colony forming units (CFU) were directly used in the PCR assays.

32.2.5 Assessment of the Specificity of the PCR assays

Specificity was evaluated by using DNA from *Yersinia pseudotuberculosis*, and *Yersinia enterocolitica*, and spleen samples from non-infected mice. Additionally positive (20 ng of *Y. pestis* DNA) and negative (non template) controls were included each time PCR was undertaken to detect variation in sensitivity or false positive results due to contamination.

32.2.6 Procedures with Animals

Seven Swiss-Webster albino mice provided by the animal facilities of the CPqAM were inoculated subcutaneously with 0.2 ml of a bacterial suspension in normal saline (about 2×10^5 viable bacteria cells). The animals were kept in an acclimatized room with water and food *ad libitum,* and observed daily until death. Spleens from six mice were triturated, suspended in saline and plated on BAB for *Y. pestis* recovery. Fragments from spleen, liver, lung and heart from one mouse were introduced into the shipment medium of Cary-Blair (Difco) and kept at room temperature for subsequent analysis.

For PCR analysis the samples, either fresh or Cary-Blair preserved, were triturated and suspended in saline, centrifuged, and the supernatants submitted to partial purification for neutralization of PCR inhibitors as previously described (Leal and Almeida 1999). Ten µl of each sample were used for PCR, NPCR and STNPCR. Simultaneously, an aliquot of the supernatants was plated on BAB for detection of *Y. pestis*.

32.2.7 Identification of *Y. pestis* in Samples from Plague Surveillance Areas

Rodent and flea samples collected during surveillance activities on Brazilian plague areas of the Sates of Pernambuco (PE), Ceará (CE) and Rio Grande do Norte (RN) were analyzed by culture and STNPCR. Sixty-two samples were from rodents: seven *Bolomys*, one *Calomys*, seven *Oryzomys*, two *Oligoryzomys*, three *Trichomys*, one

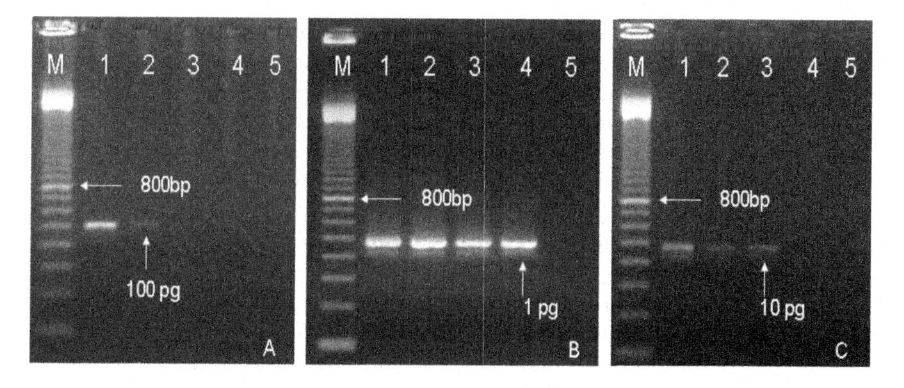

Fig. 1. Agarose gel electrophoresis showing the detection limit of PCR (A), NPCR (B) and STNPCR (C). The amounts of *Y. pestis* DNA are: lane 1, 1 ng; lane 2, 100 pg; lane 3, 10 pg; lane 4, 1 pg; lane 5, 100 fg. M, 100 bp DNA ladder (Invitrogen).

Galea, 35 *Rattus*, three *Mus*, and three from unidentified rodents from CE (35), PE (18) and RN (9), preserved on Cary-Blair medium. One hundred twenty-two samples were pools of fleas containing one to 40 specimens preserved in 2% saline: 58 *Xenopsylla*, 52 *Polygenis*, one *Ctenocephalides* and 11 without information from CE (52) and PE (70).

32.3 Results

32.3.1 Detection Limit of the PCR Approaches

When a dilution series of genomic DNA was used the detection limit obtained by conventional PCR, NPCR and STNPCR was 100 pg, one pg and 10 pg, respectively (Fig. 1). With regard to the dilution series of bacterial cells the detection limit for conventional PCR, STNPCR and NPCR was 2000, 20 and two CFU, respectively.

32.3.2 Evaluation of the PCR-based Assays in Experimentally Infected Mice

Spleens from six experimentally infected mice were plated immediately after collection. All samples were positive by culture. However, growth of other bacteria was found in five out of six cultures analyzed. NPCR and STNPCR were able to produce positive results in all samples, although simple PCR was positive in only three samples.

After storage in Cary-Blair medium for 11 months, samples from spleen, liver, lung and heart from one mouse were plated on BAB. Lung and heart cultures displayed pure *Y. pestis* growth. The spleen and liver produced multi-contaminated cultures: typical and specific phage susceptible *Y. pestis* colonies were also present on the spleen culture. On the liver culture no *Y. pestis* colonies were seen. NPCR and STNPCR yielded positive results (specific amplification) in all four samples tested, despite contamination with other organisms. In contrast simple PCR yielded positive results only in the two samples with pure *Y. pestis* cultures.

32.3.3 Detection of *Y. pestis* in Samples from Plague Area

Amplification products were observed in three samples from rodents (2 *Oryzomys* and 1 *Rattus*) from the State of Ceará. All the cultures were negative for *Y. pestis*; 49% presented growth of other organisms that surpassed the growth of any *Y. pestis* that was present.

32.3.4 Specificity of the Procedures

When DNA of other species of *Yersinia*, and spleen samples from normal non-infected mice were tested with the PCR based techniques no amplification was observed. Positive and negative controls were also used to make sure the PCR systems were working appropriately.

32.4 Discussion

In Brazil, most of the human plague cases and epizootics occur in remote areas and affect communities with precarious sanitary conditions, far from diagnostic facilities (Aragão et al. 2002; Vieira et al. 1993). The usual procedures for shipment of samples to the diagnostic reference laboratories may lead to microbiological contamination of the samples with other organisms and death of the plague bacilli, leading to false-negative results, or even hampering the test (Leal and Almeida 1999). By contrast, molecular techniques do not require culturing of the samples and are feasible even when the bacteria are not alive.

In the present study, we compared three sensitive and highly specific *Y. pestis* detection systems: conventional PCR, NPCR and STNPCR. In experiments using *Y. pestis* cells or genomic DNA we demonstrated that STNPCR is more sensitive than conventional PCR, but less sensitive than NPCR. This drawback is compensated by a lower risk of cross-contamination. The immobilization of the internal primers onto the inside of the microtube cap avoids opening the microtubes, reducing the possibility of false-positive results by contamination during the manipulation of the amplicons obtained in the first reaction. Conversely, an advantage of conventional NPCR over STNPCR is that potential DNA polymerase inhibitors are diluted during transfer of amplification products from the first step of amplification to new reactions in the second stage of amplification (Picken et al. 1996). Thus, in our assays, to achieve maximal efficiency in STNPCR the samples were submitted to an efficient purification procedure to remove DNA polymerase inhibitors. The STNPCR approach has been previously used successfully for the detection of *Schistosoma mansoni* and *Plasmodium* DNA (Abath et al. 2002; Montenegro et al. 2004; Melo et al. 2006).

Our experiments with infected animals showed that NPCR and STNPCR were able to produce positive results in all samples tested, despite contamination with other organisms. In contrast, simple PCR was less sensitive, yielding positive results in a smaller number of samples. Testing other Yersiniae species and spleens from non-infected mice did not result in PCR products indicating that the primers used do not anneal nonspecifically.

During routine plague survey, the time between the collection of rodent samples and the transfer to the reference diagnosic laboratory is usually long (at least three weeks). In the present work, samples preserved in Cary-Blair for 11 months were still appropriate for PCR analysis, although, NPCR and STNPCR had better performances than conventional PCR as more samples were amplifiable. Similarly, even when attempts to isolate *Y. pestis* by culture were unsuccessful due to contamination with other organisms or for other reasons, the PCR approaches described herein, particularly NPCR and STNPCR provided useful information.

In Brazil, since 1986, *Y. pestis* has not been isolated from rodents or fleas but a few laboratory-confirmed cases among humans were reported. Furthermore, serological surveys have detected the presence of antiplague antibodies among sentinel animals in some plague foci confirming the circulation of *Y. pestis* in these areas (Aragão et al. 2002). Positive results were obtained by STNPCR among rodents from the State of Ceará where the last human plague case was registered. Thus, detection of specific *Y. pestis* DNA by STNPCR in one of these areas (Ceará State) is an indication that this technique is potentially useful for monitoring plague transmission. In addition, provided a thermocycler is available, the method is suitable for field use, as the primer immobilization for STNPCR is simple and takes only 30 minutes. Finally, because the immobilized primers are stable for at least three months at –20°C (data not shown), the tubes can be prepared beforehand and stored until use.

32.5 Conclusion

In conclusion the PCR approaches evaluated, particularly NPCR and STNPCR, have potential to be used as alternative tools in epidemiological surveys of plague. Furthermore, as the results can be obtained quickly (less than 24 hours), these techniques could be useful in emergency situations in which the rapidity of diagnosis is essential for adoption of immediate measures of control.

32.6 Acknowledgments

Financial support for this work was provided by FIOCRUZ internal funds (PDTIS), SVS/MS (Project 433) and CNPq (Process 302584/2005-6). Gerlane Tavares was the recipient of a scholarship from CAPES and Alexandra Farias from CNPq. We thank Yara Nakazawa and Silvana Santos for technical assistance, and Roberto Werkauser and Fabio Melo for helpful advice. Animal experiments were carried out with the approval of the local ethics committee (CEUA/FIOCRUZ P0049-00).

32.7 References

Abath, F.G.C., Melo, F.L., Werkhauser, R.P., Montenegro, L., Montenegro, R. and Schindler, H.C. (2002) Single-tube nested PCR using immobilized internal primers. BioTechniques 33, 1210-1214.

Aragão, A.I., Seoane, A.C., Leal, T.C.A., Leal, N.C. and Almeida, A.M.P. (2002) Vigilância da peste no Estado do Ceará: 1990-1999. Rev. Soc. Bras. Med. Trop. 35, 143-148.

Campbell, J., Lowe, J., Walz, S. and Ezzell, J. (1993) Rapid and specific identification of *Yersinia pestis* by using a nested polymerase chain reaction procedure. J. Clin. Microbiol. 31, 758-759.

Chu, M.C. *Laboratory Manual of Plague Diagnosis Tests* (2000), World Health Organization, Geneva.

Engelthaler, D.M., Gage, K.L., Montenieri, J.A., Chu, M. and Carter, L.G. (1999) PCR detection of *Yersinia pestis* in fleas: comparison with mouse inoculation. J. Clin. Microbiol. 37, 1980-1984.

Galyov, E.E., Smirnov, O.Y., Karlishev, A.V., Volkovy, K.I., Denesyuk, A.I., Nazimov, I.V., Rubtsov, K.S., Abramov, V.M., Dalvadyanz, S.M. and Zav'yalov, V.P. (1990) Nucleotide sequence of the *Yersinia pestis* gene encoding F1 antigen and the primary structure of the protein. Putative T and B cell epitopes. FEBS Lett 277, 230-232.

Gookin, J.L., Birkenheuer, A.J., Breitschwerdt, E.B. and Levy, M.G. (2002) Single-tube nested-PCR for detection of *Thitrichomonas foetus* in feline faeces. J. Clin. Microbiol. 40, 4126-4130.

Herrmann, B., Nystron, T. and Wessel, H. (1996) Detection of *Neisseria gonorrhoeae* from air-dried genital samples by single-tube nested-PCR. J. Clin. Microbiol. 34, 2548-2551.

Hinnebush, J. and Schwan, T.G. (1993) New method for plague surveillance using polymerase chain reaction to detect *Yersinia pestis* in fleas. J. Clin. Microbiol. 31, 1511-1514.

Leal, N.C., Abath, F.G.C., Alves, L.C. and Almeida, A.M.P. (1996) A simple PCR-based procedure for plague diagnosis. Rev. Inst. Med. Trop. São Paulo 38, 371-375.

Leal, N.C. and Almeida, A.M.P. (1999) Diagnosis of plague and identification of virulence markers in *Yersinia pestis* by multiplex-PCR. Rev. Inst. Med. Trop. São Paulo, 41, 339-342.

Melo, F.L., Gomes, A.L.V., Barbosa, C.S., Werkhauser, R.P. and Abath, F.G.C. (2006) Development of molecular approaches for the identification of transmission sites of schistosomiasis. Trans. R. Soc. Trop. Med. Hyg. 100, 1049-55.

Montenegro, L.M.L., Montenegro, R.A., Lima, S.A., Carvalho, A.B., Schindler, H.C. and Abath, F.G.C. (2004) Development of a single tube hemi-nested PCR for genus-specific detection of *Plasmodium* in oligoparasitemic. Trans. R. Soc. Trop. Med. Hyg. 98, 619-625.

Nagl, J.B., Mühlbauer, G and Stekel, H. (2001) Single-tube two-round polymerase chain reaction using the LightCycler™ instrument. J. Clin. Virol. 20, 71-75.

Norkina, O.V., Kulichenko, A.N., Gintsburg, A.L., Tuchkov, I.V., Popov, YuA, Akenov, M.U. and Drosdov, I.G. (1994) Development of a diagnostic test for *Yersinia pestis* by the polymerase chain reaction. J. Appl. Bacteriol. 76, 240-245.

Picken, M.M., Picken, R.N., Han, D., Cheng, Y. and Strle, F. (1996) Single-tube nested polymerase chain reaction assay based on flagellin gene sequences for detection of *Borrelia burgdorferi* sensu lato. Eur. J. Clin. Microbiol. Infect. Dis. 15, 489-98.

Rahalison, L., Vololorinina, E., Ratsitorahina, M. and Chanteau, S. (2000) Diagnosis of bubonic plague by PCR in Madagascar under field conditions. J. Clin. Microbiol. 38, 260-263.

Tsukano, H., Itoh, K., Suzuki, S. and Watanabe, H. (1996) Detection and identification of *Yersinia pestis* by polymerase chain reaction (PCR) using multiplex primers. Microbiol. Immunol. 40, 773-775.

Vieira, J.B.F., Almeida, A.M.P. and Almeida, C.R. (1993) Epidemiologia e Controle da Peste no Brasil. Rev. Soc. Bras. M. Trop. Supl. III, 51-58.

Picture 29. People enjoy the music of Impact. Photograph by R. Perry.

Part VI – Vaccine and Antimicrobial Therapy Development

Picture 30. Stephen Smiley presents research on the role of cell-mediated defenses against plague. Photograph by R. Perry.

Picture 31. Discussions at a poster session. Photograph by R. Perry.

33
Therapeutic Potential of *Yersinia* Anti-Inflammatory Components

Benoit Foligné[1,2], Rodrigue Dessein[1], Michaël Marceau[1], Sabine Poiret[2], Joëlle Dewulf[2], Bruno Pot[2], Michel Simonet[1], and Catherine Daniel[2]

[1] Inserm U801, Faculté de Médecine-Institut Pasteur de Lille, Université de Lille, benoit.foligne@ibl.fr and michel.simonet@ibl.fr
[2] Laboratoire des Bactéries Lactiques et Immunité des Muqueuses, Institut Pasteur de Lille, catherine.daniel@ibl.fr

Abstract. Microbial pathogens have developed various stratagems for modulating and/or circumventing the host's innate and adaptive immunity. Hence, certain virulence factors can be viewed as potential therapeutic agents for human immunopathological diseases. This is the case for virulence plasmid-encoded proteins from pathogenic Yersiniae that inhibit the host's inflammatory response by interfering with various cellular signaling pathways.

33.1 Introduction

Crohn's disease (CD) and ulcerative colitis (UC) represent the two most prevalent forms of chronic inflammatory bowel disease (IBD) in Western countries, with about 2.5 million patients in Europe and North America. Both illnesses result from an inappropriate and ongoing inflammation of the mucosal immune system driven by the presence of normal luminal flora in genetically predisposed individuals. CD is a patchy, transmural inflammation that can potentially affect any part of the gastro-intestinal tract, although the disease generally concerns only the small intestine or colon (or both). In contrast, UC is a diffuse, mucosal inflammation limited to the colon. Mucosal inflammation is mediated by excessive production of IL-12, IFN-γ and TNF-α (a T_H1-type cell response) in CD and increased secretion of IL-4, IL-5 and/or IL-13 (T_H2-type cell response) in UC. However, this dichotomy is very schematic, and the immunological profile of UC patients is notably more complex (Bouma and Strober 2003; Podolsky 2002).

Until recently, there were few drugs available for IBD treatment; these essentially comprise salicylic acid derivatives, steroids, immunosuppressive agents and (since the late 1990s) monoclonal antibodies directed against TNF-α. However, given that all these compounds have various adverse effects and indeed fail to induce or maintain remission of IBD in about 30% of patients, there is a need for new anti-inflammatory drugs. The introduction of monoclonal anti-TNF-α antibodies marked the arrival of biological therapies for IBD. Many clinical trials based on antibodies that block a variety of key cytokines required for T cell activation are underway, and a number of novel approaches that act on other steps in the pathophysiological process are also being developed (Korzenik and Podolsky 2006). One of these involves the use of microbial virulence factors that modulate and/or circumvent the host's innate and adaptive immune systems.

33.2 *Yersinia pseudotuberculosis* Anti-inflammatory Components are Candidates for IBD Therapy

A number of murine models of experimental colitis have been developed in order to investigate the immunopathogenesis of IBD, and some can also be used to evaluate new anti-inflammatory strategies (Bouma and Strober 2003). The haptenating agent 2,4,6-trinitrobenzene sulfonic acid (TNBS) is frequently used to induce colitis in such investigations. When administered intrarectally in the mouse, 100 mg/kg of TNBS (dissolved in 50% ethanol, in order to disrupt the mucosal barrier) elicits loss of body weight (between 10 to 20%) within a few days, together with acute, transmural inflammation of the distal colon. The severity of inflammation can be estimated by the Wallace score, which rates the macroscopic lesions over a 4 cm-length of colon on a scale from 0 to 10, based on the following features: hyperemia, thickening of the bowel and the degree of ulceration (Wallace et al. 1989). Gene expression analysis has revealed that TNBS administration in the mouse increases colonic transcription of the proinflammatory genes *cox-2* and *tnf-α* and, above all, *il-12* and *ifn-γ* (Abad et al. 2005), corresponding to a T_H1-type cell response.

Pathogenic Yersiniae synthesize factors (encoded by the virulence plasmid pYV) that avoid or suppress host immune response (Brubaker 2003; Viboud and Bliska 2005), prompting us to wonder whether these factors might prevent experimentally-induced colitis. To address this question, we constructed a *phoP* mutant in wild-type *Y. pseudotuberculosis* (the PhoP deficiency has no impact on *yop* and *lcrV* gene expression) and challenged mice with this attenuated strain. On day 5 post-challenge, animals were given TNBS. We subsequently observed that colitis (as evaluated by the Wallace score and colonic transcription of the TNF-α gene) was reduced in these animals. Prevention of TNBS-induced colitis was *Yersinia*-specific, since it was not observed when mice were challenged with a nonpathogenic *Escherichia coli* (Marceau et al. 2004). These preliminary studies showed that anti-inflammatory components produced by *Y. pseudotuberculosis* were effective in reducing experimental colitis in mice and appear to be promising candidates for use in IBD therapy.

33.3 LcrV-Secreting *Lactococcus lactis* Reduces the Severity of TNBS-Induced Colitis in Mice

The hydrophilic LcrV protein is potentially one of the *Yersinia* components involved in the above-mentioned protective effect. Indeed, it has been previously reported that LcrV injection *in vivo* inhibits TNF-α and IFN-γ production following infection with LcrV-deficient *Y. pestis* (Nakajima et al. 1995). Additionally, LcrV upregulates IL-10 secretion in a TLR2- and CD14-dependent fashion, which in turn suppresses TNF-α secretion *in vitro* (Sing et al. 2002).

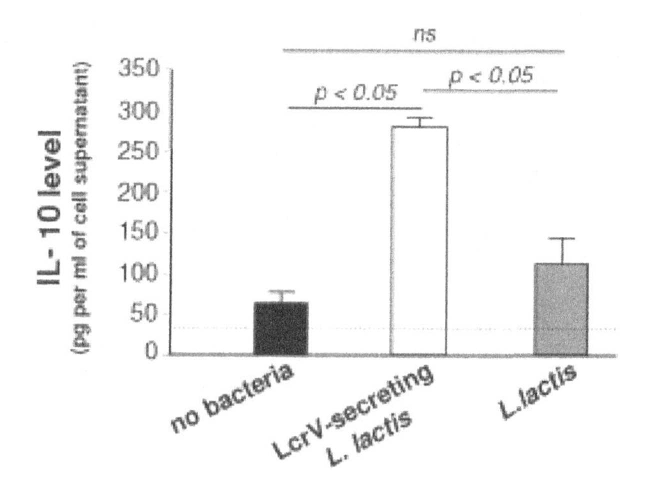

Fig. 1. The biological activity of recombinant LcrV. Peritoneal macrophages (2 x 10^6 cells/well) from BALB/c mice were incubated for 3 hours with PBS containing stationary-phase *L. lactis* (10^6 CFU/well) secreting LcrV (or not) or with carbonate buffer alone, prior to a 24-hour contact period with 2.5 μg/ml LPS from *E. coli*. Levels of the cytokine IL-10 in the cell culture supernatants were measured using an ELISA. Bars represent mean values ± SEM of three independent assays. The baseline level of IL-10 produced by untreated cells (i.e. in the absence of any stimulation) is indicated by the dashed line. As can be seen here, cellular production of the anti-inflammatory cytokine was significantly enhanced by the LcrV-secreting *L. lactis* strain (a 5-fold increase). ns: not significant.

To test the hypothesis that LcrV delivered to the intestinal mucosa can protect against colitis, we engineered a lactic acid bacterium to secrete the pYV-encoded protein. *Lactococcus lactis* (strain MG1363) - a non-pathogenic, non-invasive microorganism - was used because (i) it has no anti-inflammatory properties by itself (Foligné et al. 2006), (ii) it allows secretion of high levels of biologically active heterologous proteins (Le Loir et al. 2005) and (iii) its oral administration to humans is clinically safe (Braat et al. 2006). The resulting recombinant lactococcal strain released high amounts of LcrV, which was found to be functional as an IL-10 inducer (Fig. 1).

The recombinant *Lactococcus*' ability to prevent TNBS-induced colitis in mice was then investigated. For five consecutive days, three groups of ten BALB/c mice were intragastrically given 2 x 10^8 live *L. lactis* cells (expressing LcrV or not) or carbonate buffer, respectively. Acute colitis was triggered on day 5 by intra-rectal

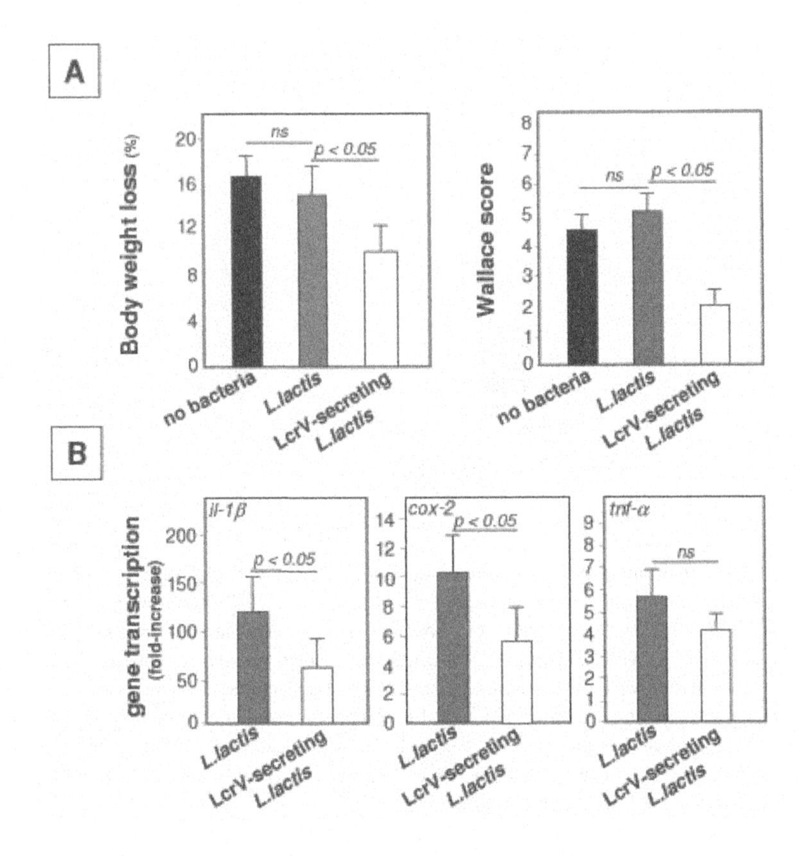

Fig. 2. The preventive effect of recombinant *L. lactis* strains on TNBS-induced acute colitis in BALB/c mice. Panel A. Loss of body weight and colonic inflammation scored according to the Wallace criteria on day 3 following intra-rectal TNBS administration. Mice were pre-treated (or not, as the case may be) with recombinant lactococcal strains. Bars represent mean values of ten animals ± SEM. ns: not significant. Panel B**.** The relative increase in levels of *il-1β*, *cox-2* and *tnf-α* transcripts (as quantified by real-time PCR) in colon samples collected at necropsy. Bars represent mean values of ten animals ± SEM. ns: not significant.

instillation of TNBS. Three days later, loss of body weight was assessed and, after sacrifice, macroscopic lesions of the colon were scored. As expected (because *L. lactis* alone does not have probiotic properties), intra-gastric gavage with $LcrV^-$-lactococci did not influence the severity of colitis. In contrast, a significant reduction in body weight loss and intestinal damage was observed in animals inoculated with $LcrV^+$-bacteria (Fig. 2A). Colonic expression of the inflammatory *il-1β*, *cox-2*, and *tnf-α* genes was quantified using real-time PCR in lactococci-treated mice. As shown in Fig. 2B, transcription of the former two genes was significantly reduced when animals received intra-gastric doses of LcrV-secreting *L. lactis*.

33.4 Conclusion

This demonstration of the protective effect of *in situ* lactococcal secretion of *Y. pseudotuberculosis* LcrV highlights the potential for using pathogen-derived immunomodulating molecules as therapeutics for human IBD. Other *Yersinia* proteins could also have anti-inflammatory potential. The MAPK and NF-κB signaling pathway inhibitor YopJ might be another good candidate for further exploration (although it would require intracellular delivery to be active) and might act synergistically with LcrV to prevent IBD.

33.5 References

Abad, C., Juarranz, Y., Martinez C., Arranz A., Rosignoli, F., Garcia-Gomez, M., Leceta, J. and Gomariz, R.P. (2005) cDNa array analysis of cytokines, chemokines, and receptors involved in the development of TNBS-induced colitis: homeostatic role of VIP. Inflamm. Bowel Dis. 11, 674-684.

Bouma, G. and Strober, W. (2003) The immunological and genetic basis of inflammatory bowel disease. Nat. Rev. Immunol. 3, 521-533.

Braat, H., Rottiers, P., Hommes, D.W., Huyghebaert, N., Remaut, E., Remon, J.P., van Deventer, S.J., Neirynck, S., Peppelenbosch, M.P. and Steidler, L. (2006) A phase I trial with transgenic bacteria expressing interleukin-10 in Crohn's disease. Clin. Gastroenterol. Hepatol. 4, 754-759.

Brubaker, R.R. (2003) Interleukin-10 and inhibition of innate immunity to Yersiniae: roles of Yops and Lcrv (V antigen). Infect. Immun. 71, 3673-3681.

Foligné, B., Nutten, S., Steidler, L., Dennin, V., Goudercourt, D., Mercenier, A. and Pot, B. (2006) Recommendations for improved use of the murine TNBS-induced colitis model in evaluating anti-inflammatory properties of lactic acid bacteria: technical and microbiological aspects. Dig. Dis. Sci. 51, 390-400.

Korzenik, J.R. and Podolsky, D.K. (2006) Evolving knowledge and therapy of inflammatory bowel disease. Nat. Rev. Drug Discov. 5, 197-209.

Le Loir, Y., Azevedo, V., Oliveira, S.C., Freitas, D.A., Miyoshi, A., Bermudez-Humaran, L.G., Nouaille, S., Ribeiro, L.A., Leclercq, S., Gabriel, J.E., Guimaraes, V.D., Oliveira M.N., Charlier, C., Gautier, M. and Langella P. (2005) Protein secretion in *Lactococcus lactis*: an efficient way to increase the overall heterologous protein production. Microb. Cell. Fact. 4, 2.

Marceau, M., Dubuquoy, L., Caucheteux-Rousseaux, C., Foligné, B., Desreumaux, P. and Simonet, M. (2004) *Yersinia pseudotuberculosis* anti-inflammatory components reduce trinitrobenzene sulfonic acid-induced colitis in the mouse. Infect. Immun. 72, 2438-2441.

Nakajima, R., Motin, V.L. and Brubaker, R.R. (1995). Suppression of cytokines in mice by protein A-V antigen fusion peptide and restoration of synthesis by active immunization. Infect. Immun. 63, 3021-3029.

Podolsky, D.K. (2002) Inflammatory bowel disease. N. Engl. J. Med. 347, 417-429.

Sing, A., Rost, D., Tvardovskaia, N., Roggenkamp, A., Wiedemann, A., Kirschning, C.J., Aepfelbacher, M. and Heesemann, J. (2002) *Yersinia* V-antigen exploits toll-like recep tor 2 and CD14 for interleukin 10-mediated immunosuppression. J. Exp. Med. 196, 1017-1024.

Viboud, G.I. and Bliska J. B. (2005) *Yersinia* outer proteins: role in modulation of host cell signaling responses and pathogenesis. Annu. Rev. Microbiol. 59, 69-89.

Wallace, J.L., MacNaughton, W.K., Morris, G.P. and Beck, P.L. (1989) Inhibition of leukotriene synthesis markedly accelerates healing in a rat model of inflammatory bowel disease. Gastroenterology 96, 29-36.

Picture 32. Michel Simonet presents research on the use of LcrV as an anti-inflammatory therapy. Photograph by R. Perry.

34
High Throughput Screening for Small-Molecule Inhibitors of Type III Secretion in *Yersinia pestis*

Ning Pan, Chrono Lee, and Jon Goguen

Department of Molecular Genetics and Microbiology, University of Massachusetts Medical School, ning.pan@umassmed.edu

Abstract. *Yersinia pestis*, *Yersinia pseudotuberculosis* and *Yersinia enterocolitica*, utilize a plasmid encoded type III secretion system (T3SS) to promote infection by delivering Yersinia outer proteins (Yops) into the cytosol of mammalian cells. This T3SS is absolutely required for *Yersinia* virulence, which makes T3SS an attractive target in the development of novel therapeutics for treatment of plague and other *Yersinia* infections. In this study, a new method for high throughput screening (HTS) of small molecules for the ability to inhibit type III secretion (T3S) in *Y. pestis* has been developed. In comparison with screening assays employed by others, this method is very simple and rapid, and thus well suited for examining very large compound sets. Using this method, we screened a diverse collection of libraries at the US National Screening Laboratory. The initial examination of 70,966 compounds and mixtures from 13 libraries resulted in 431 primary hits. Strong positive indications of inhibition were observed at a rate of 0.01%, while moderate and weak but potentially meaningful signals were observed at rates of 0.056% and 0.54% respectively. Further characterizations were conducted on selected primary hits in *Y. pestis*. Of the eight compounds examined in secondary assays, four show good promise as leads for structure activity relationship studies. They are a diverse group, each having chemical scaffolds not only distinct from one another, but also distinct from previously described candidate T3S inhibitors.

34.1 Introduction

34.1.1 Type III Secretion System (T3SS) in *Yersinia* spp. and the Low Calcium Response

Yersinia pestis, the causative agent of plague, along with the other yersinae pathogenic for humans (*Yersinia pseudotuberculosis* and *Yersinia enterocolitica*), utilizes type III secretion (T3S) to promote infection via transfer of effector proteins into the host cell cytoplasm. This type III secretion system (T3SS) is composed of a secretion machine termed the injectisome, composed primarily of proteins designated Ysc, which spans the inner and outer membranes and terminates in a needle-like structure. The system is capable of exporting the effector proteins into the surrounding medium, but requires the ancillary action of three additional proteins (LcrV, YopB and YopD) referred to as the translocon, to deliver the effectors into host cells. LcrV sits on the needle tip (Mueller et al. 2005), and YopB and YopD associate with the target cell membrane (Boland et al. 1996; Hakansson et al. 1996). Six effector proteins, Yops H, E, J, M, O, and P, some of which are paired with specific chaperones (the Syc proteins), are delivered by this system. The system also incorporates complex regulatory functions, which respond to temperature, cell contact, and calcium concentration, and altogether comprises about 50 genes, all of which reside on the 70 kb plasmid pCD1 in *Y. pestis*, and on highly homologous plasmids in the other two virulent yersiniae. The cocktail of Yops delivered by this system disarms the host cell antibacterial activities, and disrupt their signaling cascades, preventing production of proinflammatory cytokines and thus blocking the response of the host innate immune system to infection (Cornelis 2002).

T3S, which has since proven to be essential to virulence in at least 25 gram-negative species and is certainly among the most important discoveries in pathogenic microbiology, was first recognized by the remarkable insight of Hans Wolf-Watz and collaborators during study of *Y. pseudotuberculosis* in the early 1990s (Rosqvist et al. 1991; Rosqvist et al. 1994). However, the first *in vitro* phenotype directly related to T3S was observed more than 40 years earlier. Unfortunately, while rather useful as a tool, this phenotype was quite confusing and bore no obvious relationship to virulence. Virulent *Y. pestis* strains were observed to grow well in common laboratory media when incubated at temperatures below 30°C, but not when incubated at 37°C. Curiously, prolonged incubation at 37°C did lead to growth, but the bacteria recovered from such cultures were avirulent. Addition of skim milk to the medium allowed the growth of virulent strains at 37°C, and the key component of skim milk was ultimately found to be calcium (Higuchi et al. 1959). Later studies showed that this requirement for calcium, along with the production of certain antigens, was dependent on the presence of a 70 kb plasmid (Ben-Gurion and Shafferman 1981; Ferber and Brubaker 1981). Spontaneous loss of, or formation of deletions within, this plasmid during prolonged incubation both permitted growth at 37°C in the absence of calcium, but also lead to avirulence. We now know that this requirement for calcium at 37°C is directly related to the activation of T3S: in the absence of calcium at 37°C massive induction of T3S occurs in all three *Yersinia* species and growth is simultaneously arrested or greatly slowed. Growth after prolonged incubation without calcium results from loss or mutation of pCD1, resulting in dysfunction of the T3S system, and therefore loss of virulence.

Why the lack of calcium results in induction, and exactly why induction prevents growth, remain mysteries even today. Nonetheless, this phenomenon—called the low calcium response (LCR)—has been a useful tool in the study of T3S because it provides a handy *in vitro* marker for activation of secretion, allows selection schemes for a variety of mutants with altered T3S regulation or function, and permits induction of secretion on demand in the laboratory. Its significance *in vivo*, if any, remains obscure: cell contact is now thought to serve as the primary activator of *Yersinia* T3S *in vivo*. Here, we describe yet another use for the LCR phenomenon: a high throughput screening strategy for the discovery of small molecule T3S inhibitors.

34.1.2 Small-Molecule Inhibitors of T3S

Small molecule inhibitors of T3S are potentially useful both as therapeutics and as probes of T3S function. The former use is somewhat speculative given that no antibiotics designed to inhibit specific virulence-related functions, as opposed to essential functions such as protein and cell wall biosynthesis, are currently in use. Moreover, such inhibitors will be effective against a relatively narrow range of pathogens, requiring more certain diagnosis than is required with broad-spectrum antibiotics. On the other hand, use of broad-spectrum antibiotics has selected for resistance, which is now a serious problem for treatment of some pathogens, a problem that is likely to become more widespread. For certain organisms, like *Y. pestis*, which have marked potential for misuse in biological terrorism or warfare, and for which engineering of multiple antibiotic resistance is relatively simple, development of alternative therapeutics may be prudent. For these reasons, in addition to their utility as research tools, development and evaluation of specific inhibitors of key virulence functions like T3S on an exploratory basis seems justified. Several efforts to develop small molecule inhibitors and related technology specifically targeting T3SS have been reported. Abe et al. described a T3S inhibitor screening assay (Abe 2002). Caminoside A, from the extract of the marine sponge *Caminus sphaeroconia*, showed potent activity in another screen for bacterial T3S inhibitors. Bioassay guided fractionation of the extract led to the isolation of the novel antimicrobial glycolipid caminoside A (Linington et al. 2002) Kauppi et al. developed a reporter gene screening assay, and screened a 9,400 compound-library (Kauppi et al. 2003; Nordfelth et al. 2005). Screening for T3S inhibitors in pathogenic *E. coli* (Gauthier et al. 2005) has also been reported.

Inhibitors identified in previous screening projects with *Yersinia* have also been tested in other species. INP0400, a *Yersinia* T3S inhibitor had some efficacy in treatment of mice infected with the intracellular pathogen *Chlamydia trachomatis* (Muschiol et al. 2006). Wolf et al. found that another *Yersinia* T3SS inhibitor N'-(3,5-dibromo-2-hydroxybenzylidene)-4-nitrobenzo-hydrazide (Kauppi et al. 2003), disrupts progression of the chlamydial developmental cycle (Wolf et al. 2006). A limitation of these studies is that neither confirmed specific inhibition of T3S in *Chlamydia*.

34.1.3 A Novel High Throughput Screening for Inhibitors of T3S in *Y. pestis*

Here, we describe a very rapid, convenient, and inexpensive method for high-throughput screening for T3S inhibitors in *Y. pestis*. In comparison with previously reported techniques, this method is faster, simpler to execute, and requires no reagents other than bacteriological media and the compounds to be tested. This method is based on the expectation that at least some classes of T3S inhibitors will suppress the LCR, permitting growth at 37°C in the absence of calcium. A specially constructed avirulent *Y. pestis* strain that meets the requirements for exemption from Select Agent status—and thus can be used outside of BSL-3 laboratories and secure facilities—but retains the intact T3S system is used. Key properties of this strain with respect to the assay *per se* include strong luminescence, addition of a selectable marker to pCD1 to permit selection for retention of this plasmid, and stabilization of pCD1 against spontaneous deletion. In combination, these features permit detection of growth, and hence of potential inhibitors, as increased luminescence, while drastically reducing the background of false positives caused by spontaneous genetic lesions that result in loss of T3S function.

34.2 Materials and Methods

34.2.1 *Y. pestis* Strains, Growth Conditions, and Mammalian Cell Culture

A luminescent *Y. pestis* JG401 strain was constructed by introducing the *lux* operon from *Photorhabdus luminescence* contained in plasmid pML001 (Amp^R), and a pCD1 derivative, pCD1K22, marked with a Mariner transposon derivative conferring kanamycin resistance, but lacking the Mariner transposase function and thus incapable of transposition in the absence of transposase provided in trans. The location of this insertion was selected to reduce the rate of spontaneous internal deletions in pCD1. This was essential to the success of this method, because such deletions normally occur at a rate almost as high as spontaneous pCD1 segregation (10^{-4}-10^{-5} in overnight cultures). When grown as indicated, the rate of colony formation for JG401 on calcium-free medium at 37°C was less than $5x10^{-7}$. To ensure avirulence, JG401 also lacks the Pla-encoding plasmid pPCP1 and carries the 100 kb chromosomal Δ*pgm* deletion, which includes genes required for iron acquisition during infection. A T3S deficient luminescent *Y. pestis* strain JG406 (JG401 $pCD1^-$) was also constructed. *Y. pestis* strains were cultured in TB broth (1% bacto tryptose, 0.3% bacto beef extract, 0.5% NaCl) at 26°C unless otherwise indicated. HeLa cells were maintained in Dulbecco's modified Eagle's medium (DMEM) supplemented with 10% heat inactivated fetal bovine serum (FBS), 100 units/ml penicillin and 100 μg/ml streptomycin at 37°C with 5% CO_2.

34.2.2 The Primary High Throughput Screening Assay

The high-throughput screening (HTS) was conducted at the National Screening Laboratory (NSRB) located at Harvard Medical School, Boston, MA. For screening, a JG401 overnight culture was diluted into TB broth supplemented with 20 mM $MgCl_2$, 100 μg/ml ampicillin, and 50 μg/ml kanamycin. Thirty microliters of JG401 diluted with the same medium to a density of 10^4/ml was added into each well of 384-well cell culture plates (Nalge Nunc International, USA) to give 300 bacteria/well. A Bio-Tek μFill plate dispenser with Bio-Tek Bio-Stack (Bio-Tek Instruments) was used in this operation. The compounds in the screening libraries, at a concentration of 5 mg/ml (in DMSO), were arrayed in the 384 well plates. One hundred nanoliters of each compound to be screened was applied to each well of the assay plates using an Epson compound transfer robot with standard volume pin arrays. 2.5 mM $CaCl_2$ was added to the bacterial culture in two wells on each plate as the assay positive control and as an intra-plate control. The plates were incubated overnight at 37°C. The relative luminescence unit (RLU) was measured by using Perkin Elmer EnVision plate readers equipped with a 40-plate stack loader (Fig. 1). The screening was performed in duplicate.

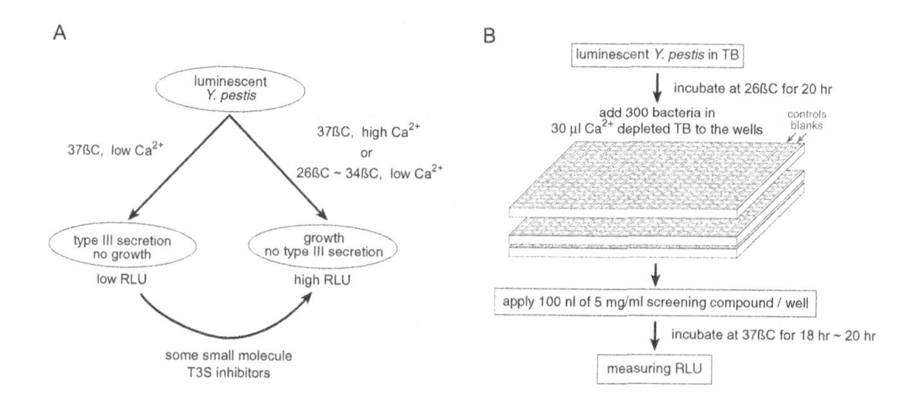

Fig. 1. The high throughput screening for inhibitors of T3S in *Yersinia*. (A) Illustration of the screening strategy. At 37°C with low or no Ca^{2+} in bacterial culture, *Y. pestis* will perform T3S with no growth. Presumably some classes of small molecules will convert the phenotype to growth with no or reduced T3S. (B) Flowchart of the HTS method to screen libraries of small molecules for their ability to promote growth of *Y. pestis* under the conditions, with which the T3S would be induced otherwise.

34.2.3 The Secondary Screening Assay

In secondary confirmation screening, different concentrations of selected primary hits in DMSO were added to the bacterial culture in the 384-well plates, with the same dilution factors as above. The plates were then incubated overnight at 37°C. The RLU was measured in an EnVision microplate reader. In addition to JG401, JG406 was also used in the secondary screening to ensure that the tested compounds did not act substantially as bacterial growth inhibitors. The experiments were performed in triplicate with both strains. JG406 was cultured in TB broth supplemented with 100 µg/ml ampicillin.

34.2.4 Statistical Analysis of Screening Results

The activity of small molecules and extracts in both sets of each screening run were calculated separately by using Eq. 1.

$$Inhibition\ (\%) = (RLU - RLU_{bkg}) \times 100 / (RLU_{CT} - RLU_{bkg}) \quad (1)$$

RLU: relative luminescence unit (luminescence unit/sec)
RLU_{bkg}: background RUL
RLU_{CT}: positive control RUL

The same analysis was applied to secondary screening data. The relative noise and dynamic range of the screening data were evaluated using the *Z'* statistic (Zhang et al. 1999).

34.2.5 Assay for Inhibition of T3S in *Y. pestis* Culture

A *Y. pestis* overnight culture in TB was subcultured into brain-heart infusion (BHI) broth, then incubated at 30°C with aeration until mid-log phase. Test compounds at the desired concentrations were then added to the bacterial cultures. The cultures were shifted to 37°C for two to four hours to induce T3S, centrifuged, and the resulting supernatant assayed for secreted Yops via antibody-based enzyme-linked immunosorbent assay (ELISA) and immunoblot. Specific anti-Yop antibodies were used in the assays.

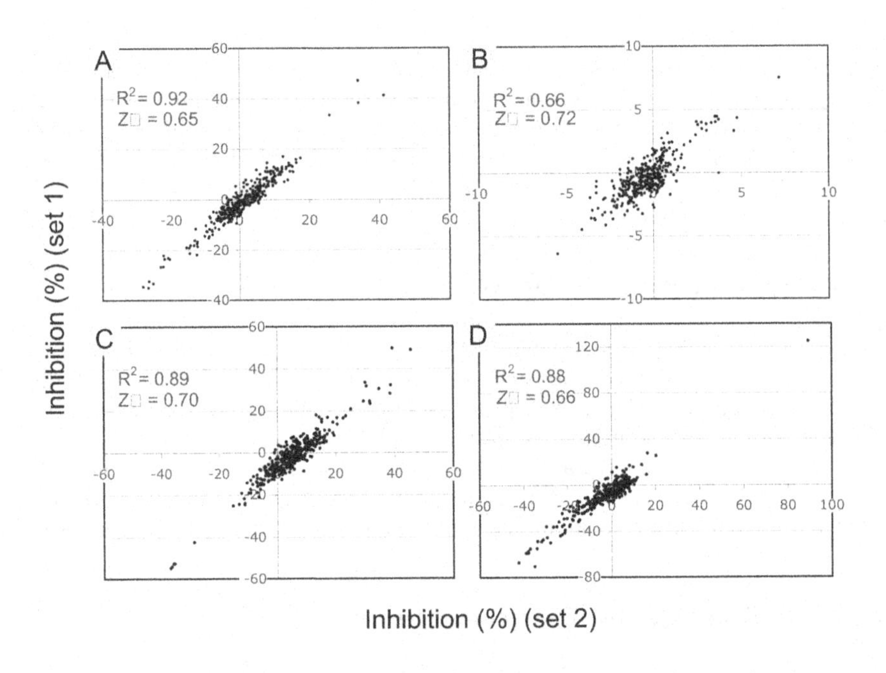

Fig. 2. Assay variability and *Z'* value. Data from four sets of 384-well duplicate plate pairs (A-D) were selected at random for analysis. The variability of the duplicate data, was evaluated by plotting the inhibition (%) values obtained in Plate 1 versus that of Plate 2 of each set. The correlations between the two plates for each of the four pairs were plotted. Correlation coefficients (R^2) were 0.92, 0.66, 0.89, and 0.88 for plate A, B, C, and D respectively. The *Z'* statistic, which evaluates signal-to-noise ratio and dynamic range, was also calculated for these four data sets, yielding values of 0.65, 0.72, 0.70, and 0.66 respectively. Z' values between 0.5 and 1 are generally regarded as excellent for HTS methods.

34.3 Results and Discussion

34.3.1 High Throughput Screening (HTS)

Using the assay described above, 70,966 compounds were screened in a period of two weeks with two people traveling daily between University of Massachusetts Medical School (UMMS) and NSRB. Use of the luminescent JG401 allowed very low-noise and sensitive detection of growth, at least 10-fold less noisy than possible by measurement of optical density in the 384 well format. The sensitivity was many orders of magnitude greater than OD measurements, allowing detection of even the low initial inoculums. In practice, a dwell time of 0.1 second/well was sufficient for luminescence readings, allowing rapid processing of plates after overnight incubation. In a two-day processing cycle, as many as 19,200 compounds were tested in duplicate. To evaluate the quality of the screening assay, four randomly selected duplicate plate pairs were analyzed (Fig. 2). Results of these analyses indicated good assay performance.

Based on this evaluation, we set the cutoff value for selecting positives at 10% of the luminescence produced by positive controls. Compounds whose tested activities were greater than this threshold (in duplicates) were considered as our primary HTS hits. Strong positive indications of inhibition were observed at a rate of 0.01%, while moderate and weak but potentially meaningful signals were observed at rates of 0.056% and 0.54% respectively. Based upon the intensity of the inhibition, chemical characteristics of the potential inhibitors, and correlation with results from other unrelated screens run by NRSB investigators (to exclude compounds positive in several unrelated assays), we selected 223 cherry-picks from 431 primary HTS hits for secondary confirmation (data not shown).

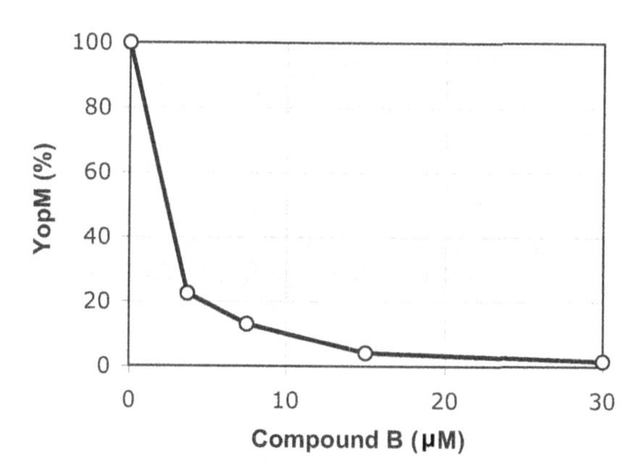

Fig. 3. Compound B inhibits YopM secretion. Compound B was added to *Y. pestis* cultures at a final concentration of 60 (not shown), 30, 15, 7.5, 3.75, and 0 μM. The samples were prepared as described in the Materials and Methods. The ELISA was performed in duplicate in a 96-well high-binding plate. Ca^{2+} was added as a control, which showed 0% YopM secretion.

34.3.2 Selected Compounds Inhibit Yop Secretion

We selected eight compounds (compound A, B, C, D, E, F, H, and I) from among the strong and moderate hits for further analysis. Four of the eight compounds (compound A, B, C, and H) effectively inhibited secretion of YopH at μM concentrations. Results for one of these compounds, compound B, is presented in Fig. 3 and shows inhibition of YopM secretion with an IC_{50} of 1 μM in the ELISA assay. With some of these compounds, differential inhibition among Yop species was observed. None of these compounds inhibited bacterial growth at the concentrations used in secretion inhibition assays (data not shown).

34.4 Conclusions

Although it is indirect (i.e. measuring luminescence as a surrogate for growth and using growth as a surrogate for inhibition of secretion) and depends upon a poorly understood phenomenon—the inhibition of growth *in vitro* when T3S is induced under low calcium conditions—the assay presented here is not only rapid, inexpensive, and convenient but also clearly yields bonifide T3S inhibitors. Because positive results depend on growth of the bacteria, this assay has the substantial advantage of efficiently excluding compounds that inhibit bacterial growth, and hence those that inhibit essential bacterial functions. This presumably excludes a wide range of compounds (e.g. protein or RNA synthesis inhibitors) that might yield false positives in more direct secretion assays. The inhibitors so far identified by this method are chemically diverse, each having chemical scaffold distinct from one another, and also distinct from previously described candidate type III inhibitors. Of the eight compounds examined in secondary assays, four show good promise as leads for structure-activity relationship studies.

34.5 Acknowledgments

We thank Karen Gingras at UMMS and Christina Anderson of HMS for expert technical assistance and Dr. Su Chiang at NRSB for facilitating our interaction with the screening facility. We thank Dr. Susan Straley, for the kind gift of anti-Yop antibodies.

34.6 References

Abe, A. (2002) Development of the screening system for the bacterial type III secretion apparatus inhibitor. Jpn. J. Antibiot. 55, 331-336.

Ben-Gurion, R. and Shafferman, A. (1981) Essential virulence determinants of different *Yersinia* species are carried on a common plasmid. Plasmid 5, 183-187.

Boland, A., Sory, M.P., Iriarte, M., Kerbourch, C., Wattiau, P. and Cornelis, G.R. (1996) Status of YopM and YopN in the *Yersinia* Yop virulon: YopM of *Y.enterocolitica* is internalized inside the cytosol of PU5-1.8 macrophages by the YopB, D, N delivery apparatus. EMBO J. 15, 5191-5201.

Cornelis, G.R. (2002) The *Yersinia* Ysc-Yop 'type III' weaponry. Nat. Rev. Mol. Cell Biol. 3, 742-752.

Ferber, D.M. and Brubaker, R.R. (1981) Plasmids in *Yersinia pestis*. Infect. Immun. 31, 839-841.

Gauthier, A., Robertson, M.L., Lowden, M., Ibarra, J.A., Puente, J.L. and Finlay, B.B. (2005) Transcriptional inhibitor of virulence factors in enteropathogenic *Escherichia coli*. Antimicrob. Agents Chemother. 49, 4101 -4109.

Hakansson, S., Schesser, K., Persson, C., Galyov, E.E., Rosqvist, R., Homble, F. and Wolf-Watz, H. (1996) The YopB protein of *Yersinia pseudotuberculosis* is essential for the translocation of Yop effector proteins across the target cell plasma membrane and displays a contact-dependent membrane disrupting activity. EMBO J. 15, 5812-5823.

Higuchi, K., Kupferberg, L.L. and Smith, J.L. (1959) Studies on the nutrition and physiology of *Pasteurella pestis*. III. Effects of calcium ions on the growth of virulent and avirulent strains of *Pasteurella pestis*. J. Bacteriol. 77, 317-321.

Kauppi, A.M., Nordfelth, R., Uvell, H., Wolf-Watz, H. and Elofsson, M. (2003) Targeting bacterial virulence: inhibitors of type III secretion in *Yersinia*. Chem. Biol. 10, 241-249.

Linington, R.G., Robertson, M., Gauthier, A., Finlay, B.B., van Soest, R. and Andersen, R.J. (2002) Caminoside A, an antimicrobial glycolipid isolated from the marine sponge *Caminus sphaeroconia*. Org. Lett. 4, 4089 -4092.

Mueller, C.A., Broz, P., Muller, S.A., Ringler, P., Erne-Brand, F., Sorg, I., Kuhn, M., Engel, A. and Cornelis, G.R. (2005) The V-antigen of *Yersinia* forms a distinct structure at the tip of injectisome needles. Science 310, 674-676.

Muschiol, S., Bailey, L., Gylfe, A., Sundin, C., Hultenby, K., Bergstrom, S., Elofsson, M., Wolf-Watz, H., Normark, S. and Henriques-Normark, B. (2006) A small-molecule inhibitor of type III secretion inhibits different stages of the infectious cycle of *Chlamydia trachomatis*. Proc. Natl. Acad.Sci. USA 103, 14566-14571.

Nordfelth, R., Kauppi, A.M., Norberg, H.A., Wolf-Watz, H. and Elofsson, M. (2005) Small-molecule inhibitors specifically targeting type III secretion. Infect. Immun. 73, 3104-3114.

Rosqvist, R., Forsberg, A. and Wolf-Watz, H. (1991) Intracellular targeting of the *Yersinia* YopE cytotoxin in mammalian cells induces actin microfilament disruption. Infect. Immun. 59, 4562-4569.

Rosqvist, R., Magnusson, K.E. and Wolf-Watz, H. (1994) Target cell contact triggers expression and polarized transfer of *Yersinia* YopE cytotoxin into mammalian cells. EMBO J. 13, 964-972.

Wolf, K., Betts, H.J., Chellas-Gery, B., Hower, S., Linton, C.N. and Fields, K.A. (2006) Treatment of *Chlamydia trachomatis* with a small molecule inhibitor of the *Yersinia* type III secretion system disrupts progression of the chlamydial developmental cycle. Mol. Microbiol. 61, 1543-1555.

Zhang, J.H., Chung, T.D. and Oldenburg, K.R. (1999) A simple statistical parameter for use in evaluation and validation of high throughput screening assays. J. Biomol. Screen 4, 67-73.

35 Cell-Mediated Defense Against *Yersinia pestis* Infection

Stephen T. Smiley

Trudeau Institute, ssmiley@trudeauinstitute.org

Abstract. *Yersinia pestis* (*Yp*) - one of the world's most deadly human pathogens - is the gram-negative bacterium that causes pneumonic plague. Virulent antibiotic-resistant *Yp* strains exist and Cold War scientists devised means to effectively aerosolize *Yp*. These facts raise grave concern that *Yp* will be exploited as a bioweapon. To counter that possibility, it is essential that we develop a safe and effective pneumonic plague vaccine. Recent studies suggest that the leading vaccine candidate, which primarily stimulates antibody-mediated humoral immunity, may not suffice. T cell-dependent cellular immunity comprises a second means by which vaccines prime long-lived protection against virulent bacterial pathogens. However, a plasmid carried by virulent *Yp* encodes factors that dampen inflammation and debilitate phagocytes, thereby compromising cellular defense mechanisms. As such, plague vaccine researchers have devoted relatively little attention to cellular immunity. Here we review our recent work demonstrating that the passive transfer of primed T cells can suffice to protect mice against lethal intranasal *Yp* infection, a model of pneumonic plague. We also demonstrate that key elements of cellular immunity even play critical roles during antibody-mediated defense against plague. We conclude that next-generation plague vaccines should strive to prime both cellular and humoral immunity.

35.1 Pneumonic Plague Vaccines

Yersinia pestis (*Yp*) is the gram-negative bacterium that causes bubonic plague, a zoonotic disease transmitted to humans by the bite of an infected flea (Perry and Fetherston 1997; Brubaker 2000). *Yp* infections can disseminate to pulmonary tissues, thereby causing pneumonic plague, a rapidly progressing and often fatal disease that can be transmitted between humans. Virulent antibiotic-resistant strains of *Yp* exist (Galimand et al. 1997) and Cold War scientists developed means to aerosolize large quantities of virulent *Yp* (Alibek 1999). Intentionally aerosolized, antibiotic-resistant *Yp* would constitute a formidable bioweapon.

Over 100 years of research have yet to generate a safe and effective pneumonic plague vaccine. Killed "whole cell" vaccines effectively protect against bubonic plague, but do not adequately protect against pneumonic plague (Meyer 1970; Titball and Williamson 2004). Live attenuated vaccines are not sufficiently safe for wide-scale use, although they reduce pneumonic plague mortality in both rodent and primate models (Girard 1963; Meyer 1970; Titball and Williamson 2004).

Current research efforts are focused on the development of subunit vaccines comprised of recombinant *Yp* proteins. The United Kingdom's defense department established that an alum formulation of the *Yp* F1 and V proteins protects mice against pulmonary *Yp* challenge (Williamson et al. 1997; Jones et al. 2000). In parallel studies, the United States Army (USAMRIID) demonstrated that an alum formulation of an engineered F1-V fusion protein also protects mice against pulmonary *Yp* challenge (Anderson et al. 1998; Heath et al. 1998). At a Plague Vaccine Workshop sponsored by the Federal Drug Administration's Center for Biologics Evaluation and Research, USAMRIID presented data from a series of primate vaccine trials (Pitt 2004). The overall conclusion was that F1/V-based vaccines provide cynomolgus macaques with significant protection against aerosolized *Yp* challenge, but fail to adequately protect African green monkeys. Presently, we lack a satisfactory explanation for the variable efficacy of F1/V-based vaccines in non-human primates. As such, there is substantial concern that F1/V vaccines may fail to protect humans against weaponized plague.

Virulent F1-negative *Yp* strains exist (Friedlander et al. 1995) and pathogenic *Yersinia* spp. express V protein variants, including some that do not confer cross-protective immunity (Roggenkamp et al. 1997). Thus, bioweapon engineers could likely circumvent vaccines based exclusively on F1 and V, even if the efficacy of F1/V-based vaccines can be improved. Clearly, additional research effort needs to be devoted to the development of safe and effective pneumonic plague vaccines. Since vaccination with live attenuated pigmentation (pgm)-negative *Yp* provides both rodents and primates with effective protection (Girard 1963; Meyer 1970; Titball and Williamson 2004), we are studying how these live vaccines protect against pneumonic plague.

35.2 How Do Vaccines Defend Against Pneumonic Plague?

Serum from animals vaccinated with live pgm-negative *Yp* passively transfers protection to naïve rodents (Girard 1963). The passive transfer of F1- or V-specific antibodies likewise protects mice against aerosolized *Yp* (Green et al. 1999). Given this documented efficacy of humoral immunity, subunit plague vaccine efforts have aimed to prime high-titer antibody responses.

In the primate vaccine studies described above, F1 antibody titers correlated with protective efficacy in cynomolgus macaques, but not in African green monkeys (Pitt 2004). V antibody titers were highly variable and did not correlate with protective efficacy in either species (Pitt 2004). These findings indicate that F1/V antibody titers are not reliable correlates of vaccine efficacy in primates. They also suggest that humoral immunity alone may not suffice to protect humans against pneumonic plague.

Vaccinating guinea pigs with live pgm-negative *Yp* solidly protects against plague, without eliciting significant protective antibody titers (Girard 1963). T cell-dependent cellular immunity comprises another means by which live vaccines can protect against bacterial infection. Cytokine products of T cells, most notably interferon-gamma (IFNγ) and tumor necrosis factor-alpha (TNFα), are pleiotropic mediators of cellular immunity. These cytokines promote the formation of granulomas - organized clusters of immune cells that function to restrain bacterial growth and dissemination. They also up-regulate expression of phagocyte antimicrobial activities, including production of reactive oxygen and reactive nitrogen.

Pre-injecting mice with IFNγ and TNFα protects against plague (Nakajima and Brubaker 1993) and pre-treating phagocytes with IFNγ and TNFα prevents the intracellular replication and survival of *Yp* in vitro (Lukaszewski et al. 2005). Thus, key products of acquired cellular immunity have the capacity to combat plague.

Yp actively counters cellular immunity by adaptively up-regulating expression of factors that reduce the antimicrobial impacts of reactive nitrogen (Sebbane et al. 2006). In addition, the pCD plasmid carried by virulent *Yp* encodes numerous factors that dampen inflammation and debilitate phagocytes, thereby compromising cell-mediated defense. Specifically, the V protein stimulates production of immunosuppressive interleukin (IL)-10 and also facilitates the translocation of pCD-encoded *Yersinia* outer proteins (Yops) into host cells, where they disrupt intracellular signaling pathways (Brubaker 2003; Viboud and Bliska 2005; Heesemann et al. 2006). Intriguingly, mutant *Yp* strains lacking expression of V protein and/or certain Yops seem to be cleared by classical cell-mediated defense mechanisms, including the formation of granulomas (Brubaker 2003; Straley and Cibull 1989; Kerschen et al. 2004). Vaccination with V protein likewise appears to defend against plague by enabling the formation of granulomas (Brubaker 2003), while also promoting phagocytosis of extracellular *Yp* organisms (Weeks et al. 2002; Cowan et al. 2005). These findings suggest that classical cellular defense mechanisms effectively combat *Yp* infection when pCD-encoded virulence factors are neutralized.

The dogma in the *Yersinia* field is that the pCD-encoded virulence factors also suppress T cell-mediated acquired immune responses, both by directly targeting T cells and by targeting dendritic cells (DC) - specialized antigen presenting cells (APC) that are particularly important for the activation of naive T cells (Brubaker 2003; Viboud and Bliska 2005; Heesemann et al. 2006). Nevertheless, here we will demonstrate that suppression of T cell responses during pulmonary *Yp* infection is, at most, incomplete.

35.3 Cellular Immunity Contributes to Antibody-Mediated Defense Against Pulmonary *Yp* Infection

We began our studies of the mechanisms underlying protection conferred by vaccination with live pgm-negative *Yp* by infecting wild type mice with a sub-lethal dose of pgm-negative *Yp* strain KIM D27 ($3x10^2$ CFU, intraperitoneally). Thirty days later, we collected serum from convalescent animals. Passive transfer of this serum protected naïve mice against lethal intranasal challenge with 10 LD_{50} *Yp* KIM D27 (Parent et al. 2005). Consistent with antibody-mediated protection, exposing the serum to Protein G-coupled sepharose removed its protective capacity.

The serum from mice infected with live pgm-negative *Yp* even provided solid protection when passively transferred 18 hours after *Yp* challenge (Parent et al. 2005; Parent et al. 2006). As such, we used this post-exposure serotherapy protocol to investigate whether cellular immunity contributes to antibody-mediated clearance of

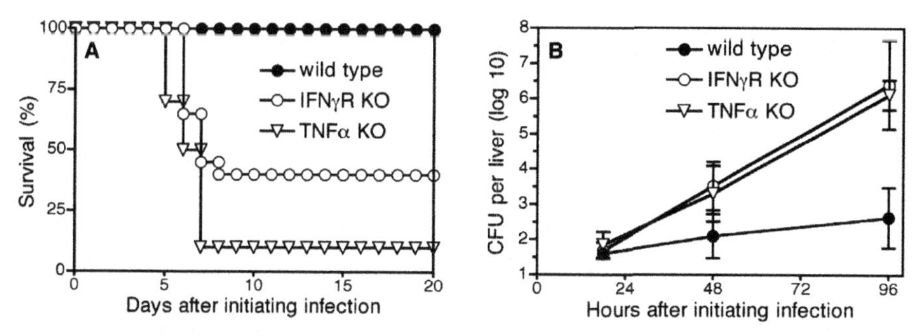

Fig. 1. IFNγ and TNFα contribute to humoral protection against pulmonary *Yp* infection. (A) Wild type, IFNγR-deficient (KO), or TNFα-deficient mice were infected intranasally with 10 LD_{50} ($2x10^5$ CFU) *Yp* KIM D27. Eighteen hours after initiating infection, the mice received 10 μl convalescent serum intraperitoneally. IFNγR- or TNFα-deficiency significantly reduced survival (both $p<0.001$). (B) At the indicated times, bacterial burden was measured in mice that were infected and treated with serum as described in (A). IFNγR- or TNFα-deficiency dramatically increased bacterial burden in hepatic tissue by 96 hr post-infection (both $p<0.002$). Adapted from Parent et al. 2006.

established *Yp* infection. Specifically, we evaluated the capacity of post-exposure serotherapy to protect mice that lacked key elements of cellular immunity. We found that genetic deficiency in either the IFNγ receptor (IFNγR) or TNFα significantly impaired the capacity of post-exposure serotherapy to reduce mortality (Fig. 1A) and suppress bacterial growth in the lung and liver (Fig. 1B and Parent et al. 2006). NOS2-deficiency, which impairs production of reactive nitrogen, also significantly reduced the efficacy of post-exposure serotherapy (Parent et al. 2006). These findings revealed a previously unrecognized role for key elements of cellular immunity during antibody-mediated defense against pulmonary *Yp* infection. Most importantly, they suggest that pneumonic plague vaccines should strive to prime both humoral and cellular immunity.

35.4 T Cell-Mediated Defense Against Pulmonary *Yp* Infection

Having found that vaccination with live pgm-negative *Yp* elicits antibodies that defend against plague in collaboration with cellular immunity, we next investigated whether vaccination with live pgm-negative *Yp* primes protective T cells. To ensure that T cells were responsible for any observed protection, we vaccinated B cell-deficient μMT mice, which cannot produce antibodies. To safely vaccinate the immunodeficient μMT mice, we took advantage of the above-described post-exposure serotherapy protocol: first we intranasally vaccinated μMT mice with 10 LD_{50} live *Yp* KIM D27 and then we prevented vaccine-induced mortality by administering a minimally protective dose of serotherapy 18 hours after initiating infection. We then waited 60 days before evaluating the capacity of *Yp*-vaccinated μMT mice to resist a lethal intranasal *Yp* challenge. As shown in Fig. 2A, vaccination with live *Yp* conferred

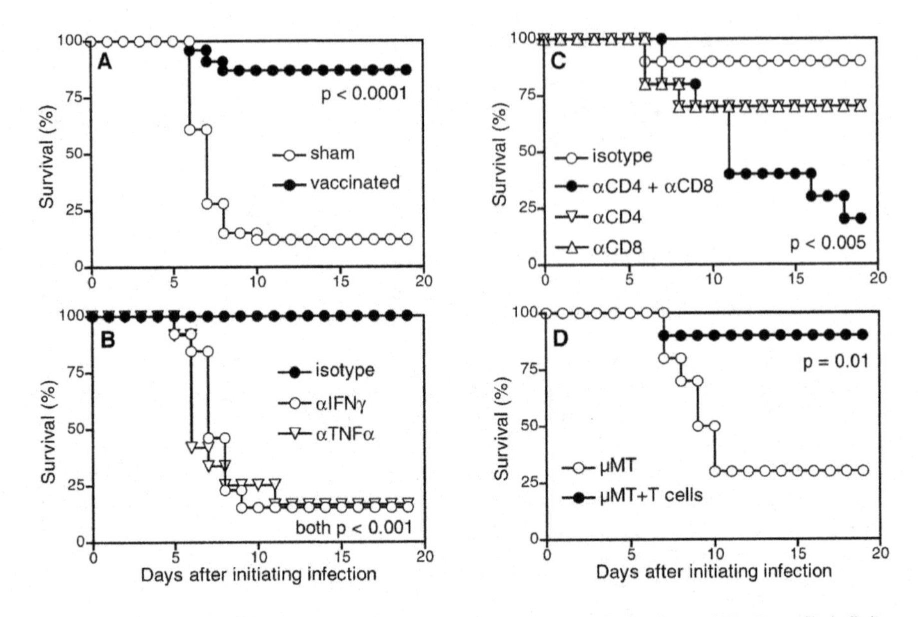

Fig. 2. Cell-mediated protection against pulmonary *Yp* infection. (A) B cell-deficient C57BL/6-backcrossed μMT mice were vaccinated by intranasal infection with 10 LD_{50} CFU *Yp* KIM D27 ($2x10^5$ CFU) followed 18 hours later by 10 μl of convalescent serum. Control mice (sham) received only convalescent serum (i.e. no *Yp* infection). Sixty days later, both groups of mice were challenged intranasally with 10 LD_{50} *Yp* KIM D27 ($2x10^5$ CFU). Day 0 refers to the day of challenge infection. Vaccination with live *Yp* significantly protected μMT mice against secondary *Yp* challenge (n = 23-29 mice/group; $p < 0.0001$). Data are pooled from three independent experiments. (B, C) μMT mice were vaccinated and challenged as described in (A). At the time of challenge, TNFα-specific, IFNγ-specific, CD4-specific, CD8-specific, or isotype-matched control mAb were administered. Treatment with TNFα- or IFNγ-specific mAb significantly reduced survival as compared with control mAb (n = 12-13 mice/group; $p < 0.001$). Treatment with both CD4- and CD8-specific mAb also significantly reduced survival (n = 10 mice/group; $p < 0.005$). Data for (B) and (C) are each pooled from two independent experiments. (D) Cells isolated from μMT mice that had been vaccinated and challenged as in (A) were stimulated in vitro with heat-killed *Yp* and splenic APC and expanded for 10 days. Then, the cells were transferred ($5x10^6$ per mouse) to naïve μMT mice, which were challenged intranasally with 10 LD_{50} *Yp* KIM D27 ($2x10^5$ CFU) one day later. The passively transferred T cells conferred significant protection (n = 10 mice/group; $p = 0.01$). Data are pooled from two independent experiments. For details, see Parent et al. 2005.

μMT mice with significant protection. By day 3 post-challenge, vaccination reduced bacterial burden in the lung, spleen, and liver and also increased the number of activated pulmonary T cells 5 to 10-fold (Parent et al. 2005). Consistent with T cell-mediated protection, depletion of IFNγ, TNFα, or both CD4 and CD8 cells abrogated vaccine efficacy (Fig. 2B and 2C). Moreover, T cells isolated from the *Yp*-vaccinated mice conferred significant protection when they were expanded in vitro and then passively transferred to naïve μMT mice (Fig. 2D). We conclude that vaccination with live pgm-negative *Yp* primes T cells that defend against lethal pulmonary *Yp* challenge (Parent et al. 2005).

Recently, we performed analogous studies using wild type mice, rather than μMT mice, as vaccine recipients (Philipovskiy and Smiley 2007). In this model, vaccination did not require serotherapy. Rather, we intranasally vaccinated wild type mice with a near-lethal dose of pgm-negative *Yp* KIM D27 and then "boosted" immunity by re-vaccinating with *Yp* KIM D27 14 days later. On day 28 after the initial vaccination, we purified splenic CD4 and CD8 T cells. Upon immediate transfer to naïve wild type mice, without any in vitro expansion, these cells protected against lethal intranasal *Yp* challenge. As shown in Fig. 3A, CD8 cells alone provided significant protection ($p<0.002$), which was enhanced by the presence of CD4 cells ($p<0.0001$), although the latter failed to confer any measurable protection on their own. We conclude that CD4 and CD8 T cells synergistically protect against lethal pulmonary *Yp* infection.

Using the live pgm-negative *Yp* prime/boost vaccination model described above, we have also begun to define the antigens recognized by protective T cells (Philipovskiy and Smiley 2007). When examined in vitro, naïve T cells produce low and similar levels of IFNγ in response to splenic APC exposed to either live *Yp* or live *Escherichia. coli* (Fig. 3B). T cells isolated from mice vaccinated with live *Yp* also produce low levels of IFNγ in response to *E. coli*. By contrast, these *Yp*-primed T cells produce large amounts of IFNγ in response to *Yp* (Fig. 3B). Both *Yp*-primed CD4 T cells and CD8 T cells respond strongly to *Yp* KIM D27. Moreover, these cells also respond strongly to bacteria lacking the capacity to express F1 protein (KIM D27/caf1-), V protein (KIM D28), or both F1 and V (KIM10+/caf1-) (Fig. 3B). These observations suggest that neither F1 nor V are dominant antigens recognized by T cells primed by vaccination with live pgm-negative *Yp*.

35.5 Discussion

The pCD plasmid-encoded V protein and Yops suppress innate cellular immune responses, and the prevailing dogma contends that these factors also suppress acquired T cell responses (Viboud and Bliska 2005; Heesemann et al. 2006). Relevant in vivo studies include demonstrations (i) that DC are targeted by *Yp* Yops (Marketon et al. 2005), (ii) that *Yp* YopJ suppresses DC migration (Velan et al. 2006), and (iii) that the *Yersinia enterocolitica* YopJ homolog suppresses the activation and expansion of naive CD8 T cells (Trulzsch et al. 2005). While certainly suggestive, these prior studies did not explicitly demonstrate suppression of T cells by *Yp* in vivo.

We isolated protective T cells from the spleens of vaccinated mice. Thus, they may have been activated at extra-pulmonary sites after bacteria disseminated from the lung. They may even have been activated after antibodies and/or neutrophils had cleared all viable bacteria. As such, future studies will need to directly investigate the impact of pulmonary *Yp* infection on the activation, expansion, and differentiation of naïve T cells. Nevertheless, our findings indicate that pCD-mediated suppression of naïve T cell responses is, at most, incomplete.

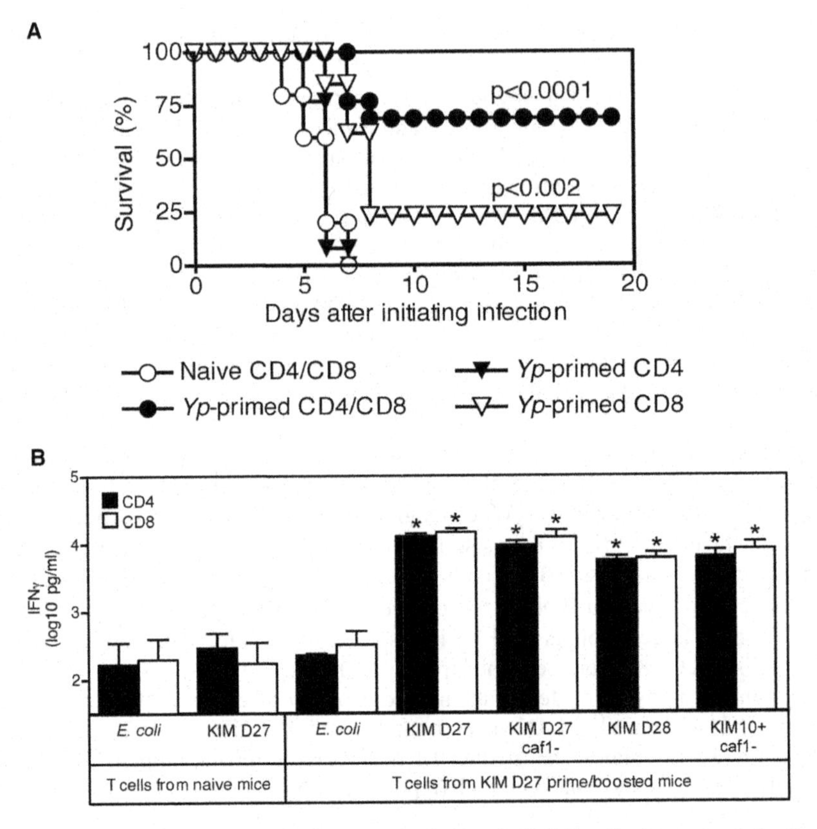

Fig. 3. (A) Vaccination with live *Yp* generates CD4 and CD8 T cells that synergistically protect against lethal pulmonary *Yp* challenge. CD4 and CD8 T cells were individually purified from wild type C57BL/6 mice that were prime-boost vaccinated with live *Yp* KIM D27 ($5x10^4$ CFU followed by $1x10^5$ CFU 14 days later). Control T cells were purified from naïve mice. The CD4 and CD8 cells were transferred intravenously ($5x10^6$ per mouse; either individually or after pooling at a 1:1 ratio). The next day, all recipient animals were challenged intranasally with 10 LD_{50} KIM D27 ($2x10^5$ CFU). Pooled *Yp*-primed CD4 and CD8 T cells provided significant protection ($p<0.0001$; n = 13 mice) as compared with pooled naïve T cells (n = 5 mice). *Yp*-primed CD8 T cells also provided modest but significant protection ($p<0.002$; n = 13 mice). *Yp*-primed CD4 T cells failed to provide measurable protection (n = 13 mice). Data is combined from three independent experiments. (B) Vaccination with live *Yp* KIM D27 generates T cells that respond strongly to *Yp* strains lacking the capacity to express the F1 and V proteins. CD4 and CD8 T cells were purified from naïve mice and mice that were prime-boost vaccinated with KIM D27 as in (A). Splenocyte APC from naïve mice were exposed to antibiotic-free media containing the indicated live bacteria for 2 hr. The purified T cells and antigen-exposed APC were then combined in the presence of antibiotics. Supernatants were assayed for levels of IFNγ protein after 48 hr of culture. Bacterial strains included KIM D27, KIM D27/caf1- (F1-negative), KIM D28 (pCD-negative), KIM10+/caf1- (F1-negative, pCD-negative, pPCP-negative) and *E. coli*. The figure depicts levels of IFNγ protein in culture supernatants and is presented as the average and standard deviation of triplicate cultures (* $p<0.001$ as compared with *E. coli*). For details, see Philipovskiy and Smiley 2007.

One possible explanation for the potent suppression of naïve T cells by *Y. enterocolitica*, but not *Yp*, is that primary infection by *Yp* leads to overwhelming bacteremia within just a few days time. Given that naïve T cell responses typically require several days to develop, the evolution of *Yp* into an extremely effective suppressor of innate immunity may have relaxed selective pressure on virulence mechanisms that diminish T cell responses. Consistent with this model, the *Y. enterocolitica* YopJ homolog was recently shown to suppress DC responses more effectively than *Yp* YopJ (Velan et al. 2006).

Upon activation, naïve T cells proliferate and differentiate into effector and memory T cells. A primary goal of T cell-based vaccines is to establish pools of these effector/memory T cells. Upon subsequent infection with the relevant pathogen, these cells rapidly exert their effector functions (i.e. cytolysis, cytokine production) and also proliferate, thereby generating new pools of effector/memory T cells. Importantly, the activation criteria for effector/memory T cells are less stringent than that for naïve T cells; whereas naïve T cells are primarily activated by DC, many cell types can act as APC for effector/memory T cells (Croft 1994). Thus, even if *Yp* possesses mechanisms that delay the activation of naïve T cells, *Yp* may nevertheless remain susceptible to vaccine-primed effector/memory T cells.

Our findings indicate that effector/memory T cells can effectively combat pulmonary infection by pCD-positive *Yp* (Fig. 2 and 3). At present, we do not know whether *Yp* virulence mechanisms partially suppress the activities of these T cells. We also do not know whether these T cells are activated in the lung or only after bacteria (or bacterial antigens) disseminate to extra-pulmonary tissues. Notably, even extra-pulmonary T cell activation could limit bacterial growth in pulmonary tissues. For example, extra-pulmonary T cell activation may generate sufficiently high levels of cytokines to systemically activate phagocyte defense mechanisms (ex. NOS2 up-regulation). Systemic increases in cytokine levels should also deplete niches for intracellular *Yp* survival, since macrophages exposed to TNFα and IFNγ no longer support *Yp* replication (Lukaszewski et al. 2005). A full understanding of the extent to which *Yp* virulence mechanisms impact effector/memory T cells will be critical for the development of pneumonic plague vaccines that optimally harness the protective capacities of cellular immunity.

The studies described herein employed *Yp* KIM D27 as both vaccine and challenge. This pgm-negative strain is only modestly attenuated by the intranasal route: the LD_{50} in mice is approximately 10^4 CFU (Parent et al. 2005). Nevertheless, we recognize the importance of extending our challenge studies to pgm-positive *Yp* strains. If we observe differences between pgm-negative and positive strains, it will suggest that the pgm locus actively combats cellular immunity. Interestingly, a recent study found that the pgm-encoded ripA gene suppresses production of reactive nitrogen, thereby allowing pgm-positive *Yp* to replicate within macrophages exposed to IFNγ after their infection (Pujol et al. 2005). Optimal cell-mediated defense against plague may thus require both vaccine-mediated priming of protective T cells and therapeutic targeting of *Yp* virulence mechanisms that actively suppress cellular defense pathways.

Perhaps most importantly, our studies demonstrate that key elements of cellular immunity (IFNγ, TNFα, and NOS2) also contribute to antibody-mediated defense against pulmonary *Yp* infection (Fig. 1). These findings add to a growing literature on interdependencies between humoral and cellular immunity (Igietseme et al. 2004). Since V-specific antibodies facilitate the phagocytosis of *Yp*, prevent Yop translocation, and suppress production of immunosuppressive IL-10 (Weeks et al. 2002; Brubaker 2003; Cowan et al. 2005), we anticipate that V-specific antibodies will also enhance antigen presentation to T cells, promote the development of anti-bacterial type 1 T cell responses, and generally buy time for the activation, expansion and recruitment of protective T cells. In turn, these T cells will activate cellular defense mechanisms, thereby facilitating the formation of granulomas and depleting intracellular niches for *Yp* survival. Given such potentially synergistic interactions, and given that humoral and cellular immunity often deploy complementary defense mechanisms, our findings strongly suggest that next-generation F1/V-based subunit vaccines should include protective T cell antigens. To achieve that goal, we must first identify specific *Yp* antigens that prime robust, long-lived, protective T cell responses. Since fully virulent F1-deficient strains exist, we are particularly interested in T cell antigens that prime cellular immunity that complements the humoral immunity conferred by V protein vaccination. Our laboratory is presently working to identify such antigens, thereby enabling next-generation pneumonic plague vaccines to harness the protective capacities of both humoral and cellular immunity.

35.6 Conclusion

There is grave concern that *Yp* will be exploited as a bioweapon because virulent antibiotic-resistant *Yp* strains exist, Cold War scientists devised means to intentionally aerosolize *Yp*, and we currently lack an effective pneumonic plague vaccine. Recombinant subunit vaccines containing the *Yp* F1 and V proteins elicit humoral immunity but may not adequately protect humans against aerosolized *Yp*. The findings presented here indicate that cellular defense mechanisms also combat pulmonary *Yp* infection and that key elements of cellular immunity even contribute to antibody-mediated defense. Thus, incorporating T cell antigens into F1/V vaccines should improve efficacy by harnessing the protective capacities of both humoral and cellular immunity.

35.7 Acknowledgements

We are indebted to Drs. Michelle A. Parent and Alexander V. Philipovskiy for performing the experiments described herein, along with excellent technical assistance from Lindsey B. Wilhelm, Debbie Duso and members of the Trudeau Institute Animal Facilities. We also thank Drs. Robert R. Brubaker, Susan C. Straley, and James B. Bliska for generously providing access to bacterial strains from their laboratories. Finally, we thank Trudeau Institute, the National Institutes of Health, and the Northeast Biodefense Center for funding these studies.

35.8 References

Alibek, K. (1999) *Biohazard.* Random House, Inc., New York.

Anderson, G.W., Jr., Heath, D.G., Bolt, C.R., Welkos, S.L. and Friedlander, A.M. (1998) Short- and long-term efficacy of single-dose subunit vaccines against *Yersinia pestis* in mice. Am. J. Trop. Med. Hyg. 58, 793-799.

Brubaker, R.R. (2000) *Yersinia pestis* and bubonic plague. In *The prokaryotes, an evolving electronic resource for the microbiological community.* Dworkin, M., Falkow, S., Rosenberg, E., Schleifer, K.-H., and Stackebrandte, E. (eds.) Springer-Verlag, New York, http://141.150.157.117:8080/prokPUB/chaprender/jsp/showchap.jsp?chapnum=265.

Brubaker, R.R. (2003) Interleukin-10 and inhibition of innate immunity to Yersiniae: roles of Yops and LcrV (V antigen). Infect. Immun. 71, 3673-3681.

Cowan, C., Philipovskiy, A.V., Wulff-Strobel, C.R., Ye, Z. and Straley, S.C. (2005) Anti-LcrV antibody inhibits delivery of Yops by *Yersinia pestis* KIM5 by directly promoting phagocytosis. Infect. Immun. 73, 6127-6137.

Croft, M. (1994) Activation of naive, memory and effector T cells. Curr. Opin. Immunol. 6, 431-437.

Friedlander, A.M., Welkos, S.L., Worsham, P.L., Andrews, G.P., Heath, D.G., Anderson, G.W., Jr., Pitt, M.L., Estep, J. and Davis, K. (1995) Relationship between virulence and immunity as revealed in recent studies of the F1 capsule of *Yersinia pestis.* Clin. Infect. Dis. 21 Suppl 2, S178-181.

Galimand, M., Guiyoule, A., Gerbaud, G., Rasoamanana, B., Chanteau, S., Carniel, E. and Courvalin, P. (1997) Multidrug resistance in *Yersinia pestis* mediated by a transferable plasmid. N. Engl. J. Med. 337, 677-680.

Girard, G. (1963) Immunity in plague. Results of 30 years of work on the "*Pasteurella pestis* Ev" (Girard and Robic) strain. Biol. Med. (Paris). 52, 631-731.

Green, M., Rogers, D., Russell, P., Stagg, A.J., Bell, D.L., Eley, S.M., Titball, R.W. and Williamson, E.D. (1999) The SCID/Beige mouse as a model to investigate protection against *Yersinia pestis.* FEMS Immunol. Med. Microbiol. 23, 107-113.

Heath, D.G., Anderson, G.W., Jr., Mauro, J.M., Welkos, S.L., Andrews, G.P., Adamovicz, J. and Friedlander, A.M. (1998) Protection against experimental bubonic and pneumonic plague by a recombinant capsular F1-V antigen fusion protein vaccine. Vaccine. 16, 1131-1137.

Heesemann, J., Sing, A. and Trulzsch, K. (2006) *Yersinia*'s stratagem: targeting innate and adaptive immune defense. Curr. Opin. Microbiol. 9, 55-61.

Igietseme, J.U., Eko, F.O., He, Q. and Black, C.M. (2004) Antibody regulation of T cell immunity: implications for vaccine strategies against intracellular pathogens. Expert Rev. Vaccines. 3, 23-34.

Jones, S.M., Day, F., Stagg, A.J. and Williamson, E.D. (2000) Protection conferred by a fully recombinant sub-unit vaccine against *Yersinia pestis* in male and female mice of four inbred strains. Vaccine. 19, 358-366.

Kerschen, E.J., Cohen, D.A., Kaplan, A.M. and Straley, S.C. (2004) The plague virulence protein YopM targets the innate immune response by causing a global depletion of NK cells. Infect. Immun. 72, 4589-4602.

Lukaszewski, R.A., Kenny, D.J., Taylor, R., Rees, D.G., Hartley, M.G. and Oyston, P.C. (2005) Pathogenesis of *Yersinia pestis* infection in BALB/c mice: effects on host macrophages and neutrophils. Infect. Immun. 73, 7142-7150.

Marketon, M.M., DePaolo, R.W., DeBord, K.L., Jabri, B. and Schneewind, O. (2005) Plague bacteria target immune cells during infection. Science. 309, 1739-1741.

Meyer, K.F. (1970) Effectiveness of live or killed plague vaccines in man. Bull. World Health Organ. 42, 653-666.

Nakajima, R. and Brubaker, R.R. (1993) Association between virulence of *Yersinia pestis* and suppression of gamma interferon and tumor necrosis factor alpha. Infect. Immun. 61, 23-31.

Parent, M.A., Wilhelm, L.B., Kummer, L.W., Szaba, F.M., Mullarky, I.K. and Smiley, S.T. (2006) Gamma interferon, tumor necrosis factor alpha, and nitric oxide synthase 2, key elements of cellular immunity, perform critical protective functions during humoral defense against lethal pulmonary *Yersinia pestis* infection. Infect. Immun. 74, 3381-3386.

Parent, M.A., Berggren, K.N., Kummer, L.W., Wilhelm, L.B., Szaba, F.M., Mullarky, I.K. and Smiley, S.T. (2005) Cell-mediated protection against pulmonary *Yersinia pestis* infection. Infect. Immun. 73, 7304-7310.

Perry, R.D. and Fetherston, J.D. (1997) *Yersinia pestis*-etiologic agent of plague. Clin. Microbiol. Rev. 10, 35-66.

Philipovskiy, A.V. and Smiley, S.T. (2007) Vaccination with live *Yersinia pestis* primes CD4 and CD8 T cells that synergistically protect against lethal pulmonary *Y. pestis* infection. Infect. Immun. In press.

Pitt, M.L. (2004) Non-human primates as a model for pneumonic plague. *Animals Models and Correlates of Protection for Plague Vaccines Workshop*. Gaithersburg, MD, http://www.fda.gov/cber/minutes/plague101304t.pdf.

Pujol, C., Grabenstein, J.P., Perry, R.D. and Bliska, J.B. (2005) Replication of *Yersinia pestis* in interferon gamma-activated macrophages requires ripA, a gene encoded in the pigmentation locus. Proc. Natl. Acad. Sci. USA. 102, 12909-12914.

Roggenkamp, A., Geiger, A.M., Leitritz, L., Kessler, A. and Heesemann, J. (1997) Passive immunity to infection with *Yersinia* spp. mediated by anti-recombinant V antigen is dependent on polymorphism of V antigen. Infect. Immun. 65, 446-451.

Sebbane, F., Lemaitre, N., Sturdevant, D.E., Rebeil, R., Virtaneva, K., Porcella, S.F. and Hinnebusch, B.J. (2006) Adaptive response of *Yersinia pestis* to extracellular effectors of innate immunity during bubonic plague. Proc. Natl. Acad. Sci. USA. 103, 11766-11771.

Straley, S.C. and Cibull, M.L. (1989) Differential clearance and host-pathogen interactions of YopE- and YopK- YopL- *Yersinia pestis* in BALB/c mice. Infect. Immun. 57, 1200-1210.

Titball, R.W. and Williamson, E.D. (2004) *Yersinia pestis* (plague) vaccines. Expert Opin. Biol. Ther. 4, 965-973.

Trulzsch, K., Geginat, G., Sporleder, T., Ruckdeschel, K., Hoffmann, R., Heesemann, J. and Russmann, H. (2005) *Yersinia* outer protein P inhibits CD8 T cell priming in the mouse infection model. J. Immunol. 174, 4244-4251.

Velan, B., Bar-Haim, E., Zauberman, A., Mamroud, E., Shafferman, A. and Cohen, S. (2006) Discordance in the effects of *Yersinia pestis* on the dendritic cell functions manifested by induction of maturation and paralysis of migration. Infect. Immun. 74, 6365-6376.

Viboud, G.I. and Bliska, J.B. (2005) *Yersinia* outer proteins: role in modulation of host cell signaling responses and pathogenesis. Annu. Rev. Microbiol. 59, 69-89.

Weeks, S., Hill, J., Friedlander, A. and Welkos, S. (2002) Anti-V antigen antibody protects macrophages from *Yersinia pestis*-induced cell death and promotes phagocytosis. Microb. Pathog. 32, 227-237.

Williamson, E.D., Eley, S.M., Stagg, A.J., Green, M., Russell, P. and Titball, R.W. (1997) A sub-unit vaccine elicits IgG in serum, spleen cell cultures and bronchial washings and protects immunized animals against pneumonic plague. Vaccine. 15, 1079-1084.

36

Oral Vaccination with Different Antigens from *Yersinia pestis* KIM Delivered by Live Attenuated *Salmonella* Typhimurium Elicits a Protective Immune Response Against Plague

Christine G. Branger[1], Jacqueline D. Fetherston[2], Robert D. Perry[2], and Roy Curtiss III[1]

[1] Center for Infectious Diseases and Vaccinology, The Biodesign Institute, Arizona State University, branger@asu.edu
[2] Department of Microbiology, Immunology and Molecular Genetics, University of Kentucky

Abstract. The use of live recombinant *Salmonella* attenuated vaccine (RASV) encoding *Yersinia* proteins is a promising new approach for the vaccination against *Yersinia pestis*. We have tested the efficacy of 2 proteins, Psn and a portion of LcrV in protecting mice against virulent *Yersinia pestis* challenge. To remove the immunosuppressive properties of LcrV protein, the *lcrV* gene, without the TLR2 receptor sequence, was cloned into a β-lactamase secretion vector. Immunizations were performed with RSAV expressing LcrV or Psn. Challenge with a virulent *Y. pestis* strain was performed 4 weeks after the last immunization. Our results show that the truncated LcrV protein delivered by RASV is sufficient to afford a full protective immune response in a mouse model of bubonic plague and the Psn protein afforded partial protection in a non-optimized system. This finding should facilitate the design and development of a new generation of vaccines against *Y. pestis*.

36.1 Introduction

Yersinia pestis causes a vector-borne disease infecting rodents and fleas; humans become infected when exposed to zoonotic reservoirs. Infection in humans usually occurs in the form of bubonic plague when they are bitten by fleas that have previously fed on plague-infected rodents (Perry and Fetherston 1999). Secondary pneumonic plague may then occur if the infection spreads to the lungs. Persons with secondary pneumonic plague can transmit the disease to others by the respiratory route, causing primary pneumonic plague, potentially leading to an epidemic. Antibiotic treatment of bubonic plague is usually effective; however pneumonic plague requires prompt antimicrobial treatment. Symptoms develop rapidly and the disease is nearly always fatal if untreated. *Y. pestis* has been responsible for three human pandemics and has caused over 200 million human deaths in the past. Plague still occurs throughout the world today, circulating in various mammalian species on most continents. In addition, the recent discovery of antibiotic resistant strains of *Y. pestis* (Galimand et al. 1997; Guiyoule et al. 2001) and its potential use by bioterrorists pose potential therapeutic and prophylactic problems (Calhoun et al. 1996).

Different vaccines have been elaborated against plague: a formaldehyde-killed whole bacilli vaccine has been used in the past but was discontinued in 1999 and is no longer available: recent studies in animals have shown that that vaccine offered poor protection against pneumonic disease and was reactogenic (Titball and Williamson 2001). A live attenuated vaccine is also available and provides protection against both bubonic and pneumonic plague. While this vaccine is effective, it retains some virulence (a fatality rate of 1% of vaccinated mice has been reported) and some side effects have been reported (Russell et al 1995). The ability of the live attenuated vaccine to protect against pneumonic plague makes the delivery of *Y. pestis* antigens by attenuated live bacteria an attractive strategy for a plague vaccine.

Live attenuated bacterial vectors such as *Salmonella* were first developed as vaccines to prevent disease caused by *Salmonella* infections of both humans and animals (Germanier and Fürer 1975). Subsequently, genetically modify attenuated *Salmonella* strains were constructed for delivery of heterologous antigens. The advantages to using *Salmonella* as a vaccine include a relatively low cost of production and an oral delivery route which abrogates the need for needles and stimulates both a mucosal and systemic immune response.

Ongoing research has been directed at improving the immunogenecity and stability of the *Salmonella* vaccine vector. To maintain the plasmid expressing the antigen of interest and avoid the use of antibiotics, a balanced-lethal host-vector system has been designed. In this system, the gene encoding for aspartate β-semialdehyde dehydrogenase (*asd*) is used to maintain plasmids co-expressing the protective antigens in Asd negative *Salmonella* strains (Nakayama et al. 1988; Galan et al. 1990). In addition, the deletion of the gene (Δ*crp*) encoding for the cyclic AMP receptor protein involved in regulating many biological functions, results in attenuation but not loss of ability to induce mucosal and systemic immune responses (Curtiss and Kelly 1987).

The virulence of *Y. pestis* is due, in part, to the expression of factors encoded on the 70 kb low-calcium response plasmid (pCD1 in KIM) (Perry et al. 1998). These determinants include the secreted LcrV antigen. LcrV is a multifunctional protein that is central to the activity of the type III secretion system apparatus. LcrV affects effector secretion by binding the negative regulator LcrG and is essential along with Yops B and D for the translocation of the effectors into eukaryotic cells (Sarker et al. 1998). In addition, LcrV suppresses the immune response by interaction with TLR2 to up-regulate Il10 levels (Brubaker 2003; Sing et al. 2002 & 2005). LcrV antigen is highly immunogenic and a major component of most plague vaccines.

Psn is the outer membrane receptor for the siderophore yersiniabactin. Expression of *psn* is regulated by iron and Fur protein (Ferric Uptake Regulator) and is required for the virulence of plague by a subcutaneous route of infection in mice (Fetherston et al. 1996).

In this study, we evaluated the efficacy of oral immunization of mice with recombinant attenuated *Salmonella enterica* serovar Typhimurium vaccines (RASV) synthesizing *Y. pestis* LcrV or Psn antigens in vivo against challenge with virulent *Y. pestis*.

36.2 Materials and Methods

36.2.1 Strains

Yersinia pestis C092 (biovar orientalis, strain CO92) was isolated from patient 92 in Colorado that died from the pneumonic form of the disease. *Y. pestis* was cultured in HIB medium at 28°C and grown to approximately 10^8 cells/ml, as estimated by OD. *Salmonella* χ8501 (*hisG* Δ*crp-28* Δ*asdA16*) and *E. coli* χ6212 (F$^-$ λ$^-$ ϕ80 Δ(*lacZYA-argF*) *endA1 recA1 hsdR17 deoR thi-1 glnV44 gyrA96 relA1* Δ*asdA4*) (Kang et al. 2002), χ6097(F$^-$ *ara* Δ(*lac-pro*) *rpsL* Δ*asdA4* Δ(*zhf-2::Tn10*) *thi* ϕ*80dlacZ* Δ*M15*) (Nakayama et al. 1988) and BL21DE3 (F$^-$*ompT hsdS*B(r_B^- m_B^-) *gal dcm* (DE3)) strains were cultured in LB broth, or on LB agar or Mac Conkey agar (supplemented with 1% maltose).

36.2.2 Cloning *lcrV* into pYA3620

To enhance the expression of LcrV in RASV, codons in the *lcrV* gene used with less than 2% frequency in highly expressed *Salmonella* genes were corrected. A fragment of the *lcrV* gene (corresponding to aa131-aa327) was amplified from the plasmid pJIT7 (Perry et al. 1986) using the forward primer, 5'-GCTCTAGAGAATTCG AT GATGATATTTTGAAAGTG-3'and the reverse primer, 5'-CCCAAGCTTTCATT T ACCAGACGTGTCATCGAG-3'. The 613 bp PCR product was cloned into pCRII.blunt (Invitrogen Co.) generating pCRII-*lcrV*. Both strands of the DNA insert were sequenced (IDT, USA). *Xba*I/*Kpn*I 664 pb fragment was isolated from pCRII-*lcrV* and ligated into the *Xba*I and *Kpn*I sites of a pUC19 generating pUC19-*lcrV*.

A 323bp PCR product was amplified from pUC19-*lcrV*, using the forward primer 5'-CCATTCAGGTGGATGGGAGCGAGAAAAAAATTGTCTCGATTAAGG-3' and the reverse primer above to optimize codons in *lcrV* at positions 698 and 705, and then cloned by TA cloning into pCRII.1 (Invitrogen Co.), generating pCRII.1-*Bst*XI-*lcrV* (698-705). Both strands of the DNA insert were sequenced (IDT, USA). A *Bst*XI/*HindIII* fragment prepared from pCRII.1-*Bst*XI-*lcrV*(698-705) was ligated into the *Bst*XI/*HindIII* site of pUC19-*lcrV* generating pUC19-*lcrV*(698-705).

The QuikChange® II Site-Directed Mutagenesis Kit was used to optimize codons in *lcrV* at positions 552 using the primer 5'-GTCTAGTAGTGGCACCATTAAT ATCCATGATAAATCC-3' and with its complement and to change codons at positions 723, 734 and 736 using the primers 5'-GGACTTTCTTGGCAGTGAGAAT AAACGCACCGGGGCGTTG-3' with its complement generating pUC19-*lcrV*$_{opt}$. pUC19-*lcrV*$_{opt}$ was used as a template for PCR with primers 5'-GCTCTAGAG AATTCATCGATGATGATATTTTGAAAGTG-3' and 5'-ACCTGCAGTTTACCA GACGTGTCATCGAGCAGACGTTG-3'. The 610 bp PCR fragment was digested with *Eco*RI and *Pst*I and ligated into the same sites of pYA3620 generating pYA3841 which was transformed into *Escherichia coli* strain χ6212 and then into *Salmonella* strain χ8501. Both strands of the DNA insert were sequenced (IDT, USA).

36.2.3 Cloning *psn* into pYA3332

The *psn* gene was amplified using RCpsn-1 (5'-GGAATTCGCATGACACGGCT TTATCCTCTG-3') and psn-his2 (5-CGGGATCCTAATCAGAAGAAATCAAT TCGC-3') primers with pPSN4 as a template (Fethertson et al. 1995). The PCR product was digested with *Bam*HI and *Eco*RI, cloned into the same sites in pWSK29 (Wang et al. 1991) and sequenced. The *Bam*HI/*Eco*RI fragment containing the *psn* gene was subcloned from pWSK29 into pYA3332, generating pYA3996 which was transformed into *Escherichia coli* strain χ6097 and then into *Salmonella* strain χ8501.

36.2.4 Expression of Recombinant Proteins in RASV

To confirm that the various constructs were functional and to determine the efficiency of protein expression in *Salmonella*, strain χ8501 harboring pYA3342 (Kang et al. 2002), pYA3841 and pYA3996 were cultured in LB broth at 37°C and harvested when they reached 0.8 OD_{600}. The bacteria were pelleted and resuspended in Laemmli sample buffer containing 2% 2-mercaptoethanol. The supernatants were filtered (0.22 μm) and precipitated overnight with 20% TCA (w/v). After centrifugation, the pellet was resuspended in cold PBS, and acetone precipitated. The precipitates were washed with acetone, resupended in 250μl of PBS and stored at –20°C.

36.2.5 Recombinant Protein Production and Purification

The plasmids pHT-V (Fields et al. 1999) and pPPH-2 containing respectively the LcrV-$_{Histag}$ or Psn-$_{Histag}$ coding sequences were amplified in *E. coli* BL21(DE3) competent cells (Stratagene). Bacteria were harvested 180 min after induction with 1 mM β-D-thiogalactopyranoside, and washed with PBS and French pressed. After centrifugation, the supernatant was directly loaded onto a Hi-Trap chelating column (Amersham Pharmacia Biotech) equilibrated with phosphate buffer. Proteins were eluted with a 40 to 250 mM imidazole gradient and then extensively dialyzed against PBS.

36.2.6 Western Blot Analysis of Recombinant Proteins

The proteins were separated by SDS-polyacrylamide gel electrophoresis as previously described (Laemmli 1970). Equal amounts of lysates or supernatants were loaded onto 12% or 7.5% polyacrylamide gels containing SDS for rLcrV and rPsn expression analysis, respectively and transferred onto nitrocellulose sheets (Biorad) using a semi-dry system with Tris buffer (48 mM Tris, pH 9.2, 39 mM glycine, 1.3 mM SDS, 20% methanol). After overnight blocking at 4°C with 3% BSA in TBST (10 mM Tris, pH 8, 150 mM NaCl, 0.05% Tween 20), recombinant proteins were identified using anti-histidine antibody (Novagen) followed by alkaline phosphatase-conjugated goat anti-mouse IgG (Sigma) or with rabbit anti-LcrV or anti-Psn serum followed by alkaline phosphatase-conjugated goat anti-rabbit IgG (Sigma). Antibody complexes were detected with the NBT-BCIP liquid substrate (Amaresco).

36.2.7 Animal Experiments

Female BALB/c mice, 6-8 weeks of age, were purchased from Harlan. Mice were deprived of food and water for 4 h prior to immunization and resupplied 30 min after. RASVs were grown in LB broth to an OD_{600} of 0.9 and concentrated to a final concentration of 5 x 10^{10} CFU/ml in phosphate-buffered saline containing 0.01% gelatin (BSG). Mice were orally immunized with 20 μl of RASV suspensions two times with a one week interval between immunizations. Blood samples were collected on days 0, 7, 35, and 49.

A subcutaneous challenge with virulent *Y. pestis* strain, CO92, (biovar orientalis) was performed 28 days after the second immunization. Each animal received by subcutaneous injection 3,000 CFU or 600 CFU in the first experiment and 1300 or 60 CFU in a second experiment. Mice were observed daily, and mortality was recorded for 14 days after the challenge. The surviving animals were euthanized after 14 days to obtain blood samples for serological analysis

36.2.8 Enzyme-Linked Immunosorbent Assay (ELISA)

Nunc Immunoplate Maxisorb F96 plates (Nalge Nunc. Rochester, NY, USA) were coated with LcrV (500ng/well), Psn (500ng/well) or LPS (100ng/well) and incubated overnight at 4°C. The plates were washed three times with PBS-t (0.1% Tween 20), blocked with PBS containing 10% (v/v) Sea BLOCK Blocking (Pierce) for 1 h at 37°C and then washed three times with PBS. The sera were diluted in PBS and added to the plates. The plates were then incubated for 1 h and washed. After 1 h incubation, the plates were washed and then incubated for 1 h at 37°C with goat anti-mouse IgG(H+L), biotin conjugated (SouthernBiotech) (1:5000). Strepavidine conjugated to AP (SouthernBiotech) (1:4000), was added and the plates were incubated for 1 h at 37°C, washed five times, and the *p*-nitrophenyl phosphate chromogenic substrate (Sigma) for alkaline phosphatase was added. After 10 min, the reaction was stopped with 2 M H_2SO_4. The optical density at 405 nm was measured with a Labsystems Multiskan MCC/340.

36.2.9 Statistical Analysis

Statistical significance was determined by log rank test, with $p < 0.05$ considered to be statistically significant.

36.3 Results

The constructions introduced into RASV strains were tested for plasmid stability, growth and the ability to colonize mouse liver, spleen and Peyer's patches (data not shown) as previously described (Kang et al. 2002). The LPS profile was verified.

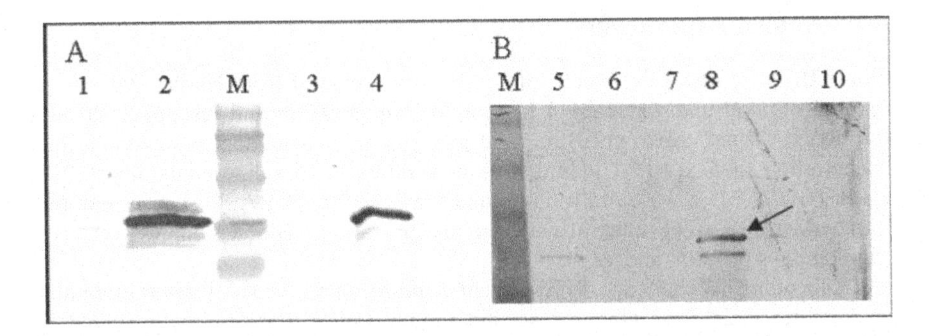

Fig. 1. Western blot performed with a polyclonal antibody against rLcrV (A) or rPsn (B). (A) Lanes 1 and 2: cells pellet from χ8501 (pYA3620) and (pYA3841) respectively. Lanes 3 and 4: culture supernatant from χ8501(pYA3620) and (pYA3841) respectively. (B) Lanes 1 and 4: cells pellet from χ8501 (pYA3342) and (pYA3996). Lanes 2 and 4: culture supernatant from χ8501(pYA3342) and (pYA3996). Lanes 3 and 6: culture supernatants after precipitation χ8501 from (pYA3342) and χ8501(pYA3996).

36.3.1 In Vivo Expression of rLcrV and rPsn Protein in *Salmonella* Cells

Expression of rLcrV and rPsn was demonstrated for each construct (pYA3841, pYA3996) by western blot analysis (Fig. 1A and Fig. 1B). As seen in Fig. 1A, samples from χ8501(pYA3841) contained LcrV protein in both the pellet and the supernatant. Analysis of the samples from χ8501(pYA3996) showed the presence of the rPsn protein only in the pellet. No bands were detected with either antisera in the χ8501(pYA3620) samples.

36.3.2 Antibody Response

The antibody response to *Salmonella* LPS was measured as previously described (data not shown). After the second immunization, all the mice developed a strong IgG antibody response against LPS with no significant difference between RASV with the empty carrier plasmid and those that expressed rLcrV or rPsn.

The IgG immune response of the mice immunized with rLcrV was analyzed by an ELISA against the recombinant protein (Fig. 2). No rLcrV antibody was detected in the non immunized mice or those receiving RASV with the empty plasmid before challenge. An anti-rLcrV response was not detected until after the second immunization at day 7, the immune response increased rapidly and was boosted by the challenge. Moreover, there was no significant decrease in the humoral response throughout the whole experiment. The challenge induced an increase in the humoral response against rLcrV in the surviving mice of each control group (mice receiving BSG only and χ8501with the empty plasmid). No significant difference was observed between the two experiments (data not shown).

The humoral immune response of mice immunized with RASV-Psn was analyzed by IgG ELISA against the recombinant rPsn (Fig. 2). A weak antibody response was detected in the mice immunized by χ8501 (pYA3342) before challenge. In groups

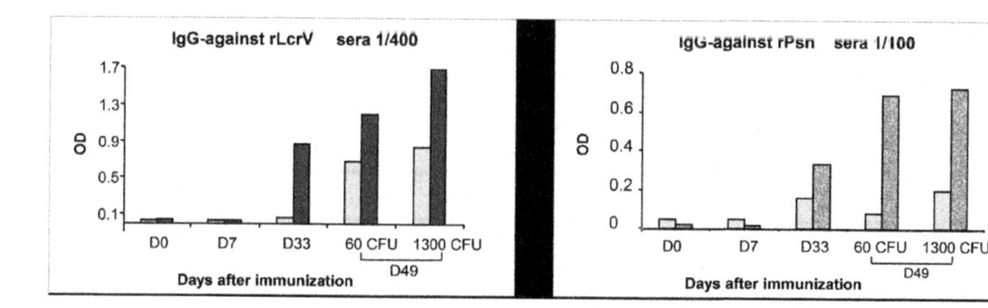

Fig. 2. Antibody responses in mice vaccinated with rLcrV or rPsn. IgG Anti-rLcrV ELISA antibody responses in mice orally immunized with RASV-LcrV (▮), or RASV-empty (Te) ( ), followed by subcutaneous *Y. pestis* challenge. IgG anti-rPsn ELISA antibody responses in mice orally immunized with RASV-rPsn ( ), or RASV-empty (Te) ( ), followed by subcutaneous *Y. pestis* challenge. Each point corresponds to the pool of sera.

immunized by RASV expressing rPsn, a weak antibody response was observed after the second immunization. The challenge induced an increase in the immune response against rPsn in the surviving animals.

36.3.3 Protection of Mice Immunized Against Plague Challenge

Mice were challenged with *Y. pestis* 28 days after the second immunization, and mortality rates were recorded for 14 days (Table 1). The statistical analysis of mortality incidence is based on the Logrank test. It is worth noting that in previous experiments, the mortality rate for the same challenge dose was lower in groups receiving RASV with an empty plasmid compared to the control group that received BSG. Therefore, in this study we compare each immunized group versus its control group that received the same *Salmonella* strains carrying the empty plasmid (RASV control). During the two trials, all the non immunized-mice (BSG controls) die within 5 days regardless the challenge dose. For the two trials, some animals immunized with the RASV control were able to survive (Table 1). Therefore, the mortality rates were significantly different between the RASV control groups and BSG control group (Table 1), indicating that some protection was provided by immunization with *Salmonella* alone. Still, the delay in onset of death in the RASV groups compared to the BSG control groups was identical.

Following the challenge with *Y. pestis* C092, almost complete protection was obtain for animals immunized by RASV-LcrV. In trial 1 (Fig. 3A), 2/10 (3000 CFU) and 1/10 (600CFU) of the vaccinated animals died versus 7/10 and 8/10, respectively of the RASV controls ($p<0.001$). In addition, we observed a delay in the time to death in the mice immunized by LcrV. In the second experiment (Fig. 3B), regardless of the challenge dose, all the animals immunized by RASV-LcrV survived (Table 1). The survival rate of mice vaccinated with RASV-*lcrV* was significantly higher than that of RASV controls ($p<0.001$). Combined analysis of the results from the two experiment showed significant protection with RASV-expressing LcrV, 14 days after the challenge ($p<0.0001$). The surviving animals immunized by RASV-LcrV did not show any symptoms of disease. The mortality rates for the animals vaccinated with

Table 1. Protective effect of immunization with RASV followed by subcutneous *Yersinia pestis* challenge. Animals were vaccinated twice at 7 day intervals and challenged with *Yersinia pestis* strain CO92, 28 days after the last immunization.

Vaccination with	Experiment 1[a]		Experiment 2[a]	
	challenge dose			
	3000 CFU	600CFU	1300 CFU	60CFU
RASV-LcrV	8/10	9/10	8/8	8/8
RASV-Psn		-	6/8	4/8
RASV-empty	3/10	1/10	2/8	3/8
BSG control	0/4	0/4	0/4	0/4

[a]Number of surviving animals at 14 days after challenge/ number of animals challenged

RASV expressing LcrV were not significantly different between the two challenge doses.

For mice immunized with RASV expressing rPsn (Fig. 3B), 2/8 (1,300 CFU) and 4/8 (60 CFU) vaccinated animals died versus 6/8 and 5/8 respectively of the RASV controls, with a delay in time to death in the mice immunized with rPsn. The survival rate of mice vaccinated with RASV-rPsn was significantly higher than that of the RASV controls at the two challenge doses ($p<0.045$). A combined analysis of the results from groups immunized by RASV-Psn showed significant protection 14 days after challenge ($p<0.004$). In terms of the onset of death, there was a significant difference between the RASV-rPsn group and the RASV control over all the experiments.

36.4 Discussion

LcrV is a *Y. pestis* virulence factor that is required for the delivery of effector proteins into the host cells via the type III secretion system (Fields et al. 1999). LcrV is well known to protect mice against *Y. pestis* infection: immunization with recombinant LcrV or passive immunization with anti-LcrV serum confers a high degree of protection (Motin et al. 1994). The protective epitope(s) of LcrV (327aa) are located in the central region of the protein (residues 135 to 275) (Hill et al. 1997). Thus, recombinant variants of LcrV lacking 30 amino acid segments within the first 150 residues in the N-term part of the protein (1-150) had no impact in terms of protecting mice against bubonic plague (Overheim et al. 2005). In addition, Nakajima and Brubaker (1993) showed that LcrV can suppress the immune response by inhibiting cytokine production. LcrV is able to suppress TNF-α production, via expression of Il10 (Sing et al. 2002). Presumably, residues 31-57 of LcrV interact with TLR2 receptor to suppress the innate immune response (Sing et al. 2002 and 2005). Therefore to obtain a safer vaccine, it may be necessary to remove the immune-modulatory residues of LcrV. In our studies, we cloned and expressed by RASV a fragment of the *lcrV* gene encoding the 131-327 residues, without the TLR2 interacting sequence but with the central part of the protein that contributes to protection.

Other studies have shown that LcrV alone or fused with F1 and expressed by *Salmonella* was able to induce only partial protection against *Y. pestis* challenge whereas F1 alone afforded full protection. One explanation for these results was that LcrV and F1-LcrV were localized in the cytoplasm whereas F1 was surface exposed (Oyston et al. 1995; Titball et al. 1997; Garmory et al. 2003).

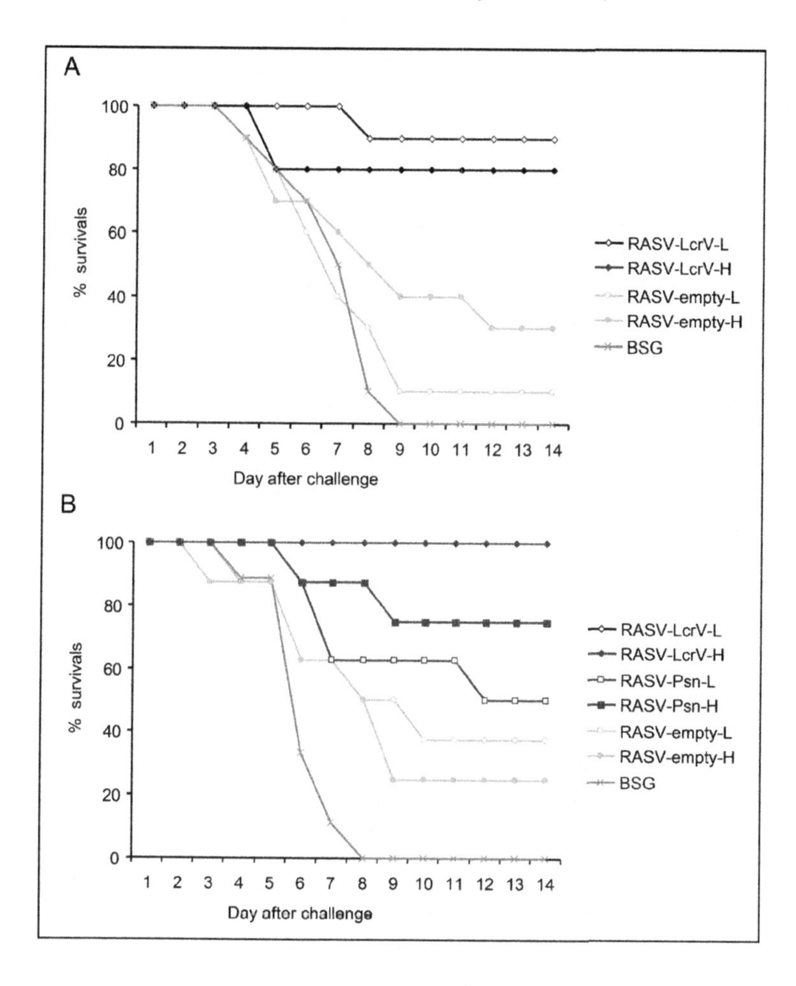

Fig. 3. Survival curves of mice after oral immunization with RASV-LcrV (L), RASV-LcrV (H), RASV-Psn (L), RASV-Psn (H), RASV control (L), RASV control (H), BSG (L) and BSG (H) after subcutaneous *Y. pestis* challenge. (A) Experiment 1 - High Dose (H)=3,000CFU; Low Dose (L)=600 CFU; (B) Experiment 2 - High Dose(H) =1,300 CFU; Low Dose(L)=60 CFU.

Previous data showed that use of β-lactamase secretion signals in RASV results in translocation of the protein to the periplasm. In addition, some of the protein is found in the supernatant of the culture possibly due to secretion of vesicles that encapsulated the accumulated periplasmic protein (Kang et al. 2002). The localization of proteins expressed in *Salmonella* affects the immune response and thus protection: antigens in the periplasm or secreted in the supernatant are more immunogenic than those in the cytoplasm (Kang et al. 2003). For these reasons, the fragment of the *lcrV* gene encoding aa131-327 was cloned into β-lactamase secretion vector (pYA3620). To enhance the expression of LcrV in *Salmonella*, any codon in the *lcrV* gene that is used less than 2% as a codon in highly expressed *Salmonella* genes has been corrected. Analysis of the expression of rLcrV showed that the protein was present as expected in the periplasm and in the supernatant, mimicking the natural expression of LcrV.

Previous attempts at developing a *Salmonella* vaccine based on LcrV have had limited success in affording protection, despite high immune responses (Th1 and Th2) that required 5 immunizations with 10^9CFU of RASV-rLcrV (Garmory et al. 2003). The poor results of those studies may be due to the choice of the attenuated *Salmonella* strain used for the vaccine and/or the vector system used. The strain must be sufficiently attenuated to be safe but at the same time must colonize the gut-associated lymphoid tissue and the visceral lymphoid tissue after oral immunization in order to induce a mucosal, humoral and cellular immune response. In our experiments, two immunizations with *S. enterica* serovar Typhimurium χ8501 (*hisGΔcrp-28ΔasdA16*) were enough to stimulate a strong immune response. Furthermore, it has been shown that mice immunized with this strain are able to survive a *Salmonella* challenge (10,000 LD_{50}) and maintain the plasmid carrying a heterologous antigen (Kang et al. 2002). It is worth noting that in our experiments some mice immunized by RASV with the empty plasmid vectors were able to survive (9 survivors/36) whatever the challenge dose (60,300, 1,300 and 3,000 CFU), whereas all the mice receiving only BSG died. This corresponds to a 25% survival rate. In comparison, immunizations with a*roA* attenuated *S. enterica* serovar Typhimurium (Garmory et al. 2003) induce protection against a 97 CFU of *Y. pestis* GB infection in only one of 40 mice.

All mice orally immunized by RASV-rLcrV (10^9CFU) developed a strong immune response against rLcrV; the profile of the IgG_1 and IgG_{2A} antibody response is similar to those of the IgG, never the less the IgG_{2A} immune response is higher than the IgG_1, despite a strong IgG_1 antibody response (data not shown). A dominant Th1 response is common after immunization with *Salmonella.* The immune response affords almost complete protection against *Y. pestis* subcutaneous infection. This level of protection shows that RASV could be an efficient vector to express rLcrV which is able to enhance a protective immune response. In addition, the truncated protein used in this study was enough to afford full protection against a 500 LD_{50} challenge, confirming that the central region of LcrV contains the principal epitope(s) involved in protection. On the other hand, secretion of the protein into the supernatant may cause the protein to fold such that the protective epitopes are better exposed and then can induce a strong Th2 response that is essential for protection as seen before after passive immunization with antisera against LcrV (Motin et al.

1994). Furthermore, the mucosal immune response elicited after oral immunization (data not shown) may be important for providing protection against pneumonic plague. We can not draw any conclusions about the role of the deletion of the TLR2 interacting sequence. For this, it will be necessary to compare the level of Il10 production in mice immunized with RASV expressing either the entire LcrV or the truncated protein to determine the immuno-modulatory properties of the truncated protein.

Psn is the outer membrane receptor for the siderophore, yersiniabactin. It confers sensitivity to the bacteriocin, pesticin, and is an integral component of an inorganic iron transport system that functions at 37°C (Fetherston et al. 1996). The expression of Psn is regulated by iron and repressed by Fur. *Y. pestis* strains with a *psn* mutation are avirulent in mice after infection by subcutaneous route (Geoffroy et al. 2000). The entire *psn* gene has been cloned without optimization of the sequence to avoid the cytotoxicity encountered with over-expressing outer membrane proteins in *Salmonella*. Immunization by RASV-rPsn was able to induce a weak IgG immune response that afforded partial protection. A delay of mortality was observed with the RASV control. An improvement in the expression of Psn may be necessary to enhance a stronger immune response.

In conclusion, our results show that rLcrV expression mediated by attenuated *Salmonella* strain χ8501 is able to elicit a Th1/Th2 immune response which affords strong protection against fully virulent *Y. pestis*. Psn seems an interesting candidate but follow-up studies are needed to determine if the immunity resulting from RASV-Psn immunization can be improved via codon and secretion optimization. Such studies will be helpful for designing and developing new generations of vaccines based on RASV.

36.5 Acknowledgements

We thank S. Straley for providing pHT-V. S. Forman, A.G. Bobrov, O. Kirillina, B. Gunn are thanked for assistance with animal experiments. W. Kong, A. Torres Escobar are thanked for suggestions during the course of this work. This work was supported by National Institutes of Health grant 5R01 AI057885.

36.6 References

Brubaker, R.R. (2003) Interleukin-10 and inhibition of innate immunity to *Yersiniae*: roles of Yops and LcrV (V antigen). Infect. Immun. 71, 3673-3681.

Calhoun L.N. and Kwon Y.M. (2006) *Salmonella*-based plague vaccines for bioterrorism. J. Microbiol. Immunol. Infect. 39, 92-97.

Curtiss, R, 3rd. and Kelly, SM. (1987) *Salmonella typhimurium* deletion mutants lacking adenylate cyclase and cyclic AMP receptor protein are avirulent and immunogenic. Infect. Immun. 55, 3035-3043.

Fetherston, J.D, Bearden, S.W. and Perry, R.D. (1996) YbtA, an AraC-type regulator of the *Yersinia pestis* pesticin/yersiniabactin receptor. Mol. Microbiol. 22, 315-325.

Fetherston, J.D, Lillard, J.W., Jr. and Perry, RD. (1995) Analysis of the pesticin receptor from *Yersinia pestis*: role in iron-deficient growth and possible regulation by its siderophore. J. Bacteriol. 177, 1824-1833.

Fields, K.A., Nilles, M.L., Cowan, C. and Straley, S.C. (1999) Virulence role of V antigen of *Yersinia pestis* at the bacterial surface. Infect. Immun. 67; 5395-408.

Galan, J.E., Nakayama, K. and Curtiss, R., 3rd. (1990) Cloning and characterization of the *asd* gene of *Salmonella typhimurium*: use in stable maintenance of recombinant plasmids in *Salmonella* vaccine strains. Gene 94, 29-35.

Galimand, M., Guiyoule, A., Gerbaud, G., Rasoamanana, B., Chanteau, S., Carniel, E. and Courvalin, P. (1997) Multidrug resistance in *Yersinia pestis* mediated by a transferable plasmid. N. Engl. J. Med. 337, 677-680.

Garmory, H.S., Griffin, K.F., Brown, K.A. and Titball, R.W. (2003) Oral immunisation with live *aroA* attenuated *Salmonella enterica* serovar Typhimurium expressing the *Yersinia pestis* V antigen protects mice against plague. Vaccine 21, 3051-3057.

Geoffroy, V.A., Fetherston, J.D., Perry, R.D. (2000) *Yersinia pestis* YbtU and YbtT are involved in synthesis of the siderophore yersiniabactin but have different effects on regulation. Infect. Immun. 68, 4452-4461.

Germanier, R. and Fürer, E. (1975) Isolation and characterization of Gal E mutant Ty 21a of *Salmonella typhi*: a candidate strain for a live, oral typhoid vaccine. J. Infect. Dis. 131, 553-558.

Guiyoule, A., Gerbaud, G., Buchrieser, C., Galimand, M., Rahalison, L., Chanteau, S., Courvalin, P. and Carniel, E. (2001) Transferable plasmid-mediated resistance to streptomycin in a clinical isolate of *Yersinia pestis*. Emerg. Infect. Dis. 7, 43-48.

Hill, J., Leary, S.E., Griffin, K.F., Williamson, E.D. and Titball, R.W. (1997) Regions of *Yersinia pestis* V antigen that contribute to protection against plague identified by passive and active immunization. Infect. Immun. 65, 4476-4482.

Kang, H.Y. and Curtiss, R., 3rd. (2003) Immune responses dependent on antigen location in recombinant attenuated *Salmonella typhimurium* vaccines following oral immunization. FEMS Immunol. Med. Microbiol. 37, 99-104.

Kang, H.Y., Srinivasan, J. and Curtiss, R., 3rd. (2002) Immune responses to recombinant pneumococcal PspA antigen delivered by live attenuated *Salmonella enterica* serovar Typhimurium vaccine. Infect. Immun. 70, 1739-1749.

Laemmli, U.K. (1970) Cleavage of structural protein during the assembly of the head bacteriophage T4. Nature 227, 680-685.

Morton, M., Garmory, H.S., Perkins, S.D., O'Dowd, A.M., Griffin, K.F., Turner, A.K., Bennett, A.M. and Titball, R.W. (2004) A *Salmonella enterica* serovar Typhi vaccine expressing *Yersinia pestis* F1 antigen on its surface provides protection against plague in mice. Vaccine 22, 2524-2532.

Motin, V.L., Nakajima, R., Smirnov, G.B. and Brubaker, R.R. (1994) Passive immunity to yersiniae mediated by anti-recombinant V antigen and protein A-V antigen fusion peptide. Infect. Immun. 62, 4192-4201.

Nakajima, R. and Brubaker, R.R. (1993) Association between virulence of *Yersinia pestis* and suppression of gamma interferon and tumor necrosis factor alpha. Infect. Immun. 61, 23-31.

Nakayama, K., Kelly, S.M. and Curtiss, R., 3rd. (1988) Construction of an *asd*$^+$ expression-cloning vector: stable maintenance and high level expression of cloned genes in a Salmonella vaccine strain. Biotechnology 6, 693–697.

Overheim, K.A., Depaolo, R.W., Debord, K.L., Morrin, E.M., Anderson, D.M., Green, N.M., Brubaker, R.R., Jabri, B. and Schneewind, O. (2005) LcrV plague vaccine with altered immunomodulatory properties. Infect. Immun. 73, 5152-5159.

Oyston, P.C., Williamson, E.D., Leary, S.E., Eley, S.M., Griffin, K.F. and Titball, R.W. (1995) Immunization with live recombinant *Salmonella typhimurium aroA* producing F1 antigen protects against plague. Infect. Immun. 63, 563-568.

Perry, R.D., Balbo, P.B., Jones, H.A., Fetherston, J.D. and DeMoll, E. (1999) Yersiniabactin from *Yersinia pestis*: biochemical characterization of the siderophore and its role in iron transport and regulation. Microbiology 145, 1181-1190.

Perry, R.D. and Fetherston, J.D. (1997) *Yersinia pestis* - etiologic agent of plague. Clin. Microbiol. Rev. 10, 35-66.

Perry, R.D., Harmon, P.A., Bowmer, W.S. and Straley, S.C. (1986) A low-Ca^{2+} response operon encodes the V antigen of *Yersinia pestis*. Infect. Immun. 54, 428-434.

Perry, R.D., Straley, S.C., Fetherston, J.D., Rose, D.J., Gregor, J. and Blattner, F.R. (1998) DNA sequencing and analysis of the low-Ca^{2+}-response plasmid pCD1 of *Yesinia pestis* KIM5. Infect. Immun. 66, 4611-4623.

Russell, P., Eley, S.M., Hibbs, S.E., Manchee, R.J., Stagg, A.J. and Titball, R.W. (1995) A comparison of Plague vaccine, USP and EV76 vaccine induced protection against *Yersinia pestis* in a murine model. Vaccine 13, 1551-6.

Sarker, M.R., Neyt, C., Stainier, I. and Cornelis, G.R. (1998) The *Yersinia* Yop virulon: LcrV is required for extrusion of the translocators YopB and YopD. J. Bacteriol. 180, 1207-1214.

Sing, A., Reithmeier-Rost, D., Granfors, K., Hill, J., Roggenkamp, A. and Heesemann, J. (2005) A hypervariable N-terminal region of *Yersinia* LcrV determines Toll-like receptor 2-mediated IL-10 induction and mouse virulence. PNAS 102, 16049-16054.

Sing, A., Rost, D., Tvardovskaia, N., Roggenkamp, A., Wiedemann, A., Kirschning, C.J., Aepfelbacher, M. and Heesemann, J. (2002) *Yersinia* V-antigen exploits toll-like receptor 2 and CD14 for interleukin 10-mediated immunosuppression. J. Exp. Med. 196, 1017-1024.

Titball, R.W., Howells, A.M., Oyston, P.C. and Williamson, E.D. (1997) Expression of the *Yersinia pestis* capsular antigen (F1 antigen) on the surface of an *aroA* mutant of *Salmonella typhimurium* induces high levels of protection against plague. Infect. Immun. 65, 1926-1930.

Titball, R.W. and Williamson, E.D. (2001) Vaccination against bubonic and pneumonic plague. Vaccine 19, 4175-4184.

Wang, R.F. and Kushner, S.R. (1991) Construction of versatile low-copy-number vectors for cloning, sequencing and gene expression in *Escherichia coli.* Gene 100, 195-199.

Picture 33. Christine Branger presents results on oral immunization with live attenuated *Salmonella* expressing *Y. pestis* antigens. Photograph by R. Perry.

37
Yersinia pestis YadC: A Novel Vaccine Candidate Against Plague

Brian S. Murphy[1], Christine R. Wulff[2], Beth A. Garvy[1,2], and Susan C. Straley[2]

[1] Department of Internal Medicine, University of Kentucky, bsmurp1@uky.edu
[2] Dept. of Microbiology, Immunology, and Molecular Genetics, University of Kentucky

Abstract. Current subunit vaccines provide partial protection against pneumonic plague if the infecting *Y. pestis* strain is encapsulated ($F1^+$). Here we describe YadC, a novel *Y. pestis* outer membrane protein that provides partial protection against a $F1^-$ *Y. pestis* strain. Swiss-Webster mice were immunized subcutaneously with glutathione S-transferase (GST) or His_6-tagged (HT) purified fusion proteins ($GST\text{-}YadC_{137\text{-}409}$ or HT-LcrV) or buffer emulsified with Alhydrogel. Intravenous challenge with 1 x 10^4 $F1^-$ *Δpgm Y. pestis* CO99-3015 revealed no protection for those mice immunized with GST-Alhydrogel alone, full protection for HT-LcrV-immunized mice, and partial protection for $GST\text{-}YadC_{137\text{-}409}$ -immunized mice. Similarly, C57BL/6 mice were immunized with $GST\text{-}YadC_{137\text{-}409}$, HT-LcrV, or GST all with Alhydrogel adjuvant. After intranasal challenge with 3 x 10^3 $F1^-$ *Y. pestis* CO99-3015, 87% of $GST\text{-}YadC_{137\text{-}409}$-immunized mice survived pneumonic plague. This is compared to the GST control group (0 surviving mice) and the LcrV-immunized group where 50% survived the challenge. This protection was correlated with a predominantly IgG1 response in LcrV-immunized mice and an IgG1/IgG3 antibody response in YadC-immunized mice. Additionally, we report the cytokine response from HT-LcrV- and $GST\text{-}YadC_{137\text{-}409}$-stimulated peripherally derived macrophages. YadC-stimulated cells demonstrated a predominant pro-inflammatory cytokine production. This mixed Th1/Th2 response suggests that YadC's protection may involve a different adaptive immune response than the LcrV protein that currently is part of plague vaccines.

37.1 Introduction

There is currently no effective plague vaccine. Vaccination with whole-cell plague vaccines using formalin-inactivated *Yersinia pestis*, although protective versus bubonic plague, has limited effectiveness for protection from pneumonic challenge and has been associated with significant side effects (Marshall et al. 1974; Russell et al. 1995; Williamson et al. 1997; Inglesby et al. 2000; Gruchalla and Jones 2003; Williamson et al. 2005). To address these concerns, subunit vaccines composed of the fibrillar capsular protein Caf1 or the V antigen (LcrV) of *Y. pestis* have been developed and are in phase II clinical trials (Simpson et al. 1990; Motin et al. 1994; Andrews et al. 1996; Anderson et al. 1997). Vaccines combining F1 and LcrV (Eyles et al. 1998; Eyles et al. 2004) or recombinant F1/V fusion proteins (Leary et al. 1997; Heath et al. 1998; Glynn et al. 2005) have been found protective when appropriate adjuvants are utilized.

Still, the current formulations and the planned third-generation vaccines do not provide adequate protection against a non-encapsulated ($F1^-$) *Y. pestis* strain (British and US vaccine research groups 2002). F1 is not required for the pathogenesis of pneumonic plague and studies have shown that *Y. pestis* with mutations in the *caf1* pilus operon remains fully virulent in mice and non-human primates (Friedlander et al. 1995; Welkos et al. 1995; Andrews et al. 1996; Davis et al. 1996).

The other component that is currently being incorporated, LcrV, is a potent protective antigen that is exposed at the tip of the type III secretion needle on the bacterial surface. However, several new studies have raised the concern that full-length LcrV may be immunosuppressive (Sing et al. 2002a; Sing et al. 2003b; DeBord et al. 2006) and cause a Th2 cytokine polarization (Nakajima and Brubaker 1993; Nakajima et al. 1995; Nedialkov et al. 1997). Further, both LcrV and the F1 capsular protein induce an IgG1/IgG2a antibody mixture that is consistent with a Th2 response (Williamson et al. 1999; Glynn et al. 2005). Consequently, additional work is required to achieve a more balanced adaptive immune response as well as provide protection against $F1^-$ *Y. pestis*.

We have been identifying and characterizing surface proteins that sponsor adhesion and invasion of *Y. pestis* into mammalian cells. *Y. pestis* requires tight adherence to mammalian cells for the type three secretion mechanism to work, but the responsible adhesins are unknown (Perry and Fetherston 1997). *Y. pestis* can also invade epithelial cells and macrophages avidly, mediated by more than one invasin (Cowan et al. 2000). The surface aspartyl protease Pla is one of these invasins (Cowan et al. 2000) that mediates adherence to epithelial cell lines in addition to its role in promoting dissemination, but the identities of the major invasins are unknown. Moreover, a Pla^- mutant is not significantly decreased in adherence to type 1 pneumocytes (Cowan et al. 2000; Lahteenmaki et al. 2001). Consequently, additional adhesins must be present. In this study, we tested the protective efficacy of a potential surface protein with weak sequence similarity but striking predicted structural relatedness to the anchor, linker, stalk, and neck regions of YadA, a major adhesin for *Y. pseudotuberculosis* and *Y. enterocolitica.* Here we report that this protein appears to be protective against $F1^-$ *Y. pestis*, does not suppress innate immune responses, and may be useful in a new generation vaccine.

37.2 Materials and Methods

37.2.1 Bacterial Strains and Plasmids

Y. pestis CO99-3015 strains were grown in Heart Infusion Broth (Difco laboratories, Detroit, MI) supplemented with 2.5mM $CaCl_2$ and 0.2% xylose (sHIB) or on sHIB agar. *Escherichia coli* strains were grown in Luria-Bertani broth (LB) or on LB-based agar. Antibiotics were used at the following concentrations during genetic constructions: carbenicillin (Cb), 100 µg/ml; kanamycin, 25 µg/ml chloramphenicol, 25 µg/ml. Cb was used at 50 ug/ml for growth of *Y. pestis* strains containing pCD2Ap. The presence of the pigmentation locus (Perry and Fetherston 1997) was confirmed by the formation of red colonies on Congo Red agar (Surgalla and Beesley 1969). The presence of a functional Lcr virulence plasmid was confirmed

Table 1. Bacterial Strains and Plasmids Used in This Study.

Strain or Plasmid	Key Properties	Source or Reference
Y. pestis		
CO99-3015	Wildtype strain; subclone of CO92; molecular grouping 1.ORI.	CDC Ft. Collins
CO99-3015.S2	CO92 Pgm^+ Lcr^- Δ*caf1*, pPst; $F1^-$ derivative made by allelic exchange into avirulent Lcr^- $F1^+$ *Y. pestis* CO99-3015 obtained from Scott Bearden (CDC Ft. Collins); entire *caf1* coding sequence and 12 upstream bp deleted.	Spencer Leigh
CO99-3015.S5	CO92 Δ*pgm* Lcr^+ pFra, pPst; conditionally virulent[a] Δ*pgm* derivative of CO99-3015.	Scott Bearden
CO99-3015.S6	CO92 Δ*pgm* Lcr^+ pFra, pPst; pCD2 *yadA::bla* (pCD2Ap; Ap^r) made from CO99-3015.S5 by allelic exchange; used as source of pCD2Ap for electroporation into Lcr^- strains to reconstitute potential virulence.	Forman et al. 2006
CO99-3015.S7	CO92 Δ*pgm* Lcr^+ Δ*caf1*, pPst; $F1^-$ Δ*pgm* attenuated strain made from CO99-3015.S5 by allelic exchange.	This Study
CO99-3015.S9	CO92 Pgm^+ Lcr^+ Δ*caf1*, pPst; reconstituted Lcr^+ $F1^-$ conditionally virulent[b] strain made by allelic exchange.	This Study
E. coli		
DH5α	end-1 hsdR17(rk-mk+) supE44 thi-1 recA1 *gyrA* (Nalr) *relA1* Δ(lacIZYA-argF)U169 deoR (Φ80dlacΔ (lacZ)M15).	Life Technologies, Gaithersburg MD
Plasmids		
pKD3	Template plasmid; *cat* flanked by FRT sites; Cm^rAp^r.	Datsenko and Wanner 2000
pKD46	Red recombinase expression plasmid; Ap^r.	Datsenko and Wanner 2000
pCP20	Suicide plasmid with temperature-sensitive replication and thermally induced expression of FLP recombinase; Ap^rCm^r	Datsenko and Wanner 2000
pHT-V	PCR-amplified *lcrV* cloned into pProEX-1; expresses full-length LcrV with a His_6-containing 23-residue N-terminal leader sequence (HT-LcrV).	Fields and Straley 1999
pGEX-$YadC_{137-409}$	Sequence encoding amino acids 137- 409 of YadC in pGEX-3X; expresses GST-$YadC_{137-409}$.	Forman et al. 2006

[a] Full virulence is conditional on intravenous route of infection (Perry and Fetherston 1997).
[b] Full virulence is conditional on intranasal route of infection (Davis1996).

by absence of growth at 37°C on sHIB agar containing 20mM $MgCl_2$ and 20mM sodium oxalate (MgOx plates) and by assaying expression and secretion of Yops during growth in the defined medium TMH (Straley and Bowmer 1986) with and without 2.5mM $CaCl_2$.

Production of the protein capsule F1 was abolished by deleting the *caf1* gene. Primers 449 [ACAGGACACAAGCCCTCTCTACGAATTTGTTCGTGGATTGG ATTATTCGAGTGTAGGCTGGAGCTGCTTC] and 450 [ATTAAAGGAGGGC ATAATAGCCCTCCTTTATCTATTATCTATATGGATTACATATGAATATCCT CCTTAGTTCCT] were used to generate a PCR fragment containing the FRP site-flanked *cat* gene from pKD3 flanked by *caf1* sequence, and *caf1* was deleted by allelic exchange mediated by the Red and Flp recombinases as previously described

(Datsenko and Wanner 2000). The resulting deletion started at −12 bp upstream of the *caf1* translation initiation codon through the last translated codon. The absence of F1 expression was verified by immunoblot of 37°C-grown yersiniae using the monoclonal antibody YPF1 (Research Diagnostics, Inc., Flanders, NJ), and the absence of *cat* in the final strain was confirmed by PCR and by inability to grow on chloramphenicol-containing plates.

Virulence was reconstituted in BSL3 containment with select-agent security by introducing pCD2Ap by electroporation. Briefly, *Y. pestis* CO99-3015.S2 was electroporated with plasmid DNA from *Y. pestis* CO99-3015.S6, that contained the *Y. pestis* plasmids pFra (110 kb) and pPst (9 kb) (Perry and Fetherston 1997) in addition to pCD2Ap. Their presence in the transformation would not pose a problem since the recipient strains already contained copies of these two plasmids.

Transformed strains were selected for Cb-resistance and screened for the Pgm^+ and Lcr^+ phenotypes on agar media. Individual isolates that passed these tests were then subcultured two times in 3 ml TMH + Cb at 26°C for a total of 24 hours prior to the experiment. The cultures were subcultured a third time into TMH with or without 2.5 mM $CaCl_2$ and allowed to grow for an additional two hours at 26°C. Following this, each culture was shifted to 37°C and permitted to grow another four hours. A 500-µl sample of each culture was removed and briefly placed on ice to aid cellular aggregation. Cells were pelleted in a microcentrifuge at room temperature (RT) for 10 min. The top 250 µl of supernatant was removed, transferred to a microcentrifuge tube containing trichloroacetic acid to a final concentration of 5%, and incubated overnight at 4°C to precipitate secreted proteins. The rest of the supernatant was discarded. The precipitated proteins were collected by centrifugation at RT. Pellets of whole cells and of secreted proteins were suspended in lysis loading buffer (400 mM TrisCl, pH 6.8, 25% glycerol, 8% SDS, 5% dithiothreitol, 0.4% bromophenol blue) to a concentration corresponding to a culture OD_{620} of 2.0 and heated to 95°C for 15 minutes. Half of each heat-treated sample was plated and incubated for 5 days at 28°C to ensure that no viable cells remained prior to removing the samples from the containment facility.

The resultant cellular extracts and supernatant proteins were then analyzed by immunoblot for the presence of secreted LcrV and YopM as described previously (Wulff-Strobel et al. 2001) to determine whether the type three secretion system was fully functional (data not shown). Several isolated colonies were then pooled and stocked.

37.2.2 Preparation of Protein Antigens

To produce GST-$YadC_{137\text{-}409}$ (Forman et al. 2006), *E. coli* DH5α containing plasmid pGEX-$YadC_{137\text{-}409}$ was grown to an OD_{620} of 0.5, IPTG was added to a final concentration of 1.0 mM and the bacteria were incubated for an additional 3 hours. Cells were harvested by centrifugation at 4°C, and resuspended at 0.05 of the original volume in column buffer (20 mM Tris, 200 mM NaCl, 1 mM EDTA pH 7.4). Samples were passed twice through a French press to lyse bacteria, cleared by centrifugation at 9000 x g for 30 minutes at 4°C, diluted 1:4 with column buffer and passed over a glutathione sepharose column. Bound protein was eluted using column buffer containing 10 mM glutathione and collected in 1 mL aliquots. Purity was assessed on SDS-PAGE gels stained with Coomasie brilliant blue. Protein-containing fractions were pooled and dialyzed, and the protein concentration measured by the bicinchoninic acid assay (Pierce Chemical Co., Rockford, IL).

N-terminally His_6-tagged LcrV (HT-LcrV) was expressed in *E. coli* DH5α from plasmid pHT-V and purified on Talon metal-affinity resin (Clonetech Laboratories, Inc, Palo Alto, CA) as described previously (Fields and Straley 1999).

37.2.3 Protection Tests

Sets of 5 female outbred Swiss-Webster mice were immunized subcutaneously with 40 μg amounts of purified fusion proteins (GST-$YadC_{137\text{-}409}$, or HT-LcrV) or buffer emulsified with Alhydrogel (aluminum hydroxide gel) (Sigma Chemical Company, St. Louis, MO; 6.25% vol/vol final concentration) on days 0, 8, 14, and 22. On day 35, they were challenged intravenously with 10,000 CFU (5 LD_{50} doses) of *Y. pestis* CO99-3015.S7.

Similarly, groups of 8 C57BL/6 mice were immunized with 80 μg of HT-LcrV, GST-$YadC_{137\text{-}409}$, or GST (Sigma Chemical Company, St. Louis, Mo.) adjuvanted with Alhydrogel as in the previous experiment. Two weeks after the 3rd immunization, the mice were challenged intranasally with 3000 CFU of CO92.S2 (6 times the LD_{50} dose) in BSL3 conditions.

37.2.4 Determination of IgG Isotypes

Measurement of antibody titer and isotype response was performed by a modified ELISA on serum samples taken 1 week prior to bacterial challenge. 96-well ELISA plates were coated with 10 μg/ml of GST-$YadC_{137\text{-}409}$, HT-LcrV, GST or with buffer alone. Mouse sera were then used at serial dilutions to a maximum dilution of 1:5,120,000. HRP conjugates against mouse IgG, IgG1, IgG2a, IgG2b, and IgG3 (BD Pharmingen, San Diego, CA) were used as secondary antibodies. Protein-specific antibody titer was estimated as the maximum dilution of serum giving an absorbance at 450 nm reading 0.1 units over the non-immunized mouse background sera (pre-immunization sera). Mean titers were determined per treatment group.

37.2.5 Human Macrophages and Cytokine Analysis

We evaluated the cytokine response of human peripheral blood monocytes (PBMs) to HT-LcrV, GST-$YadC_{137\text{-}409}$, and the negative control GST. Blood was obtained from two healthy volunteers. Red blood cells were lysed with ACK buffer (155 mM NH_4Cl/1 mM $KHCO_3$/0.5 mM EDTA, pH 7.2), the samples were washed, and PBMs were isolated on Ficoll-Paque density gradients. PBMs were cultured for one week in RPMI with 12.5% human serum and 10 μg/ml of penicillin, streptomycin, and neomycin at a concentration of 1 x 10^6 cells/ml in 5% CO_2 to differentiate into macrophages.

Differentiated PBMs were stimulated with 10 μg of F1, HT-LcrV, GST, or GST-$YadC_{137\text{-}409}$ in 96-well ELISA plates. A 100 μl aliquot of supernatant was removed at 3 hours and then at 24 hours. Supernatants were kept at –80°C until processed for cytokine analysis. All analyses were done in triplicate in 2 different experiments.

Cytokine levels were measured by Cytokine Bead Analysis (BDBiosciences) and evaluated by flow cytometry.

37.2.6 Statistical Analysis

Unpaired Student *t* tests were used to determine the probability that differences in test results would occur by chance alone. Log Rank survival curves were used to determine statistical significance in virulence and immunization challenge experiments.

37.3 Results

37.3.1 *yadC* Operon and Predicted Properties of YadC

The adhesin YadA of the enteropathogenic yersiniae is encoded by the same Lcr plasmid that encodes LcrV. YadA contributes to colonization through binding to extracellular matrix components such as fibronectin, collagen, and laminin that in turn are bound to β1 integrins (Tahir and Skurnik 2001; Hudson et al. 2005). YadA belongs to a family of oligomeric coiled-coil (Oca) adhesins that are a subgroup of autotransporters composed of trimers with short outer membrane-anchor domains. The trimeric character may provide increased avidity for tight adherence (Hoiczyk et al. 2000; Cotter et al. 2005). Each YadA monomer is composed of an N-terminal head that contains a collagen binding motif, a neck that serves as a platform for the head and contributes to stability of the complex, a stalk composed of seven 15-mer units, a coiled-coil linker of two to three 7-mer units that is necessary for transport of the head, neck, and stalk through the outer membrane, and a membrane anchor (Hoiczyk et al. 2000; Tahir and Skurnik 2001; Roggenkamp et al. 2003; Nummelin et al. 2004).

In *Y. pestis*, the *yadA* gene contains a frameshift mutation that prevents the expression of a functional YadA protein. To identify surface proteins of potential pathogenic importance for *Y. pestis*, we examined the *Y. pestis* CO92 genome for genes encoding proteins similar to YadA and identified the 1869 bp ORF YPO1388 that was subsequently named *yadC*. *yadC* is likely part of a bicistronic operon with the upstream 1095 bp ORF encoding another protein with homology to YadA and called *yadB* (YPO1387). *yadBC* is fairly isolated on the chromosome with the nearest upstream gene 300 base pairs away and the nearest downstream gene 366 base pairs away.

The *yadB* and *yadC* genes in the chromosome of *Y. pestis* CO92 also encode members of the Oca family (denoted as 653a/b in Hoiczyk et al. 2000)) and share the structural arrangement seen in YadA, although the amino-acid sequence similarity is

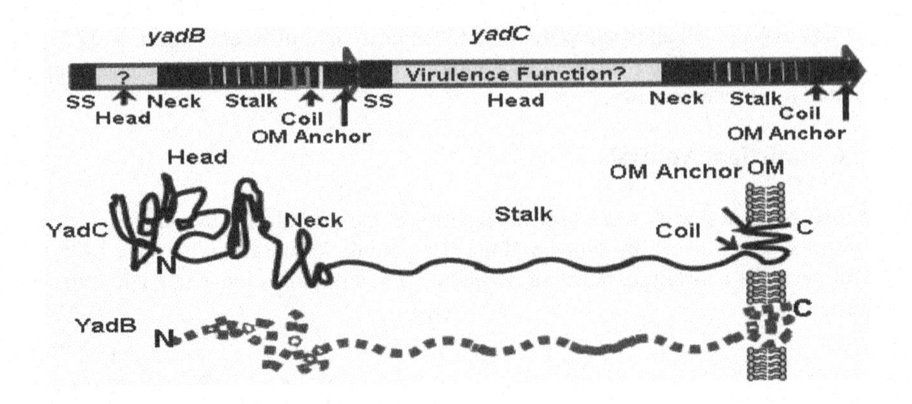

Fig. 1. Predicted YadB and YadC structure.

not strong (Fig. 1). There is essentially no similarity between YadA and the two predicted *Y. pestis* proteins in the head regions. YadC is 25% identical and 45% similar in 285 amino acids in the C-terminus of YadA. *yadC* encodes a putative 622 amino acid preprotein while yadB encodes a putative 364 residue preprotein. The predicted YadC protein (61.6kDa) has a much larger head domain than does YadB (35 kDa), which carries only a short extension N-terminal to the neck domain. The isoelectric point of YadC is 4.0 while YadB's PI is 4.5. Both YadB and YadC have signal sequences as found by SignalP 3.0 (Bendtsen et al. 2004). These observations, considered with the *yadBC* bicistronic genetic arrangement, suggest that YadB and YadC may function together.

37.3.2 Importance of YadC for Virulence in Bubonic Plague

We have found that some human plague convalescent sera recognize YadC (unpublished data), and a *yadBC* mutant *Y. pestis* has been found to show decreased adherence to epithelioid cells (Forman et al. 2006), suggesting that YadB and YadC might have a virulence role in plague. Further, the *yadBC* mutant was attenuated from the subcutaneous route (mimicking bubonic plague) as compared to the virulent *Y. pestis* parent strain. This finding shows that *yadBC* encodes a new virulence determinant for bubonic plague. This attenuation was not demonstrated when mice were challenged intranasally (mimicking pneumonic plague), showing that YadB and YadC are dispensable for the lung infection (Forman et al. 2006).

37.3.3 Protection Tests

One of our lab's interests is to improve the subunit plague vaccine by identifying components that will provide protection against pneumonic plague due to an $F1^-$ *Y. pestis* strain or against bubonic and systemic plague due to an $F1^+$ strain. Since

YadC was determined to be a virulence factor for bubonic plague and is detectable by antibody, it likely is displayed on the bacterial surface where it would be accessible to antibody. Therefore, we tested its role as a protective antigen. We made two tests for protective efficacy, using different challenge strains and infection routes. Mice were immunized and challenged either intravenously or intranasally according to the protocol outlined in 2.3.

Negative-control mice received either Alhydrogel only or GST-alhydrogel immunization. HT-LcrV-immunized mice were used as the positive control. From an intravenous route, a test with Swiss-Webster mice challenged with 5 LD_{50} doses (10,000 CFU) of *Y. pestis* CO99-3015.S7 revealed no protection for those immunized with buffer alone, full protection for HT-LcrV-immunized mice, and partial

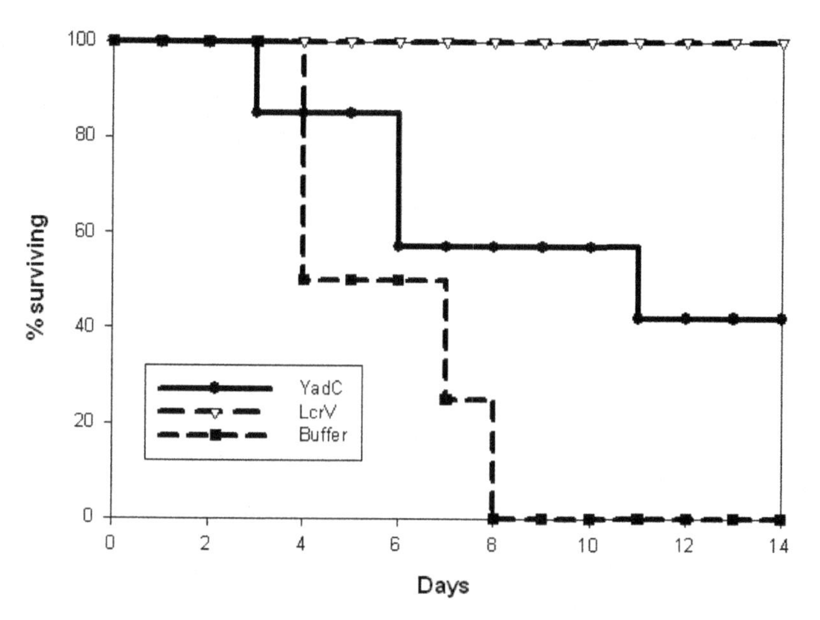

Fig. 2. Intravenous challenge after immunization. Immunized Swiss-Webster mice were challenged with 5 LD_{50} doses (10,000 CFU) of Y. pestis CO99-3015.S7. There was no protection for those immunized with buffer alone, full protection for HT-LcrV-immunized mice, and partial protection for GST-$YadC_{137\text{-}409}$ -immunized mice (with 40% of mice surviving).

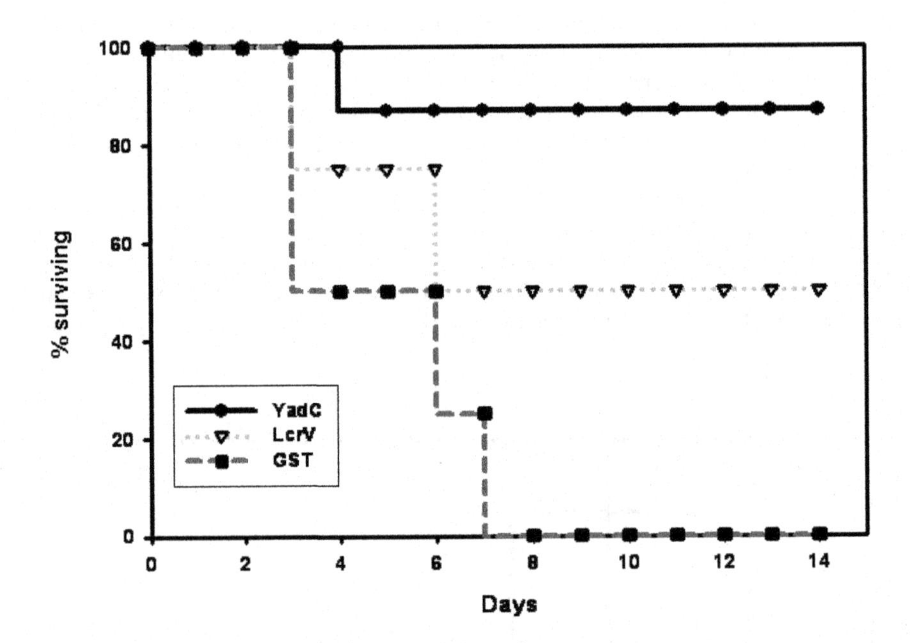

Fig. 3. Intranasal challenge after immunization. 87% of GST-YadC$_{137-409}$-immunized C57BL/6 mice challenged intranasally with 6 LD_{50} doses (3,000 CFU) of *Y. pestis* CO99-3015.S9 survived pneumonic plague. This is compared to the GST-immunized control group (0 surviving mice) and the HT-LcrV-immunized group where 50% survived.

protection for GST-YadC$_{137-409}$ -immunized mice (with 40% of mice surviving) (Fig. 2). After intranasal challenge, 87% of GST-YadC$_{137-409}$ immunized C57BL/6 mice given 6 LD_{50} doses (3,000 CFU) of *Y. pestis* CO99-3015.S9 survived pneumonic plague. This is compared to the GST control group (0 surviving mice) and the HT-LcrV-immunized group where 50% survived the challenge (Fig. 3). There was not a significant difference in survival between mice immunized with GST-YadC$_{137-409}$ or with HT-LcrV in either experiment; however, there was a significant difference between results for both proteins when compared to the GST- or Alhydrogel-only negative control groups. Accordingly, YadC appears to be partially protective against F1$^-$ *Y. pestis* from both intravenous and intranasal routes.

This test also supports the prediction that YadC is accessible to antibody *in vivo* and is a true surface-localized protein. The results are even more promising since the GST-YadC$_{137-409}$ we used was not the full part of the molecule that may be surface-exposed: the complete YadC may contain additional protective epitopes. If, in future experiments, YadB and YadC are found to function as a heteromeric unit at the bacterial surface, the YadBC complex may provide better protection.

37.3.4 IgG Isotypes

Protein-specific IgG1, IgG2A, IgG2B, and IgG3 antibodies were measured by ELISA in the sera of the mice immunized with GST-Alhydrogel, GST-YadC$_{137\text{-}409}$ or HT-LcrV. As has been reported, protection with the F1-LcrV antigen is associated with predominantly a Th2 IgG1 response (Williamson et al. 1999). Likewise, in our experiment, mice immunized with HT-LcrV demonstrated a similar IgG isotype distribution (IgG1>IgG2a~IgG2b~IgG3, $p<0.005$ when IgG1 compared to other isotypes). Mice immunized with YadC demonstrated a more balanced Th1/Th2 response (IgG1=IgG3>IgG2a>IgG2b, $p<0.005$ when IgG1and IgG3 compared to other isotypes). The total IgG titers to GST-YadC$_{137\text{-}409}$ (1:5.6 x 10^6) were significantly higher than those for HT-LcrV (1:1 x 10^6) and total IgG titers to the GST component in the negative control mice (1:6.4 x 10^5).

37.3.5 Anti-inflammatory/Pro-inflammatory Cytokines

In response to stimulation with either GST-YadC$_{137\text{-}409}$, HT-LcrV, or GST, PBMs produced little or no IL-10. PBMs stimulated with YadC showed a significant TNF-α and IL-6 production that was much higher than that due to LPS or HT-LcrV (Fig. 4). IL-12 production was not detectable in response to either GST-YadC$_{137\text{-}409}$ or HT-LcrV (not shown) but a small amount of IL-1 was produced in response to GST-YadC$_{137\text{-}409}$ but not to HT-LcrV. Overall, YadC provided more of a pro-inflammatory

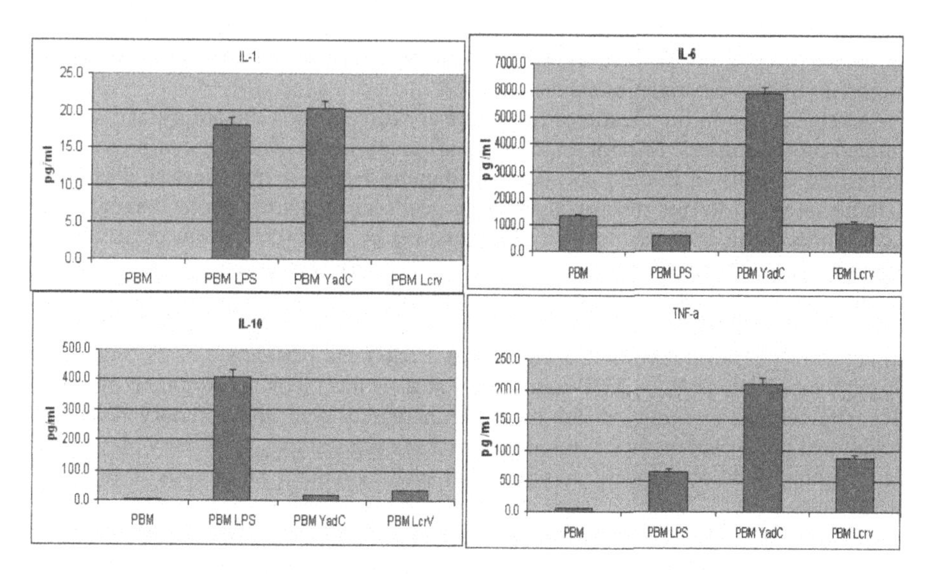

Fig. 4. Peripheral macrophage (PBM) cytokine response. GST-YadC$_{137\text{-}409}$ leads to increased IL-1, IL-6 and TNF-α production when compared to HT-LcrV.

response than did LcrV (i.e., increased IL-1and IL-6). This finding suggests that a more balanced Th1/Th2 plague vaccine is possible. We have done one experiment using human alveolar macrophages (AM) from bronchoalveolar lavage fluid. This test revealed a different cytokine profile with significantly more IL-10 produced in response to HT-LcrV than to GST-YadC$_{137\text{-}409}$ and significant IL-12 production after stimulation with GST-YadC$_{137\text{-}409}$ but not with HT-LcrV or GST alone (data not shown). Again, the overall picture was that GST-YadC$_{137\text{-}409}$ was more pro-inflammatory and HT-LcrV was anti-inflammatory. These different cytokine profiles suggest that the antigens presented during bubonic and pneumonic plague have a very different immune effect. These data, combined with the previous challenge experiments, provide evidence that surface proteins like YadC would be beneficial in inducing a more robust adaptive immune response in future plague vaccines. Further work with PBMs and AMs will help to delineate the immune responses between bubonic and pneumonic plague.

37.4 Discussion

It is clear from previous studies (Cowan et al. 2000) that *Y. pestis* has multiple adhesins and invasins to substitute for the absence of proteins such as YadA. We hypothesize that these and other *Y. pestis* surface proteins could alter the current vaccine to elicit a more balanced Th1/Th2 phenotypic profile. Our hope in developing a new vaccine is to find antigens that induce production of antibody that blocks a crucial surface function such as epithelial adhesion and invasion. Ideally, genes for such crucial proteins could not be deleted from a potential bioterroristic weapon without loss of virulence, unlike F1. This study examined *Y. pestis* YadC which, as an acidic protein, could have a significant impact on the bacterial surface and be important in host-pathogen interactions.

A critical step in the clearance of pulmonary infections is the ability of the lung antigen-presenting cell to migrate to regional lymph nodes where it can present the antigen to T cells to initiate an adaptive immune response. Previous studies have demonstrated a role for macrophage-lineage cells in mediating protection by anti-LcrV antibody in at least the liver (Philipovskiy et al. 2005; Cowan et al. 2005). Recently it has been shown that when Lcr$^-$ *Y. pestis* was given to mice intratracheally, the bacteria were taken up by a characteristic population of DCs (CD11c$^+$ DEC205$^+$CD11b$^-$) (Bosio et al. 2005). Data presented here suggest that PBMs when stimulated by a more favorable antigen (i.e., YadC) can participate in the induction of cytokines such as IL-1, TNF-α, and IL-6 that would favor development of adaptive Th1 cellular immunity in the lung and increased local inflammatory response. Additional pro-inflammatory cytokines from AMs suggest that YadC may play a role in recruiting additional innate immune cells such as natural killer cells to the lung environment.

Finally, as YadC has shown to be required for virulence from a subcutaneous infection and is partially protective against pneumonic and systemic plague, we conclude that YadC is a surface protein and deserves to be studied further for potential use in a new generation plague vaccine.

37.5 Acknowledgments

This study was funded in part from an NIH K30 program seed grant 5K30HL004163-05 (PI Steven Shedlofsky) to BSM, sabbatical support to BSM from NIAID-funded 1 UF4 AI057175 (Region IV Center of Excellence for Biodefense and Emerging Infectious Diseases; Bart Haynes PI), from Project 4.1 of UF4 AI057175 SCS PI, and funds from the University of Kentucky. The investigators also thank Stanislav Forman for providing antibodies, pGST-$YadC_{137-409}$, and background information from his manuscript in preparation. Additionally, we would like to thank Heather Hoy, Clarissa Cowan, Tanya Myers-Morales, and Annette Uittenbogaard for their expertise in assisting with these projects. Spencer Leigh (present address USDA ARS Poultry Science Research Unit, Mississippi State, MS) created the $F1^-$ *Y. pestis* CO99-3015.S2.

37.6 References

Anderson, G.W., Jr., Worsham, P.L., Bolt, C.R., Andrews, G.P., Welkos, S.L., Friedlander, A.M. and Burans, J.P. (1997). Protection of mice from fatal bubonic and pneumonic plague by passive immunization with monoclonal antibodies against the F1 protein of *Yersinia pestis*. Am. J. Trop. Med. Hyg. 56, 571-573.

Andrews, G., Heath, D.G., Anderson, Jr., G.W., Welkos, S.L. and Friedlander, A.M. (1996). Fraction 1 capsular antigen (F1) purification from *Yersinia pestis* CO92 and from an *Escherichia coli* recombinant strain and efficacy against lethal plague challenge. Infec. Immun. 64, 2180-2187.

Bendtsen, J.D., Nielsen, H., von Heijne, G. and Brunak, S. (2004). Improved detection of signal peptides: SignalP 3.0. J. Mol. Biol. 340: 783-795.

Bosio, C., Goodyear, A.W. and Dow, S.W. (2005). Early interaction of *Yersinia pestis* with APCs in the lung. J. Immunol. 175, 6750-6756.

British and US vaccine research groups. (2002). Abstracts O-35 and O-40, 8th International Symposium on Yersinia September 2002. Turky, Finland.

Cotter, S., NSurana, N.K. and St. Geme III, J.W. (2005). Trimeric autotransporters: a distinct subfamily of autotransporter proteins. Trends Microbiol. 13, 199-205.

Cowan C., Philipovskiy, A.V., Wulff-Strobel, C.R., Ye, Z. and Straley, S.C. (2005). Anti-LcrV antibody inhibits delivery of Yops by *Yersinia pestis* KIM5 by directly promoting phagocytosis. Infect. Immun. 73, 6127-6137.

Cowan, C., Jones, H.A., Kaya, Y.H., Perry, R.D. and Straley, S.C. (2000). Invasion of epithelial cells by *Yersinia pestis*: evidence for a Y. pestis-specific invasin. Infect. Immun. 68, 4523-4530.

Datsenko, K.A. and Wanner, B.L. (2000). One-step inactivation of chromosomal genes in *Escherichia coli* K-12 using PCR products. PNAS USA 97, 6640-6645.

Davis, K, Fritz, D.L., Pitt, M.L.M., Welkos, S.L., Worsham, P.L. and Friedlander, A.M. (1996). Pathology of experimental pneumonic plague produced by fraction 1-positive and fraction 1-negative Yersinia pestis in African green monkeys (Cercopithecus aethiops). Arch. Pathol. Lab. Med. 120, 156-163.

DeBord, K., Anderson, D.M., Marketon, M.M., Overheim, K.A., DePaolo, R.W., Ciletti, N.A., Jabri, B. and Schneewind, O. (2006). Immunogenicity and protective immunity against bubonic plague and pneumonic plague by immunization of mice with the recombinant V10 antigen, a variant of LcrV. Infect. Immun. 74, 4910-4914.

Eyles, J., Spiers, I.D., Williamson, E.D. and Alpar, H.O. (1998). Analysis of local and systemic immunological responses after intra-tracheal, intra-nasal and intra-muscular administration of microsphere co-encapsulated *Yersinia pestis* sub-unit vaccines. Vaccine 16, 2000-2009.

Eyles, J., Elvin, S.J., Westwood, A., LeButt, C.S., Alpar, H.O., Somavarapu, S. and Williamson, E.D. (2004). Immunisation against plague by transcutaneous and intradermal application of subunit antigens. Vaccine 22, 4365-4373.

Fields, K.A. and Straley, S.C. (1999). LcrV of *Yersinia pestis* enters infected eukaryotic cells by a virulence plasmid-independent mechanism. Infect. Immun. 67, 4801-4813.

Forman, S., Wulff, C.R., Perry, R.D. and Straley, S.C. (2006). Manuscript in preparation.

Friedlander, A., Welkos, S.L., Worsham, P.L., Andrews, G.P., Heath, D.G., Anderson, Jr., G.W., Pitt, L.M., Estep, J. and Davis, K. (1995). The relationship between virulence and immunity as revealed in recent studies of the F1 capsule of Yersinia pestis. Clin. Infect. Dis. 21, S178-S181.

Glynn, A., Roy, C.J., Powell, B.S., Adamovicz, J.J., Freytag, L.C. and Clements, J.D. (2005). Protection against aerosolized *Yersinia pestis* challenge following homologous and heterologous prime-boost with recombinant plague antigens. Infect. Immun. 73, 5256-5261.

Glynn, A., Freytag, L.C. and Clements, J.D. (2005). Effect of homologous and heterologous prime-boost on the immune response to recombinant plague antigens. Vaccine 23, 1957-1965.

Gruchalla, R.S. and Jones, J. (2003). Combating high-priority biological agents: what to do with drug-allergic patients and those for whom vaccination is contraindicated? J. Allergy Clin. Immunol. 112, 675-682.

Heath, D., Anderson, Jr., G.W., Mauro, J.M., Welkos, S.L., Andrews, G.P., Adamovicz, J. and Friedlander, A.M. (1998). Protection against experimental bubonic and pneumonic plague by a recombinant capsular F1-V antigen fusion protein vaccine. Vaccine 16, 1131-1137.

Hoiczyk, E., Roggenkamp, A., Reichenbecher, M., Lupas, A. and Heeseman, J. (2000). Structure and sequence analysis of *Yersinia* YadA and Moraxella UspAs reveal a novel class of adhesins. EBMO J. 19, 5989-5999.

Hudson, K., Bliska, J.B. and Bouton, A.H. (2005). Distinct mechanisms of integrin bindings by *Yersinia pseudotuberculosis* adhesins determine the phagocytic response of host macrophages. Cell. Microbiol. 7, 1471-1489.

Inglesby T.V., Dennis, D.T., Henderson, D.A., Bartlett, J.G., Ascher, M.S., Eitzen, E., Fine, A.D., Friedlander, A.M., Hauer, J., Koerner, J.F., Layton, M., McDade, J., Osterholm, M.T., O'Toole, T., Parker, G., Perl, T.M., Russell, P.K., Schoch-Spana, M. and Tonat, K. (2000). Plague as a biological weapon: Medical and public health management. Working Group on Civilian Biodefense. J. Am. Med. Assoc. 283, 2281-2290.

Lahteenmaki, K., Kukonen, M. and Korhonen, T.K. (2001). The Pla surface protease/adhesin of Yersinia pestis mediates bacterial invasion into human endothelial cells. FEBS Lett. 504, 69-72.

Leary, S., Griffin, K.F., Garmory, H.S., Williamson, E.D. and Titball, R.W. (1997). Expression of an F1/V fusion protein in attenuated *Salmonella typhimurium* and protection of mice against plague. Microb. Pathog. 23, 167-179.

Marshall J.D., Jr., Bartelloni, P.J., Cavanaugh, D.C., Kadull, P.J. and Meyer, K.F. (1974). Plague immunization. II. relation of adverse clinical reactions to multiple immunizations with killed vaccine. J. Infect Dis. 129(Suppl), S19-S25.

Motin, V., Nakajima, R., Smirnov, G.B. and Brubaker, R.R. (1994). Passive immunity to yersiniae mediated by anti-recombinant V antigen and protein A-V antigen fusion peptide. Infect. Immun. 62, 4192-4201.

Nakajima, R., Motin, V.L. and Brubaker, R.R. (1995). Suppression of cytokines in mice by protein A-V antigen fusion peptide and restoration of synthesis of active immunization. Infect. Immun. 63, 3021-3029.

Nakajima, R. and Burbaker, R.R. (1993). Association between virulence of *Yersinia pestis* and suppression of gamma interferon and tumor necrosis factor alpha. Infect. Immun. 61, 23-31.

Nedialkov, Y., Motin, V.L. and Brubaker, R.R. (1997). Resistance to lipopolysaccharide mediated by the *Yersinia pestis* V antigen-polyhisitidine fusion peptide: amplification of interluekin-10. Infect. Immun. 65, 1196-1203.

Nummelin H., Merckel, M.C., Leo, J.C., Lankinen, H., Skurnik, M. and Goldman, A. (2004). The Yersinia adhesin YadA collagen-binding domain structure is a novel left-handed parallel beta-roll. EBMO J 23, 701-711.

Perry, R. and Fetherston, J.D. (1997). *Yersinia pestis*--etiologic agent of plague. Clin. Microbiol. Rev. 10, 35-66.

Philipovskiy, A.V., Cowan, C., Wulff-Strobel, C.R., Burnett, S.H., Kerschen, E.J., Cohen, D.A., Kaplan, A.M. and Straley, S.C. (2005). Antibody against V antigen prevents Yop-dependent growth of *Yersinia pestis*. Infect. Immun. 83, 1532-1542.

Roggenkamp A., Ackerman, N.A., Joacobi, C.A., Truelzsch, K., Hoffman, H. and Heeseman, J. (2003). Molecular analysis of transport and oligomerization of the*Yersinia enterocolitica* adhensin YadA. J. Bacteriol. 185, 3735-3744.

Russell, P., Eley, S.M., Hibbs, S.E., Manchee, R.J., Stagg, A.J. and Titball, R.W. (1995). A comparison of plague vaccine, USP and EV76 vaccine induced protection against *Yersinia pestis* in a murine model. Vaccine 13, 1551-1556.

Simpson, W., Thomas, R.E. and Schwan, T.G. (1990). Recombinant capsular antigen (fraction 1) from *Yersinia pestis* induces a protective antibody response in BALB/c mice. Am. J. Trop. Med. Hyg. 43, 389-396.

Sing, A., Roggenkamp, A., Geiger, A.M. and Heesemann, J. (2002). *Yersinia enterocolitica* evasion of the host immune response by V antigen-induced IL-10 production of macrophages is abrogated in IL-10-deficient mice. J. Immunol. 168, 1315-1321.

Sing, A., Tvardovaskaia, N., Rost, D., Kirschning, C., Wagner, H. and Heesemann, J. (2003). Contribution of Toll-like receptors 2 and 4 in an oral Yersinia enterocolitica mouse infection model. Int. J. Med. Microbiol. 293, 341-348.

Sing, A., Tvardovskaia, N., Rost, D., Kirschning, C.J., Wagner, H. and Heeseman, J. (2003). Yersinia V-antigen exploits toll-like receptor 2 and CD14 for interleukin 10-mediated immunosuppression. J. Exp. Med. 197, 1017-1024.

Straley, S.C. and Bowmer, W.S. (1986). Virulence genes regulated at the transcriptional level by Ca^{2-} in *Yersinia pestis* include structural genes for outer membrane proteins. Infect. Immun. 51, 445-454.

Surgalla, M. and Beesley, E.D. (1969). Congo red-agar plating medium for detecting pigmentation in *Pasturella pestis*. Appl. Microbiol. 18, 834-837.

Tahir E.Y. and Skurnik, M. (2001). YadA, the multifaceted *Yersinia* adhesin. Int. J. Med. Microbiol. 291, 209-218.

Welkos, S., Davis, K.M., Pitt, L.M., Worsham, P.L. and Friedlander, A.M. (1995). Studies on the contribution of the F1 capsule-associated plasmid pFra to the virulence of *Yersinia pestis*. Contrib. Microbiol. Immunol. 13, 299-305.

Williamson, E., Flick-Smith, H.C., LeButt, C., Rowland, C.A., Jones, S.M., Waters, E.L., Gwyther, R.J., Miller, J., Packer, P.J. and Irving, M. (2005). Human immune response to a plague vaccine comprising recombinant F1 and V antigens. Infect. Immun. 2005, 3598-3608.

Williamson, E., Vessey, P.M., Gillhespy, K.J., Eley, S.M., Green, M. and Titball, R.W. (1999). An IgG1 titre to the F1 and V antigens correlates with protection against plague in the mouse model. Clin. Exp. Immunol. 116, 107-114.

Williamson, E., Eley, S.M., Stagg, A.J., Green, M., Russell, P. and Titball, R.W. (1997). A sub-unit vaccine elicits IgG in serum, spleen cell cultures and bronchial washings and protects immunized animals against pneumonic plague. Vaccine 15, 1079-1084.

Wulff-Strobel, C. R., Williams, A.W. and Straley, S.C. 2001. LcrQ and SycH function together at the Ysc type III secretion system in *Yersinia pestis* to impose a hierarchy of secretion. Mol. Microbiol. 43, 411-423.

Picture 34. Susan Straley introduces a speaker. Photograph by A. Anisimov.

38
Protective Immunity Against Plague

Claire Cornelius, Lauriane Quenee, Deborah Anderson, and Olaf Schneewind

Department of Microbiology, University of Chicago, oschnee@bsd.uchicago.edu

Abstract. Plague, an infectious disease that reached catastrophic proportions during three pandemics, continues to be a legitimate public health concern worldwide. Although antibiotic therapy for the causative agent *Yersinia pestis* is available, pharmaceutical supply limitations, multi-drug resistance from natural selection as well as malicious bioengineering are a reality. Consequently, plague vaccinology is a priority for biodefense research. Development of a multi-subunit vaccine with Fraction 1 and LcrV as protective antigens seems to be receiving the most attention. However, LcrV has been shown to cause immune suppression and *Y. pestis* mutants lacking F1 expression are thought to be fully virulent in nature and in animal experiments. The LcrV variant, rV10, retains the well documented protective antigenic properties of LcrV but with diminished inhibitory effects on the immune system. More research is required to examine the molecular mechanisms of vaccine protection afforded by surface protein antigens and to decipher the host mechanisms responsible for vaccine success.

38.1 Global Perspective

Yersinia pestis, the causative agent of bubonic and pneumonic plague, infects a wide range of mammalian hosts via flea vectors or aerosol droplets. Throughout history, plague has devastated human populations, notably during the three major cyclical pandemics: Justinian's plague (AD 541 to 767); Black Death (1346 to the early 19th century); and the modern plague (since 1894). Plague continues to be widespread throughout the world, particularly in environments that favor close contact between human and rodent populations. Active plague foci have been identified in the Democratic Republic of the Congo (DRC), Madagascar, Mozambique, Uganda, United Republic of Tanzania, Kazakhstan, Mongolia, Brazil, Bolivia, Ecuador, Peru and the United States. Of these countries, Madagascar and the DRC are considered the most endemic for the disease. The average annual incidence of plague cases in Madagascar figures around 900 with a case-fatality rate of approximately 19 deaths per 100 cases. About 1,000 cases are reported annually in the DRC, primarily in the north-eastern Orientale Province. It is important to note, though, that only a fraction of the reported cases are actually confirmed through laboratory testing due to insufficient resources and civil unrest. Although the majority of the patients present with bubonic plague, both the pneumonic and the septicemic forms of plague do occur. For example, in May 2006, DRC recorded 100 patients that contracted pneumonic plague. In China, at least 19 provinces have been identified as plague foci. Most patients infected with plague in the western and northern provinces of China, where transmission occurs during the slaughtering and skinning of animals such as marmots, develop either septicemic or primary pneumonic plague (WHO 2006).

Various antimicrobials such as streptomycin, tetracycline, chloramphenicol, and sulfonamides have been employed to treat the plague infected patient. However, large scale use of antibiotics in many of the endemic countries is already not feasible for various reasons and so, by extension, reliance on pharmaceutical remediation in the face of a severe pandemic whether by natural or artificial means would be foolhardy. Furthermore, multi-drug resistance is already occurring naturally. In 1997, Galimand et al. described a *Y. pestis* strain-isolated from a teenage boy with bubonic plague in Madagascar-that possessed a multi-drug resistant plasmid which under scrutiny *in vitro* was readily transferred between *Y. pestis* 17/95 and *Escherichia coli* as well as other *Y. pestis* strains (Galimand et al. 1997). Finally, there is additional speculation that genetically engineered strains resistant to a variety of antimicrobials were constructed and stockpiled in the past when biological warfare industries were not regulated.

38.2 Early Research on *Y. pestis* Whole Cell Vaccines

Various plague vaccine formulations have been devised, employed and discarded over the last century. In the first acknowledged attempt, W. M. Haffkine used a killed whole-cell preparation. However, inconsistent dosing suggestions and severe reactions in the vaccinated led to its cessation of use in India and abroad (Haffkine 1897; Meyer 1974). On October 22, 1941, the Subcommittee on Tropical Diseases, National Research Council Committee on Medical Research, passed the following resolution, 'Resolved that, even though the available knowledge does not seem to afford definite evidence of the benefits from the use of plague vaccine, it is considered advisable to vaccinate with killed plague bacilli of an approved strain all military and navy personnel under serious threat of exposure to bubonic plague' (Meyer et al. 1974). The "Army Vaccine", now commercially known as Plague Vaccine, USP (Cutter Laboratory, Inc./Greer Laboratory), was a formaldehyde-killed preparation of the highly virulent strain 195/P *Y. pestis,* administered to military personnel during both World War II and the Vietnam War. Unfortunately, this vaccine is thought to be less effective against the pneumonic form of plague, annual boosters are required and there is a high propensity for both local and systemic reactions (Meyer et al. 1974; Russell et al. 1995).

A live-attenuated plague vaccine was first experimented in humans by R. Strong in 1908 and introduced to the general populace as E.V. by Girard and Robic in 1933 and as Tjiwidej by Otten in 1934 (Meyer et al. 1974). The modified EV76 strain vaccine, though still used in Russia, has not been adopted for worldwide use due to inconsistent immune responses, adverse reactions in vaccinated individuals, laborious booster schedules for full effectiveness and the legitimate concern that it might revert to a fully virulent form. In their study comparing the Plague Vaccine, USP and the EV76 vaccine, Russell et al. (1995) found that both vaccines induced protec tion against a subcutaneous challenge with *Y. pestis* GB strain. However, mice challenged by the intranasal route were only partially protected by the Plague Vaccine USP and fully protected by EV76. Interestingly, side effects were not detected in

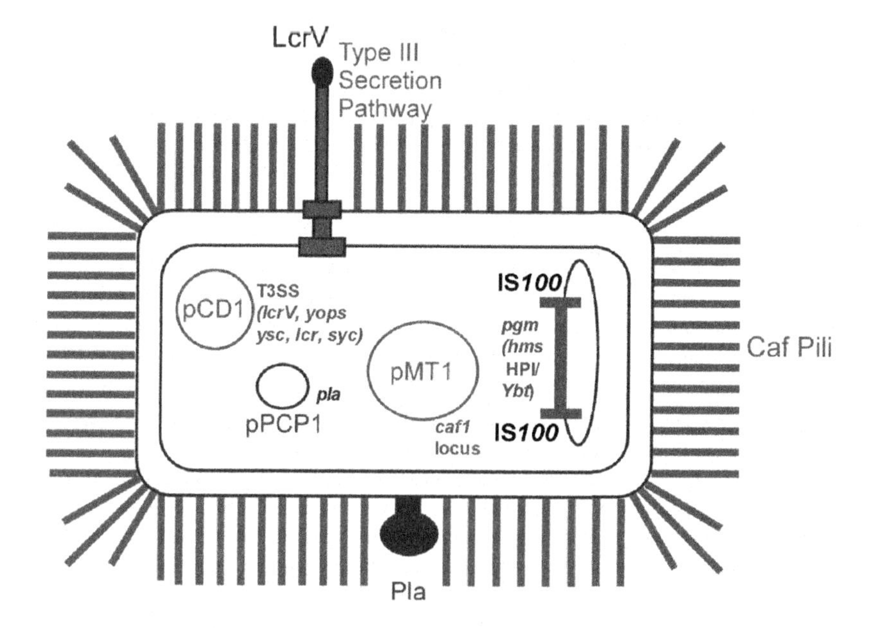

Fig. 1. Plasmid and chromosomal components of the *Y. pestis* genome.

mice immunized with the Plague Vaccine, USP but they were detected-some being quite severe-in EV76 immunized mice.

38.3 Advent of Sub-unit Plague Vaccines

Failure in these vaccine attempts shifted research interests to sub-unit type vaccines. *Y. pestis* requires three virulence plasmids (pCD1, pPCP1, and pMT1/pFra) to cause mammalian disease (Fig. 1). The pCD1 (calcium dependence) plasmid provides for Yersiniae type III secretion machinery such as V/LcrV antigen and secretion substrate genes (*yop*, Yersinia outer proteins, Yop). The pPCP1 (pesticin, coagulase, plasminogen activator) encodes for *Yersinia* Pla, a surface protease that facilitates the spread of bacteria throughout the host. And finally, the PMT1/pFra plasmid encodes for murine toxin (Ymt) and the capsular fraction 1 antigen (F1). Prepara tions with the capsular F1 antigen and/or V antigen (LcrV) have been investigated for nearly half a century. At the present time, defense scientists at CBDE, Porton Down, UK and USAMRIID, USA are engaged in clinical trials with their respective recombinant F1 and LcrV combination sub-unit vaccines. However, successful licensure of either of these vaccines for global use may be limited due to the reported negative immuno-modulatory effects of LcrV and the possibility that anomalous F1 minus strains would not be neutralized by F1 antibody *in vivo* (Nakajima and Brubaker 1993; Winter et al. 1960).

Y. pestis has evolved a very clever strategy to evade the host immune response. In fact, this microorganism appears to manipulate the natural host immune system in order to facilitate its own survival. In 1993, Nakajima and Brubaker found that wild type *Y. pestis* but not LcrV mutants suppressed the expression of TNF-α and IFN-γ *in vivo* using a mouse model. Subsequent experiments demonstrated similar results and also found that LcrV (to include its recombinant purifications) is responsible for the upregulation of interleukin 10 (IL-10) which inhibits NF-κB activation and toll-like receptor (TLR) synthesis thus suppressing the synthesis of such pro-inflammatory cytokines as TNF-α and IFN-γ (Brubaker 2003; Sing et al. 2002). While recombinant LcrV (rLcrV) appears to subdue innate immunity and inflammatory responses, it does incite an appropriate, if not beneficial, adaptive immune response in the mammalian host. In several animal models, immunization with rLcrV or polyclonal antibodies against rLcrV is sufficient to generate protective immunity against infections with highly virulent *Y. pestis* (Une and Brubaker 1984; Titball and Williamson 2001; Motin et al. 1994; Leary et al. 1995).

The next challenge for yersiniologists, then, became defining those regions of rLcrV that are responsible for immuno-modulatory and/or protective immune responses. Overheim et al. designed and expressed 11 rLcrV (rV1-rV11) variants with staggered 30-amino-acid deletions (from the KIM5 coding sequence *lcrv*) in *E. coli* strain BL21 (DE3). Although all of the variants demonstrated a lower IL-10 secretion than rLcrV, the rV7 (lacking 181-210) and rV10 (lacking 271-300) variants displayed a five- or four-fold decrease, respectively. Of the variants, only the rV7 peptide suppressed TNF-α release by LPS-stimulated macrophages. Furthermore, immunization of mice with rV10 protected against lethal plague infections caused by 1,000 mean lethal doses (MLD) of *Y. pestis* KIM5, a Δ(*pgm*) (pigmentation defective), attenuated strain that causes plague infections only when inoculated intravenously (Overheim et al. 2005). DeBord et al. demonstrated that rV10 vaccination can also prevent plague disease in animal challenge studies using bubonic (100,000 MLD or 100,000 cfu) and pneumonic (2,570 MLD or 1,000,000 cfu) models of challenge with the fully virulent isolate *Y. pestis* CO92 (Fig. 2A and 2B, respectively). The pneumonic model was simulated with intranasal application of the microorganism versus utilizing aerosolization. Furthermore, it was shown that the rV10 vaccine stimulated a stronger immune response in mice than rLcrV; rV10 stimulated IgG1, IgG2a or IgG2b isotypes as well as elevated T-cell responses (DeBord et al. 2006).

The molecular mechanism(s) whereby antibodies directed against either rLcrV or rV10 provide protection against plague infection *in vivo* has yet to be conclusively demonstrated. It has been hypothesized that some antibodies to rLcrV may improve the efficiency of host polymorphonuclear cell phagocytosis of *Y. pestis*, whereas others serve to block bacterial type III injection of effector proteins into host cells (Philipovskiy et al. 2005; Weeks et al. 2002). Debord et al. examined the effect of both rV10 and rLcrV on type III injection of effector Yops into immune cells using

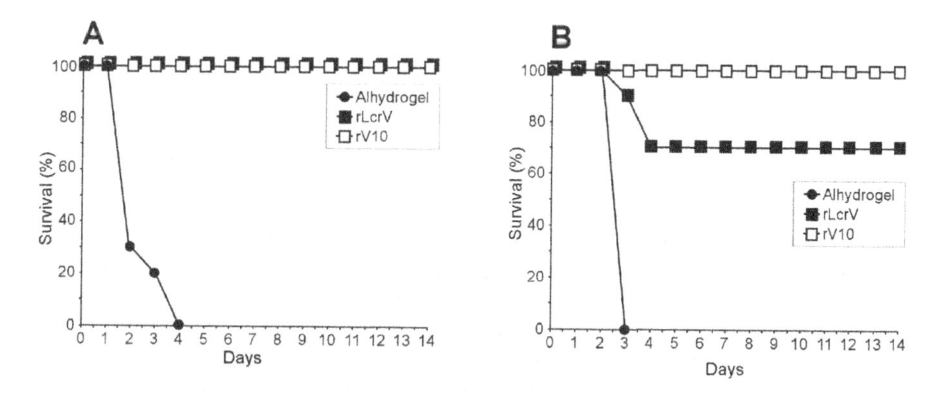

Fig. 2. Protection against bubonic (A) and pneumonic (B) plague.

Y. pestis strain KIM5 (pMM83) as a reporter for Gene BLAzer technology (Invitrogen). This strain of *Y. pestis* secretes YopM-Bla, a hybrid between the YopM effector with a C-terminal fusion to β-lactamase. When YopM-Bla is injected via the typeIII pathway into host cells, β-lactamase cleaves the fluorogenic compound CCF2-AM and emission of blue fluorescence can be measured by fluorescence microscopy or flow cytometry. Mice immunized with rLcrV or rV10 were infected intravenously with 1,000 CFU of *Y. pestis* KIM5 expressing either YopM-Bla or GST-Bla, a control hybrid that cannot travel the type III pathway. Two days post-infection, spleens were harvested from mice. Splenic homogenates were cultured to determine bacterial load or treated with CCF2-AM and analyzed by flow cytometry for injection of YopM-Bla into immune cells. Mice that received adjuvant developed high bacterial titers in the spleen, ranging from 10^6 to 10^7 cfu. Only one out of seven mice immunized with LcrV developed colonization of the spleen. However, clinically, this animal appeared healthy. Spleens from mice immunized with rV10 did not have any detectable bacteria. When analyzed by flow cytometry, mice immunized with adjuvant alone and then infected with *Y. pestis* expressing YopM-Bla displayed significant injection. As a control, adjuvant mice infected with GST-Bla *Yersinia* displayed no type III injection. Only one mouse that had been vaccinated with LcrV showed about 1% injection. On the other hand, rV10 vaccinated mice did not show any injection. These results suggest that rV10 vaccination blocks plague infection and may prevent type III injection into immune cells (Debord et al. 2005).

38.4 Pili and Other Surface Proteins

Taylor and colleagues hypothesize that there are protective antigens (besides F1 and LcrV), shared between *Yersinia pseudotuberculosis* and *Y. pestis*, that are responsible for generating protective immunity (Taylor et al. 2005). The researchers demonstrated that Balb/c mice immunized either intravenously with 10^5 cfu or orally with 10^8 cfu of a pYV-cured derivative of the *dam* mutant of *Y. pseudotuberculosis*, were protected against subcutaneous challenge with 100 MLD of the highly virulent *Y. pestis* GB strain. Protection cannot be attributed to the F1 antigen since it is not produced by *Y. pseudotuberculosis* strains, nor is it attributable to lipopolysaccharide (LPS) due to known LPS dissimilarities between *Y. pestis* and *Y. pseudotuberculosis* (Skurnik et al. 2000; Prior et al. 2001).

Bioinformatic analysis of *Y. pestis* and *Y. pseudotuberculosis* genome sequences using Pfam searches, topology predictions and a newly developed algorithm to detect ß-barrel outer membrane domains, identified 52 genes that fulfilled the criteria of surface or secreted antigens that are conserved in *Y. pestis* KIM5, *Y. pestis* CO92 and *Y. pseudotuberculosis*. In our laboratory, these antigens have been expressed as recombinant proteins in *E. coli*. Purified products have been tested for protective properties in a mouse model of plague disease, using intravenous and subcutaneous lethal challenges of *Y. pestis* strain KIM5 and CO92, respectively. Using LcrV and F1 antigens as positive controls, immunization of mice with these antigens did not generate complete protective immunity against plague (unpublished data).

38.5 Conclusions

Plague vaccine design continues to be a challenge. There is much to learn about the *in vivo* virulence mechanisms of LcrV, F1 as well as other surface antigens. There is also much to learn about host immune responses to *Y. pestis* infection, particularly those that generate protective immunity. At this time, rV10 represents a promising vaccine candidate because of the high levels of protection afforded by rV10 immunization in mice that correlate with the development of specific humoral immune responses. Ideally, efforts to create a plague vaccine should strive for the most far-reaching impact; hence, it should be effective in multiple animal models and in humans with robust or weakened immune systems.

38.6 Acknowledgements

All authors acknowledge membership within and support from the Region V "Great Lakes" Regional Center of Excellence in Biodefense and Emerging Infectious Diseases Consortium (GLRCE; NIH Award 1-U54-AI-057153).

38.7 References

Achtman, M., Zurth, K., Morelli, G., Torrea, G., Guiyoule, A. and Carniel, E. (1999) *Yersinia pestis*, the cause of plague, is a recently emerged clone of *Yersinia pseudotuberculosis*. Proc. Natl. Acad. Sci. USA 96, 14043-14048.

Anderson, G.W., Jr., Leary, S.E.C., Williamson, E.D., Titball, R.C., Welkos, S.C., Worsham, P.L. and Friedlander, A.M. (1996) Recombinant V antigen protects mice against pneumonic and bubonic plague caused by F1-capsule-positive and -negative strains of *Yersinia pestis*. Infect. Immun. 64, 4580-4585.

Andrews, G.P., Heath, D.G., Anderson, G.W, Jr., Welkos, S.L and Friedlander, A.M. (1996) Fraction 1 capsular antigen (F1) purification from *Yersinia pestis* CO92 and from an *Escherichia coli* recombinant strain and efficacy against lethal plague challenge. Infect. Immun. 64, 2180-2187.

Andrews, G.P., Strachan, S.T., Benner, G.E., Sample, A.K., Anderson, G.W. Jr., Adamovicz, J.J., Welkos, S.L., Pullen, J.K. and Friedlander, A.M. (1999) Protective efficacy of recombinant *Yersinia* outer proteins against bubonic plague caused by encapsulated and nonencapsulated *Yersinia pestis*. Infect. Immun. 67, 1533-1537.

Baker, E.E., Somer, H., Foster, L.W., Meyer, E. and Meyer, K.F. (1952) Studies on immunization against plague. I. The isolation and characterization of the soluble antigen of *Pasteurella pestis*. J. Immunol. 68, 131-145.

Benner, G.E., Andrews, G.P., Byrne, W.R., Strachan, S.D., Sample, A.K., Heath, D.G. and Friedlander, A.M. (1999) Immune response to *Yersinia* outer proteins and other *Yersinia pestis* antigens after experimental infection in mice. Infect. Immun. 67, 1922-1928.

Boyd, A.P., Grosdent, N., Totemeyer, S., Geuijen, C., Bleves, S., Iriarte, M., Lambermont, I., Octave, J-N. and Cornelis, G.R. (2000) *Yersinia enterocolitica* can deliver Yop proteins into a wide range of cell types, development of a delivery system for heterologous proteins. Eur. J. Cell Biol. 79, 659-671.

Brubaker, R. R. (2003) Interleukin-10 and the inhibition of innate immunity to yersiniae: roles of Yops and LcrV (V antigen). Infect. Immun. 71, 3673-3681.

Brubaker, R.R. (1969) Mutation rate to nonpigmentation in *Pasteurella pestis*. Journal of bacteriology. 98(3), 1404-1406.

Burrows, T.W. (1957) Virulence of *Pasteurella pestis*. Nature 179, 1246-1247.

Burrows, T. W. (1963) Virulence of *Pasteurella pestis* and immunity to plague. Ergebn. Mikrobiol. 37, 59-113.

Burrows, T.W. and Bacon, G.A. (1956) The basis of virulence in *Pasteurella pestis*, antigen determining virulence. Br. J. Exp. Pathol. 37, 481-493.

Burrows, T.W. and Bacon, G.A. (1958) The effect of loss of different virulence determinants on the virulence and immunogenicity of strains of *Pasteurella pestis*. Br. J. Exp. Pathol. 39, 278-291.

Cornelis, G. R. (1998) The *Yersinia* deadly kiss. J. Bacteriol. 180, 5495-5504.

Cornelis, G.R., Boland, A., Boyd, A.P., Geuijen, C., Iriarte, M., Neyt, C., Sory, M.-P and Stainier, I. (1998) The virulence plasmid of *Yersinia*, an antihost genome. Microbiol. Mol. Biol. Rev. 62, 1315-1352.

Crook, L.D. and Tempest, B. (1992) Plague, a clinical review of 27 cases. Arch. Intern. Med. 152, 1253-1256.

Davis, K.J., Fritz, D.L., Pitt, M.L.M., Welkos, S.L., Worsham, P.L. and Friedlander, A.M. (1996) Pathology of experimental pneumonic plague produced by fraction 1-positive and fraction 1-negative *Yersinia pestis* in African green monkeys (*Cercopitheus aethiops*). Arch. Pathol. Lab. Med. 120,156-163.

DeBord K.L., Anderson, D.M., Marketon, M.M., Overheim, K.A., DePaolo, R.W., Ciletti, N.A, Jabri, B. and Schneewind, O. (2006) Immunogenicity and protective immunity against bubonic plague and pneumonic plague by immunization of mice with the recombinant V10 antigen, a variant of LcrV. Infect Immun. 74, 4910-4914.

Deng, W., Burland, V., Plunkett, G.R., Boutin, A., Mayhew, G.F., Liss, P., Perna, N.T., Rose, D.J., Mau, B., Zhou, S., Schwartz, D.C., Fetherston, J.D., Lindler, L.E., Brubaker, R.R., Plano, G.V., Straley, S.C., McDonough, K.A., Nilles, M.L., Matson, J.S., Blattner, F.R. and Perry, R.D. (2002) Genome sequence of *Yersinia pestis* KIM. J. Bacteriol. 184, 4601-4611.

Eyles, J.E., Williamson, E.D., Spiers, I.D. and Alpar, H.O. (2000) Protection studies following bronchopulmonary and intramuscular immunization with yersinia pestis F1 and V subunit

vaccines coencapsulated in biodegradable microspheres, a comparison of efficacy. Vaccine 18, 3266-3271.
Ferber, D.M. and Brubaker, R.R. (1981) Plasmids in *Yersinia pestis*. Infect. Immun. 31, 839-841.
Friedlander, A.M., Welkos, S.L., Worsham, P.L., Andrews, G.P., Heath, D.G., Anderson, G.W., Pitt, M.L.M., Estep, J. and Davis, K. Relationship between virulence and immunity as revealed in recent studies of the F1 capsule of *Yersinia pestis*. Clin. Infect. Dis. 21 (Suppl. 2), S178-181.
Galimand, M., Carniel, E. and Courvalin, P. (2006) Resistance of *Yersinia pestis* to antimicrobial agents. Antimicrob Agents Chemother. 50, 3233-3236
Galimand, M., Guiyoule, A., Gerbaud, G., Rasoamanana, B., Chanteau, S., Carniel, E. and Courvalin, P. (1997) Multidrug resistance in *Yersinia pestis* mediated by a transferable plasmid. N Engl J Med. 337:677-680.
Haffkine, W.M. (1897) Remarks on the plague prophylactic fluid. Br. Med. J. 1, 1461.
Heath, D.G., Anderson, G.W., Jr., Mauro, J.M., Welkos, S.L., Andrews, G.P., Adamovicz, J.J. and Friedlander, A.M. (1998) Protection against experimental bubonic and pneumonic plague by a recombinant capsular F1-V antigen fusion protein vaccine. Vaccine. 16, 1131-1137.
Hu, P., Elliott, J., McCready, P., Skowronski, E., Garnes, J., Kobayashi, A., Brubaker, R. and Garcia, E. (1998) Structural organization of virulence-associated plasmids of *Yersinia pestis*. J. Bacteriol. 180, 5192-5202.
Inglesby, T.V., Dennis, D.T., Henderson, D.A., Bartlett, J.G., Ascher, M.S., Eitzen, E., Fine, A.D., Friedlander, A.M., Hauer, J., Koerner, J.F., Layton, M., McDade, J., Osterholm, M.T., O'Toole, T., Parker, G., Perl, T.M., Russell, P.K., Schoch-Spana, M. and Tonat, K. (2000) Plague as a biological weapon, medical and public health management. JAMA 283, 2281-2290.
Jones, S.M., Day, F., Stagg, A.J. and Williamson, E.D. (2001) Protection conferred by a fully recombinant sub-unit vaccine against *Yersinia pestis* in male and female mice of four inbred strains. Vaccine. 19, 358-366.
Lawto, W.D., Erdman, R.L. and Surgalla, M.J. (1963) Biosynthesis and purification of V and W antigen in *Pasteurella pestis*. J. Immunol. 91, 179-184.
Marketon, M.M., DePaolo, R.W., DeBord, K.L., Jabri, B. and Schneewind, O. (2005) Plague bacteria target immune cells during infection. Science. 309, 1739-1741.
Meyer, K. (1961) Pneumonic plague. Bacteriol. Rev. 25, 249-261.
Meyer, K., Cavanaugh, D.C., Bartelloni, P.J. and Marshall, J.D. Jr. (1974) Plague Immunization I. Past and Present Trends. J. Infect. Dis. 129, S13-S18.
Meyer, K.F., Hightower, J.A. and McCrumb, F.R. (1974) Plague immunization. VI. Vaccination with the fraction 1 antigen of *Yersinia pestis*. J. Infect. Dis. 129, S41-S45.
Michiels, T., Wattiau, P., Brasseur, R., Ruysschaert, J.-M and Cornelis, G. (1990) Secretion of Yop proteins by yersiniae. Infect. Immun. 58, 2840-2849.
Motin, V.L., Nedialkov, Y.A. and Brubaker, R.R. (1994) Passive immunity to yersiniae mediated by anti-recombinant V antigen and protein A-V antigen fusion peptide. Infect. Immun. 62, 3021-3029.
Nakajima, R., Motin, V.L. and Brubaker, R.R. (1993) Association between virulence of *Yersinia pestis* and suppression of gamma interferon and tumor necrosis factor alpha. Infect. Immun. 63, 3021-3029.
Nakajima, R., Motin, V.L. and Brubaker, R.R. (1995) Suppression of cytokines in mice by protein A-V antigen fusion peptide and restoration of synthesis by active immunization. Infect. Immun. 63, 3021-3029.

Nedialkov, Y.A., Motin, V.L. and Brubaker, R.R. (1997) Resistance to lipopolysaccharide mediated by the *Yersinia pestis* V antigen-polyhistidine fusion peptide, amplification of interleukin-10. Infect. Immun. 65, 1196-1203.

Overheim, K.A., Depaolo, R.W., Debord, K.L., Morrin, E.M., Anderson, D.M., Green, N.M., Brubaker, R.R., Jabri, B. and Schneewind, O. (2005) LcrV plague vaccine with altered immunomodulatory properties. Infect. Immun. 73, 5152-5159.

Parkhill, J., Wren, B.W., Thompson, N.R., Titball, R.W., Holden, M.T., Prentice, M.B., Sebaihia, M., James, K.D., Churcher, C., Mungall, K.L., Baker, S., Dasham, D., Bentley, S.D., Brokks, K., Cerdeno-Tarraga, A.M., Chillingworth, T., Cronin, A., Davies, R.M., Davis, P., Dougan, G., Feltwell, T., Hamlin, N., Holroyd, S., Jagels, K., Karlyshev, A.V., Leather, S., Moule, S., Oyston, P.C., Quail, M., Rutherford, K., Simmonds, M., Skelton, J., Stevens, K., Whitehead, S. and Barrell, B.G. (2001) Genome sequence of *Yersinia pestis*, the causative agent of plague. Nature 413, 523-527.

Perry, R.D. and Fetherston, J.D. (1997) *Yersinia pestis* - etiologic agent of plague. Clin. Microbiol. Rev. 10, 35-66.

Philipovskiy, A.V., Cowan, C.R., Wulff-Strobel, S.H., Burnett, S.H., Kerschen, E.J., Cohen, D.A., Kaplan, A.M. and Straley, S.C. (2005) Antibody against V antigen prevents Yop-dependent growth of *Yersinia pestis*. Infect. Immun. 73, 1532-1542.

Prior J.L., Hitchen P.G., Williamson D.E., Reason A.J., Morris H.R., Dell A., Wren B.W. and Titball R.W. (2001) Characterization of the lipopolysaccharide of Yersinia pestis. Microb. Pathog. 30, 49-57.

Russell, P., Eley, S.M., Hibbs, S.E., Manchee, R.J., Stagg, A.J. and Titball, R.W. (1995) A comparison of Plague Vaccine, USP and EV76 vaccine induced protection against *Yersinia pestis* in a murine model, Vaccine. 13, 1551-1556.

Sing, A., Roggenkamp, A., Geiger, A.M. and Heesemann, J. (2002) *Yersinia enterocolitica* evasion of the host immune response by V antigen-induced IL-10 production of macrophages is abrogated in IL-10-deficient mice. J. Immunol. 168, 1315-1321.

Sing, A., Rost, D., Tvardovaskaia, N., Roggenkamp, A., Wiedemann, A., Kirschning, C., Aepfelbacher, J.M. and Heesemann, J. (2002) *Yersinia* V-antigen exploits Toll-like receptor 2 and CD14 for interleukin 10-mediated immunosuppression. J. Exp. Med. 196, 1017-1024.

Skurnik, M., Peippo, A. and Ervela, E. (2000) Characterization of the O-antigen gene clusters of Yersinia pseudotuberculosis and the cryptic O-antigen gene cluster of Yersinia pestis shows that the plague bacillus is most closely related to and has evolved from Y. pseudotuberculosis serotype O:1b. Mol. Microbiol. 37, 316-330.

Taylor, V.L., Titball, R.W. and Oyston, P.C.F. (2005) Oral immunization with a dam mutant of *Yersinia pseudotuberculosis* protects against plague. Microbiol. 151, 1919-1926.

Titball, R.W. and Williamson, E.D. (2001) Vaccination against bubonic and pneumonic plague. Vaccine. 19, 4175-4184.

Une, T. and Brubaker, R.R. (1984) Role of V antigen in promoting virulence and immunity in yersiniae. J. Immunol. 133, 2226-2230.

Weeks, S., Hill, J., Friedlander, A. and Welkos, A. (2002) Anti-V antigen antibody protects madcrophages from Yersinia pestis-induced cell death and promotes phagocytosis. Microb. Pathog. 32, 227-237.

WHO (2006) Wkly. Epidemiol. Rec. 81, 278-84.

Williamson, E.D (2001) Plague vaccine research and development. J. Appl. Microbiol. 91, 606-608.

Williamson, E.D., Eley, S.M., Stagg, A.J., Green, M., Russell, P. and Titball, R.W. (2000) A single dose sub-unit vaccine protects against pneumonic plague. Vaccine. 19, 566-571.

Williamson, E.D., Eley, S.M., Stagg, A.J., Green, M., Russell, P. and Titball, R.W. (1997) A sub-unit vaccine elicits IgG in serum, spleen cell cultures and bronchial washings and protects immunized animals against pneumonic plague. Vaccine. 15, 1079-1084.

Winter, C.C., Cherry, W.B. and Moody, M.D. (1960) An unusual strain of *Pasteurella pestis* isolated from a fatal case of human plague. Bull. W.H.O. 23, 408-409.

Picture 35. Conversations at the banquet. Photograph by R. Perry.

Index

Page numbers are the first page of an article in which the term is discussed.

Page numbers are the first page of an article in which the term is discussed.

Page numbers are the first page of an article in which the term is discussed.

Page numbers are the first page of an article in which the term is discussed.

Printed in the United States
By Bookmasters